IUTAM Symposium on Theoretical and Numerical Methods in Continuum Mechanics of Porous Materials

SOLID MECHANICS AND ITS APPLICATIONS

Volume 87

Series Editor: G.M.L. GLADWELL
Department of Civil Engineering
University of Waterloo
Waterloo, Ontario, Canada N2L 3GI

Aims and Scope of the Series

The fundamental questions arising in mechanics are: *Why?*, *How?*, and *How much?* The aim of this series is to provide lucid accounts written by authoritative researchers giving vision and insight in answering these questions on the subject of mechanics as it relates to solids.

The scope of the series covers the entire spectrum of solid mechanics. Thus it includes the foundation of mechanics; variational formulations; computational mechanics; statics, kinematics and dynamics of rigid and elastic bodies: vibrations of solids and structures; dynamical systems and chaos; the theories of elasticity, plasticity and viscoelasticity; composite materials; rods, beams, shells and membranes; structural control and stability; soils, rocks and geomechanics; fracture; tribology; experimental mechanics; biomechanics and machine design.

The median level of presentation is the first year graduate student. Some texts are monographs defining the current state of the field; others are accessible to final year undergraduates; but essentially the emphasis is on readability and clarity.

For a list of related mechanics titles, see final pages.

IUTAM Symposium on

Theoretical and Numerical Methods in Continuum Mechanics of Porous Materials

Proceedings of the IUTAM Symposium
held at the University of Stuttgart, Germany,
September 5–10, 1999

Edited by

WOLFGANG EHLERS
Institute of Applied Mechanics,
University of Stuttgart, Germany

SPRINGER-SCIENCE+BUSINESS MEDIA, B.V.

A C.I.P. Catalogue record for this book is available from the Library of Congress.

ISBN 978-94-017-3887-3 978-0-306-46953-4 (eBook)
DOI 10.1007/978-0-306-46953-4

Printed on acid-free paper

CONTENTS

Poster Session A

Session B1: Homogenization

Session B2: Biot's Theory

Session B3: Flow in Porous Media

Session B4: Waves in Porous Media I

Poster Session B

Session C1: Localization

Session C2: Extended Models

Session D1: Micromechanics

Session D2: Fracture and Damage

Session D3: Swelling, Drying and Shrinkage

Session D4: Waves in Porous Media II

Session D5: Applications

Poster Session D

Preface

During the last decades, continuum mechanics of porous materials has achieved great attention, since it allows for the consideration of the volumetrically coupled behaviour of the solid matrix deformation and the pore-fluid flow. Naturally, applications of porous media models range from civil and environmental engineering, where, e. g., geotechnical problems like the consolidation problem are of great interest, via mechanical engineering, where, e. g., the description of sinter materials or polymeric and metallic foams is a typical problem, to chemical and biomechanical engineering, where, e. g., the complex structure of living tissues is studied. Although these applications are principally very different, they basically fall into the category of multiphase materials, which can be described, on the macroscale, within the framework of the well-founded Theory of Porous Media (TPM).

With the increasing power of computer hardware together with the rapidly decreasing computational costs, numerical solutions of complex coupled problems became possible and have been seriously investigated. However, since the quality of the numerical solutions strongly depends on the quality of the underlying physical model together with the experimental and mathematical possibilities to successfully determine realistic material parameters, a successful treatment of porous materials requires a joint consideration of continuum mechanics, experimental mechanics and numerical methods. In addition, micromechanical investigations and homogenization techniques are very helpful to increase the phenomenological understanding of such media.

The intention of the IUTAM Symposium on "Theoretical and Numerical Methods in Continuum Mechanics of Porous Materials", held at the University of Stuttgart, Germany, from September 5 to 10, 1999, was to bring together scientists from all over the world to discuss and to exchange their ideas and findings in the field of porous materials. The Scientific Committee selected the participants to be invited and to present their papers during the symposium. Altogether, 86 scientists followed the invitation, 59 of them from West European and 14 from East European countries, 6 from Northern and 1 from Southern America, 2 from the Near East, 2 from Africa and 1 each from Asia and Australia.

During the Symposium, the state of the art of modelling empty, partially saturated and fully saturated porous materials, such as soil, concrete, sinter materials, metallic and polymeric foams, glacier and rock ice, living tissues, *etc.* has been presented. In addition, microscopic and macroscopic approaches including homogenization methods have been discussed. Furthermore, the theoretical models and the material parameters therein have been compared to experimental data, generally based on parameter identification and optimization methods applied to geometrically linear and finite approaches for the description of the elastic, viscous and plastic properties of various porous matrix materials. Apart from the solid deformation, fluid flow in porous media has carefully been studied. In particular, the flow-deformation interaction of the volumetrically strongly coupled solid-fluid problem has been investigated and applied to quasi-static and fully dynamic situations. In addition to these general topics, several contributions have been presented in the fields of wave propagation, localization phenomena, *Biot*'s approach to porous media, fracture and damage, swelling, drying and shrinkage as well as composite materials.

The present volume contains the printed contributions to the IUTAM Symposium on "Theoretical and Numerical Methods in Continuum Mechanics of Porous Materials" having reached the editor and having successfully passed the review process. The time and effort spent by the authors for preparing their manuscripts for this volume is greatly appreciated. Beneath the scientific part of any symposium, financial support plays an important role in organizing a successful meeting. Our thanks are due to the IUTAM Bureau, the German Research Foundation (Deutsche Forschungsgemeinschaft, DFG), the State Government of Baden-Württemberg through the Ministry of Science and Art, the University of Stuttgart, and some further institutions and companies for sponsoring the symposium and for providing funds for international participants. The organization of the Symposium was in the hands of the local organizing committee. The work of this team, especially the work of the Conference Secretary, Dr.-Ing. Stefan Diebels, is gratefully acknowledged. Thanks also to René Ott who carefully assembled the manuscript and to Kluwer Academic Publishers, Doordrecht, The Netherlands, for their efforts in producing this attractive Proceedings Volume.

July 2000 *Wolfgang Ehlers*

Sponsors of the IUTAM Symposium on Theoretical and Numerical Methods in Continuum Mechanics of Porous Materials

Stuttgart, September 5–10, 1999

BMW Niederlassung Stuttgart
City of Stuttgart
Deutsche Forschungsgemeinschaft (DFG)
Ed. Züblin AG
International Union of Theoretical and Applied Mechanics (IUTAM)
IBM Deutschland
Kluwer Academic Publishers
Landesbank Baden-Württemberg
Ministerium für Wissenschaft, Forschung und Kunst Baden-Württemberg
Silicon Graphics Inc.
Universität Stuttgart

International Scientific Committee

Wolfgang Ehlers (Germany)
Olivier Coussy (France)
Reint de Boer (Germany)
René de Borst (The Netherlands)
Stefan Kowalski (Poland)
Hans-Bernd Mühlhaus (Australia)
Jean-Herve Prévost (USA)
Bernhard Schrefler (Italy)
Iannis Vardoulakis (Greece)
Ren Wang (China)

Local Organizing Committee

Wolfgang Ehlers (Chairman)
Stefan Diebels (Secretary)
Martin Ammann
Peter Blome
Alexander Droste
Peter Ellsiepen
Bernd Markert
Günter Thomas

List of Participants

Prof. Y. Abousleiman, Ph. D., The Rock Mechanics Institute, Sarkeys Energy Center, University of Oklahoma, Norman, Oklahoma 73019-0628, U.S.A.
yabousle@ou.edu

M. Ammann, Institut für Mechanik (Bauwesen), Universität Stuttgart, 70550 Stuttgart, GERMANY
ammann@mechbau.uni-stuttgart.de

Prof. R. A. Anandarajah, Ph. D., Department of Civil Engineering, The Johns Hopkins University, 34th and Charles Streets, Baltimore, MD 21218, U.S.A.
rajah@jhu.edu

Prof. Y. C. Angel, Ph. D., Centre de Mechanique, Bât 721, Université Claude Bernard, 43 Bd du 11 Novembre 1918, 69622 Villeurbanne Cedex, FRANCE
angel@mecamat.univ-lyon1.fr

J. Banaszak, Polish Academy of Sciences, Poznań University of Technology, Institute of Technology and Chemical Engineering, pl. Marii Sklodowskiej-Curie 2, 60-965 Poznań, POLAND
gmm@rose.man.poznan.pl

Dr. E. Bauer, Institute of Mechanics, Technical University Graz, Kopernikusgasse 24, 8010 Graz, AUSTRIA
bauer@mech.tu-graz.ac.at

Dr. S. C. Baxter, University of South Carolina, 300 S. Main St., Columbia, S.C. 29208, U.S.A.
baxter@engr.sc.edu

Dr. H. Besserer, Institut für Computeranwendungen (ICA-1), Universität Stuttgart, 70550 Stuttgart, GERMANY
besserer@ica1.uni-stuttgart.de

P. Blome, Institut für Mechanik (Bauwesen), Universität Stuttgart, 70550 Stuttgart, GERMANY
blome@mechbau.uni-stuttgart.de

PD Dr. J. Bluhm, Institut für Mechanik, FB 10, Universität-GH Essen, 45117 Essen, GERMANY
bluhm@mechanik.bauwesen.uni-essen.de

Prof. Dr. R. de Boer, Institut für Mechanik, FB 10, Universität-GH Essen, 45117 Essen, GERMANY
deboer@mechanik.bauwesen.uni-essen.de

Prof. Dr. R. de Borst, Faculty of Civil Engineering, TU Delft, P.O. BOX 5048, 2600 GA Delft, THE NETHERLANDS
R.deBorst@LR.TUDelft.nl

Dr. J. Brauns, Riga Technical University, 16 Azenes St., Riga, 1048, LETTLAND
braun@acad.latnet.lv

Dr. J. Carmeliet, Lab. of Bilding Physics, Celestijnenlaan 131, 3001 Heuerlee, BELGIUM
jan.carmeliet@bwk.kuleuven.ac.be

X. Chateau, Laboratoire des Matériaux et des Structures du Génie Civil, U.M.R.113 C.N.R.S.-L.C.P.C., Cité Descartes, 2 Allée Kepler, 77420 Champs sur Marne, FRANCE
Xavier.Chateau@lcpc.fr

Dr. M. Cieszko, Department of Environmental Mechanics, Pedagogical University, Chodkiewicza 30, 85-064 Bydgoszcz, POLAND
mkk@rose.man.poznan.pl

A. Cortis, Faculty of Applied Earth Sciences, Delft University of Technology, Mijnbouwstraat 120 P.O. Box 5028, 2600 GA Delft, THE NETHERLANDS
a.cortis@ta.tudelft.nl

Prof. Dr. O. Coussy, Labroratoire Central des Ponts et Chaussées, Modélisation Mécanique, 58,bd Lefèbvre, 75732 Paris, FRANCE
coussy@lcpc.fr

R. Cudmani, Institut für Bodenmechanik und Felsmechanik, Abteilung Erddammbau und Deponiebau , Universität Karlsruhe , 76128 Karlsruhe, GERMANY
cudmani@ibf-tiger.bau-verm.uni-karlsruhe.de

Dr. P. Dangla, Labroratoire Central des Ponts et Chaussées, Modélisation Mécanique, 58,bd Lefebvre, 75732 Paris, FRANCE
Patrick.Dangla@lcpc.fr

Dr. A. I. M. Denneman, Applied Geophysics & Petrophysics, Department of Applied Earth Sciences, Delft University of Technology, P.O. Box 5028, 2600 GA Delft, THE NETHERLANDS
arthur@akst12.tn.tudelft.nl

Dr. A. K. Didwania, Department of Applied Mechanics, University of California, San Diego, 9500 Gilman Drive, Dept 0411, La Jolla, CA 92093-0411, U.S.A.
didwania@ames.ucsd.edu

Dr. S. Diebels, Institut für Mechanik (Bauwesen), Universität Stuttgart, 70550 Stuttgart, GERMANY
diebels@mechbau.uni-stuttgart.de

Prof. Dr. L. Dormieux, CERMMO, ENPC, 6-8 av. Blaise Pascal, 77 240 Champus sur Marne, FRANCE
Dormieux@cermmo.enpc.fr

A. Droste, Institut für Mechanik (Bauwesen), Universität Stuttgart, 70550 Stuttgart, GERMANY
droste@mechbau.uni-stuttgart.de

Prof. Dr. W. Ehlers, Institut für Mechanik (Bauwesen), Universität Stuttgart, 70550 Stuttgart, GERMANY
ehlers@mechbau.uni-stuttgart.de

Dr. D. Elata, Faculty of Mechanical Engineering, Technion - Israel Institute of Technology (I.I.T.), Haifa 32000, ISRAEL
elata@techunix.technicon.ac.il

P. Ellsiepen, Institut für Mechanik (Bauwesen), Universität Stuttgart, 70550 Stuttgart, GERMANY
ellsiepen@mechbau.uni-stuttgart.de

A. J. H. Frijns, Eindhoven University of Technology, Department of Mathematics, P. O. Box 513, 5600 MB Eindhoven, THE NETHERLANDS
frijns@win.tue.nl

S. Grasberger, Lehrstuhl für Statik und Dynamik, Ruhr-Universität Bochum, 44780 Bochum, GERMANY
Stefan.Grasberger@sd.ruhr-uni-bochum.de

Prof. Dr. A. A. Gubaidullin, Tyumen Institute of Mechanics of Multiphase Systems (TIMMS), Taymirskaya St. 74, 625000 Tyumen, RUSSIA
timms@sbtx.tmn.ru

J. Hartikainen, Laboratory of Theoretical and Applied Mechanics, Helsinki University of Technology, P.O.Box 1100, 02015 HUT, FINLAND
Juha.Hartikainen@csc.fi

Prof. C. T. Herakovich, Ph. D., Applied Mechanics Program, University of Virginia, Charlottesville, VA 22903-2442, U.S.A.
herak@virginia.edu

PD Dr. R. Hilfer, Institut für Computeranwendungen (ICA-1), 70550 Stuttgart, GERMANY
hilfer@ica1.uni-stuttgart.de

Dr. U. Hunsche, Bundesanstalt für Geowissenschaften und Rohstoffe, Stilleweg 2, 30655 Hannover, GERMANY
udo.hunsche@bgr.de

Dr. J. M. Huyghe, Engineering Mechanics Institute, Eindhoven University of Technology, P.O. Box 513, 5600 MB Eindhoven, THE NETHERLANDS
jacques@wfw.wtb.tue.nl

L. Jacobsson, Department of Solid Mechanics, Chalmers University of Technology, 41296 Göteborg, SWEDEN
laja@solid.chalmers.se

T. Jakpa, Department of Chemical Engineering, University of Lagos, Akoka-Yaba, NIGERIA
jakpa@hotmail.com

Dr. A. G. Kolpakov, Siberian State University of Telecommunications and Informatics, 324, Bld. 95, 9th November str, 630009 Novosibirsk, RUSSIA
AGK@neic.nsk.su

Prof. Dr. S. J. Kowalski, Poznań University of Technology, Institute of Technology and Technical Engineering, pl. Marii Skladowskiej-Curie 2, 60-965 Poznań, POLAND
kowal@rose.man.poznan.pl

Prof. Dr. S. Krenk, Institute of Structural Mechanics, Dept. of Structural Engineering and Materials, Building 118, Technical University of Denmark, 2800 Lynby, DENMARK
sk@bkm.dtu.dk

Prof. Dr. J. Kubik, Pedagogical University, Chodkiewicza 30, 85-064 Bydgoszcz, POLAND
kubik@rose.man.poznan.pl

Dr. D. Kuhl, Institute for Structural Mechanics, Ruhr-University Bochum, 44780 Bochum, GERMANY
detlef.kuhl@ruhr-uni-bochum.de

Dr. L. Kyziol, Akademia Marynarki Hojenney, ul. Smidowicza, 81-919 Gdynia, POLAND
lkyz@amw.gdynia.pl

Prof. Dr. R. Lancellotta, Dipart. di. Ingegneria Strutturale, Politecnico di Torino, Corso Duca degli Abruzzi, 24, 10129 Torino, ITALIA
lancellotta@polito.it

J. Larsson, Departement of Solid Mechanics, Chalmers University of Technology, 412 96 Göteborg, SWEDEN
jonasla@solid.chalmers.se

Dr. R. Larsson, Departement of Solid Mechanics, Chalmers University of Technology, 412 96 Göteborg, SWEDEN
ragnar@solid.chalmers.se

A. J. Lempinen, Helsinki University of Technology, Laboratory of Theoretical and Applied Mechanics, P.O. Box 1100, 02015 HUT, FINLAND
antti.lempinen@hut.fi

Dr. Z. Lu, Beijing University of Aeronautics and Astronautics, Research Center of Solid Mechanics, Beijing 100083, CHINA

PD Dr. R. Mahnken, Institut für Baumechanik und Numerische Mechanik, Universität Hannover, 30160 Hannover, GERMANY
mahnken@ibnm.uni-hannover.de

B. Markert, Institut für Mechanik (Bauwesen), Universität Stuttgart, 70550 Stuttgart, GERMANY
markert@mechbau.uni-stuttgart.de

Prof. Dr. C. Miehe, Institut für Mechanik (Bauwesen), Universität Stuttgart, 70550 Stuttgart, GERMANY
miehe@mechbau.uni-stuttgart.de

Prof. D. Sc. M. J. Mikkola, Helsinki University of Technology, Laboratory of Theoretical and Applied Mechanics, P.O. Box 1100, 02015 HUT, FINLAND
martti.mikkola@hut.fi

A. Millard, CEA / DMT / LAMS, CEN Saclay, 91191 GIF / Yvette Cedex, FRANCE
am@semt2.smts.cea.fr

M. M. Molenaar, Eindhoven University of Technology, Department of Mechanical Engineering, Section MAterials TEchnology, P.O. Box 513, 5600 MB Eindhoven, THE NETHERLANDS
mathieu@wfw.wtb.tue.nl

Dr. H.-B. Mühlhaus, Chief Research Scientist, Rock Mechanics Research Centre, CSIRO Division of Exploration and Mining, 39 Fairway, Nedlands WA 6009, AUSTRALIA
h.muhlhaus@dem.csiro.au

H. Müllerschön, Institut für Mechanik (Bauwesen), Universität Stuttgart, 70550 Stuttgart, GERMANY
müllerschön@mechbau.uni-stuttgart.de

Dr. M. A. Murad, Laboratorio Nacional de Computacao Cientifica, Petropolis, RJ, C. Postal 95113, and Instituto Politecnico/UERJ, Nova Friburgo, RJ, 2860, BRAZIL
murad@iota.lncc.br

Dr. G. Musielak, Poznań University of Technology, Institute of Technology and Technical Engineering, pl. Marii Skladowskiej-Curie 2, 60-965 Poznań, POLAND
gmm@rose.man.poznan.pl

Prof. Dr. R. Nova, Milan University of Technology, Department of Structural Engineering, Piazza L. da Vinci, 32, 20133 Milano, ITALIA
nova@rachele.stru.polimi.it

Dr. A. Papastavrou, Institut für Techno- und Wirtschaftsmathematik e. V. (ITWM), 67663 Kaiserslautern, GERMANY

M. Paul, Institut für Computeranwendungen im Bauingenieurwesen, TU Braunschweig, 38106 Braunschweig, GERMANY
m.paul@TU-BS.de

Prof. J.-H. Prévost, Ph. D., Department of Civil Engineering and Operations Research, School of Engineering/Applied Science, Princeton University, Princeton, New Jersey 08544-5263, U.S.A.
prevost@princeton.edu

A. Pudewills, Institut für Nukleare Entsorgungstechnik, Forschungszentrum Karlsruhe GmbH, Postfach 3640, 76021 Karlsruhe, GERMANY
pudewill@ine.fzk.de

K. Rajewska, Polish Academy of Sciences, Poznań University of Technology, Institute of Technology and Chemical Engineering, pl. Marii Sklodowskiej-Curie 2, 60-965 Poznań, POLAND
gmm@rose.man.poznan.pl

Dr. P. Redanz, Cambridge Centre for Micromechanics, Cambridge University, Engineering Department, Trumpington Street, Cambridge CB2 1PZ, U.K.
pr227@eng.cam.hc.uk

T. Ricken, Institut für Mechanik, FB 10, Universität-GH Essen, 45117 Essen, GERMANY
tricken@mechanik.bauwesen.uni-essen.de

Prof. Dr. K. Runesson, Dept. of Solid Mechanics, Chalmers University of Technology, 41296 Göteborg, SWEDEN
keru@solid.chalmers.se

Dr. A. Rybicki, Poznań University of Technology, Institute of Technology and Technical Engineering, pl. Marii Skladowskiej-Curie 2, 60-965 Poznań, POLAND
ryban@rose.man.poznan.pl

Dr. T. Sadowski, Department of Applied Mechanics, Faculty of Mechanical Engineering, Technical University of Lublin, ul. Nadbystrzycka 40, 20-618 Lublin, POLAND
tskms@archimedes.pol.lublin.pl

Dr. L. Sanavia, Dipartimento di Costruzioni e Trasporti, Università degli Studi di Padova, via F. Marzolo 9, 35131 Padova, ITALIA
sanavia@caronte.dic.unipd.it

Prof. Dr. A. Sawicki, Institute of Hydro-Engineering, IBW PAN, ul. Koscierska 7, 80-953 Gdansk-Oliwa, POLAND
as@hapcio.ibwpan.gda.pl

A. Scheuermann, University of Karlsruhe, Institute of Soil and Rock Mechanincs, Division of Earth Dams and Landfill Technology, Engler-Bunte-Ring, 76131 Karlsruhe, GERMANY
Alexander.Scheuermann@bau-verm.uni-karlsruhe.de

Prof. B. A. Schrefler, Ph. D., Dipartmento di Costruzioni e Trasporti, Università degli Studi di Padova, Via marzolo 9, 35131 Padova, ITALY
bas@caronte.dic.unipd.it

Dr. J. Schröder, Institut für Mechanik (Bauwesen), Universität Stuttgart, 70550 Stuttgart, GERMANY
schroeder@mechbau.uni-stuttgart.de

Dr. S. Sellers, School of Mathematics, UEA, Norwich NR4 7TJ, U.K.
sellers@uea.ac.uk

J. Skolnik, Institut für Mechanik, FB 10, Universität-GH Essen, 45117 Essen, GERMANY
judith@mechanik.bauwesen.uni-essen.de

Dr. D. M. J. Smeulders, Delft University of Technology, Faculty of Applied and Earth Sciences, P.O. Box 5028, 2600 GA Delft, THE NETHERLANDS
D.M.J.Smeulders@MP.TUDelft.NL

Dr. A. H. Soliman, 4 Habib St, Port Said St, 11411 Cairo, EGYPT
azzasoliman@hotmail.com

Prof. Dr. S. Sorek, Ben-Gurion University of the Negev, J. Blaustein Institute for Dessert Research, Dryland Environmental Water Resources, Sde Boker Campus, 84990, ISRAEL
shaul@hydro.boker.bgu.ac.il

Prof. Dr. P. Steinmann, Lehrstuhl für Technische Mechanik, Universität Kaiserslautern, Postfach 3049, 67653 Kaiserslautern, GERMANY
ps@rhrk.uni-kl.de

PD Dr. G. Thomas, Institut für Mechanik (Bauwesen), Universität Stuttgart, 70550 Stuttgart, GERMANY
thomas@mechbau.uni-stuttgart.de

Prof. Dr. R. Uklejewski, Department of Environmental Mechanics, Pedagogical University, Chodkiewicza 30, 85-064 Bydgoszcz, POLAND
mkk@rose.man.poznan.pl

Prof. Dr. I. Vardoulakis, Nat. Techn. Univ. of Athens, 5 Heroes of Polytechnion Ar., Dept. of Engineering Science, 15773 Zographou, GREECE
I.Vardoulakis@mechan.ntua.gr

J. Widjajakusuma, Institut für Computeranwendungen (ICA-1), Universität Stuttgart, 70550 Stuttgart, GERMANY
jack@ica1.uni-stuttgart.de

T. Wilhelm, Universität Innsbruck, Institut für Geotechnik und Tunnelbau, Technikerstr. 13, 6020 Innsbruck, AUSTRIA
c81315@uibk.ac.at

Prof. Dr. S. Sorek, Ben-Gurion University of the Negev, J. Blaustein Institute for Desert Research, and Environmental Water Resources, Sde Boker Campus, 84990 ISRAEL

Session A1: Opening

Chairman: W. Ehlers

Introduction to the Porous Media Theory

R. de Boer
Institute of Mechanics, FB-10,
University of Essen, D-45117 Essen, Germany

1. Introduction

In the last decade, particularly in recent years, the macroscopic porous media theory has made a decisive progress concerning the fundamentals of the theory (kinematics, balance equations, and constitutive theory) and the development of mathematical models in various fields of engineering and biomechanics. This progress attracted some attention and therefore conferences (colloquia, symposia, etc.) were organized in the last three years devoted exclusively to the macroscopic porous media theory. Also in national and international journals a great number of important contributions have been published which has brought the porous media theory, in some parts, to a close.

The goal of this paper is to give an introduction to the macroscopic porous media theory, whereby also some aspects of the microscopic scale will be touched. The introduction to the porous media theory must, however, be restricted, due to space limitations, to some important features of the complex porous media theory.

The modern macroscopic porous media theory is a combination of the volume fraction concept and elements of the mixture theory. The volume fraction concept contains the introduction of volume fractions n^α which relate the volume elements of the individual constituents ρ^α to the bulk volume element, where α denotes the individual constituents (e.g., $\alpha = \mathrm{S}$: solid, $\alpha = \mathrm{F}$: fluid). Due to the volume fraction concept, all geometric and physical quantities, such as motion, deformation, and stress, are defined in the total control space, which is formed by the porous solid. Thus, the geometric and physical quantities can be interpreted as the statistical average values of the real quantities. Within the framework of the porous media theory, a saturated porous medium will be treated as an immiscible mixture of all constituents, with particles X_α. This immiscible mixture is, of course, a substitute model; it can be treated with the methods of continuum mechanics, especially with those of the mixture theory which had been developed for the description of the thermodynamical behavior of heterogenously composed continua.

Porous media theory has a long tradition. The purely mechanical theory had already been founded by Fillunger in 1936. Recently, some

W. Ehlers (ed.), IUTAM Symposium on Theoretical and Numerical Methods in Continuum Mechanics of Porous Materials, 3–12.

important new findings in the porous media theory have been worked out concerning, in particular, the constitutive theory (Readers who are interested in the historical development and in the fundamentals of the current state of porous media are referred to de Boer, [5]).

In this place the content of this paper should be indicated. The next section is devoted to kinematics and balance equations. Section 3 gives an introduction to the constitutive theory. Section 4 shows the usefulness of the macroscopic theory in its application to various problems in engineering and biomechanics. Finally, some conclusions shall be drawn.

2. Kinematics and balance equations

If the motion of the constituent is understood as a chronological succession of placements χ_α, then for the spatial position vector $\mathbf{x}$ of the material points X_α, which can be identified with the reference position vector $\mathbf{X}_\alpha$ at time $t = t_0$, the following relations hold at time t:

$$\mathbf{x} = \chi_\alpha(\mathbf{X}_\alpha, t) \, , \quad \mathbf{X}_\alpha = \chi_\alpha^{-1}(\mathbf{x}, t) \, . \tag{1}$$

Eqns. $(1)_1$ and $(1)_2$ represent the Lagrange and Euler description of motion. The function χ_α is postulated to be unique, and uniquely invertible, at any time t. A mathematically necessary and sufficient condition for the existence of Eqn. $(1)_2$ is given, if the Jacobian

$$J_\alpha = \det \mathbf{F}_\alpha \quad (\text{ with the deformation gradient } \mathbf{F}_\alpha = \operatorname{Grad}_\alpha \chi_\alpha) \tag{2}$$

differs from zero. The differential operator "$\operatorname{Grad}_\alpha$" denotes a partial differentiation with respect to the reference position $\mathbf{X}_\alpha$ of the constituents φ^α. The inverse of the deformation gradient $\mathbf{F}_\alpha$ is given with the differential operator "grad" referring to the spatial point $\mathbf{x}$ by

$$\mathbf{F}_\alpha^{-1} = \operatorname{grad} \mathbf{X}_\alpha \, . \tag{3}$$

With the Lagrange and Euler description of the motion (1), the velocity and the acceleration of a material point of a constituent φ^α are defined by

$$\begin{aligned} & \frac{\partial \chi_\alpha(\mathbf{X}_\alpha, t)}{\partial t} \, , \quad \mathbf{x}''_\alpha = \frac{\partial^2 \chi_\alpha(\mathbf{X}_\alpha, t)}{\partial t^2} \, , \\ \mathbf{v}_\alpha = \mathbf{x}'_\alpha = \mathbf{x}'_\alpha(\mathbf{x}, t) \, , \quad & \mathbf{a}_\alpha = \mathbf{x}''_\alpha = \mathbf{x}''_\alpha(\mathbf{x}, t) \, . \end{aligned} \tag{4}$$

With (4), the material velocity gradient and the spatial velocity gradient of the constituent φ_α are obtained:

$$(\mathbf{F}_\alpha)'_\alpha = \mathrm{Grad}_\alpha \, \mathbf{x}'_\alpha \, , \quad \mathbf{L}_\alpha = \mathrm{grad}\, \mathbf{x}'_\alpha = \mathrm{grad}\, \mathbf{v}_\alpha \, ,$$
$$\mathbf{L}_\alpha = (\mathbf{F}_\alpha)'_\alpha \mathbf{F}_\alpha^{-1} \, . \tag{5}$$

The additive decomposition of $\mathbf{L}_\alpha$ yields the symmetrical part $\mathbf{D}_\alpha$ of the spatial velocity gradient and the skew-symmetric spin tensor $\mathbf{W}_\alpha$:

$$\mathbf{L}_\alpha = \mathbf{D}_\alpha + \mathbf{W}_\alpha \;\text{ with }\; \mathbf{D}_\alpha = \frac{1}{2}(\mathbf{L}_\alpha + \mathbf{L}_\alpha^T) \, , \; \mathbf{W}_\alpha = \frac{1}{2}(\mathbf{L}_\alpha - \mathbf{L}_\alpha^T) \, . \tag{6}$$

The symmetric strain tensors $\mathbf{E}_\alpha$ and $\mathbf{A}_\alpha$, the Green strain tensor, and the Almansi strain tensor depend on the deformation gradient $\mathbf{F}_\alpha$ in the following way:

$$\mathbf{E}_\alpha = \frac{1}{2}(\mathbf{C}_\alpha - \mathbf{I}) \, , \quad \mathbf{A}_\alpha = \frac{1}{2}(\mathbf{I} - \mathbf{B}_\alpha^{-1}) \, ,$$

where

$$\mathbf{C}_\alpha = \mathbf{F}_\alpha^T \mathbf{F}_\alpha \quad \text{and} \quad \mathbf{B}_\alpha = \mathbf{F}_\alpha \mathbf{F}_\alpha^T \tag{7}$$

denote the right and left Cauchy-Green deformation tensors, respectively. In order to describe properties of the microscale, namely compressibility and incompressibility, it is useful to multiplicatively decompose the deformation gradient $\mathbf{F}_\alpha$ (see the extensive discussion of this problem and the consequences concerning the kinematics in Bluhm and de Boer, [3]):

$$\mathbf{F}_\alpha = \mathbf{F}_{\alpha N} \hat{\mathbf{F}}_{\alpha R} \, . \tag{8}$$

The part $\hat{\mathbf{F}}_{\alpha R}$ is interpreted as that part of $\mathbf{F}_\alpha$ which describes the deformation of the real material, whereas $\mathbf{F}_{\alpha N}$ describes the remaining part of the deformation of the control space, namely the change of the pores in size and shape. The parts $\mathbf{F}_{\alpha N}$ and $\hat{\mathbf{F}}_{\alpha R}$ are to be understood as local mappings of tangent (vector) spaces in each material point of the body. In the case of homogeneous deformations, the multiplicative decomposition (8) leads to an intermediate state $(\hat{\ldots})$. In analogy to (5) through (6), material time derivatives of the deformation tensor $\hat{\mathbf{F}}_{\alpha R}$ as well as of the spherical part of the time derivative of $\mathbf{F}_{\alpha N}$ can be introduced:

$$\hat{\mathbf{D}}_{\alpha R} = \frac{1}{2}(\hat{\mathbf{L}}_{\alpha R} + \hat{\mathbf{L}}_{\alpha R}^T) , \quad \hat{\mathbf{L}}_{\alpha R} = \frac{\partial(\hat{\mathbf{x}}_\alpha)'_\alpha}{\partial \mathbf{X}_\alpha}\frac{\partial \mathbf{X}_\alpha}{\partial \hat{\mathbf{x}}_\alpha} = \frac{\partial(\hat{\mathbf{x}}_\alpha)'_\alpha}{\partial \hat{\mathbf{x}}_\alpha}$$
$$\text{(homogeneous deformations)} , \quad \mathbf{D}_{\alpha N} \cdot \mathbf{I} = \mathbf{D}_\alpha \cdot \mathbf{I} - \mathbf{D}_{\alpha R} \cdot \mathbf{I} . \quad (9)$$

In the mixture and porous media theory, balance equations – balance of mass, balance of momentum, and moment of momentum, as well as balance of energy – must be established for each constituent φ^α in consideration of all interaction and external agencies. The interaction effects (supply terms) have to be in the sum equal to zero. The following balance equations will be formulated in local forms for the individual constituents in consideration of all interaction effects (see de Boer, [5]).

2.1. Balance of mass

The balance of mass for the individual constituents φ^α in the local form is given by:

$$(\rho^\alpha)'_\alpha + \rho^\alpha \operatorname{div} \mathbf{v}_\alpha = \hat{\rho}^\alpha , \qquad \sum_{\alpha=1}^{\kappa} \hat{\rho}^\alpha = 0 , \qquad (10)$$

where $\hat{\rho}^\alpha$ is the mass supply per volume element.

2.2. Balance of momentum and moment of momentum

In this section, the consequences of applying the axioms of the *balance of momentum* to each individual constituent will be discussed. Cauchy's first equation of motion (balance of momentum) for φ^α is obtained as:

$$\operatorname{div} \mathbf{T}^\alpha + \rho^\alpha \mathbf{b}^\alpha + \hat{\mathbf{p}}^\alpha = \rho^\alpha \mathbf{a}_\alpha + \hat{\rho}^\alpha \mathbf{v}_\alpha , \qquad \sum_{\alpha=1}^{\kappa} \hat{\mathbf{p}}^\alpha = \mathbf{0} . \qquad (11)$$

Here, $\mathbf{T}^\alpha$ is Cauchy's stress tensor, $\rho^\alpha \mathbf{b}^\alpha$ the external volume force and $\hat{\mathbf{p}}^\alpha$ the momentum supply. The *balance of moment of momentum* for nonpolar materials results in the statements that the partial stresses $\mathbf{T}^\alpha$ and the sum of all partial stresses $\mathbf{T}$ are symmetric.

2.3. Balance of energy

The first law of thermodynamics (balance of energy) states that the sum of the material time derivatives of the internal and kinetic energies equals the rates of the mechanical work and heat. This balance equation is transferred to the individual constituents:

$$\rho^\alpha(\varepsilon^\alpha)'_\alpha - \mathbf{T}^\alpha \cdot \mathbf{L}_\alpha - \rho^\alpha r^\alpha + \operatorname{div} \mathbf{q}^\alpha = \hat{e}^\alpha - \hat{\mathbf{p}}^\alpha \cdot \mathbf{v}_\alpha -$$
$$- \hat{\rho}^\alpha(\varepsilon^\alpha - \frac{1}{2}\mathbf{v}_\alpha \cdot \mathbf{v}_\alpha), \quad \sum_{\alpha=1}^{\kappa} \hat{e}^\alpha = 0 . \tag{12}$$

Here, $\varepsilon^\alpha = \varepsilon^\alpha(\mathbf{x}, t)$ is the specific internal energy, $r^\alpha = r^\alpha(\mathbf{x}, t)$ the partial energy source, and $\mathbf{q}^\alpha = \mathbf{q}^\alpha(\mathbf{x}, t)$ the partial heat flux vector, which is positive when entering the body, and $\hat{e}^\alpha$ the energy supply. Eqn $(12)_2$ represents a constraint on the energy supplies.

3. Constitutive theory

Mixture theory – the basis of porous media theory – is closed, i.e., the number of unknown fields is equal to the sum of the balance equations and the constitutive relations. This can easily be proven. However, by the introduction of the volume fractions n^S and n^F for the real constituents φ^S and φ^F in a binary porous media model (in order to obtain "smeared" continua which can be treated by continuum mechanical methods), the problem arises that two field equations are missing.

Therefore, one has to look for additional equations in order to close the fields. It is, however, difficult to gain additional fields since the volume fractions touch quantities of the microscale for which balance or constitutive equations are not contained in the macroscopic mixture theory. Therefore, in order to close the system of field equations, it is necessary to introduce constitutive equations. These equations connect certain mechanical or thermodynamic quantities via material-dependent constants. The constitutive relations must reflect the main phenomena gained by test observations and they are restricted by the entropy inequality. The evaluation of the entropy inequality in the porous media theory differs considerably from the evaluation in the theory of one-component materials. One must, namely, be aware of an important constraint (saturation condition), which states that the sum of all volume fractions is equal to unity. This constraint is valid for incompressible and compressible materials and restricts, in the rate formulation, the rates of the volumetric changes and must, therefore be considered in the evaluation of the entropy inequality. This evaluation is rather cumbersome, lengthy and laborious, see de Boer ([5]). Here, only the constitutive equations for a binary model (compressible and incompressible elastic solid and viscous fluid phases) will be listed.

$$
\begin{aligned}
\mathbf{T}^S &= -\,n^S p\mathbf{I} - n^S \rho^S \frac{\partial \psi^S}{\partial n^S}\mathbf{I} + 2\rho^S J_S^{-2/3}(\mathbf{F}_S \frac{\partial \psi^S}{\partial \breve{\mathbf{C}}_S}\mathbf{F}_S^T)^D \\
&= n^S p^{SR}\mathbf{I} + 2\rho^S J_S^{-2/3}(\mathbf{F}_S \frac{\partial \psi^S}{\partial \breve{\mathbf{C}}_S}\mathbf{F}_S^T)^D\,, \qquad (13)\\
p^{SR} &= -(\rho^{SR})^2 \frac{\partial \psi^S}{\partial \rho^{SR}}\,,\\
\mathbf{T}^F &= -\,n^F p\mathbf{I} + 2\mu^F \mathbf{D}_F^D
\end{aligned}
$$

where in the incompressible case p^{SR} and p^{FR} are indetermined and in the compressible case determind by the following constitutive equation

$$
p^{\alpha R} = -\,(\rho^{\alpha R})^2 \frac{\partial \psi^\alpha}{\partial \rho^{\alpha R}}\,. \qquad (14)
$$

In (13) and (14) ψ^α is the free Helmholtz energy function, $p^{\alpha R}$ the hydrostatic stress state in the corresponding phase, $\breve{\mathbf{C}}_S$ the volume-preserving part of the deformation tensor of the solid phase, μ^F the shear modulus of the viscous fluid, and $\rho^{\alpha R}$ denotes the real density of the constituents.

It should be mentioned that $p^{\alpha R}$ can be expressed by the fluid hydrostatic stress state p^{FR} and the so-called configuration pressure (the second term on the right-hand side of $(13)_1$).

4. Applications in engineering and biomechanics

With the development of kinematics, balance equations, and the constitutive theory for saturated and empty porous solids in the last years an ensured base has been provided for the investigation of special problems in different fields of engineering and biomechanics. Great progress has been achieved in different domains such as `soil mechanics`, `process engineering`, `material science`, and `environmental engineering`, as well as in `biomechanics`. In this section some problems in the various fields will be discussed and the current state of the treatment of these problems will be addressed.

The vast domain of `soil mechanics` is a preferred field for the application of the porous media theory. Saturated and empty sandy bodies or clay fulfill, to a high extent, the basic assumption of the porous media theory regarding statistically distributed pores. Therefore, it is not surprising that the porous media theory has its roots in soil mechanics (see the extensive review of de Boer, [4]). Based on experimental observations – in particular, in soil mechanics – fundamental findings, e.g., the effective stress concept, have been recognized. In the meantime, modern porous media theory has been successfully applied to static and dynamic problems in soil mechanics.

In **process engineering**, many applications of the porous media theory are known. In the following paragraphs we will discuss, however, only two problems.

For aerospace and other high-tech applications, many powder-metallurgy alloys have been developed in order to achieve better combinations of strength, toughness, fatigue resistance, and resistance to stress-corrosion cracking than those found in alloys produced by ingot metallurgy (see Doraivelu *et al.*, [8]). The success of powder-metallurgy processing depends heavily upon the ability to economically produce a near-net-shape. The formation of this shape is dependent on the success of the die compaction process in delivering defect-free, uniform density green parts (Lewis *et al.*, [14]). The compaction process for compressing the powder is, without any doubt, the main process in manufacturing engineered products in powder metallurgy. Therefore, this process should be clearly understood from the mechanical and thermodynamic point of view, the more so as there are many difficulties which exist in the compaction process for powders. This concerns, for example, the non-homogeneous density distribution and considerably large residual stresses in the green end product. Hence, there is a need to develop a mathematical model which can predict mechanical phenomena for the compaction process.

Another example for the application of the porous media theory in process engineering are drying processes. These processes (phase transitions which are accompanied by a liquid transport) occur in many fields of engineering, e.g., in soil mechanics, agriculture and, in particular, in chemical engineering. Thus, the problem of liquid transport in porous solids, and its removal from them, has an important practical feature. As indicated, the drying process consists of a change of fluid into moisture and the transport of the liquid in capillary-porous solids.

The drying process is very complex and not all problems have been solved till yet. The thermodynamics of phase transition of water into gas (steam) has come to a primary close only recently (see de Boer and Bluhm, [6]). The capillary rise of fluids in capillary-porous bodies is still under study concerning the continuum mechanical description of this phenomenon.

It seems that Kowalski ([12]) was the first scientist who tried a consequent continuum mechanical approach to the complex drying process. Kowalski ([12]) pointed out: "This paper corresponds to the review article by Luikov ([15]), and stands for the generalization of the concept presented there by taking into consideration the deformability of the capillary-porous body."

Porous media theory is an essential part of `biomechanics`. Since Mow's *et al.* ([17]) introduction of a binary model (similar to Bowen's, [7], approach) with incompressible constituents to investigate creep and stress relaxation of articular cartilage, a vast amount of contributions in biomechanics, based on the fundamentals of porous media theory, have been published. Within the framework of this review, it is impossible to mention and discuss all of these valuable papers. We will restrict our review, therefore, to some selected papers and will discuss only the mechanical feature. During the 1980s, much effort was made to investigate single problems in biomechanics on the basis of the so-called linear Kuei, Lai, and Mow (KLM) biphasic theory for cartilage (see Armstrong *et al.*, [1]). In this paper, analytical solutions for the deformations, flux of the fluid, creep, and stress relaxations of fluid-filled elastic porous solids (e.g., cartilage) were developed. In the contribution of Holmes *et al.* ([10]), viscoelastic behavior of the solid phase was taken into account.

An excellent review article *Cartilage and diarthrodial joints as paradigms for hierarchical materials and structures* was written by Mow and Ratcliffe ([18]). This article gives a clear overview on the most essential mechanical properties of cartilage and diarthrodial joints. In particular, Mow *et al.* explained many notions of biomechanics and classical mechanics. Thus, this review article is excellently suited to give an introduction to that part of biomechanics concerned with the description of the behavior of loaded cartilage.

A decisive step towards a consistent theory to describe the mechanical behavior of articular cartilage was done by Lai *et al.* ([13]). They developed a ternary model for the swelling and deformation behaviors of articular cartilage.

Little has been done to derive the description of some physical phenomena, important in `building physics`, from the fundamentals of porous media theory. Most of the describing equations for e.g., moisture transport and heat conduction in saturated porous solids, are obtained purely intuitively.

A field of growing interest is `environmental mechanics`. Typical problems in this domain are transport of contaminants in clayey soils, debris flow, and flow of an *ice-till mixture.*

Another field of application is `agriculture (soil physics)`, see, e.g., Raats *et al.* [19].

Well-known transport phenomena occur also in the field of the `petroleum industry`. Oil and natural gas, driven by, in parts, high pressure are gained from porous reservoirs. Recently, attention has been focused on foamy oil flows consisting of two phases: oil and gas. There is a special issue of the journal Transport in Porous Media (Maini and Hayes, [16]) devoted exclusively to this problem.

Moreover, the field of `material science` will be mentioned. The powder compaction described in Section 8.1 can be considered as a part of material science. Other materials, which can be treated by the methods of porous media theory, are metallic foams (see, e.g., Ehlers and Eipper, [9]) and ceramic composites (see, e.g., Besmann *et al.*, [2]). All mentioned materials are used for high-tech applications and just in this field it can be expected that other porous materials with special properties will be developed.

The porous media theory is also used in the field of `local water supplies`. Zimmer ([21]) applied the Richard equations (a combination of the partial balance equations of mass and momentum) to the retention and filtration problem of rainwater.

Finally, the mechanical behavior of plants, in particular, the analysis of `plant growth` in the continuum mechanical context, will be mentioned. First approaches have already been developed, see, e.g., Karalis ([11]) with a chapter devoted to plant growth. In this chapter Silk ([20]) remarked, however: "Much work remains to understand the dynamic of growth."

5. Conclusions

In this paper the current state of the macroscopic porous media theory has been stated along with the citation of numerous contributions. The investigations concerning the fundamentals of the theory of porous media have revealed that in the last decade a consistent theory has been derived, consistent with the basic principles of continuum mechanics, in particular, the dissipation principles.

References

1. Amstrong, C. G., Lai, W. M. and Mow, V. C. An analysis of the unconfined compression of articular cartilage, *Journal of Biomechanical Engineering* (J Biomech Eng Trans), **106**, 165–173, 1984.
2. Besmann, T. M., Sheldon, B. W., Lowden, R. A. and Stinton, D. P. Vapor-phase fabrication and properties of continuous-filament ceramic composites *Science*, **253**, 1104–1108, 1991.
3. Bluhm, J. and de Boer, R. The volume fraction concept in the porous media theory, *Zeitschrift für angewandte Mathematik und Mechanik* (ZAMM), **77**, 563–577, 1997.
4. de Boer, R. Highlights in the historical development of the porous media theory – toward a consistent macroscopic theory, *Applied Mechanics Review* (Appl Mech Rev), **49**, 201–262, 1996.

5. de Boer, R. Theory of Porous Media: Highlights in the historical development and current state, Springer-Verlag, Berlin • Heidelberg • New York, 1999.
6. de Boer, R. and Bluhm, J. Phase transitions in gas- and liquid-saturated porous solids, *Transport in PorousMedia* (T I P M), **34**, 249–267, 1999.
7. Bowen, R. M. Incompressible porous media models by use of the theory of mixtures, *International Journal of Engineering Science* (Int J Eng Sci), **18**, 1129–1148, 1980.
8. Doraivelu, S. M., Gegel, H. L., Grunasekara, J. S., Malas, J. C., Morgan, J. T. and Thomas, Y. F. A new yield function for compressible P/M materials, *International Journal of Mechanics Science* (Int J Mech Sci), **26**, 527–535, 1984.
9. Ehlers, W. and Eipper, G. Finite elastic deformations in liquid-saturated and empty porous solids, *Transport in Porous Media* (T I P M), **34**, 179–191, 1999.
10. Holmes, M. H., Lai, W. M. and Mow, V. C. Singular pertubation analysis of the nonlinear, flow-dependent, compressive stress-relaxation behavior of articular cartilage, *Journal of Biomechanical Engineering* (ASME), **107**, 206–218, 1985.
11. Karalis, T. K. Thermodynamics of soils swelling non-hydrostatically, NATO ASI Series, Vol. H 64, Mechanics of Swelling (ed.: by T. K. Karalis) Springer-Verlag Berlin • Heidelberg, 1991.
12. Kowalski, S. J. Thermomechanics of constant drying rate period, *Archives of Mechanics* (Arch Mech), **39**, 157–176, 1987.
13. Lai, W. M., Hou, J. S. and Mow, V. C. A triphasic theory for the swelling and deformation behaviors of articular cartilage, *Journal of Biomechanical Engineering* (J Biomech Eng), **113**, 245–258, 1991.
14. Lewis, R. W., Jinka, A. G. and Getin, D. T. Computer-aided simulation of metal powder die compaction processes, PMI Vol. 25, No. 6, 287–293, 1993.
15. Luikov A. V. Systems of differential equations of heat and mass transfer in capillary-porous bodies, *International Journal of Heat and Mass Transfer* (Int J Heat Mass Trans), **18**, 1–14, 1975.
16. Maini, B. B. and Hayes, R. E. (eds) Foamy oil flow, *Transport in Porous Media* (T I PM), **35** (special issue), 1999.
17. Mow, V. C., Kuei, S. C., Lai, W. M. and Amstrong, C. G. Biphasic creep and stress relaxation of articular cartilage in compression: theory and experiments, *Journal of Biomechanical Engineering* (J Biomech Eng) (ASME), **102**, 73–84, 1980.
18. Mow, V. C. and Ratcliffe, A. Cartilage and diarthrodial joints as paradigms for hierarchical materials and structures, *Biomaterials*, **13**, 67–97, 1991.
19. Raats, P. A. C., Rogoar, H. and van den Heuvel-Pieper (eds.) Soil structure and transport processes, The Netherlands Integrated Soil Research Programme Reports **6**, Graphish Service Centrum Van Gils B.V., Wageningen, The Netherlands, 1996.
20. Silk, W. K. On the kinematics and dynamics of plant growth. NATO ASI Series, Vol. H 64, Mechanics of Swelling, (ed.: by T. K. Karalis) Springer-Verlag, Berlin • Heidelberg, 1991.
21. Zimmer, U. Durchströmung poröser Medien, Einsatz numerischer Modellrechnungen zur Beschreibung und Bewertung von Anlagen zur Retention und Versickerung von Regenwasser. Fachbereich Bauwesen Universität-GH Essen, Dissertation, 1997.

2-D Localisation Analysis of Saturated Porous Media

B. A. Schrefler and H. W. Zhang
Department of Structural and Transportation Engineering
University of Padua, Italy
Department of Engineering Mechanics, Dalian
University of Technology, P. R. China

Abstract. Conditions for localisation of deformation into a planar shear band in the incremental response of elastic-plastic saturated porous media are studied in the case of small strains and rotations. The critical modulus for localisation of both undrained and drained conditions are given in terms of the discontinuous bifurcation analysis. Further loss of uniqueness of the response of the coupled problem is investigated by means of positiveness of the second order work density. As an example, the critical hardening moduli and shear band angle for localisation are derived for the case of generalised plasticity. Quantitative results are given for different sands, characterised by appropriate material parameters of the Pastor Zienkiewicz model.

Keywords: Strain localisation, fully saturated porous media, generalised plasticity, critical hardening moduli

1. Introduction

The deformations within soils are often observed to concentrate in well defined zones called shear bands. A suitable tool for describing localisation in terms of continuum theory is the discontinuous bifurcation analysis. The theory was first proposed by Hadamard [1], and later developed by Thomas [2], Hill [3], Mandel [4], Rudinicki [5], Rice [6] and Vardoulakis [7]. Strain localisation implies loss of uniqueness in the incremental elastoplastic response of a homogeneously strained body and, as described by Rice [6], also implies an imaginary speed of acceleration waves. Based on discontinuous bifurcation analysis and investigation of positiveness of the second order work density the critical hardening moduli and shear band angle for localisation of saturated porous media have been derived analytically in Zhang and Schrefler [9]. The aim of the present work is the prediction of the loss of uniqueness and localisation in saturated porous sands which exhibit either strain softening or non associativity during the loading process. The generalized plasticity constitutive model developed by Pastor *et al.* [10] is used for our investigations. Parameters such as critical hardening moduli for shear band generation, loss of second order work positiveness and shear band angle when shear band takes place are obtained numerically for several cases of saturated sand materials. A detailed description of the Pastor Zienkiewiz model and of the equations for the analysis of

W. Ehlers (ed.), IUTAM Symposium on Theoretical and Numerical Methods in Continuum Mechanics of Porous Materials, 13–18.

saturated porous media can be found in Lewis and Schrefler [11] and Zienkiewicz *et al.* [12] and is omitted here for sake of brevity. Similar investigations for dry sand only where carried out by Imposimato and Nova [13].

2. Uniqueness and localisation analysis of saturated porous media under undrained conditions

Based on undrained assumptions, the critical hardening modulus for shear band generation in a saturated porous medium has been obtained analytically by Zhang and Schrefler [9] choosing the principal axes of stress denoted by 1,2,3 as reference system. We show here the case when one of the components of the vector n orthogonal to the band is null. The shear band normal lies in one of the planes formed by two of the principal axes of stress. The corresponding critical modulus is

$$H_{TPC}(n_i^2) = H_{TSC}(n_i^2) - H^*_{TWC}(n_i^2) \quad (1)$$

where

$$\begin{aligned} H_{TSC}(n_i^2) &= 4G(n_1^2F_1P_1 + n_2^2F_2P_2 + n_3^2F_3P_3) - \\ &-2G(n_1^2P_1 + n_2^2P_2 + n_3^2P_3)(n_1^2F_1 + n_2^2F_2 + n_3^2F_3) - \\ &-2G(F_1P_1 + F_2P_2 + F_3P_3) - \\ &-\frac{2G\Lambda}{\Lambda + 2G}[(\mathrm{tr}P) - n_1^2P_1 - n_2^2P_2 - n_3^2P_3][(\mathrm{tr}F) - \\ &-n_1^2F_1 - n_2^2F_2 - n_3^2F_3] \end{aligned} \quad (2)$$

$$\begin{aligned} H^*_{TWC}(n_i^2) &= \frac{Q^*}{(\Lambda + 2G)(\lambda + 2G + Q^*)} \\ &[\Lambda(\mathrm{tr}P) + 2G(n_1^2P_1 + N_2^2P_2 + N_3^2P_3)] \\ &[\Lambda(\mathrm{tr}F) + 2G(n_1^2F_1 + n_2^2 + n_3^2F_3)] \end{aligned} \quad (3)$$

and F and P are the gradients of yield function and plastic potential respectively and Λ and G are Lame's constants. Note also that, in order to yield an admissible solution, the components of vectors F_i and P_i must be such as to give a value of n_i^2 internal to the interval [0,1]. When this condition is not met, the extrema of hardening modulus will be searched in the case where two components of the vector n are null. To separate the critical hardening modulus of the single phase problems from eq. (1), we change this equation to the following form

$$H_{TPC}(n_i^2) = H_{SC}(n_{is}^2) - H^*_{TWC}(n_i^2, n_{is}^2, n_{iw}^2) \quad (4)$$

where $H_{SC}(n_{is}^2)$ is the critical hardening modulus for shear band generation of the single phase material without fluid

$$\frac{H_{SC}}{2G} = n_i^2\Big(2(P_iF_i - P_jF_j) - \frac{1}{1-\nu}[(P_i - P_j)F_j +$$

$$+(F_i - F_j)P_j] + \frac{1}{1-\nu}[\mathrm{tr}F(P_i - P_j) + \mathrm{tr}P(F_i - F_j)]\Big) -$$
$$-\frac{n_i^4}{1-\nu}(P_i - P_j)(F_i - F_j) - P_iF_i - P_kF_k -$$
$$-\frac{1}{1-\nu}(F_i - F_k)(P_i - P_k) \tag{5}$$

and

$$H_{TWC}(n_i^2, n_{is}^2, n_{iw}^2) = H^*_{TWC}(n_i^2) + H_{TSC}(n_{is}^2) - H_{TSC}(n_i^2) \tag{6}$$

is the additional part due to the action of fluid phase in the media. For plane strain situations we assume without loss of generality, that two of the principal directions i and j are located in the plane of interest, and the direction k is out of the plane and $n_k = n_3 = 0$. In this case solution (3) and the case where two components of n are null, obtained for three dimensional deformations still hold true. The results for associative plasticity can be immediately obtained from the above solutions. It can easily be shown that in this case the hardening modulus corresponding to shear band generation and strain localisation is never positive in saturated porous media. For comparison also the critical hardening modulus corresponding to the zero second order work has been derived. The interested reader is referred to Zhang and Schrefler [9].

3. Examples for a particular constitutive model

The above equations are now evaluated numerically for different cases of the Pastor Zienkiewicz model for sand. This model is strongly parameter and stress history dependent. The used parameters are taken from Pastor *et al.* [10]. The initial stresses for Pastor-Zienkiewicz model are: $\sigma_{01} = 1.3\sigma_0$, $\sigma_{02} = \sigma_0 = \sigma_{03}$ and $\sigma_0 = -1.0 * 10^{+5}$ Pa. Only loading process is considered in this paper. During this process the following loading conditions are assumed: $\sigma_1 = (1.3 + \zeta)\sigma_0$, $\sigma_2 = (10 + \nu\zeta)\sigma_0$, $\sigma_3 = \sigma_0$, where z¿0 is the loading factor. The results at bifurcation and localisation points are shown in Tables 1 and 2, respectively without consideration of fluid phase and with the effect of the liquid at undrained conditions. The cases of 11 to 18 do not present associative features. The corresponding critical hardening modulus for shear band formation in Table 1 is larger than zero. For the associative case, the critical hardening modulus is not only zero, but also the same as the modulus when second order work becomes zero. This was pointed out by several authors (Rice [6], Bigoni and Hueckel [8]) in single phase problems. For non-associative constitutive behaviour the critical hardening modulus at which the loss of positiveness of the second order work occurs is always greater than the one when shear band formation takes place. Further, the critical hardening modulus without consideration of fluid phase is generally greater than that with the liquid phase effects in the materials. This shows that the existence of fluid will delay the occurrence of loss of uniqueness and localisation. However, this does not hold in some non-associative cases, particularly when the non-associated behaviour is strong. This can be seen in the cases 11(a), 11(b) and 16 shown in Tables 1 and 2.

Table I. The results without consideration of liquid phase

	Θ_L(deg)	H_{SC}(MPa)	H_{S2}(MPa)	g_1	g_3	Dilatancy
11(a)	47.7	1.510	4.270	-0.826	0.564	No
11(b)	41.2	1.010	2.870	-0.730	0.684	Yes
11(c)	38.3	0.586	1.680	-0.674	0.739	Yes
11(d)	35.4	0.325	0.769	-0.611	0.791	Yes
13(a)	31.6	1.320	2.700	-0.580	0.814	Yes
13(b)	39.3	1.290	3.560	-0.713	0.702	Yes
14	33.9	0.008	0.024	-0.564	0.826	Yes
15	38.2	0.774	1.980	-0.667	0.745	Yes
16	47.4	1.520	5.000	-0.823	0.569	No
17	28.8	1.790	3.470	-0.558	0.830	Yes
18	44.4	0.477	1.390	-0.675	0.738	Yes
0	34.7	0.000	0.000	-0.569	0.822	Yes

The value of hardening modulus in P-Z model depends on the loading history and stress state of the material. The variation of the hardening modulus H_L with loading factor ζ for associative P-Z model is shown in Figures 1a and 1b where the material parameters of the cases 0 and 11(c) are used. The corresponding critical hardening moduli H_{TPC} or H_{SC} and loading plastic hardening modulus H_L varying with shear band angle Θ_L are plotted in Figures 1a and 1b for associative behaviour. Figure 1a shows the behaviour of the sand with a fluid phase, Figure 1b without the consideration of this effect. It is clear that the cross-over point between the diagrams of H_{TPC} or H_{SC} and H_L is the critical point for occurrence of localisation: there the hardening modulus H_L starts to be smaller than H_{TPC} or H_{SC} in the loading process.

Table II. The results with consideration of liquid phase

	Θ_L(deg)	H_{TPC}(MPa)	H_{W2P}(MPa)	g_1	g_3	Dilatancy
11(a)	45.2	2.470	2.580	-0.718	0.696	No
11(b)	44.9	2.580	2.820	-0.713	0.701	No
11(c)	23.1	-29.90	-19.00	-0.444	0.896	Yes
11(d)	23.1	-4.060	-2.670	-0.422	0.907	Yes
13(a)	25.4	-114.0	-72.30	-0.484	0.875	Yes
13(b)	27.1	-437.0	-384.0	-0.533	0.846	Yes
14	29.4	-443.0	-443.0	-0.497	0.868	Yes
15	26.0	-118.0	-75.60	-0.488	0.873	Yes
16	45.1	4.000	4.290	-0.717	0.697	No
17	25.3	-82.30	-41.80	-0.501	0.865	Yes
18	44.9	0.987	1.240	-0.699	0.715	Yes
0	25.2	-580.0	-580.0	-0.425	0.905	Yes

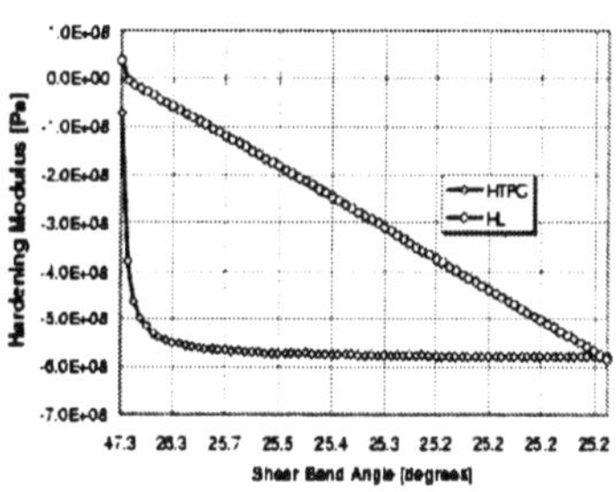

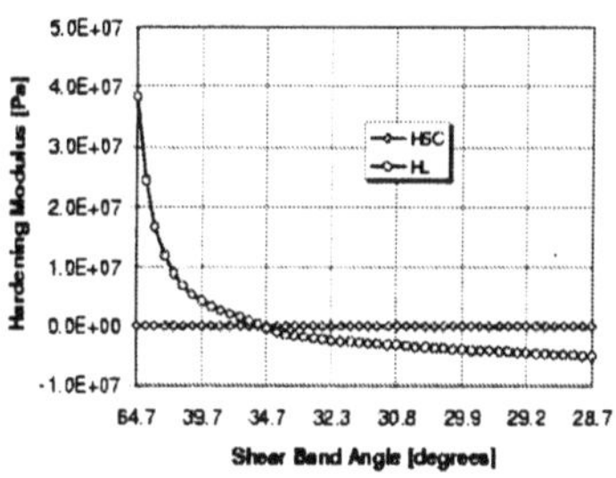

Figure 1. Variation of hardening modulus H_L and critical hardening $HTPC$ with shear band angle Θ_L with (left) and without (right) consideration of liquid phase.

4. Conclusions

The uniqueness and localisation analysis for the Pastor-Zienkiewicz saturated sand constitutive model based on the generalized plasticity theory is presented in this paper. It appears that there are two critical hardening moduli which correspond respectively to single phase material (large permeability) and to undrained conditions (small permeability). In between there is a transition where the critical moduli and hence the onset of localisation are dependent on the permeability.

References

1. Hadamard, J. Leçons sur la propagation des ondes et les equations de lhydrodynamique. Librarie Scientifique. A. Hermann, Paris, 1903.
2. Thomas, T. Plastic flow and fracture in solids. *Academic Press*, New York, 1961.
3. Hill, R. Acceleration waves in solids. *J. Mech. Phys. Solids*, **10**, 1–16, 1962.
4. Mandel, J. Conditions de stabilite et postulate de Drucker. IUTAM Symposium on Rheology and Soil Mechanics (Edited by J. Kravtchenko and P.M. Sirieys), Springer, Berlin, 58, 1966.
5. Rudnicki, J. W. and Rice, J. R. Conditions for the localisation of the deformation in pressure-sensitive dilatant material. *J. Mech. Phys. Solids*, **23**, 371–394, 1975.
6. Rice, J. R. The localisation of plastic deformation. Proc. 14th Int. Congress on Theoretical and Applied Mechanics (Edited by W.T. Koiter), 1976.
7. Vardoulakis, I. Equilibrium theory of shear bands in plastic bodiesblock *Mech. Res. Commun.*, **3**, 209–214, 1903.
8. Bigoni, D. and Hueckel, T. Uniqueness and localisation-I. associative and non-associative elastoplasticity. *Int. J. Solids and Structures*, **28**, 197–213, 1991.
9. Zhang, H. W. and Schrefler, B. A. Uniqueness and localisation analysis of elasto-plastic saturated porous media, to appear.
10. Pastor, M., Zienkiewicz, O. C. and Chan, A. H. C. Generalized plasticity and the modelling of soil behaviour. *Int. J. Numer. Anal. Meths. Geomech.*, **14**, 151–190, 1990.
11. Lewis, R. W. and Schrefler, B. A. The Finite Element Method in the Static and Dynamic Deformation and Consolidation of Porous Media.2nd Ed. Wiley, Chichester, 1998.
12. Zienkiewicz, O. C., Chan, A. H. C., Pastor M., SchreflerB. A. and Shiomi, T. Computational Geomechanics, with special reference to Earthquake Engineering. Wiley, Chichester, 1999.
13. Imposimato, S. and Nova, R. An investigation on the uniqueness of the incremental response of elastoplastic models for virgin sand. *Mechanics of Cohesive-Frictional Material*, **3**, 65–87, 1998.

Evolution of the Volume Fractions in Compressible Porous Media

S. Diebels
Institute of Applied Mechanics (Civil Engineering), University of Stuttgart, D-70550 Stuttgart, Germany
e-mail: Diebels@mechbau.uni-stuttgart.de

Abstract. In porous media theories the volume fractions are introduced as internal variables. While for incompressible constituents the evolution of the volume fractions follows from the balance of mass, for mixtures with compressible constituents additional equations are required, which must be given constitutively. In the present contribution, an evolution equation is postulated and a related two-phase model is examined.

Keywords: Compressible porous media, evolution of the volume fractions, saturation, effective stress

1. Introduction

In porous media theories the local composition of the mixture is described by the volume fractions. They were introduced as internal variables which may be determined by additional evolution equations while the saturation condition acts as a constraint. Porous media with compressible constituents were discussed by Bowen [1], Ehlers [5, 6], Svendsen & Hutter [8], and others.

In the present contribution an evolution equation for the volume fraction of the solid matrix material of a two-phase mixture is motivated from a split of the balance of mass into two parts as proposed by Diebels [3, 2]. The resulting model may easily be reduced to the so-called hybrid model or to the incompressible model (Ehlers, [5, 6]) where either the solid skeleton or both, the skeleton and the pore fluid, respectively, are incompressible.

2. Kinematics and balance equations

Following the concept of superimposed continua as usual in the theory of porous media, each constituent follows its own function of motion χ_α which relates the reference position $\mathbf{X}_\alpha$ of a particle of constituent φ^α to the actual position $\mathbf{x}$

$$\mathbf{x} = \chi_\alpha(\mathbf{X}_\alpha, t). \tag{1}$$

W. Ehlers (ed.), IUTAM Symposium on Theoretical and Numerical Methods in Continuum Mechanics of Porous Materials, 21–26.

In a two-phase model, $\alpha = S$ represents the solid skeleton and $\alpha = F$ the fluid. Line elements $\mathrm{d}\mathbf{x}$ are mapped by the deformation gradient $\mathbf{F}_\alpha$ from the reference configuration onto the actual configuration

$$\mathrm{d}\mathbf{x} = \mathrm{Grad}_\alpha\, \chi_\alpha(\mathbf{X}_\alpha, t)\, \mathrm{d}\mathbf{X}_\alpha = \mathbf{F}_\alpha(\mathbf{X}_\alpha, t)\, \mathrm{d}\mathbf{X}_\alpha. \tag{2}$$

Therein, the symbol $\mathrm{Grad}_\alpha(\cdot)$ means differentiation with respect to $\mathbf{X}_\alpha$. The corresponding velocities are obtained from (1) by differentiation with respect to time t

$$\mathbf{x}'_\alpha = \frac{\partial \chi_\alpha(\mathbf{X}_\alpha, t)}{\partial t}. \tag{3}$$

In general, $(\cdot)'_\alpha$ represents the material time derivative following the motion of the constituent φ^α.

In porous media theories, the volume fractions n^α are introduced as local ratio of the volume $\mathrm{d}v^\alpha$ occupied by the constituent φ^α to the volume $\mathrm{d}v$ occupied by the whole mixture φ

$$n^\alpha = \frac{\mathrm{d}v^\alpha}{\mathrm{d}v}. \tag{4}$$

The saturation condition

$$n^S + n^F = 1 \tag{5}$$

guaranties that each spatial point is occupied with matter. According to (4), two different density functions may be introduced, namely the bulk density ρ^α relating the element of mass dm^α of constituent φ^α to the volume occupied by the mixture and the effective density $\rho^{\alpha R}$ relating the same element of mass to the volume occupied by φ^α, respectively:

$$\rho^\alpha = \frac{\mathrm{d}m^\alpha}{\mathrm{d}v}, \quad \rho^{\alpha R} = \frac{\mathrm{d}m^\alpha}{\mathrm{d}v^\alpha}, \quad \rho^\alpha = n^\alpha \rho^{\alpha R}. \tag{6}$$

In the following, incompressibility is understood microscopicaly in a material sense, i. e. the effective density is assumed to be constant, $\rho^{\alpha R} = \text{const.}$

The corresponding balance equations of a two-phase model are the balances of mass and the balances of momentum, c. f. Ehlers [5, 6],

- Balance of mass:

$$(\rho^\alpha)'_\alpha + \rho^\alpha \,\mathrm{div}\, \mathbf{x}'_\alpha = 0. \tag{7}$$

- Balance of momentum (quasi-static):

$$\mathbf{0} = \mathrm{div}\, \mathbf{T}^\alpha + \rho^\alpha \mathbf{b} + \hat{\mathbf{p}}^\alpha, \quad \hat{\mathbf{p}}^S = -\hat{\mathbf{p}}^F. \tag{8}$$

For simplicity, an exchange of mass between the solid skeleton and the pore fluid is not assumed here. Furthermore, $\mathbf{T}^\alpha$ are the stress tensors, $\rho^\alpha \mathbf{b}$ the body forces, $\hat{\mathbf{p}}^\alpha$ are the momentum exchange terms [4].

3. Constitutive equations

While for incompressible constituents due to $(6)_3$ the balance of mass (7) degenerates to a balance of volume, for compressible constituents an additional evolution equation is required. For a gas filling the pore space such an equation follows from the saturation condition, because it is the property of a gas to fill up all the available space. For the matrix material an evolution equation may be motivated from (7) if it is re-written in the following form

$$\begin{aligned} \rho^{SR} & \left((n^S)'_S + z^S n^S \operatorname{div} \mathbf{x}'_S \right) + \\ + n^S & \left((\rho^{SR})'_S + (1 - z^S) \rho^{SR} \operatorname{div} \mathbf{x}'_S \right) = 0. \end{aligned} \tag{9}$$

In a thermodynamical process both, ρ^{SR} and n^S, are independent variables. Therefore, (9) may be separated into a balance of the efffective density and into the evolution equation for n^S

$$(n^S)'_S + z^S n^S \operatorname{div} \mathbf{x}'_S = 0. \tag{10}$$

On the other hand, (10) may be understood as the defintion of z^S and, in this case, the balance of effective density according to $(9)_2$ follows as a consequence. The factor z^S determines which percentage of changes of the partial density is governed by changes in volume fraction. For $z^S = 1$, the well-known result of an incompressible skeleton is recovered. Based on (10) a thermodynamically consistent model may be derived (Diebels, [3, 2]). Here, the essential results are summarized:

– Stress tensor of the solid phase:

$$\mathbf{T}^S = - z^S n^S [p^{FR} + p^K] \mathbf{I} + \rho^S \mathbf{F}_S \frac{\partial \psi^S}{\partial \mathbf{C}_S} \mathbf{F}_S^T. \tag{11}$$

– Stress tensor of the fluid phase:

$$\mathbf{T}^F = - n^F p^{FR} \mathbf{I} + \mathbf{T}_E^F, \qquad \mathbf{T}_E^F = \overset{4}{\mathbf{T}} \mathbf{D}_F \approx \mathbf{0}. \tag{12}$$

– Configuration pressure:

$$p^K = \rho^S \frac{\partial \psi^S}{\partial n^S}. \tag{13}$$

– Fluid pressure:

$$p^{FR} = (\rho^{FR})^2 \frac{\partial \psi^F}{\partial \rho^{FR}}. \tag{14}$$

– Momentum exchange (interaction force):

$$\hat{\mathbf{p}}^F = -p^{FR} \operatorname{grad} n^S + \hat{\mathbf{p}}_E^F, \qquad \hat{\mathbf{p}}_E^F = -\frac{(n^F)^2 \gamma^{FR}}{k^F} \mathbf{w}_F. \tag{15}$$

Therein, ψ^α are the free energy functions, $\mathbf{C}_S = \mathbf{F}_S^T \mathbf{F}_S$ is the right Cauchy Green deformation tensor, and $\mathbf{w}_F = \mathbf{x}_F - \mathbf{x}_S$ is the so-called seepage velocity.

Comparing the result (11) to the principle of effective stress in the extension to compressible constituents as given e. g. by Šuklje [9], leads to an identification of the function z^S in terms of the matrix compressibility K^S and of the material compressibility K^{SR} in the following form

$$z_S^S = 1 - \frac{K^S}{K^{SR}} = 1 - n^S. \tag{16}$$

The last equality is valid under the assumption of the Voigt bound of homogenization theory, i. e. $K^S = n^S K^{SR}$.

To become more specific, an extension of the well-known Neo-Hooke elasticity law may be given for the weighted solid stresses (Kirchhoff stress tensor) in terms of the left Cauchy Green deformation tensor $\mathbf{B}_S = \mathbf{F}_S \mathbf{F}_S^T$ and the volumetric deformation $J_S = \det \mathbf{F}_S$:

$$\begin{aligned} J_S \mathbf{T}_E^S &= \mu^S (\mathbf{B}_S - \mathbf{I}) + \alpha \lambda^S \frac{\mathrm{d}W^S(J_S)}{\mathrm{d}J_S} \mathbf{I} + P_0 \mathbf{I}, \\ J_S p^K &= (1-\alpha) \lambda^S \frac{\mathrm{d}W^S(n_{0S}^S / n^S)}{\mathrm{d}n^S}. \end{aligned} \tag{17}$$

The last term $P_0 \mathbf{I}$ allows for an pressure in the undeformed reference configuration, which is required to ensure equilibrium with a compressible pore fluid. Furthermore, the material parameters are the Lamé constants μ^S and λ^S. In (17), W^S represents the volumetric part of the strain energy function. According to α, the compressibility is split in two parts depending on changes in porosity and total volumetrical deformation, respectively.

With the choice $z^S \equiv 1$, the so-called hybrid model (Ehlers, [5, 6]) is recovered, which consists of an incompressible skeleton and a compressible pore fluid. In addition, the stress strain relation should posses a point of compaction [7]. Furthermore, if also the fluid is assumed to be incompressible, the incompressible model follows, if the pore pressure p^{FR} is not given by the equation of state (14) but is interpreted as a Lagrange parameter P.

4. Examples

The differences between the three models (compressible, hybrid, incompressible) are illustrated in the simple tension test and in a consolidation test as shown in Figures 1 and 2, respectively.

While the difference between the incompressible model and the compressible model is small in the range of tension, differences in the load-deflection curves become visible in the range of compression. This is an effect according to the point of compaction, i. e. the incompressible model does not allow further compaction if all pores are closed. The hybrid model is not taken into account in this problem because under ideal drained conditions the different properties of the pore fluid do not influence the solution.

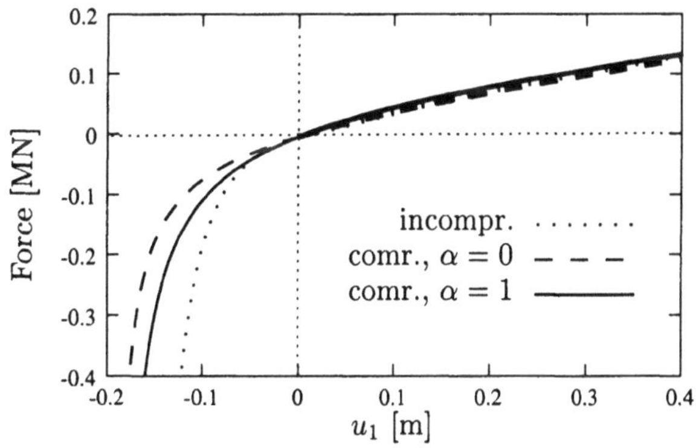

Figure 1. Tension test

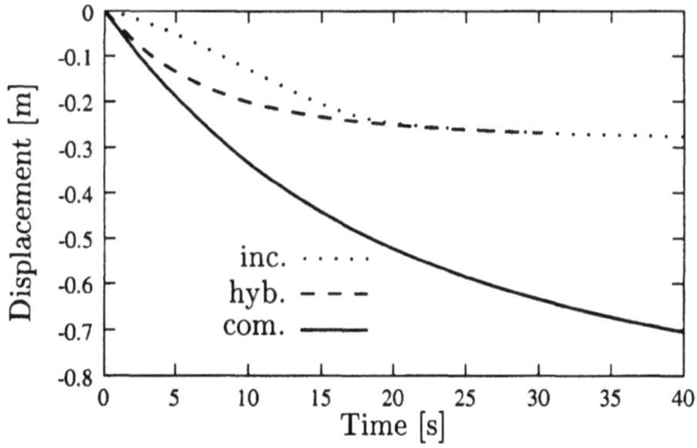

Figure 2. Consolidation test

In the second example the consolidation problem is investigated. The curves in Figure 2 show the settlement under a load which increases monotonly in time. The hybrid and the compressible model show an instantaneous response due to the fact that a pore pressure arises only due to a compression of the pore space. In contrast to this, the incompressible model allows for a compression only if the pore fluid is drained out and, therefore, it shows a strong time dependence of the solution. Under high loads the hybrid model and the incompressible model show the same settlement, which is restricted by the point of compaction. In general, the compressible model would allow a compaction to a point if the load is large enough.

5. Conclusions

Based on the theory of porous media, a two-phase model is presented which consists of a deformable porous skeleton saturated with a pore fluid. Both constituents are assumed to be compressible. Therefore, an additional evolution equation is required for the volume fraction of the solid constituent. Such an equation may be motivated from the balance of mass. The thermodynamically admissible model including this evolution equation permits the determination of the additional parameters if the effective stress principle is applied. Some numerical results show the evidence of the proposed model.

References

1. Bowen, R. M. Compressible porous media models by use of the theory of mixtures. *Int. J. Engng. Sci.*, **20**, 697–735, 1982.
2. Diebels, S. Constitutive modelling of micropolar porous media. In: J.-F. Thimus *et al.* (eds.): *Poromechanics – A Tribute to Maurice A. Biot.* A. A. Balkema, Rotterdam, pp. 71–76, 1998.
3. Diebels, S. A Micropolar Theory of Porous Media: Constitutive Modelling. *Transport in Porous Media*, **34**, 193–208, 1999.
4. Diebels, S. and Ehlers, W. On Basic Equations of Multiphase Micropolar Materials. *Technische Mechanik*, **16**, 77–88, 1996.
5. Ehlers, W. Compressible, incompressible and hybrid two-phase models in porous media theories. In: Y. C. Angel (ed.): *Anisotropy and Inhomogeneity in Elasticity and Plasticity*, AMD-Vol. 158. ASME, 25–38, 1993a.
6. Ehlers, W. Constitutive equations for granular materials in geomechanical context. In: K. Hutter (ed.): *Continuum mechanics in environmental sciences and geophysics*, CISM Courses and Lectures No. 337. Springer-Verlag, Wien, 313–402, 1993b.
7. Eipper, G. *Theorie und Numerik finiter elastischer Deformationen in fluidgesättigten porösen Festkörpern.* Dissertation, Bericht Nr. II-1, Institut für Mechanik (Bauwesen), Universität Stuttgart, 1998.
8. Svendsen, B. and Hutter, K. On the thermodynamics of a mixture of isotropic materials with constraints. *Int. J. Engng. Sci.*, **33**, 2021–2054, 1995.
9. Šuklje, L. *Rheological Aspects of Soil Mechanics.* Wiley-Interscience, London, 1969.

Constitutive Relations for Thermo-Elastic Porous Solids within the Framework of Finite Deformations

J. Bluhm
Institute of Mechanics, FB-10 Bauwesen
University of Essen, D-45117 Essen, Germany

1. Introduction

Based on the theory of porous media (TPM), a constitutive relation for the so-called "effective" stresses of the compressible or incompressible thermo-elastic solid phase of a fluid-saturated porous solid will be presented. For the description of a compressible solid within the framework of finite deformation processes, well-known constitutive laws for one-component materials can be used. For incompressible porous solids, one has to consider that the so-called compression point exists (at which point all pores are closed and no further volume compression can occur). With respect to the description of the above-mentioned effect, a constitutive relation for incompressible solids will be developed following the structure of a well-known constitutive law for compressible one-component materials.

2. Basics of the constitutive theory

The constitutive relation for the "effective" Cauchy stress tensor $\mathbf{T}_{\mathrm{E}}^{\mathrm{S}} = \mathbf{T}^{\mathrm{S}} + \mathrm{n}^{\mathrm{S}} p(1-\delta)\mathbf{I}$ ($\mathbf{T}^{\mathrm{S}}$: partial Cauchy stress tensor of the solid; $\mathrm{n}^{\mathrm{S}}p$: pore pressure of the fluid, weighted ith the volume fraction n^{S} of the solid; δ: reduction term, $\delta = 0$ for incompressible solids, $\delta \leq 1$ for compressible solids) of a compressible or an incompressible thermo-elastic porous solid is given by

$$\mathbf{T}_{\mathrm{E}}^{\mathrm{S}} = 2\,\rho^{\mathrm{S}}\,\mathbf{F}_{\mathrm{S}}\,\frac{\partial \psi^{\mathrm{S}}}{\partial \mathbf{C}_{\mathrm{S}}}\,\mathbf{F}_{\mathrm{S}}^{\mathrm{T}} \quad , \quad \psi^{\mathrm{S}} = \hat{\psi}^{\mathrm{S}}\,(\Theta^{\mathrm{S}}\,,\,\mathbf{C}_{\mathrm{S}}) \quad , \tag{1}$$

see, e.g., de Boer [1] and Bluhm [2]. In this, the free Helmholtz energy ψ^{S} of the solid phase is a function of the absolute temperature Θ^{S} and the right Cauchy-Green tensor $\mathbf{C}_{\mathrm{S}} = \mathbf{F}_{\mathrm{S}}^{\mathrm{T}}\mathbf{F}_{\mathrm{S}}$. The quantities $\mathbf{F}_{\mathrm{S}} = \mathbf{F}_{\mathrm{SN}}\mathbf{F}_{\mathrm{SR}}$ and $\rho^{\mathrm{S}} = \mathrm{n}^{\mathrm{S}}\rho^{\mathrm{SR}}$ denote the deformation gradient and the partial density of the solid body, where the deformation parts $\mathbf{F}_{\mathrm{SN}}$ and $\mathbf{F}_{\mathrm{SR}}$ describe the change of the pores in size and shape and the deformation of the real solid material; n^{S} and ρ^{SR} are the volume fraction and the real density of φ^{S}, respectively. For an incompressible

W. Ehlers (ed.), IUTAM Symposium on Theoretical and Numerical Methods in Continuum Mechanics of Porous Materials, 27–32.

solid, considering that $\rho^{SR} = \rho_{0S}^{SR} = \text{const.}$ and $J_{SR} = \rho_{0S}^{SR}/\rho^{SR} = 1$, i.e., the Jacobian $J_S = \det \mathbf{F}_S = \rho_{0S}^{S}/\rho^{S} = (n_{0S}^{S}\rho_{0S}^{SR})/(n^{S}\rho^{SR})$ of the porous solid is given by $J_S = n_{0S}^{S}/n^{S}$. The lower index $(\ldots)_{0S}$ denotes the quantities at the position $\mathbf{X}_S$ at time $t = t_0$ (reference placement of φ^S). A detailed discussion of the multiplicative decomposition of $\mathbf{F}_S$ and the corresponding expression of the determinants can be found in Bluhm [2]. Furthermore, for empty porous solids ($p = 0$) the "effective" Cauchy stress tensor $\mathbf{T}_E^S$ is identical with the partial Cauchy stress tensor $\mathbf{T}^S$.

In the following, only isotropic solid materials will be considered. For these kinds of materials the condition

$$\hat{\psi}^S(\Theta, \mathbf{C}_S) = \overset{\circ}{\psi}{}^S(\Theta, \mathbf{B}_S) \quad , \tag{2}$$

concerning the free Helmholtz energy for φ^S, must hold, see Beatty [3], where the quantity $\mathbf{B}_S = \mathbf{F}_S\mathbf{F}_S^T$ denotes the left Cauchy-Green tensor. The condition (2) is fulfilled, if the free Helmholtz energy $\hat{\psi}^S$ is an isotropic scalar valued function of the temperature and the principal invariants of $\mathbf{C}_S$ or $\mathbf{B}_S$. Thus, for the description of isotropic thermo-elastic materials, the quantity $\mathbf{C}_S$ in $(1)_2$ must be replaced by the invariants of $\mathbf{C}_S$, i.e.,

$$\begin{aligned} \psi^S = \hat{\psi}^S(\Theta^S, \mathbf{C}_S) \quad \Longrightarrow \quad \psi^S &= \hat{\hat{\psi}}{}^S(\Theta^S, \mathrm{I}_{\mathbf{C}_S}, \mathrm{II}_{\mathbf{C}_S}, \mathrm{III}_{\mathbf{C}_S}) \\ &= \hat{\hat{\psi}}{}^S(\Theta^S, \mathrm{I}_{\mathbf{C}_S}, \mathrm{II}_{\mathbf{C}_S}, J_S) \quad , \end{aligned} \tag{3}$$

where

$$\begin{aligned} \mathrm{I}_{\mathbf{C}_S} &= \mathbf{C}_S \cdot \mathbf{I} \quad , \qquad \mathrm{II}_{\mathbf{C}_S} = \frac{1}{2}[(\mathbf{C}_S \cdot \mathbf{I})^2 - \mathbf{C}_S \cdot \mathbf{C}_S] \quad , \\ \mathrm{III}_{\mathbf{C}_S} &= \det \mathbf{C}_S = J_S^2 \end{aligned} \tag{4}$$

are the three invariants of the right Cauchy-Green tensor $\mathbf{C}_S$. Replacing $\mathbf{C}_S$ with $\mathbf{B}_S$, the three invariants $\mathrm{I}_{\mathbf{B}_S}$, $\mathrm{II}_{\mathbf{B}_S}$, and $\mathrm{III}_{\mathbf{B}_S}$ of the left Cauchy-Green tensor can be calculated with the help of (4).

With (3), the "effective" stress of the solid phase can be written as

$$\begin{aligned} \mathbf{T}_E^S &= 2\rho^S \mathbf{F}_S \left(\frac{\partial \hat{\hat{\psi}}{}^S}{\partial \mathrm{I}_{\mathbf{C}_S}} \frac{\partial \mathrm{I}_{\mathbf{C}_S}}{\partial \mathbf{C}_S} + \frac{\partial \hat{\hat{\psi}}{}^S}{\partial \mathrm{II}_{\mathbf{C}_S}} \frac{\partial \mathrm{II}_{\mathbf{C}_S}}{\partial \mathbf{C}_S} + \frac{\partial \hat{\hat{\psi}}{}^S}{\partial J_S} \frac{\partial J_S}{\partial \mathbf{C}_S} \right) \mathbf{F}_S^T \\ &= \alpha_0 \mathbf{I} + \alpha_1 \mathbf{B}_S + \alpha_2 \mathbf{B}_S \mathbf{B}_S \quad . \end{aligned} \tag{5}$$

The quantities

$$\begin{aligned} \alpha_0 &= \rho^S J_S \frac{\partial \hat{\hat{\psi}}{}^S}{\partial J_S} \quad , \qquad \alpha_1 = 2\rho^S \left(\frac{\partial \hat{\hat{\psi}}{}^S}{\partial \mathrm{I}_{\mathbf{C}_S}} + \mathrm{I}_{\mathbf{C}_S} \frac{\partial \hat{\hat{\psi}}{}^S}{\partial \mathrm{II}_{\mathbf{C}_S}} \right) \quad , \\ \alpha_2 &= -2\rho^S \frac{\partial \hat{\hat{\psi}}{}^S}{\partial \mathrm{II}_{\mathbf{C}_S}} \end{aligned} \tag{6}$$

are the material or elastic response functions of the solid.

3. Constitutive relations for incompressible and compressible porous solids

For isotropic thermo-hyperelastic compressible or incompressible porous solids, the following formulation of the free Helmholtz energy ψ^S within the framework of finite deformation processes will be postulated:

$$\begin{aligned}
\psi^S &= \hat{\hat{\psi}}^S(\Theta^S, I_{C_S}, J_S = J_{SN} J_{SR}) \\
&= \frac{1}{\rho_{0S}^S}\{\frac{1}{2}\lambda_{cp}^S[(\log J_S)^2 + 2 J_{cp}^S \log J_S] + \\
&\quad + \lambda_{cp}^S \frac{1 - J_{cp}^S}{J_{cp}^S - 2}[\log \frac{J_{cp}^S - J_S}{J_S(J_{cp}^S - 1) - J_{cp}^S} - \\
&\quad - \log(1 - J_{cp}^S)] - \\
&\quad - \mu^S \log J_S + \frac{1}{2}\mu^S(I_{C_S} - 3) - \\
&\quad - 3\alpha^S k^S(\log J_S)(\Theta^S - \Theta_{0S}^S) - \\
&\quad - \rho_{0S}^S c^S(\Theta^S \log\frac{\Theta^S}{\Theta_{0S}^S} - \Theta^S + \Theta_{0S}^S)\} \quad ,
\end{aligned} \tag{7}$$

where μ^S, λ_{cp}^S, k^S, α^S, and c^S are interpreted as constant macroscopic material parameters of the constituent φ^S. The quantity J_{cp}^S ($0 \leq J_{cp}^S \leq 1$) denotes the compression point of the solid phase, i.e., for compressible solids J_{cp}^S is equal to zero ($J_{cp}^S = 0$) and for incompressible solids J_{cp}^S is equal to n_{0S}^S ($J_{cp}^S = n_{0S}^S$).

Inserting (7) into (5) yields the following constitutive relation for the "effective" Cauchy stress tensor of the solid phase:

$$\begin{aligned}
\mathbf{T}_E^S &= \frac{1}{J_S}[2\mu^S \mathbf{K}_S + \lambda_{cp}^S(\log J_S + \zeta^S)\mathbf{I} - \\
&\quad - 3\alpha^S k^S(\Theta^S - \Theta_{0S}^S)\mathbf{I}] \quad ,
\end{aligned} \tag{8}$$

where the abbreviation

$$\zeta^S = J_{cp}^S\left(1 - \frac{J_S}{J_S^2 + \frac{(J_{cp}^S)^2}{1 - J_{cp}^S}(J_S - 1)}\right) \tag{9}$$

has been used. The quantity $\mathbf{K}_S = 1/2\,(\mathbf{B}_S - \mathbf{I})$ is the Karni-Reiner strain tensor referred to the actual placement of the constituent φ^S. Furthermore, concerning the derivation of (8), the relation $J_S = \rho^S_{0S}/\rho^S$ has been used.

In the following, a constitutive relation for the "effective" stress of the solid phase, referred to the reference placement for small deformation processes, will be derived. Therefore, the "effective" second (symmetric) Piola-Kirchhoff stress $\mathbf{S}^S_E = J_S\,\mathbf{F}_S^{-1}\mathbf{T}^S_E\mathbf{F}_S^{T-1}$ will be introduced:

$$\begin{aligned} \mathbf{S}^S_E &= 2\mu^S \overset{R}{\mathbf{K}}_S + \lambda^S_{cp}\,(\log J_S + \zeta^S)\,\mathbf{C}_S^{-1} - \\ &\quad - 3\,\alpha^S k^S\,(\Theta^S - \Theta^S_{0S})\,\mathbf{C}_S^{-1} \quad . \end{aligned} \tag{10}$$

The quantity $\overset{R}{\mathbf{K}}_S = 1/2\,(\mathbf{I} - \mathbf{C}_S^{-1})$ denotes the Karni-Reiner strain tensor referred to the reference placement of φ^S.

In order to obtain Hookean-type constitutive laws referred to the reference placement, the linearization

$$\mathbf{S}^S_{E\,lin} = \mathbf{S}^S_E\Big|_{\mathcal{P}_0} + \frac{\partial \mathbf{S}^S_E}{\partial \mathbf{C}_S}\Big|_{\mathcal{P}_0} \Delta\mathbf{C}_S + \frac{\partial \mathbf{S}^S_E}{\partial \Theta^S}\Big|_{\mathcal{P}_0} \Delta\Theta^S \quad , \tag{11}$$

where

$$(\dots)\Big|_{\mathcal{P}_0} = (\dots)\Big|_{\mathbf{C}_S = \mathbf{I},\,\Theta^S = \Theta^S_{0S}} \quad , \tag{12}$$

of the stress tensor $\mathbf{S}^S_E$ at the position $\mathcal{P}_0$ will be introduced, i.e., the stress (10) will be evaluated in a Taylor series for $\mathbf{C}_S = \mathbf{I}$ and $\Theta^S = \Theta^S_{0S}$ by neglecting higher-order terms. With the elastic tangent operator $\partial\mathbf{S}^S_E/\partial\mathbf{C}_S$ and the thermal tangent operator $\partial\mathbf{S}^S_E/\partial\Theta^S$,

$$\begin{aligned} \frac{\partial \mathbf{S}^S_E}{\partial \mathbf{C}_S} &= \frac{1}{2}\,\lambda^S_{cp}\,(1 + J_S\,\frac{\partial \zeta^S}{\partial J_S})\,(\mathbf{C}_S^{-1} \otimes \mathbf{C}_S^{-1}) + \\ &\quad + \{\,\mu^S - \lambda^S_{cp}\,[\log J_S + \zeta^S] + \\ &\qquad + 3\,\alpha^S k^S\,(\Theta^S - \Theta^S_{0S})\,\}\,(\mathbf{C}_S^{-1} \otimes \mathbf{C}_S^{-1})^{\overset{14}{T}} \quad , \\ \frac{\partial \mathbf{S}^S_E}{\partial \Theta^S} &= -\,3\,\alpha^S k^S\,\mathbf{C}_S^{-1} \quad , \end{aligned} \tag{13}$$

where

$$\frac{\partial \zeta^S}{\partial J_S} = J^S_{cp}\, \frac{J_S^2 + \dfrac{(J^S_{cp})^2}{1 - J^S_{cp}}}{\left[J_S^2 + \dfrac{(J^S_{cp})^2}{1 - J^S_{cp}}\,(J_S - 1)\right]^2} \quad , \tag{14}$$

the following form of the linearized law of Hookean type, with respect to the reference placement, is obtained:

$$\mathbf{S}^{\mathrm{S}}_{\mathrm{Elin}} = 2\,\mu^{\mathrm{S}}\,\mathbf{E}_{\mathrm{S}} + \lambda^{\mathrm{S}}(\mathbf{E}_{\mathrm{S}} \cdot \mathbf{I})\,\mathbf{I} - 3\,\alpha^{\mathrm{S}}\,k^{\mathrm{S}}\,(\Theta^{\mathrm{S}} - \Theta^{\mathrm{S}}_{0\mathrm{S}})\,\mathbf{I} \quad , \tag{15}$$

where

$$\lambda^{\mathrm{S}} = \lambda^{\mathrm{S}}_{\mathrm{cp}}\,[\,1 + \mathrm{J}^{\mathrm{S}}_{\mathrm{cp}}\,(\,1 + \frac{(\mathrm{J}^{\mathrm{S}}_{\mathrm{cp}})^2}{1 - \mathrm{J}^{\mathrm{S}}_{\mathrm{cp}}}\,)\,] \quad . \tag{16}$$

In comparison with the linear theory of one component materials the material parameters μ^{S}, λ^{S} and $k^{\mathrm{S}} = 2/3\,\mu^{\mathrm{S}} + \lambda^{\mathrm{S}}$ can be interpreted as the macroscopic Lamé constants and the macroscopic compression modulus of the constituent φ^{S}. For compressible materials, i.e., $\mathrm{J}^{\mathrm{S}}_{\mathrm{cp}} = 0$, the Lamé constant λ^{S} is equal to $\lambda^{\mathrm{S}}_{\mathrm{cp}}$. Furthermore, within the framework of the linear theory in (15) the rates $\Delta\mathbf{C}_{\mathrm{S}}$ and $\Delta\Theta^{\mathrm{S}}$, see (11), have been replaced by $\Delta\mathbf{C}_{\mathrm{S}} = \mathbf{C}_{\mathrm{S}} - \mathbf{C}_{\mathrm{S}}\,|_{\mathcal{P}_0} = \mathbf{C}_{\mathrm{S}} - \mathbf{I} = 2\,\mathbf{E}_{\mathrm{S}}$ and $\Delta\Theta^{\mathrm{S}} = \Theta^{\mathrm{S}} - \Theta^{\mathrm{S}}\,|_{\mathcal{P}_0} = \Theta^{\mathrm{S}} - \Theta^{\mathrm{S}}_{0\mathrm{S}}$, where $\mathbf{E}_{\mathrm{S}} = 1/2\,(\mathbf{C}_{\mathrm{S}} - \mathbf{I})$ is the Lagrange strain tensor. The determination of the strain tensor

$$\begin{aligned} \mathbf{E}_{\mathrm{S}} = {} & \frac{1}{2\,\mu^{\mathrm{S}}}\,\mathbf{S}^{\mathrm{S}}_{\mathrm{Elin}} - \frac{\lambda^{\mathrm{S}}}{2\,\mu^{S}\,(\,2\,\mu^{\mathrm{S}} + 3\,\lambda^{\mathrm{S}}\,)}\,(\,\mathbf{S}^{\mathrm{S}}_{\mathrm{Elin}} \cdot \mathbf{I}\,)\,\mathbf{I} + \\ & + \alpha^{\mathrm{S}}\,(\,\Theta^{\mathrm{S}} - \Theta^{\mathrm{S}}_{0\mathrm{S}}\,)\,\mathbf{I} \end{aligned} \tag{17}$$

from (15) shows that the material parameter α^{S} can be interpreted as the macroscopic thermal expansion coefficient of the constituent φ^{S}. It is worth mentioning that the quantity c^{S} in (7) can be identified with the help of the balance of energy as the macroscopic specific heat of the solid.

4. Examples

In the following, the constitutive law for $\mathbf{T}^{\mathrm{S}}_{\mathrm{E}}$, see (8), for compressible and incompressible empty elastic porous solids ($p = 0$, $\mathbf{T}^{\mathrm{S}}_{\mathrm{E}} = \mathbf{T}^{\mathrm{S}}$) within the framework of finite deformation processes will be discussed. Furthermore, thermal effects will be neglected, i.e., $\Theta^{\mathrm{S}} = \Theta^{\mathrm{S}}_{0\mathrm{S}}$. In Fig. 1b the results (displacements versus load referring to μ^{S}) of the FEM-simulation of the boundary problem shown in Fig. 1a for the plain strain state for a compressible and an incompressible porous solid are presented. For a given Lamé constant μ^{S} the other material parameters are taken as $\lambda^{\mathrm{S}} = 0.6667\mu^{\mathrm{S}}$ and $k^{\mathrm{S}} = 1.3333\mu^{\mathrm{S}}$, i.e., Poison's ratio is 0.2, i.e., stresses perpendicular to the load direction are considered. The parameter $\lambda^{\mathrm{S}}_{\mathrm{cp}}$ is determined by (16) for a given $\mathrm{J}^{\mathrm{S}}_{\mathrm{cp}}$.

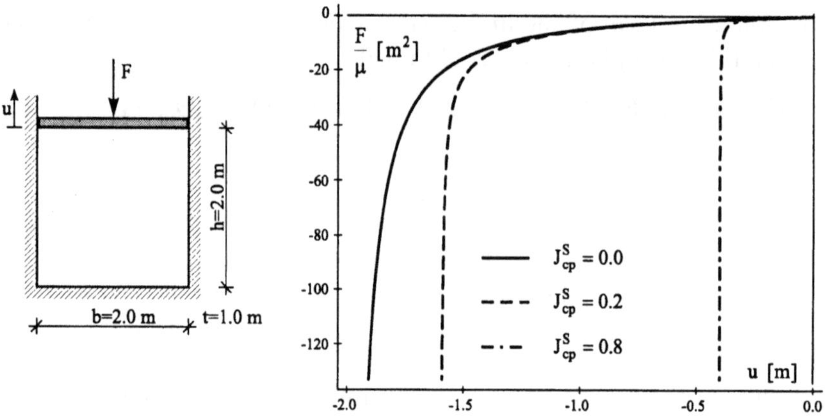

Figure 1. Displacements versus load referring to μ^S for compressible and incompressible porous solids (compression).

5. Conclusion

In this paper, a constitutive law for the description of the thermo-elastic compressible and incompressible porous solids has been presented. For compressible materials ($J^S_{cp} = 0$), the relation (8) takes on the form of the well-known law of Simo & Pister [4]. For incompressible materials ($J^S_{cp} = n^S_{0S}$), the displacement curves, see right Fig., of the law in question are almost identical with the curves calculated with the finite law of Ehlers & Eipper [5] developed for incompressible elastic porous solids without thermal effects.

References

1. de Boer, R. Highlights in the historical development of the porous media theory: Toward a consistent macroscopic theory. *Appl. Mech. Rev.*, **49**, 201–261, 1996.
2. Bluhm, J. A consistent model for saturated and empty porous media. Forschungsbericht aus dem Fachbereich Bauwesen 74, Universität-GH Essen, 1997.
3. Beatty, M. F. Topics in finite elasticity: Hyperelasticty of rubber, elastomers, and biological tissues – with examples. *Appl. Mech. Rev.*, **40**, 1699–1734, 1987.
4. Simo, J. C. and Pister, K. S. Remarks on rate constitutive equations for finite deformation problems: Computational implications. *Comput. Meth. Appl. Mech. Eng.*, **46**, 201–215, 1984.
5. Ehlers, W. and Eipper, G. Finite elastic deformations in liquid-saturated and empty porous solids. *Transport in Porous Media*, **34**, 179–191, 1999.

A Generalized Cam-Clay Model

S. Krenk
Department of Structural Engineering and Materials
Technical University of Denmark
DK-2800 Lyngby, Denmark

1. Introduction

The classical Cam-Clay model may be considered as an associated elasto-plastic material model with volumetric plastic work hardening, Schofield and Wroth [8]. Non-linear elastic properties were included by a tangent bulk modulus proportional to the mean stress. The main features of the Cam-Clay model is the identification of a flow potential via an approximate friction hypothesis and a simple hardening rule providing dilative or contractive behaviour depending on the location of the current stress state relative to a 'critical surface' corresponding to stress states with deformation at constant volume.

The concepts of the Cam-Clay model need some refinement in order to permit accurate representation of the experimentally determined behaviour of granular materials. The theory presented here introduces three essential generalizations: the friction representation of the flow rule is refined and a separate non-associated yield surface is introduced, a small contribution from shear is included in the hardening rule to permit dilation before failure, and the theory is extended in a consistent manner into three dimensions.

2. Flow potential and yield surface

The theory is first developed for a two-dimensional state of stress (σ_1, σ_3) with $\sigma_1 \geq \sigma_3$ and stress positive in compression. These results are then 'unfolded' into three-dimensional stress-space. Introduce the mean stress $\sigma = \frac{1}{2}(\sigma_1 + \sigma_3)$ and the maximum shear stress $\tau = \frac{1}{2}(\sigma_1 - \sigma_3)$. The corresponding strain components are the volumetric strain $\varepsilon = \varepsilon_1 + \varepsilon_3$ and the maximum shear strain $\gamma = \varepsilon_1 - \varepsilon_3$.

The plastic strain increment follows from the flow potential $g(\sigma, \tau)$ in the form $(d\varepsilon^p, d\gamma^p) = d\chi(\partial g/\partial\sigma, \partial g/\partial\tau)$. Curves of constant $g(\sigma, \tau)$ can be determined by using the orthogonality condition $(d\sigma, d\tau)\cdot(d\varepsilon^p, d\gamma^p) = 0$ in the plastic work relation

$$\sigma\, d\varepsilon^p \;+\; \tau\, d\gamma^p \;=\; dW^p \tag{1}$$

W. Ehlers (ed.), *IUTAM Symposium on Theoretical and Numerical Methods in Continuum Mechanics of Porous Materials*, 33–38.

leading to the differential equation

$$\frac{\tau}{\sigma} - \frac{d\tau}{d\sigma} = \frac{dW^p}{\sigma\, d\gamma^p} \tag{2}$$

For a friction material it is assumed that the plastic work increment can be expressed in the form $n\sigma_* d\gamma^p$, where σ_* represents a normal stress on a critical section, and n represents a coefficient of friction. Three hypotheses are illustrated in Fig. 1. The first is a generalized Coulomb friction theory for granular materials, in which the friction is assumed to act on a critical interface between the grains, determined by the current stress state, Krenk [5]. The second is the classical Cam-Clay assumption of $\sigma_* = \sigma$, while the third is a modified Cam-Clay hypothesis, i which $\sigma_* = \sigma_{min}$, Krenk [6]. The integrated form of these three hypotheses is

$$\frac{\tau}{\sigma} = \gamma_g(\sigma/\sigma_0) = \begin{cases} \sin\left(n \ln(\sigma_0/\sigma)\right) & , \text{ granular friction} \\ n \ln(\sigma_0/\sigma) & , \text{ mean stress, } \sigma \\ 1 - (\sigma/\sigma_0)^n & , \text{ minimum stress, } \sigma_{min} \end{cases} \tag{3}$$

Figure 1a shows the flow potential for granular friction. It blends

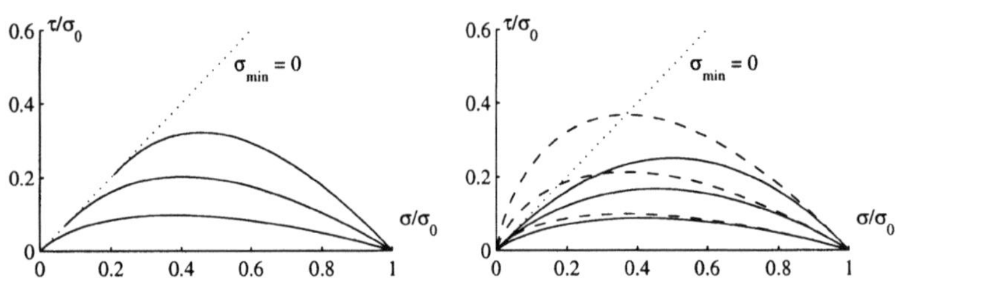

Figure 1. Plastic flow potential contours for $\varphi = 15°, 30°, 45°$, a) granular friction, b) minimum stress σ_{min}: —, mean stress σ: - -.

smoothly into a tension cut-off at $\sigma/\sigma_0 = e^{-\pi/2n}$. The classical Cam-Clay hypothesis, using the mean stress in the friction relation continues into the tension region as shown in Fig. 1b, and for friction angle $\varphi > 45°$ the highest point on the curve, indicating the characteristic state of vanishing dilation, lies in the tension region. This defect is corrected in the modified friction relation based on σ_{min}. The flow potential of the granular friction theory is seen to lie between those based on σ and σ_{min}.

The flow potential curves of two dimensional σ, τ-theory can be extended into the compression octant of three-dimensional principal stress-space, Fig. 2a, by defining the flow potential surfaces as (Krenk, [4], [6])

$$g(\boldsymbol{\sigma}) = -J_3 + pJ_2 - p^3\gamma_g(p/p_g)^2 = 0 \tag{4}$$

where $p = \frac{1}{3}(\sigma_1+\sigma_2+\sigma_3)$ is the mean stress and J_2 and J_3 the deviatoric stress invariants. $p=p_g$ defines the intersection with the isostatic axis.

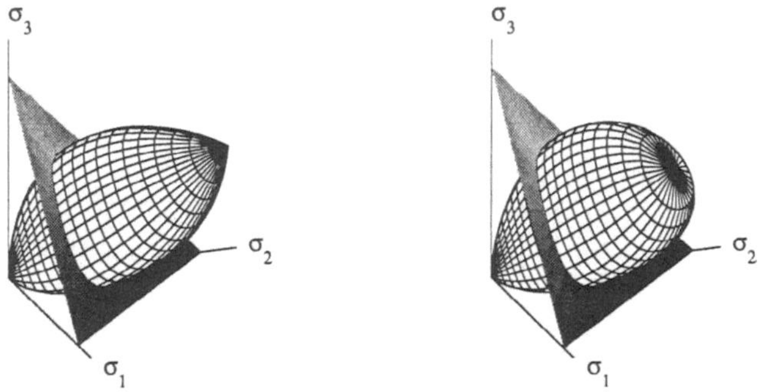

Figure 2. Principal stress representation of a) plastic flow potential, b) yield surface

Experiments indicate a rounded shape of the yield surface at the isostatic axis, as represented by the function

$$\left(\frac{\tau}{\sigma}\right)^2 = \gamma_f(\sigma/\sigma_f)^2 = 1 - \left(\frac{\sigma}{\sigma_f}\right)^m \tag{5}$$

The corresponding three-dimensional extension

$$f(\boldsymbol{\sigma}) = -J_3 + pJ_2 - p^3\gamma_f(p/p_f)^2 = 0 \tag{6}$$

gives the yield surface shown in Fig. 2b.

3. Hardening and dilation

Typical results from a standard triaxial test on normally consolidated sand are shown in Fig. 3. The stress ratio q/p increases from zero, reaching $M_c = (q/p)_c$ at the characteristic state and $M_u = (q/p)_u$ at the ultimate state. The shear strain $\varepsilon_q = \frac{2}{3}(\varepsilon_1 - \varepsilon_3)$ increases during the test, while the volumetric strain $\varepsilon_v = \varepsilon_1 + 2\varepsilon_3$ first increases (contraction), reaches the characteristic state and then decreases (dilation). The elastic and elasto-plastic volumetric stiffness are determined by the parameters κ, λ, and the elastic shear stiffness is G, like the classical Cam-Clay model,

$$d\varepsilon_v^e = \frac{\kappa}{p}dp \quad , \quad d\varepsilon_v^{ep} = \frac{\lambda}{p}dp \quad , \quad d\varepsilon_q^e = 3G\,dq \tag{7}$$

It follows from the stress curve in Fig. 3, that hardening continues, when the characterisic state is passed. This is modeled by introducing plastic work hardening as the sum of the work associated with plastic

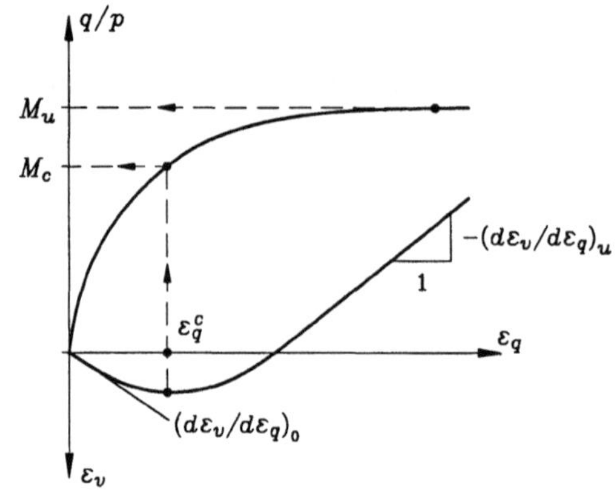

Figure 3. Normalized stress and volumetric strain versus shear strain.

volume change, $pd\varepsilon_v^p$, and a contribution from the work associated with deviatoric plastic strains, $s_{ij}de_{ij}^p$ with weight factor $w << 1$. The size of the current yield surface is determined by its intersection with the isostatic axis at $p = p_f$, and in terms of the stress components p, q the weighted plastic work hardening takes the form

$$dp_f = \frac{1}{\lambda - \kappa}\left(p\, d\varepsilon_v^p + w\, q\, d\varepsilon_q^p\right) \tag{8}$$

The non-dimensional weight factor w on the deviatoric work determines the asymptotic rate of dilation prior to failure.

4. Model calibration

The model parameters can be determined from a standard triaxial test as outlined below and described in detail by Ahadi and Krenk [1]. First the flow potential shape parameter n is determined from the stress ratio $M_c = (q/p)_c$ of the characteristic state.

The non-dimensional hardening weight parameter w is determined from the asymptotic rate of dilation at the ultimate state. At the ultimate state hardening vanishes, and it then follows from the hardening rule (8) that

$$-w\, M_u = \left.\frac{d\varepsilon_v^p}{d\varepsilon_q^p}\right|_u = \left.\frac{\partial g/\partial p}{\partial g/\partial q}\right|_{M_u} \tag{9}$$

The last equality is used to determine the stress ratio M_u by iteration, and subsequently the parameter w is determined from the first equality.

Figure 4 shows the experimentally determined direction of the plastic strain increment and the local direcion of the yield surface for

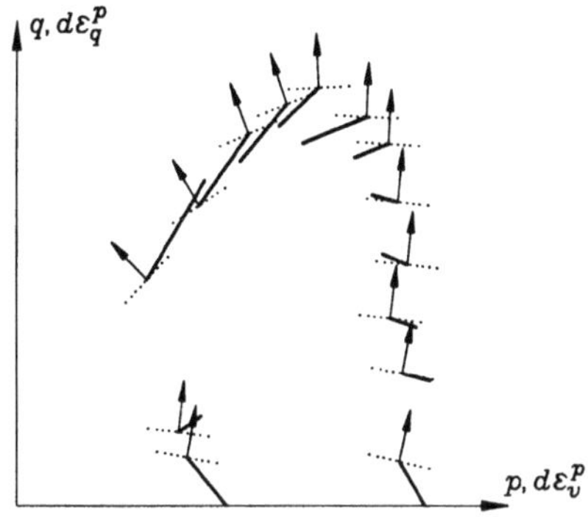

Figure 4. Direction of plastic strain increments for sand, (Andersen *et al.*, 1997).

sand. It is seen that the maximum of the yield contour, corresponding to $M_f = (q/p)_f$, lies to the right of the maximum of the flow potential. Analysis of several test series Ahadi and Krenk [1] show that for sand the shape of the yield surface can be estimated from $M_f = \min(1, 2M_c - M_u)$.

The stiffness parameters κ, λ and G are determined from the initial slopes $d\varepsilon_v/dp$, $d\varepsilon_q/dq$ and the position of the asymptote of the curve $(\varepsilon_q, \varepsilon_v)$.

5. Representative results

Figure 5 shows the ability of the model to represent drained triaxial test data for dense sand with initial specific pore volume $e = 0.55$ and confining pressure $p_0 = 0.64\,\text{MPa}$, Ibsen and Jakobsen [3]. The model represents the gradual development of dilation very well.

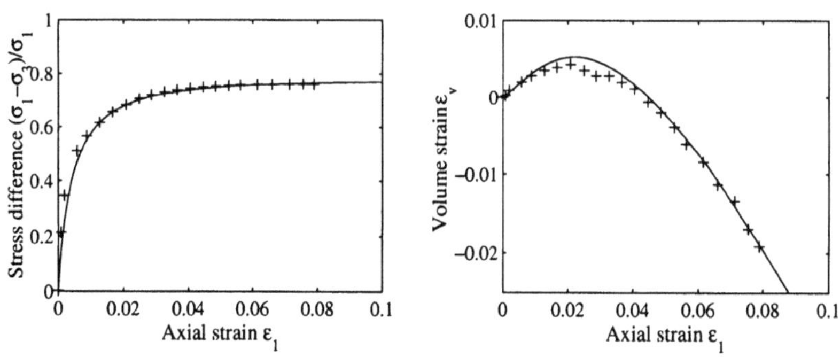

Figure 5. Normalized stress and volumetric strain for dense Lund sand.

Figure 6 shows undrained triaxial test data for loose and dense sand, Lee and Seed [7]. The model represents the difference between the two types of sand as well as the dependence on initial confining pressure very well.

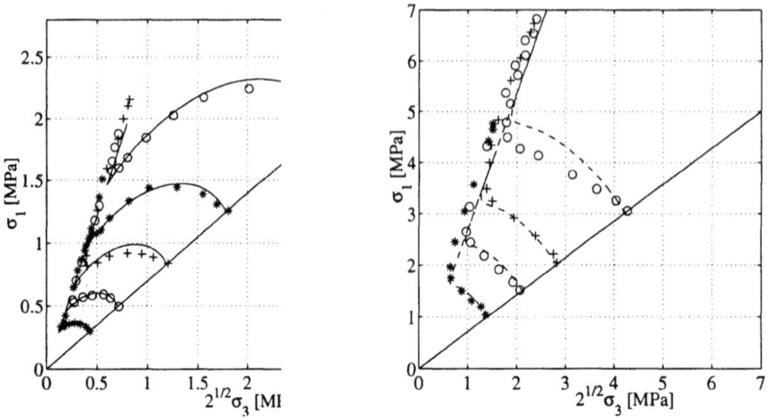

Figure 6. Constant volume triaxial compression tests. a) loose Sacramento sand, $p_0 = 0.3, 0.5, 0.85, 1.27, 2.0\,\mathrm{MPa}$, b) dense Sacramento sand, $p_0 = 1.0, 1.5, 2.0, 3.0\,\mathrm{MPa}$.

Acknowledgements

Support from the Danish and Swedish Technical Research Councils is gratefully acknowledged.

References

1. Ahadi, A. and Krenk, S. Characteristic state plasticity for granular materials. Part 2: Calibration and results. *International Journal of Solids and Structures*, **37**, 6361–6380, 1999.
2. Andersen, A. T., Madsen, E. B. and Schaarup-Jensen, A. L. Constitutive Modeling in Soil Mechanics, Master Thesis, Building Technology and Structural Engineering, Aalborg University, 1997.
3. Ibsen, L. B. and Jakobsen F. R. Data Report, Lund Sand No. 0, Part 1, Geotechnical Engineering Group, Aalborg University, 1996.
4. Krenk, S. A family of invariant stress surfaces. *Journal of Engineering Mechanics*, **122**, 201–208, 1996.
5. Krenk, S. Friction, dilation and plastic flow potential, *Physics of Dry Granular Media*, eds. H. J. Hermann, J. P. Hovi and S. Ludig, Kluwer, Dordrecht, 255-260, 1998.
6. Krenk, S. Characteristic state plasticity for granular materials. Part 1: Basic theory. *International Journal of Solids and Structures*, **37**, 6343–6360, 1999.
7. Lee, K. L. and Seed, H. B. Undrained strength characteristics of sand, *Journal of the Soil Mechanics and Foundation Division, ASCE*, **93**, 333-360, 1967.
8. Schofield, A. N. and Wroth, C. P. *Critical State Soil Mechanics*. McGraw-Hill, New York, 1968.

Theoretical and Experimental Investigation into Mechanical Properties of High Density Foam Plastics

Z. Lu and Z. Gao
Research Center of Solid Mechanics
Beijing University of Aeronautics and Astronautics, Beijing 100083

R. Wang
Dept. of Mechanics and Engineering Science
Peking University, Beijing 100871

Abstract. This paper presents the experimental results of high-density PUR foam plastics for tensile and compressive loading at different strain-rates and temperatures. The elastic properties and yield stress of foam plastics is addressed at the same time. Theoretical formulae for moduli and strength of foam plastics are proposed, which agree well with the available experimental data.

Keywords: Foam plastics, mechanical properties, mechanical models

1. Introduction

Foam plastics are important protecting materials. They are extensively used in both civilian and military industries. Since the end of the 50s, investigations into mechanical properties of foam plastics have been done extensively in the world [1]. However, most of the past works studying their mechanical properties were concentrated on the low-density materials [2,3]. For example, many cell structural models of low-density foams to describe their mechanical behavior have been proposed by Gent and Thomas [2], Matonis [2], Chan and Nakamura [2], Gibson and Ashby [3], Christensen [4], Warren and Kraynik [5], Zhu and Knott [6], respectively. In recent years, we have done some works in this field, our interest is in studying the mechanical properties of high-density PUR foam plastics. We have obtained some valuable theoretical and experimental results [7-13]. These are presented here simply, as a summary, including their mechanical properties in compression and tension and some theoretical formulae for moduli and strength of foam plastics.

W. Ehlers (ed.), IUTAM Symposium on Theoretical and Numerical Methods in Continuum Mechanics of Porous Materials, 41–50.

2. Experimental description

In our experiments, several PUR foam plastics with different densities were prepared and made into specimens. All the compression specimens are cylindrical, with size of 25mm in diameter and 25mm in height. While, the tensile specimens are all made into dumbbell shape, with section size of 10mmx5mm. The axis of all specimens is directed in the rise direction of foam. Before the tests, SEM analyses were performed. It was seen that the cell structure of higher density foam plastics is basically spherical, different from that of low-density foam plastics, which is polyhedron [7]. All compressive tests were done on an Instron-1196 Universal Testing Machine, but tensile tests were conducted on a MTS-880 Universal Testing Machine. These machines are all equipped with a heating device. Deformation of the specimens was measured by electronic extensometer, with gauge lengths being 13mm for compression and 25mm for tension. After the test, the compression specimens were made into SEM samples along the vertical and transverse sections, then placed into a X-650 model SEM for observation.

3. Experimental results

3.1. Compressive mechanical properties of PUR foam plastics [8,9]

The compressive stress-strain curves of PUR foam plastics under different velocities are plotted in Fig. 1a. A clear strain-rate effect can be seen from this figure; Furthermore, stress-strain curves at different strain-rates are almost parallel after the elastic region, thus, an overstress model can be used to describe the mechanical behavior after yielding and before densification. In addition, the comparison among compression curves for different densities foam plastics shows that foam plastics with higher density are more sensitive to strain rate. It also indicates that the strain-rate effects of foam plastics are mainly governed by the magnitude of strain-rate and volume ratio of the matrix material.
Stress-strain curves during cyclic compression were obtained under different velocities, a typical one is shown in Fig. 1b. From which, we can observe the following phenomena: (1) The unloading curves in early elastic stage do not coincide with the reloading ones (see the first cycle in Fig. 1b). This shows that PUR foam plastics have obvious viscous effects from the very beginning stage of deformation. (2) If the crossing point of loading and unloading curves is connected to the residual point, the corresponding slope of this line (dash line in Fig. 1b)

may be considered as the average elastic modulus in the course of cyclic compression. It decreases monotonically with the number of cycles. (3) If the hysteresis loop occurred in every loading-unloading cycle is used as the measurement of energy loss, its area may represent the viscous damping work dissipated in every cycle. The calculation shows that the area of hysteresis loop increases with the strain of the material.

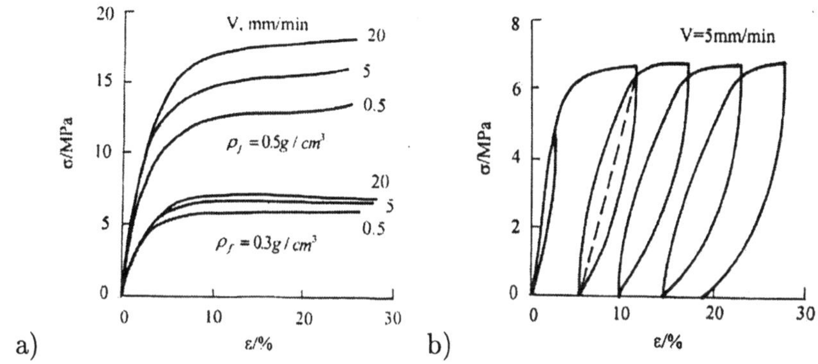

Figure 1. a) The compressive stress-strain curves of PUR foam plastics under different velocities and b) the stress-strain curve of loading-unloading cyclic compression

Apart from strain-rate effects, we have also investigated the influence of temperature on the mechanical behavior of PUR foam plastics because they are often used at different temperatures. Fig. 2 shows the comparison of stress-strain curves for two kind of foam plastics at different temperatures. From this figure, it can be seen that both strength and modulus of PUR foam plastics decrease with an increase of temperature. It is worth noting that the temperature alteration has a stronger influence on the mechanical behavior of foam plastics with higher density than those with lower density.

The compressive failure mechanism of foam plastics was investigated by means of SEM analyses with specimens compressed by quasi-static loading. Some typical SEM pictures are shown in Fig. 3. Some main conclusions can be drawn as follows: (1) Cells in the middle of a specimen and larger cells are more severely damaged under quasi-static loading. (2) Under quasi-static loading, two main failure modes are observed on the cell wall, viz. the bending breakage of the cell wall (Fig.3a) and the open rupture of the cell wall (Fig. 3b).

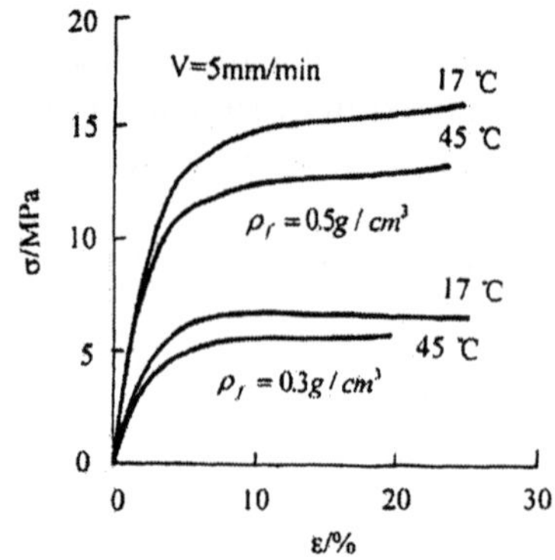

Figure 2. The stress-strain curves of foam plastics at different temperature

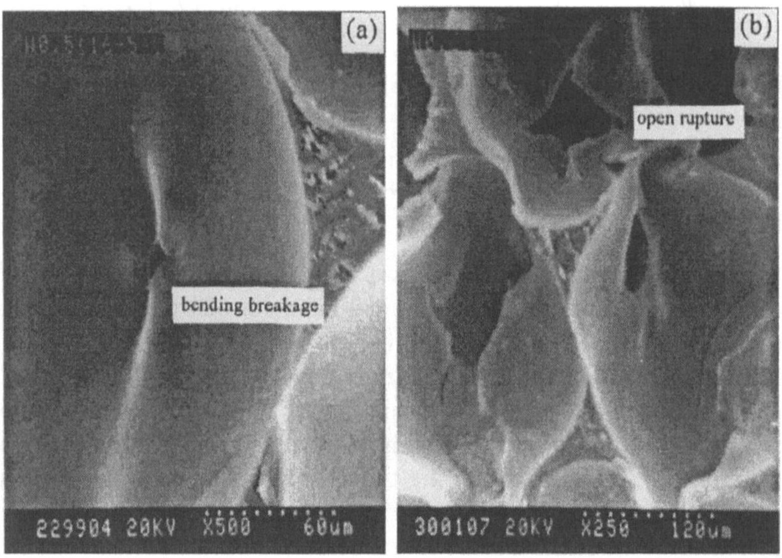

Figure 3. The typical SEM pictures of compressive failure in foam plastics

3.2. Tensile mechanical behavior of PUR foam plastics [10]

The tensile stress-strain curves of different density foam plastics are plotted in Fig. 4. From this figure, we can find that the mechanical behavior of PUR foam plastics in tension is different from that in compression. For example, the tensile deformation capacity of PUR foam plastics is poor, the rupture strain is usually smaller than 5%. Furthermore, the macroscopic fractures of high-density foam plastics exhibit brittle characteristics, whose fracture surfaces are basically a plane. Different from the compressive stress-strain curve, the tensile stress-strain curve may be divided roughly into a linear elastic region, and a nonlinear visco-elastic region.

The tensile stress-strain curves of $0.48g/cm^3$ foam plastics at different strain-rates are shown in Fig. 4b. From this figure, we see that the strain-rate effects in tension are not so obvious as those in compression. However, it is worth noting that the tensile rupture strain decreases with the increase in strain-rate. The tensile stress-strain curves at different temperatures of the same foam plastics are shown in Fig. 5, from which we can see that the temperature effect on foam plastics is obvious and its mechanical properties decrease with the increase of temperature. For example, when the temperature is increased from room temperature to 65 C, the tensile strength and elastic modulus of foam plastics are reduced by 26% and 20%, respectively. For higher density foam ($0.586g/cm^3$), the losses are 23% and 21%, respectively. Furthermore, the tensile rupture strain is increased with the increase in temperature.

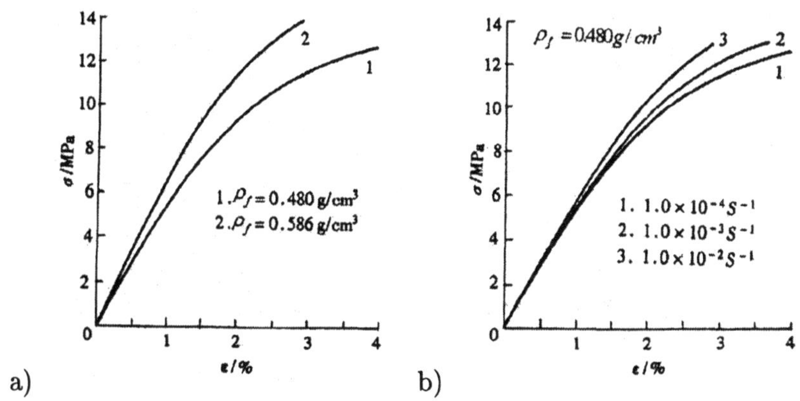

Figure 4. a) The tensile stress-strain curves of different density foam plastics and b) The tensile stress-strain curves of foam plastics under different strain-rate

4. Elastic modulus prediction of foam plastics [11,12]

The Young's modulus and the Poisson's ratio of foam plastics are important parameters in representing their mechanical properties. For lower-density foam plastics, many modulus prediction formulae have been proposed in the past [1]. But, for higher density materials, only some empirical formulae may be used [14]. In recent years, we have tried to find a theoretical prediction formula for the modulus of higher density foam plastics by means of the "Eshelby equivalent inclusion method" and the "differential scheme". We have obtained the effective moduli formulae of foam plastics with linear elastic matrix. In addition, we have also used a three phase spheroidal model in composite mechanics to determine the effective moduli of foam plastics.

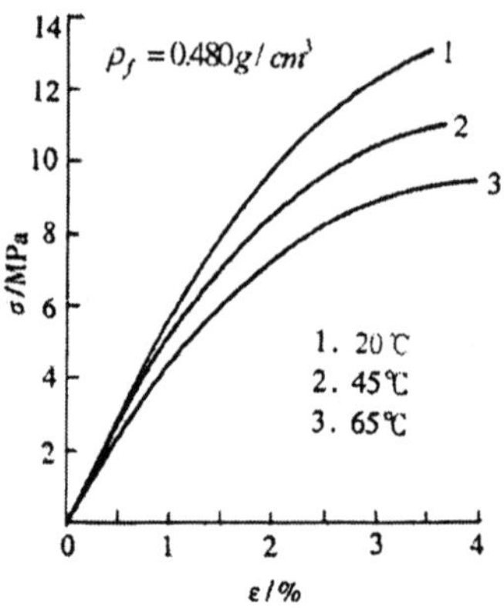

Figure 5. The tensile stress-strain curves of foam plastics under different temperatures

4.1. Differential scheme [11]

By using the Eshelby equivalent inclusion method and a differential scheme, in the case of large porosity, the Youngs modulus, E, and Poisson's ratio, ν, of foam plastics can be shown to satisfy the following differential equations:

$$\frac{dE}{E} = -(\frac{P}{1-f})df\,, \quad dv = -(\frac{\nu P - Q}{1-f})df \tag{1}$$

where

$$P = \frac{3(1-\nu)(9+5\nu)}{2()7-5\nu}\,, \quad Q = \frac{3(1-\nu)(1+5\nu)}{2(7-5\nu)} \tag{2}$$

The effective moduli of foam plastics with porosity, f, may be obtained by integrating the above formulae, and it can be shown that the following approximate expressions for E and ν are good to 92% and better:

$$E = E_0(1-f)^2\,, \quad \nu = 0.2 + (\nu_0 - 0.2)(1-f)^{\frac{5}{6}} \tag{3}$$

in which E_0 and ν_0 are the Young's modulus and Poisson's ratio of the matrix material.

4.2. THREE PHASE SPHEROIDAL MODEL [12]

The three phase spheroidal model in composite mechanics includes the inclusion, the matrix and the equivalent homogeneous media. In the case of foam plastics, the inclusion can be considered as gas. Thus, the corresponding bulk modulus and shear modulus equal zero. Based on this kind of model, the bulk modulus can be easily determined as follows:

$$K = \frac{4\,K_0\,\mu_0\,(1-f)}{4\,\mu_0 + 3\,K_0\,f} \tag{4}$$

where K_0 and μ_0 are the bulk modulus and shear modulus of the matrix material. Similar to the method used by Christensen [15], the following quadratic equation satisfied by shear modulus of foam plastics μ can be obtained:

$$A\,(\frac{\mu}{\mu_0})^2 + B\,(\frac{\mu}{\mu_0}) + C = 0 \tag{5}$$

where

$$\begin{aligned} A &= 4(7-10\nu_0)(7-5\nu_0) + 50(7-12\nu_0+8\nu_0^2)f - 252f^{\frac{5}{3}} + \\ &\quad +25(7-\nu_0^2)f^{\frac{7}{3}} + 2(4-5\nu_0)(7+5\nu_0)f^{\frac{10}{3}} \\ B &= -3(7-5\nu_0)(7-15\nu_0) - 150\nu_0(3-\nu_0)f + 504f^{\frac{5}{3}} - \\ &\quad -50(7-\nu_0^2)f^{\frac{7}{3}} + (5\nu_0-1)(7+5\nu_0)f^{\frac{10}{3}} \\ C &= -(7-5\nu_0)(7+5\nu_0) + 25(7-\nu_0^2)f - 252f^{\frac{5}{3}} + \\ &\quad +25(7-\nu_0^2)f^{\frac{7}{3}} - (7-5\nu_0)(7+5\nu_0)f^{\frac{10}{3}} \end{aligned}$$

From formula (5) and the expressions of A, B and C, we can see that the shear modulus of foam plastics depends only on ν_0 and f. If a matrix material is isotropic, the foam plastics of spherical cell structure can be considered as an isotropic material. Hence, once the bulk modulus and shear modulus are determined, the Youngs modulus and Poisson's ratio of foam plastics can be derived at the same time. In Fig. 6a, the Youngs modulus of foam plastics based on the three-phase model is compared with that by eq.(3) and the experimental results. It is easy to see that the two predictions are in agreement with the available experimental data.

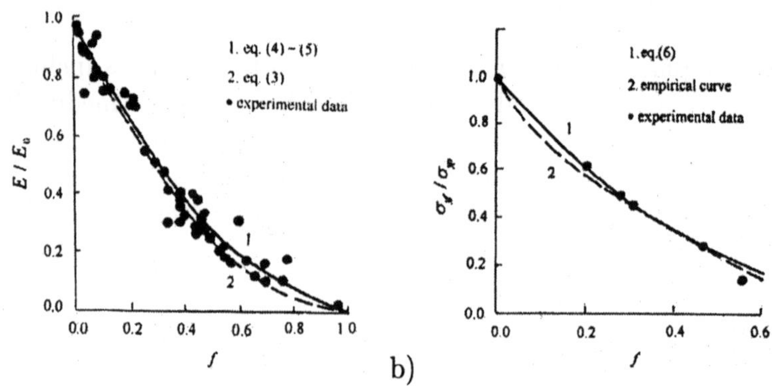

Figure 6. a) The comparison between predicted Youngs modulus and experimental results of foamed plastics and b) he comparison and prediction of yield strength for foam plastics

5. A strength formula for foam plastics [13]

So far as we know, there is no theoretical prediction for yield strength of high-density foam plastics at present, there exist only some empirical formulae of strength [16]. As discussed in the above section, the high-density foam plastics with spherical cell structure can be considered as a kind of composite composed of two phases so that it can be investigated by the method used to determine the mechanical properties of composite with the same spherical reinforced particles [14]. According to Gao and Lelievre's method in [17], dealing with the composite reinforced by spherical particles, the mechanical properties of high-density foam plastics under uniaxial load (tension) is investigated by a self-consistent model. A detailed discussion has been given in reference [13], we shall only present the basic outline and the final result here. At first, we construct a model to simulate a local region of foam plastics. It consists of a spherical cavity surrounded by a matrix spherical shell, which in turn is embedded in the equivalent homogeneous media whose properties represent the mechanical properties of foam plastics. Then, under uniaxial loading, the local model will be deformed axisymmetrically so that the basic equations can be set up in the spherical co-ordinate system. Furthermore, under the suppositions of matrix being incompressible and the hydrostatic stress in the equivalent homogeneous media being a constant, all the basic equations are solved, and the displacement, strain and stress fields determined in the matrix material and the equivalent homogeneous media. Finally, on the basis of the stress fields in the matrix material and the von-Mises yield criterion, the relation between the yield strength of foam plastics and the porosity of material can then be derived as follows:

$$\frac{\sigma_{yf}}{\sigma_{yp}} = \frac{G^*}{G}[(1 + \frac{G^*}{G}\frac{3f}{4(1-f)}^{2}) + \frac{25}{48}(\frac{f}{1-f})^2(\frac{G^*}{G})^2]^{-\frac{1}{2}} \tag{6}$$

where σ_{yf} is the yield strength of foam plastics, $\sigma_{yp}$is the yield strength of the matrix, G^* is the shear modulus of foam plastics, and $G$is the shear modulus of the matrix. Therefore, if the shear modulus ratio and the porosity of the material are given, the yield strength ratios of foam plastics can be predicted by formula (6). Fig. 6b shows the prediction for yield strength of foam plastics given by the present model are in good agreement with the experimental data as well as Masi and Nicolais'empirical curve [16], which is used to fit the experimental data.

6. Conclusions

(1) Under compressive loading, the strain-rate sensitivity for PUR foam plastics increases with its density. Furthermore, Temperature has a clear effect on the stress-strain response of PUR foam plastics, and the effect is greater for higher density foam.
(2) The cyclic compression of PUR foam plastic shows that it is a visco-elastic-plastic material and the cyclic compression deformation enhances the development of damage in the material.
(3) During quasi-static compression, two main failure modes are observed on the cell wall, viz. the bending breakage of the cell wall and the open rupture of the cell wall.
(4) Under tensile loading, the higher density foam plastics exhibit brittle characteristics, and their tensile stress-strain curves have a clear nonlinear region. In addition, the strain-rate effect of foam plastics is not so noticeable in this case, but its temperature effect is quite apparent.
(5) Some formulae for the modulus prediction (Eq. 3~5) and the strength prediction (Eq. 6) of high-density foam plastics are proposed in this paper. They all are in agreement with the experimental data.

Acknowledgements

The support of the National Natural Science Foundation of China (Grant No. 19672005) is gratefully acknowledged.

References

1. Lu, Z., Wang R., Huang Z. *et al.* A Review of Studies on the Mechanical Properties of Foam Plastics. *Advances in Mechanics*, **26**, 306–323 (in Chinese), 1996.
2. Hiyard, N. C. (ed). *Mechanics of cellular plastics.* London: Applied Sci. Publisher, 1982.
3. Gibson, L. J. and Ashby, M. F. *Cellular solids: structures & properties.* Oxford: Pergamon Press, 1988.
4. Christensen, R. M. Mechanics of Low Density Materials, *J. Mech. Phys. Solids*, **34**, 563-578, 1986.
5. Warren, W. E. and Kraynik, A. M. The Linear Elastic Properties of Open-cell Foams, *J. Appl. Mech.*, **55**, 341–346, 1988.
6. Zhu, H. X., Knott, J. F. and Mills, N. J. Analysis of the Elastic Properties of Open-Cell Foams with Tetrakaidecahedral Cells. *J. Mech. Phys. Solids*, **45**, 319–343, 1997.
7. Lu, Z., Li, H. and Tian, C. Determination of Cell Structural Characteristics of Polyurethane Foam Plastics. *Polymeric Materials Science & Engineering*, **11**, 86–91 (in Chinese), 1995.
8. Lu, Z., Wang. R., Huang, Z. *et al.* Experimental Investigation of the Mechanical Properties of PUR Foams. *Acta Mechanica Solida Sinica*, Special Issue, **8**, 535–539, 1995.
9. Lu, Z., Xie, R., Fu, C. *et al.* Visco-elastic(plastic) Effects and Failure Behavior of PUR Foamed Plastics. *Chinese Journal of Aeronautics*, **11**, 1–7, 1998.
10. Lu, Z., Kou, C. and Li, H. Investigation into Tensile Mechanical Properties of PUR Foam Plastics. *Journal of Beijing University of Aeronautics and Astronautics*, **24**, 646–649 (in Chinese), 1998.
11. Huang, Z., Lu, Z. and Wang, R. On Nonlinear Constitutive Relation of Foam Plastics. *Proceedings of IMMM95*, Beijing, Int. Academic Publishers, 65–72, 1995.
12. Lu, Z., Huang, Z. and Wang, R. Determination of Effective Moduli for Foam Plastics Based on Three Phase Spheroidal Model. *Acta Mechanica Solida Sinica*, **8**, 294–302, 1995.
13. Lu, Z. and Gao, Z. Theoretical Prediction for Young's Modulus and Yield Strength of High-density Foamed Plastics. *Science in China* (Series E), **41**, 271–279, 1998.
14. Progelhof, R. C. and Throne, J. L. Young's Modulus of Uniform Density Thermoplastic Foam, *Polym. Eng. Sci.*, **19**, 493–499, 1979.
15. Christensen, R. M. and Lo, K. H. Solution for Effective Shear Properties in Three Phase Sphere and Cylinder Models, *J. Mech. Phys. Solids*, **27**, 315–330, 1979.
16. Masi, P., Nicolais, L., Mazzola, M. *et al.* Tensile Behavior of High-Density Thermosetting Polyester Foams. *Polym. Eng. Sci.*, **24**, 469–472, 1984
17. Gao, Y. C. and Lelievre, J. A Theoretical Analysis of the Strength of Composite Gels with Rigid Filler Particles, *Polym. Eng. Sci.*, **34**, 1369–1376, 1994.

A Unified Sensitivity Analysis Approach for Parameter Identification of Material Models in Fluid-Saturated Porous Media

R. Mahnken
Institute for Structural and Computational Mechanics
University of Hannover, Appelstrasse 9a, D-30167 Hannover

P. Steinmann
Chair of Applied Mechanics, University of Kaiserslautern
Postfach 3049, D-67653 Kaiserslautern

1. Introduction

A large body of references can be found in the literature on the issue of sensitivity analysis (see e.g. the survey in [1] with 148 references). The main consequences and guidelines with regard to linear and non-linear (history dependent) single-phase problems can be summarized as follows [4, 5, 7, 8].

1. The analytical approach yields a *linear* equation for the sensitivities, and thus no iterations are necessary. This also holds for history dependent problems of continuum mechanics, where the response is the result of an iterative procedure.

2. As noted in [7] "the sensitivity formulation must use the consistent tangent operator, whether or not it is used to accelerate the convergence of the equilibrium iterations".

3. Concerning history dependent problems the design sensitivity at a given time (or load) step also necessitates results for the sensitivity at the previous time step, and thus the sensitivity formulation essentially exhibits a recursion structure. Consequently, errors at previous time steps effect the sensitivity at the actual time step.

W. Ehlers (ed.), IUTAM Symposium on Theoretical and Numerical Methods in Continuum Mechanics of Porous Materials, 51–58.

4. A derivative of (stiffness-)matrices as occasionally found in the literature is not necessary.

5. From a computational point of view, the sensitivity computations are relatively inexpensive, since no iterations are required. Basically it consists of three steps: (1) assembly of a "pseudo-load" (pre-processing), (2) solution of a linear system of equations (with consistent tangent matrix already factorized in the equilibrium iteration), (3) back-substitu-tion in order to obtain the design sensitivity of quantities such as forces, stresses, strains or history-variables (post-processing).

6. The sensitivities are obtained simultaneously to the incremental computation at each time (load) step, at the converged state of the equilibrium iteration.

It is the aim of this paper to reconsider the above guidelines towards a unified approach for the sensitivity analysis of a coupled problem modeling fluid saturated porous media.

2. Incremental weak form of the coupled IBVP

Porous media theories eventually lead to an initial boundary value problem (IBVP) consisting of a set of partial differential equations combined with appropriate boundary and initial conditions. For a fluid saturated two phase mixture the balance of linear momentum and the balance of mass in terms of the basic independent quantities, i.e. displacement field $\mathbf{u}$ and pore pressure p, are the essential ingredients of the theory, however, not specified here for brevity of representation, see e.g. [2, 3]. In order to obtain an incremental weak form of the IBVP both balance equations are multiplied with appropriate test functions, i.e. the balance of momentum with a test function $\delta\mathbf{u} \in [H_0^1]^{n_{dim}}$ and the balance of mass equation with a virtual pore pressure $\delta p \in H_0^1(\mathcal{B})$. Furthermore, upon applying the divergence theorem, taking into account the corresponding Neumann boundary conditions and doing a time discretization the following incremental weak formulation is obtained at each time step $\Delta t = t^{n+1} - t^n$:

$$\boxed{\begin{array}{ll} R_s = G_s^{int,n+1} - G_s^{ext,n+1} = 0 & \forall\, \delta\mathbf{u} \in [H_0^1(\mathcal{B})]^{n_{dim}} \\ R_f = G_f^{int,n+1} - G_f^{ext,n+1} - F_f^{n+1} = 0 & \forall\, \delta p \in H_0^1(\mathcal{B}) \end{array}} \quad (1)$$

where

$$
\begin{aligned}
&1. \quad G_s^{int,n+1} &&= \int_{\mathcal{B}} (\nabla_{\mathbf{x}}\delta\mathbf{u} : \boldsymbol{\sigma}^{n+1} - \mathrm{div}\delta\mathbf{u}p^{n+1})dV, \\
&2. \quad G_s^{ext,n+1} &&= \int_{\mathcal{B}} \rho_b \delta\mathbf{u}\cdot\mathbf{b}dV + \int_{\mathcal{A}_t} \delta\mathbf{u}\cdot\bar{\mathbf{t}}_p^{n+1}dA \\
&3. \quad G_f^{int,n+1} &&= \int_{\mathcal{B}} (\nabla_{\mathbf{x}}\delta p\cdot\mathbf{w}_f^{n+1} - \delta p\mathrm{div}\mathbf{v}_s^{n+1})dV \\
&4. \quad G_f^{ext,n+1} &&= \int_{\mathcal{A}_f} \delta p\bar{f}_p^{n+1}dA \\
&5. \quad F_f^{n+1} &&= \int_{\mathcal{B}} S\frac{1}{\Delta t}(p^{n+1}-p^n)\delta p dV
\end{aligned}
\tag{2}
$$

Here $\bar{\mathbf{t}}_p$ and $\bar{f}_p$ denote boundary loads for both balance equations, $\mathbf{b}$ is the distributed body force per unit mass, S is the storativity coefficient and $\mathbf{v}_s$ is the solid velocity. The effective stress tensor $\boldsymbol{\sigma}^{n+1}$ and the Darcy fluid velocity $\mathbf{w}_f^{n+1}$ are obtained from the incremental update schemes

$$
\boxed{
\begin{array}{ll}
\boldsymbol{\sigma}^{n+1}=\mathbf{C}[\boldsymbol{\kappa}_s]:(\boldsymbol{\varepsilon}^{n+1}-\boldsymbol{\varepsilon}_{pl}^{n+1}), & \text{where } \boldsymbol{\varepsilon}^{n+1}=\nabla_{\mathbf{x}}\mathbf{u}^{n+1} \text{ and} \\
\boldsymbol{\varepsilon}_{pl}^{n+1}=\boldsymbol{\varepsilon}_{pl}^{n}+\Delta\boldsymbol{\varepsilon}_{pl}^{n+1}, & \Delta\boldsymbol{\varepsilon}_{pl}^{n+1}=\Delta\boldsymbol{\varepsilon}_{pl}^{n+1}[\boldsymbol{\varepsilon}^{n+1},...;\boldsymbol{\kappa}_s] \\
\mathbf{q}^{n+1}=\mathbf{q}^{n}+\Delta\mathbf{q}^{n+1}, & \Delta\mathbf{q}^{n+1}=\Delta\mathbf{q}^{n+1}[\boldsymbol{\varepsilon}^{n+1},...;\boldsymbol{\kappa}_s] \\
\mathbf{w}_f^{n+1}=\mathbf{K}[\boldsymbol{\kappa}_f]\cdot\mathbf{h}^{n+1}, & \text{where } \mathbf{h}^{n+1}=\nabla_{\mathbf{x}}p^{n+1}
\end{array}
}
\tag{3}
$$

Here $\mathbf{C}$ and $\mathbf{K}$ denote the fourth order elasticity tensor and second order permeability tensor, depending on material parameters $\boldsymbol{\kappa}_s$ and $\boldsymbol{\kappa}_f$ for the solid problem and the fluid problem, respectively. Furthermore $\Delta\boldsymbol{\varepsilon}_{pl}^{n+1}$ and $\Delta\mathbf{q}^{n+1}$ denote the incremental plastic strain and additional (strain-like) internal variables, respectively, both dependent on material parameters $\boldsymbol{\kappa}_s$.

In order to obtain a more compact representation we define

$$
\mathbf{Z} := \begin{bmatrix} \boldsymbol{\varepsilon}_{pl} \\ \mathbf{q} \end{bmatrix}, \quad \mathbf{R} := \begin{bmatrix} R_s \\ R_f \end{bmatrix}, \quad \mathbf{Y} := \begin{bmatrix} \mathbf{u} \\ p \end{bmatrix}, \quad \boldsymbol{\kappa} := \begin{bmatrix} \boldsymbol{\kappa}_s \\ \boldsymbol{\kappa}_f \end{bmatrix} \tag{4}
$$

where $\mathbf{Z}$ is a generalized vector of history variables, and $\mathbf{R}$ and $\mathbf{Y}$ are generalized residual and configuration vectors, respectively. Then, the generalized equilibrium problem is summarized as

$$\boxed{\begin{array}{l}\text{For given initial conditions } \mathbf{Y}^n, \mathbf{Z}^n \text{ and material parameters } \boldsymbol{\kappa}: \\ \text{Find } \mathbf{Y}^{n+1}, \text{such that } \mathbf{R}^{n+1}[\boldsymbol{\kappa}, \mathbf{Z}^n, \mathbf{Y}^n, \mathbf{Y}^{n+1}] = \mathbf{0}\end{array}} \tag{5}$$

The solution of this problem is obtained with a Newton iteration scheme

$$\mathbf{Y}^{(j+1)} = \mathbf{Y}^{(j)} - \left[\frac{\partial \mathbf{R}^{n+1}}{\partial \mathbf{Y}^{n+1}}\right]^{-1} \cdot \mathbf{R}^{n+1(j)}, \qquad j = 0, 1, 2, ... \tag{6}$$

where, by misuse of notation, we have introduced a symbolic notation for the inverse tangent operator of the continuous equations. The individual results for a quadratic Drucker-Prager material model for the solid skeleton combined with a Darcy-model for the fluid are presented in [6].

3. Least-squares functional

Typically a least-squares functional is used as an criterion for parameter identification, such that the discrepancies between experimental data and simulated data are minimized. A possible example is

$$f[\boldsymbol{\kappa}] = \frac{1}{2}\sum_{n=0}^{N-1} ||\tilde{M}\mathbf{Y}^{n+1}[\boldsymbol{\kappa}] - \bar{\mathbf{Y}}^{n+1}||^2 \tag{7}$$

Here $\bar{\mathbf{Y}}^{n+1} = [\bar{\mathbf{u}}^{n+1}, \bar{\mathbf{p}}^{n+1}]$ denote experimental data at different observations points for an arbitrary testing object, which may be obtained e.g. from in-situ experiments. $\tilde{M}$ is an obervation operator, mapping the simulated data $\mathbf{Y}[\boldsymbol{\kappa}]$ obtained with the finite-element-method to the same observation points. In order to employ a gradient based optimization algorithm for minimization of the above functional, the first derivative of the functional

$$\frac{df[\boldsymbol{\kappa}]}{d\boldsymbol{\kappa}} = \left(\tilde{M}\mathbf{Y}^{n+1}[\boldsymbol{\kappa}] - \bar{\mathbf{Y}}^{n+1}\right) \cdot \tilde{M}\frac{d\mathbf{Y}^{n+1}[\boldsymbol{\kappa}]}{d\boldsymbol{\kappa}} \tag{8}$$

has to be determined, thus requirering the parameter sensitivity of the configuration vector, i.e.

$$d\mathbf{Y}^{n+1}[\boldsymbol{\kappa}]/d\boldsymbol{\kappa} = \left[d\mathbf{u}^{n+1}[\boldsymbol{\kappa}]/d\boldsymbol{\kappa}\ ,\ dp^{n+1}[\boldsymbol{\kappa}]/d\boldsymbol{\kappa}\right]$$

4. Sensitivity analysis for the coupled incremental weak form

The starting point for determination of $d\mathbf{Y}^{n+1}[\boldsymbol{\kappa}]/d\boldsymbol{\kappa}$ is the general representation for the residual in (5) with all independent variables, i.e. $\mathbf{R}^{n+1} = \hat{\mathbf{R}}^{n+1}\left[\boldsymbol{\kappa}, \mathbf{Z}^n, \mathbf{Y}^n, \mathbf{Y}^{n+1}\right] = \mathbf{0}$. Since, additionally the configuration vectors and the history variables are dependent on the material parameters the complete material parameter dependence is represented by

$$\mathbf{R}^{n+1} = \hat{\mathbf{R}}^{n+1}\left[\boldsymbol{\kappa}, \mathbf{Z}^n[\boldsymbol{\kappa}], \mathbf{Y}^n[\boldsymbol{\kappa}], \mathbf{Y}^{n+1}[\boldsymbol{\kappa}]\right] = \mathbf{0} \tag{9}$$

Consequently the total differential of this equation is obtained as

$$\begin{aligned}
&1.\quad \frac{d\mathbf{R}^{n+1}}{d\boldsymbol{\kappa}} = \frac{\partial^p \mathbf{R}^{n+1}}{\partial \boldsymbol{\kappa}} + \frac{\partial^Y \mathbf{R}^{n+1}}{\partial \boldsymbol{\kappa}} = \mathbf{0}, \qquad \text{where}\\
&2.\quad \frac{\partial^p \mathbf{R}^{n+1}}{\partial \boldsymbol{\kappa}} = \underbrace{\frac{\partial \mathbf{R}^{n+1}}{\partial \boldsymbol{\kappa}}}_{\text{Term 1}} + \underbrace{\frac{\partial \mathbf{R}^{n+1}}{\partial \mathbf{Z}^n}}_{\text{Term 2}} \cdot \underbrace{\frac{d\mathbf{Z}^n}{d\boldsymbol{\kappa}}}_{\text{Term 3}} + \underbrace{\frac{\partial \mathbf{R}^{n+1}}{\partial \mathbf{Y}^n}}_{\text{Term 4}} \cdot \underbrace{\frac{d\mathbf{Y}^n}{d\boldsymbol{\kappa}}}_{\text{Term 5}}\\
&3.\quad \frac{\partial^Y \mathbf{R}^{n+1}}{\partial \boldsymbol{\kappa}} = \underbrace{\frac{\partial \mathbf{R}^{n+1}}{\partial \mathbf{Y}^{n+1}}}_{\text{Term 6}} \cdot \frac{d\mathbf{Y}^{n+1}}{d\boldsymbol{\kappa}}
\end{aligned} \tag{10}$$

By the same misuse of notation as in Eq(6) this can be solved for

$$\frac{d\mathbf{Y}^{n+1}}{d\boldsymbol{\kappa}} = -\left[\frac{\partial \mathbf{R}^{n+1}}{\partial \mathbf{Y}^{n+1}}\right]^{-1} \frac{\partial^p \mathbf{R}^{n+1}}{\partial \boldsymbol{\kappa}} \tag{11}$$

which in turn has the same structure as Eq(6). Concerning the above Terms 1-6 the following remarks are noteworthy: Term 1 denotes the partial derivative of $\mathbf{R}$ w.r.t. $\boldsymbol{\kappa}$ and Term 2 the partial derivative of $\mathbf{R}$ w.r.t. $\mathbf{Z}^n$. The Term 3 is obtained from the previous time step, Term 4 denotes the partial derivative of $\mathbf{R}$ w.r.t. $\mathbf{Y}^n$, and Term 5 again is a result from the previous time step. The Term 6 denotes the tangent operator and is identical to the operator used in Eq(6). In this way, the notation $(.)^p$ for the above partial derivative in Eq(10.2) basically excludes the partial derivative w.r.t. the configuration vector $\mathbf{Y}^{n+1}$. Note, the fact that Term 3 and Term 5 are obtained from the previous time step, gives the above result a *recursion* structure. The individual results of the above Terms 1-6 for a model combining a quadratic Drucker-Prager law for the solid problem and a Darcy law for the fluid problem are presented in [6].

For the algorithmic procedure the main consequences of the recursion structure are that $d\mathbf{Z}^{n+1}/d\boldsymbol{\kappa}$ and $d\mathbf{Y}^{n+1}/d\boldsymbol{\kappa}$ have to be calculated at the actual time step to make these quantities available for the following time step. Since $d\mathbf{Y}^{n+1}/d\boldsymbol{\kappa}$ is obtained according to the above result, it remains to calculate $d\mathbf{Z}^{n+1}/d\boldsymbol{\kappa}$, which is determined next.

5. Sensitivity of history variables

From the update scheme Eq(3), one can observe the following dependency for the inelastic plastic strain $\boldsymbol{\varepsilon}_{pl}^{n+1} = \boldsymbol{\varepsilon}_{pl}^{n+1}[\boldsymbol{\varepsilon}_{pl}^{n}[\boldsymbol{\kappa}], \boldsymbol{\varepsilon}^{n+1}[\boldsymbol{\kappa}]] = \boldsymbol{\varepsilon}_{pl}^{n+1}[\boldsymbol{\varepsilon}_{pl}^{n}[\boldsymbol{\kappa}], \mathbf{u}^{n+1}[\boldsymbol{\kappa}]]$ as a result of a "strain driven" algorithm. This motivates the more general representation

$$\mathbf{Z}^{n+1} = \mathbf{Z}^{n+1}[\boldsymbol{\kappa}, \mathbf{Z}^{n}[\boldsymbol{\kappa}], \mathbf{Y}^{n+1}[\boldsymbol{\kappa}]] \tag{12}$$

from which the history sensitivity results as

$$\frac{d\mathbf{Z}^{n+1}}{d\boldsymbol{\kappa}} = \frac{\partial \mathbf{Z}^{n+1}}{\partial \boldsymbol{\kappa}} + \frac{\partial \mathbf{Z}^{n+1}}{\partial \mathbf{Z}^{n}} \cdot \frac{d\mathbf{Z}^{n}}{d\boldsymbol{\kappa}} + \frac{\partial \mathbf{Z}^{n+1}}{\partial \mathbf{Y}^{n+1}} \cdot \frac{d\mathbf{Y}^{n+1}}{d\boldsymbol{\kappa}} \tag{13}$$

This result shows, that in the algorithmic implementation $d\mathbf{Z}^{n+1}/d\boldsymbol{\kappa}$ can only be obtained in a *post-processing* step for given sensitivities $d\mathbf{Y}^{n+1}/d\boldsymbol{\kappa}$.

6. Algorithmic implementation

In the algorithmic implementation firstly $d\mathbf{Y}^{n+1}/d\boldsymbol{\kappa}$ is computed according to the formula (10) in a *pre-processing* step, where it should be noted, that the tangent operator $\partial \mathbf{R}^{n+1}/\partial \mathbf{Y}^{n+1}$ at the converged state of the equilibrium iteration can be used. Then $d\mathbf{Z}^{n+1}/d\boldsymbol{\kappa}$ is computed in a *post-processing* step. An overall structure of the algorithm for determination of the derivative of configuration vector with respect to material parameters is summarized in Table I.

Table I. Overall structure of the algorithm for determination of the derivative of configuration vector with respect to material parameters. Steps 3,4 and 5 are the additional steps in the context of a standard FEM iteration procedure.

Step 1 OBTAIN set of material parameters $\boldsymbol{\kappa}$ from least–squares optimization procedure.
INITIALIZE time step $n = 0$, history variables $\mathbf{Z}^n$, configuration vector $\mathbf{Y}^n$ and its sensitivities $\partial_{\boldsymbol{\kappa}}\mathbf{Z}^n$, $\partial_{\boldsymbol{\kappa}}\mathbf{Y}^n$.

Step 2 PERFORM standard FEM–Newton iteration for the direct problem and obtain:

- Stresses $\boldsymbol{\sigma}^{n+1}$, history variables $\mathbf{Z}^{n+1}$ and filter velocity $\mathbf{w}_f^{n+1}$ at all quadrature points by local iteration procedure.
- Configuration vector $\mathbf{Y}^{n+1}[\boldsymbol{\kappa}]$ by global iteration procedure.

Step 3 PRE–PROCESSING: Determine partial load vector:

- Calculate $\partial^p_{\boldsymbol{\kappa}}\boldsymbol{\sigma}^{n+1}$, $\partial^p_{\boldsymbol{\kappa}}\mathbf{w}_f^{n+1}$ at all quadrature points.
- Assemble $\partial^p_{\boldsymbol{\kappa}}\mathbf{R} = \mathop{\mathbf{A}}_{e=1}^{n_{elm}} \int_{v_e} \mathbf{B}_e^T \begin{bmatrix} \partial^p_{\boldsymbol{\kappa}}\boldsymbol{\sigma}^{n+1} \\ \partial^p_{\boldsymbol{\kappa}}\mathbf{w}_f^{n+1} \end{bmatrix} dv_e$

Step 4 SOLVE linear system of equations

$$\partial_{\boldsymbol{\kappa}}\mathbf{Y}^{n+1} = -[\mathbf{K}_T]^{-1}\partial^p_{\boldsymbol{\kappa}}\mathbf{R}$$

with factorized matrix $\mathbf{K}_T$ from the direct problem.

Step 5 POST-PROCESSING: Update history variables $\mathbf{Z}^{n+1}[\boldsymbol{\kappa}]$ and its derivatives $\partial_{\boldsymbol{\kappa}}\mathbf{Z}^{n+1}$. If n .eq. $N-1$, then EXIT else: Set $n \leftarrow n+1$, GOTO 2.

References

1. Adelmann, H. M., Haftka, R. Sensitivity Analysis of Discrete Structural Systems, *AIAA Journal*, **24**, 823–832, 1986.
2. de Boer, R. and Ehlers, W.: *Theorie der Mehrkomponentenkontinua mit Anwendung auf bodenmechanische Probleme, Teil I,* Forschungsberichte aus dem Fachbereich Bauwesen der Universität-GH-Essen, **40**, Essen, 1986.
3. Lewis, R. W. and Schrefler, B. A.: *The Finite Element Method in the Static and Dynamic Deformation and Consolidation of Porous Media*, Wiley, Chichester, 1998.
4. Mahnken, R. and Stein, E. The Parameter-Identification for Visco–Plastic Models via Finite–Element–Methods and Gradient–Methods, *Modelling Simul. Mater. Sci. Eng.*, **2**, 597–616, 1994.
5. Mahnken, R. and Stein, E. Gradient–Based Methods for Parameter Identification of Viscoplastic Materials, In: Inverse Problems in Engineering Mechanics, Eds.: H. D. Bui & M. Tanaka et al, A. A. Balkama, Rotterdam, 1994.

6. Mahnken, R. and Steinmann, P. A Finite Element Algorithm for Parameter Identification of Material Models for Fluid-Saturated Porous Media, Submitted to: *Mechanics of Cohesive-Frictional Materials*, 1998.
7. Vidal, C. A., Lee, H.-S. and Haber, R. B. The Consistent Tangent Operator for Design Sensitivity Analysis of History–Dependent Response, *Comp. Syst. in Engng.*, **2**, 509–523, 1991.
8. Vidal, C, Haber, R. B. Design Sensitivity for rate–independent elastoplasticity, *Comp. Meths. Appl. Mech. Eng.*, **107**, 393–431, 1993.

Numerical Solution of Soil Freezing Problem by a New Finite Element Scheme

J. Hartikainen and M. J. Mikkola
Laboratory of Theoretical and Applied Mechanics, Helsinki University of Technology, P.O.Box 1100, FIN-02015 HUT, FINLAND
e-mail: juha.hartikainen@csc.fi ; martti.mikkola@hut.fi

Abstract. A finite element scheme is presented for the numerical solution of soil freezing problem. It is based on the space-time finite element concept and on the appropriate choice of weighting functions.

Keywords: Soil freezing, thermomechanical model, space-time finite elements, proper weighting

1. Introduction

Soil freezing is an intricate physical problem of several interactive physical processes of heat and mass transfer. The problem involves propagation of a moving internal layer, the frozen fringe in which water freezes over a range of temperature [6], so that water coexists in solid, liquid and gaseous phases with mineral particles of soil skeleton. The phase change in the frozen fringe creates a strong depression of water, the cryogenic suction which induces migration of water from the unfrozen soil to the frozen fringe. The formation of ice expands pores of the frozen soil giving rise to the frost heave whereas the unfrozen soil consolidates under the negative pore-water pressure.

The mathematical model of the physical problem consists of a system of nonlinear partial differential equations coupling thermal, mechanical and hydraulic aspects, see [2] and [5]. The model is governed by the moving frozen fringe in which the physical parameters have abrupt changes and the solution is nonsmooth and sensitive to small perturbations. As a consequence, the application of the standard Galerkin method produces a poor approximation of the solution. In this paper a numerical scheme for solving the soil freezing problem is presented. It is based on the space-time finite element method and on the proper choice of weighting functions. These functions are constructed elementwise by using Taylor series for the approximation of the adjoint system of the partial differential equations to be solved. This procedure results in a family of test functions which guarantee the stability of the solution and are capable to deal with the moving frozen fringe in a regular

W. Ehlers (ed.), IUTAM Symposium on Theoretical and Numerical Methods in Continuum Mechanics of Porous Materials, 61–66.

finite element mesh. The approach is applied to some examples and comparisons are made with experimental findings of soil freezing tests.

2. Thermomechanical model

The derivation of thermomechanical model for freezing of water saturated soil has been presented in detail in the paper [5]. It is based on the mixture concept, on the conservation laws of mass, linear momentum and energy and on the entropy inequality. The model is outlined in the following.

The state variables defining the state of freezing soil are the thermodynamic temperature T, the volume fractions β^s, β^w and β^i (skeleton, water and ice) and the strains of skeleton and ice, i.e. $\epsilon^s = \epsilon^i = \epsilon$. The volume fractions satisfy the obvious conditions

$$\beta^s + \beta^w + \beta^i = 1\,,\ \beta^s \geq 0,\ \beta^w \geq 0,\ \beta^i \geq 0. \tag{1}$$

Introducing the porosity n and the unfrozen water content χ the volume fractions can be presented in the form

$$\beta^s = 1 - n, \quad \beta^w = \chi n, \quad \beta^i = (1-\chi)n. \tag{2}$$

The densities of constituents are $\rho^k = \beta^k \bar{\rho}^k$, where the bulk densities $\bar{\rho}^k$ are taken as constants. The velocity of water is denoted by $\vec{U}^w$ and the velocities of skeleton and ice by $\vec{U}^s = \vec{U}^i = \vec{U}$. The dissipative behaviour of the freezing soil is assumed to depend on the heat flow $\vec{q}^k,\ k \in \{s, w, i\}$ and on the relative flow of water with respect to the solid constituent, i.e. $\vec{V}^w = \vec{U}^w - \vec{U}$.

The balance laws of mass, linear momentum and energy of the freezing soil can be expressed in the following compact form

$$\tfrac{\partial}{\partial t}\mathbf{Q} + \nabla\cdot\mathbf{J}^{\mathrm{conv}} + \nabla\cdot\mathbf{J}^{\mathrm{diff}} = \mathcal{F} \tag{3}$$

where

$$\begin{aligned}
\mathbf{Q} &= [\ \rho^s \quad \rho^w+\rho^i \quad \vec{0} \quad \textstyle\sum_k \rho^k e^k \ ]^{\mathrm{T}},\\
\mathbf{J}^{\mathrm{conv}} &= [\ \rho^s\vec{U} \quad (\rho^w+\rho^i)\vec{U} \quad \textstyle\sum_k p^k_{\mathrm{th}}\mathbf{I} \quad \sum_k \rho^k e^k \vec{U} \ ]^{\mathrm{T}},\\
\mathbf{J}^{\mathrm{diff}} &= [\ \vec{0} \quad \rho^w\vec{V}^w \quad -\sigma^s_\epsilon - \sigma^i_\epsilon \quad \textstyle\sum_k \vec{q}^k + \rho^w e^w \vec{V}^w \ ]^{\mathrm{T}},\\
\mathcal{F} &= [\ 0 \quad 0 \quad \textstyle\sum_k \rho^k \vec{g} \quad r \ ]^{\mathrm{T}}.
\end{aligned} \tag{4}$$

Above, the specific internal energies of skeleton, water and ice are $e^s = C^sT + \mathcal{W}^s(\epsilon)/\bar{\rho}^s$, $e^w = C^wT + L$ and $e^i = C^iT + \mathcal{W}^i(\epsilon)/\bar{\rho}^i$, respectively. C^k is the specific heat capacity and the constant L is related to the latent heat of fusion ℓ by the relationship $\ell = (C^i - C^w)T_0 + L$, where T_0 is the temperature at 0 °C. $W^s(\epsilon)$ and $W^i(\epsilon)$ are the strain energy functions of skeleton and ice. r is the external heat source. The system above is completed by the following constitutive relations based on the entropy inequality:

(i) The effective stress tensors of skeleton and ice are given by

$$\sigma_\epsilon^s = \begin{cases} 2\beta^s G^s \epsilon' + \beta^k K^s \mathrm{tr}\epsilon\, \mathbf{I} & \text{if} \quad \mathrm{tr}\epsilon \le 0 \\ 0 & \text{if} \quad \mathrm{tr}\epsilon > 0 \end{cases}, \tag{5}$$

$$\sigma_\epsilon^i = 2\beta^i G^i \epsilon' + \beta^i K^i \mathrm{tr}\epsilon\, \mathbf{I}, \tag{6}$$

where ϵ' is the deviator of the strain tensor ϵ. G^k is the shear modulus and K^k the bulk modulus.

(ii) The thermodynamic pressures p_{th}^k are defined by

$$p_{\mathrm{th}}^k = \beta^k \left(\hat{B}^k + \rho^w \frac{\ell T}{T_0} \frac{\partial f}{\partial \beta^k} \right), \quad k \in \{s, w, i\}, \tag{7}$$

where the pressure components $(\hat{B}^s, \hat{B}^w, \hat{B}^i)$ are equal for partially frozen soil and $\hat{B}^s = \hat{B}^w \ge \hat{B}^i$ for unfrozen soil. These so-called subgradients are result of the constraints (1) which are taken into account in the mathematical modelling procedure by the indicator function concept [1]. The function $f(\chi = \beta^w/(1-\beta^s))$ characterizes the reduction of the chemical potential of water due to the adsorption of water to soil particles.

(iii) The heat fluxes of skeleton, water and ice are given by the usual Fourier law

$$\vec{q}^k = -\beta^k \kappa^k \nabla T, \quad k \in \{s, w, i\}, \tag{8}$$

where κ^k are the thermal conductivities.

(iv) The relative flow of water with respect to the solid constituent is defined by the generalized law of Darcy

$$\beta^w \vec{V}^w = \vec{v} = -\frac{k}{\bar{\rho}^w g} \left[\nabla \left(\frac{p_{\mathrm{th}}^w}{\beta^w} \right) - \bar{\rho}^w \vec{g} + \bar{\rho}^w \tfrac{\ell T}{T_0} \nabla f \right], \tag{9}$$

where the reduction of the permeability k due to freezing is taken into account by the unfrozen water content: $k(\chi) = k_o \chi^\alpha$, $\alpha > 0$. The subscript $()_o$ denotes the unfrozen state.

(v) The equation of phase change is the Clausius-Clapeyron equation

$$(C^w - C^i) T \ln\left(\tfrac{T}{T_0}\right) + L \tfrac{T - T_0}{T_0} + \left(\tfrac{\hat{B}^i}{\bar{\rho}^i} - \tfrac{\hat{B}^w}{\bar{\rho}^w} \right) = \tfrac{\ell T}{T_0} \left(f + \beta^w \tfrac{\partial f}{\partial \beta^w} \right). \tag{10}$$

3. Numerical formulation

The soil freezing problem which has the basic unknowns $\mathbf{U} = [n\ \hat{B}^w \vec{u}\ T]^\top$ is defined completely by the thermomechanical model presented before. The weak form of the problem is obtained by using the space-time finite element concept [4].

Consider a subdivision of the space-time $\Omega\times]0,\mathcal{T}[$ into time slabs $\Omega\times]t_n, t_{n+1}[$ with boundaries $\Gamma\times]t_n, t_{n+1}[$, where $n = 0,\dots,N-1$, and in turn, a discretization of each time slab into space-time elements $\Omega^e\times]t_n, t_{n+1}[$, $e = 1,\dots,n_{el}$, where n_{el} is the number of elements. Given the trial solution space $\mathcal{V}$ and the test function space $\tilde{\mathcal{V}}$, the weak formulation for the soil freezing problem can be stated. Within each $\Omega\times]t_n, t_{n+1}[$, $n = 0,\dots,N-1$, find $\mathbf{U}\in\mathcal{V}$ such that $\forall\tilde{\mathbf{U}}\in\tilde{\mathcal{V}}$

$$\begin{aligned}\int_{t_n}^{t_{n+1}}\int_\Omega &\left(-\tfrac{\partial}{\partial t}\tilde{\mathbf{U}}^\top\mathbf{Q} - \nabla\tilde{\mathbf{U}}^\top\cdot\,\mathbf{J}^{\text{conv}} - \nabla\tilde{\mathbf{U}}^\top\cdot\,\mathbf{J}^{\text{diff}}\right)\mathrm{d}\Omega\mathrm{d}t\\ &+\int_\Omega\left[\tilde{\mathbf{U}}^\top(t_{n+1}^-)\,\mathbf{Q}(t_{n+1}^-) - \tilde{\mathbf{U}}^\top(t_n^+)\,\mathbf{Q}(t_n^-)\right]\mathrm{d}\Omega\\ = \int_{t_n}^{t_{n+1}}\int_\Omega &\tilde{\mathbf{U}}^\top\mathcal{F}\,\mathrm{d}\Omega\mathrm{d}t - \int_{t_n}^{t_{n+1}}\int_\Gamma\tilde{\mathbf{U}}^\top\vec{\mathrm{n}}\cdot\left(\mathbf{J}^{\text{conv}}+\mathbf{J}^{\text{diff}}\right)\mathrm{d}\Gamma\mathrm{d}t.\end{aligned}\tag{11}$$

The weak form (11) is obtained in the way that the system (3) is multiplied by the test functions $\tilde{\mathbf{U}} = [\tilde{n}_h\ \tilde{\hat{B}}_h^w\ \tilde{\vec{u}}_h\ \tilde{T}_h]^\top$ and integrated over the space-time domain $\Omega\times]t_n, t_{n+1}[$ and finally, the conservation, the convection and the diffusion terms are integrated by parts. The second integral enforces weakly the continuity of the solution in time from one slab to the next, so that globally discontinuous functions in time are allowed. Correspondingly, the last integral is used to take into account the Neumann type boundary conditions.

The trial functions in each time slab are assumed to be piecewise linear polynomials in space and constants in time whereas the proper space of weighting functions is selected so that the weighting functions satisfy the adjoint associated with the original problem in each element. This is a fundamental condition to get a nodally accurate solution [3]. However, it has been turned out to be satisfied merely in an approximative sense. Firstly, the linearization of the weak form (11) after integration by parts of the diffusion terms and the terms including the velocity of solid constituent produces an approximation for the adjoint of the soil freezing problem (3). Further, assuming that the weighting functions are constants in time the following form is obtained

$$\left.\begin{aligned}\tilde{n}_{i+1} \;-\; \nabla\tilde{n}_{i+1}\cdot\vec{u}_i &= 0\\ \nabla\cdot\tilde{\vec{u}}_{i+1} \;+\; \nabla\cdot(\nabla\tilde{\hat{B}}^w{}_{i+1}\tfrac{k_i}{\hat{\rho}^w g}) &= 0\\ \nabla\tilde{\hat{B}}^w{}_{i+1} + (t_{n+1}-t_n)\nabla\cdot(\tilde{\sigma}^s_{\epsilon_{i+1}} + \tilde{\sigma}^i_{\epsilon_{i+1}}) &= \vec{0}\\ \tilde{T}_{i+1}\textstyle\sum_k \rho_i^k C_i^k - (t_{n+1}-t_n)\nabla\cdot(\nabla\tilde{T}_{i+1}\textstyle\sum_k\beta_i^k\kappa_i^k) &= 0\end{aligned}\right\}\ \text{in }\Omega^e\text{ at }t_{n+1}\tag{12}$$

where the subscript i denotes the iteration step. Secondly, the system above is solved approximately by the application of Taylor series. The idea is to construct a system of equations of finite difference type such that the partial differential equations to be solved are used to express the second and higher order derivatives in the Taylor series expansions by the values of the functions themselves and their first order derivatives.

4. Numerical example

A numerical simulation is carried out for a full-scale experiment in which a gas pipe with the diameter of 273mm and the length of 16m is buried in soil and subjected to controlled thermal and hydraulic conditions [1]. A cross section of the pipe line buried in silt is considered as a two dimensional problem. Figure 1 shows the computational frame together with the boundary and the initial conditions. The finite element mesh consists of 60 × 80 quadrilaterals or 120 × 160 triangles, respectively. The function f takes the form $f = a(1/\chi - 1)^2$, where $a = 0.00004$, so that the unfrozen water content χ corresponds to the experimental observations. The permeability of the unfrozen soil is $10^{-8}\mathrm{ms}^{-1}$ and the parameter α defining its value in frozen soil has the value of 4.75. The average values of 1MPa and 0.3 are chosen for the elastic modulus and the Poisson's ratio of the soil, respectively.

The movement of the 0 °C isotherm with time is shown in the Figure 2A while the Figure 2B shows the calculated total frost heaves at different sites together with the experimental findings. The roughness in zero isotherms is not real, rather it follows from imperfect post processing.

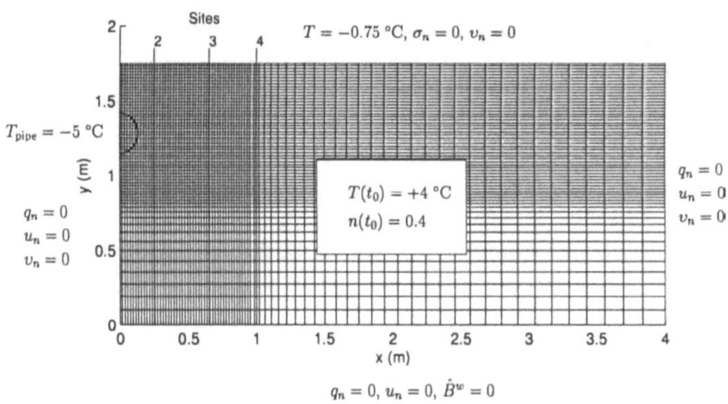

Figure 1. Computational frame.

5. Conclusions

A numerical scheme for the approximation of the solution of soil freezing problem has been presented. The proposed scheme is based on the space-time finite element formulation and on the appropriate weighting functions which are elementwise approximations for the adjoint system

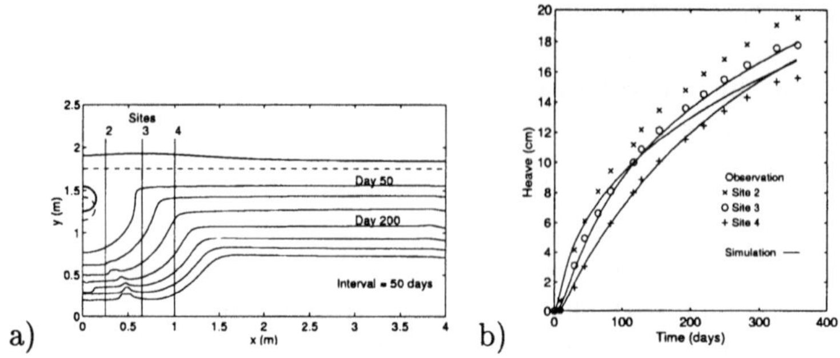

Figure 2. a) Evolution of the 0 °C isotherm. b) Evolution of total frost heave at different sites.

of the partial differential equations to be solved. The numerical results presented illustrate the performance of the method.

Notes

[1] The indicator function is a function $I : \mathbb{R}^3 \to \bar{\mathbb{R}} = \mathbb{R} \cup \{+\infty\}$

$$I(\beta^s, \beta^w, \beta^i) = \begin{cases} 0 & \text{if} \quad (\beta^s, \beta^w, \beta^i) \in \mathcal{C}, \\ +\infty & \text{otherwise}, \end{cases}$$

where the set $\mathcal{C} \subset \mathbb{R}^3$ is a convex set defined by the internal constraints (1)

$$\mathcal{C} = \left\{ (\beta^s, \beta^w, \beta^i) \in \mathbb{R}^3 | \ \beta^s + \beta^w + \beta^i = 1, \quad \beta^k \geq 0 \right\}$$

The subgradients $(\hat{B}^s, \hat{B}^w, \hat{B}^i)$ are elements of the subdifferential ∂I

$$(\hat{B}^s, \hat{B}^w, \hat{B}^i) \in \partial I(\beta^s, \beta^w, \beta^i).$$

References

1. The Formal Meetings of the Canada-Fance Seminar: Gas Pipelines, Oil Pipelines and Civil Engineering in Artic Climates. In *Proceedings of a Seminar Held in Caen and Paris, France*, Geotechnical Science Laboratories, Carleton University, Ottawa, Canada, 1993.
2. Frémond, M. and Mikkola, M. Thermomechanical modelling of freezing soil. In *Ground Freezing 91, Proceedings of the Sixth International Symposium on Ground Freezing*, A. A. Balkema, Rotterdam, Netherlands, 17–24, 1991.
3. Freund, J. and Salonen, E.-M. Notes on Petrov-Galerkin Weighting Functions in Connection with the Scalar Convection-Diffusion equation. Report 28, Institute of Mechanics, Helsinki University of Technology, Finland, 1990.
4. Hughes, T. J. R. and Hulbert, G. M. Space-Time Finite Element Methods for Elastodynamics: Formulations and Error Estimates. *Comput. Meths. Appl. Mech. Engrg.*, **66**, 339–363, 1988.
5. Mikkola, M. and Hartikainen, J. Mathematical Model of Soil Freezing and Its Numerical Implementation. In *Proceedings of the European Conference on Computational Mechanics, ECCM '99*, München, Germany, August 31 – September 3, 1999.
6. Williams, P. Properties and behaviour of freezing soils. *Norwegian Geotech. Inst.*, Publ. 72, 1967.

Space-Time Finite Elements and Adaptive Strategy for the Coupled Poroelasticity Problem

K. Runesson, F. Larsson and P. Hansbo
Department of Solid Mechanics, Chalmers University of Technology, SE-412 96 Göteborg, Sweden

Abstract. A space-time finite element method for the poroelasticity problem is outlined. In particular, this paper focuses on the error generation in time. An adaptive strategy in the time-domain is proposed, and its performance is investigated with the aid of a numerical example in 2D.

Keywords: Adaptivity, a priori error estimate, FEM, poroelasticity

1. Introduction

Very little (if any) attention has been paid to error estimation and adaptive finite element techniques for the coupled poroelasticity problem. Here, we propose a finite element formulation that is based on the discontinuous Galerkin method in time. With this formulation it is possible to establish a residual-based *a posteriori* error estimate in L_2-norm along the lines of the (now) well-established approach using a dual problem, cf. [1].

2. Coupled poroelasticity

Consider the problem determining the displacement field $\boldsymbol{u}(\boldsymbol{x},t)$ and the (excess) pore pressure field $\theta(\boldsymbol{x},t)$ within a poroelastic body occupying the spatial domain Ω:

$$-\boldsymbol{\nabla}\cdot\bar{\boldsymbol{\sigma}}(\boldsymbol{u},\theta)=\boldsymbol{0} \quad \text{with} \quad \bar{\boldsymbol{\sigma}}(\boldsymbol{u},\theta)=\boldsymbol{\sigma}(\boldsymbol{u})-\boldsymbol{\delta}\theta \quad \text{in } \Omega \tag{1}$$

$$i[\dot{\boldsymbol{u}},\dot{\theta}]+\boldsymbol{\nabla}\cdot\boldsymbol{h}(\theta)=0 \quad \text{in } \Omega \tag{2}$$

with suitable boundary conditions on $\boldsymbol{u}$, $\bar{\boldsymbol{T}}(\boldsymbol{u},\theta)\stackrel{\text{def}}{=}\boldsymbol{n}\cdot\bar{\boldsymbol{\sigma}}$, θ, and $Q(\theta)\stackrel{\text{def}}{=}\boldsymbol{n}\cdot\boldsymbol{h}$, and the initial condition

$$i[\boldsymbol{u}_0,\theta_0]=0 \quad \text{with } \boldsymbol{u}_0=\boldsymbol{u}(\boldsymbol{x},0),\ \theta_0=\theta(\boldsymbol{x},0) \tag{3}$$

We introduced the linear form $i[\boldsymbol{u},\theta]$, defined as

$$i[\boldsymbol{u},\theta]\stackrel{\text{def}}{=}\boldsymbol{\nabla}\cdot\boldsymbol{u}+c^{\mathrm{f}}\theta \tag{4}$$

W. Ehlers (ed.), IUTAM Symposium on Theoretical and Numerical Methods in Continuum Mechanics of Porous Materials, 67–72.

The formulation is based on the following assumptions: The *effective* stress $\boldsymbol{\sigma}$ is related to the (bulk) strain
$\boldsymbol{\epsilon}(\boldsymbol{u}) \stackrel{\text{def}}{=} (\boldsymbol{u} \otimes \boldsymbol{\nabla} + \boldsymbol{\nabla} \otimes \boldsymbol{u})/2$
via Hooke's law, whereas the drainage velocity $\boldsymbol{h}$ is related to the pore pressure gradient $\boldsymbol{g}(\theta) \stackrel{\text{def}}{=} \boldsymbol{\nabla}\theta$ via Darcy's law:

$$\boldsymbol{\sigma}(\boldsymbol{u}) = \boldsymbol{\mathcal{E}}^{\mathrm{e}} : \boldsymbol{\epsilon}(\boldsymbol{u}), \quad \boldsymbol{h}(\theta) = -\boldsymbol{K} \cdot \boldsymbol{g}(\theta) \tag{5}$$

Finally, we have introduced the fluid compressibility c^{f}.
We define energy products:

$$(\boldsymbol{u}, \boldsymbol{u}')_{\mathrm{EU}} \stackrel{\text{def}}{=} \int_\Omega \boldsymbol{\epsilon}(\boldsymbol{u}) : \boldsymbol{\mathcal{E}}^{\mathrm{e}} : \boldsymbol{\epsilon}(\boldsymbol{u}')\mathrm{d}\Omega, \quad (\theta, \theta')_{\mathrm{ET}} \stackrel{\text{def}}{=} \int_\Omega \boldsymbol{g}(\theta) \cdot \boldsymbol{K} \cdot \boldsymbol{g}(\theta')\mathrm{d}\Omega \tag{6}$$

Moreover, for $\boldsymbol{f}, \boldsymbol{f}' \in [L_2(\omega)]^2$, with $\omega = \Omega$ or Γ, we define

$$(\boldsymbol{f}, \boldsymbol{f}')_{L_2(\omega)} \stackrel{\text{def}}{=} \int_\omega \boldsymbol{f} \cdot \boldsymbol{f}'\mathrm{d}\omega, \quad \|\boldsymbol{f}\|_{L_2(\omega)} \stackrel{\text{def}}{=} \sqrt{(\boldsymbol{f}, \boldsymbol{f})_{L_2(\omega)}} \tag{7}$$

Consider a finite time interval $I = [0, T]$. The space-time variational format of (1) to (3) can be expressed as follows: Find $\boldsymbol{u}$ and θ such that

$$\int_I (\boldsymbol{u}, \boldsymbol{u}')_{\mathrm{EU}}\mathrm{d}t - \int_I (\theta, \boldsymbol{\nabla} \cdot \boldsymbol{u}')\mathrm{d}t = \int_I (\bar{\boldsymbol{T}}_{\mathrm{p}}(t), \boldsymbol{u}')_{L_2(\Gamma_{\mathrm{T}})}\mathrm{d}t \tag{8}$$

$$\int_I (i[\dot{\boldsymbol{u}}, \dot{\theta}], \theta')\mathrm{d}t + \int_I (\theta, \theta')_{\mathrm{ET}}\mathrm{d}t + (i[\boldsymbol{u}_0, \theta_0], \theta'(\boldsymbol{x}, 0)) = 0 \tag{9}$$

for all $\boldsymbol{u}'$ and θ' in the same function spaces as the solution $\boldsymbol{u}$ and θ, respectively.

3. Finite element formulation

3.1. The coupled space-time method cG-dG(0,0)

We introduce a partition $0 = t_0 < t_1 < \ldots < t_N = T$ of $I = [0, T]$ into time-intervals $I_n = (t_{n-1}, t_n)$ of length $k_n = t_n - t_{n-1}$. We also introduce a triangulation of Ω defined by the mesh function $h_n(\boldsymbol{x})$. We restrict our attention to the approximations: The pore pressure and displacements are piecewise *continuous* polynomial functions in space, whereas they are piecewise constant and, thus, *discontinuous* in time. The method reads: For $n = 1, 2, \ldots, N$, find $\boldsymbol{U}_n$ and Θ_n such that

$$(\boldsymbol{U}_n, \boldsymbol{U}_n')_{\mathrm{EU}} - (\Theta_n, \boldsymbol{\nabla} \cdot \boldsymbol{U}_n') = (\{\bar{\boldsymbol{T}}_{\mathrm{p}}\}_n, \boldsymbol{U}_n')_{L_2(\Gamma_{\mathrm{T}})} \tag{10}$$

$$(i[\boldsymbol{U}_n, \Theta_n], \Theta_n') + k_n(\Theta_n, \Theta_n')_{\mathrm{ET}} = (i[\boldsymbol{U}_{n-1}, \Theta_{n-1}], \Theta_n') \tag{11}$$

for all $\boldsymbol{U}_n'$ and Θ_n' in the finite element spaces to which $\boldsymbol{U}_n$ and Θ_n belong. Here, $\{\boldsymbol{f}\}_n$ is the mean value of $\boldsymbol{f}(t)$ on I_n.

Letting the *constant* column vector $\underline{\boldsymbol{Y}}_n$ contain the space-nodal values of $\boldsymbol{U}_n$ and Θ_n, we obtain the matrix equation

$$(\underline{\boldsymbol{A}}_n + k_n \underline{\boldsymbol{B}}_n)\underline{\boldsymbol{Y}}_n = \underline{\boldsymbol{A}}_{n,n-1}\underline{\boldsymbol{Y}}_{n-1} + k_n\{\underline{\boldsymbol{f}}\}_n \tag{12}$$

3.2. The decoupled method in space and time

The traditional "semi-continuous" strategy to deal with time-dependence is to first discretize in space and then in time while keeping the spatial finite element mesh fixed. This leads to the set of ODE's:

$$\underline{\boldsymbol{A}}\,\dot{\underline{\boldsymbol{y}}}(t) + \underline{\boldsymbol{B}}\,\underline{\boldsymbol{y}}(t) = \underline{\boldsymbol{f}}(t), \quad \underline{\boldsymbol{A}}\,\underline{\boldsymbol{y}}(0) = \underline{\boldsymbol{0}} \tag{13}$$

where the *time-dependent* column vector $\underline{\boldsymbol{y}}(t)$ represent the space-nodal values of $\boldsymbol{U}$ and Θ.

We now introduce the approximation $\underline{\boldsymbol{Y}}(t)$ for $\underline{\boldsymbol{y}}(t)$ such that $\underline{\boldsymbol{Y}}(t) = \underline{\boldsymbol{Y}}_n$ for $t \in I_n$, which gives the method: For $n = 1, 2, ..., N$, find $\underline{\boldsymbol{Y}}_n$ such that

$$(\underline{\boldsymbol{A}} + k_n \underline{\boldsymbol{B}})\underline{\boldsymbol{Y}}_n = \underline{\boldsymbol{A}}\,\underline{\boldsymbol{Y}}_{n-1} + k_n\{\underline{\boldsymbol{f}}\}_n \tag{14}$$

Remark: It is $\underline{\boldsymbol{y}}$ that should be considered as the *exact* solution to the finite element problem in time. □

4. Error estimates

4.1. A posteriori error estimate in space-time

Following [1], we choose to control the quantity $i[\boldsymbol{e}_{u,N}, e_{\theta,N}]$ in L_2-norm, where

$$\boldsymbol{e}_{u,N} \stackrel{\text{def}}{=} \boldsymbol{e}_u(\cdot, t_N) = \boldsymbol{u}(\cdot, t_N) - \boldsymbol{U}_N, \quad e_{\theta,N} \stackrel{\text{def}}{=} e_\theta(\cdot, t_N) = \theta(\cdot, t_N) - \Theta_N \tag{15}$$

We now introduce the dual problem to (1) and (2) as follows:

$$-\boldsymbol{\nabla} \cdot \hat{\boldsymbol{\sigma}}(\boldsymbol{\varphi}, \dot{\psi}) = 0 \quad \text{with} \quad \hat{\boldsymbol{\sigma}}(\boldsymbol{\varphi}, \dot{\psi}) = \boldsymbol{\sigma}(\boldsymbol{\varphi}) - \boldsymbol{\delta}\dot{\psi} \quad \text{in } \Omega \tag{16}$$

$$-\boldsymbol{\nabla} \cdot \boldsymbol{\varphi} - c^{\mathrm{f}}\dot{\psi} + \boldsymbol{\nabla} \cdot \boldsymbol{h}(\psi) = 0 \quad \text{in } \Omega \tag{17}$$

with homogeneous boundary conditions on φ, $\hat{\boldsymbol{T}}(\varphi,\dot{\psi})$, ψ and $Q(\psi)$, and a suitable "initial" condition expressed in $i[\boldsymbol{e}_{u,N}, e_{\theta,N}]$.

The dual problem is utilized as follows: Multiplying (16) with $\boldsymbol{e}_u$ and (17) with e_θ, and integrating, we obtain an *exact* error representation formula. Details are omitted here.

4.2. A POSTERIORI ERROR ESTIMATE IN TIME

Let us consider only the error due to discretization in time. We then *assume* that φ and ψ vary in space exactly as do $\boldsymbol{U}$ and Θ. Letting the time-dependent vector $\underline{\chi}(t)$ contain the space-nodal values of φ and ψ, we obtain the error representation

$$\begin{aligned} e^* = &- \sum_{n=1}^{N} (\underline{\boldsymbol{A}}\,\underline{\boldsymbol{Y}}_n - \underline{\boldsymbol{A}}\,\underline{\boldsymbol{Y}}_{n-1})^{\mathrm{T}} \underline{\chi}^*(t_{n-1}) \\ &- \sum_{n=1}^{N} k_n \left[(\underline{\boldsymbol{B}}\,\underline{\boldsymbol{Y}}_n)^{\mathrm{T}} \{\underline{\chi}^*\}_n - \{\underline{\boldsymbol{f}}(t)^{\mathrm{T}} \underline{\chi}^*\}_n \right] \end{aligned} \tag{18}$$

where we introduced $\underline{\chi}^* = \underline{\chi} - \underline{\boldsymbol{Y}}'_n$ and the generic error measure e^*.

The corresponding (time-continuous) dual equation reads

$$-\underline{\boldsymbol{A}}^{\mathrm{T}} \dot{\underline{\chi}} + \underline{\boldsymbol{B}}^{\mathrm{T}} \underline{\chi} = 0 \tag{19}$$

A possible choice of the "initial" condition for $\underline{\chi}(t_N)$ is

$$\underline{\chi}(t_N) = \frac{\underline{\boldsymbol{A}}\,\underline{\boldsymbol{e}}_N}{|\underline{\boldsymbol{A}}\,\underline{\boldsymbol{e}}_N|} \quad \text{or} \quad \underline{\boldsymbol{A}}^{\mathrm{T}} \underline{\chi}(t_N) = \frac{\underline{\boldsymbol{A}}^{\mathrm{T}} \underline{\boldsymbol{A}}\,\underline{\boldsymbol{e}}_N}{|\underline{\boldsymbol{A}}\,\underline{\boldsymbol{e}}_N|} \leadsto e^* = |\underline{\boldsymbol{A}}\,\underline{\boldsymbol{e}}_N| \tag{20}$$

5. Numerical performance – prototype problem

A half-space is loaded with a strip load of intensity $T_{\mathrm{p}}(t)$, cf. Figure 1, which is applied in the shape of a linear ramp up to $t = t_r$ and then held at a constant value, T_{pa}. Both the displacement and pore pressure are approximated as piecewise linear in space; however, due to finite pore compressibility ($c_{\mathrm{f}} > 0$) and finite rate of loading the LBB-condition is not relevant in this case.

Using a fixed mesh in space we compute the error as the sum of the error contributions, E_n, from each time step. The estimated error is then compared to the true error for different uniform time-meshes via the effectivity index, η,

$$\eta = \frac{e}{e_{true}} = \frac{\sum_{n=1}^{N} E_n}{|\underline{\boldsymbol{A}}\,\underline{\boldsymbol{e}}_N|} \tag{21}$$

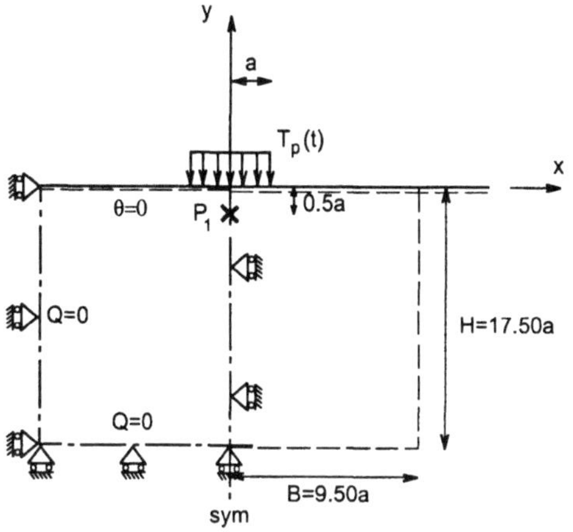

Figure 1. A two-dimensional poroelasticity problem with a strip loading on a half-space.

which is shown in Figure 2 to be close to unity (except for very small number of steps).

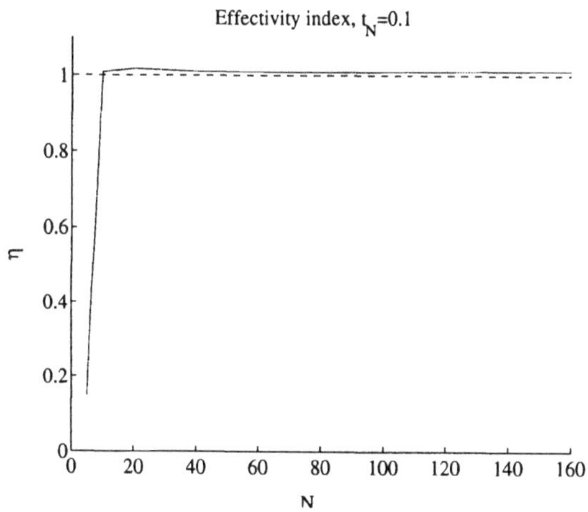

Figure 2. The effectivity index, η, versus number of uniform timesteps, N ($t_N = 0.1 > t_r = 0.01$).

In the adaptive strategy, we remesh until the estimated error is lower than the chosen tolerance. For the loading ramp, $t_r = 0.01$, Figure 3 shows the error after each remeshing. Finally, we compare the relative error for a uniform time-mesh and an adaptively chosen mesh in Figure 4. We note that the adaptive mesh is more efficient.

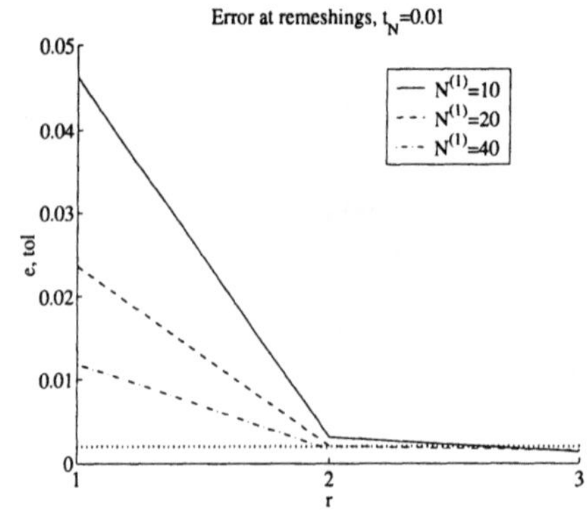

Figure 3. Relative error estimate, e, during adaptive remeshings for 4 different initial uniform time-meshes ($N^{(1)} = 10$, 20 and 40).

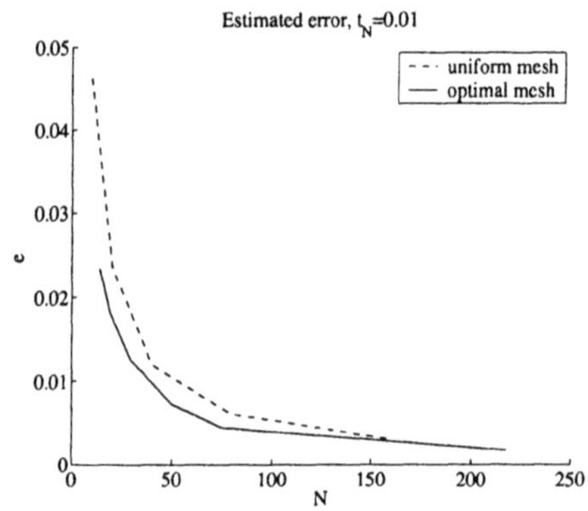

Figure 4. Relative error for adaptively chosen timesteps compared to relative error for uniform time-meshes. (This result is extremely problem-dependent.)

6. Concluding remarks

We have established a rational procedure for computing the error in time and the consequent adaptive time-meshes. It seems to be possible to achieve very accurate error estimates (effectivity index close to one) based on a dual solution that is computed on a crude time-mesh.

References

1. Eriksson, K., Estep, D., Hansbo, P. and Johnson, C. Introduction to adaptive methods for differential equations *Acta Numerica*, 105–158, 1995.

Packed-to-Fluidized Bed Transition and Origin of Particle-Free Regions

A. K. Didwania
Department of Mechanical and Aerospace Engineering
University of California, San Diego, La Jolla, CA 92093-0411, USA

Abstract. The stability of the state of "isothermal" homogeneous fluidization to small amplitude "non-isothermal" disturbances is analyzed. A new general expression for particle phase pressure based on dimensional arguments is proposed and the dispersion relation governing the growth and propagation characteristics of the disturbances is derived.

1. Introduction

When a fluid flows upwards through an assembly of solid particles resting on a porous plate, the particles experience a drag force. For low velocities of the fluid, the drag force is not strong enough to balance the weight of the particles. The particles remain in contact with each other and behave as a packed bed (porous media). As the fluid flow is increased, at some critical flow u_{mfi}, the minimum fluidization velocity, the drag force becomes just sufficient to support the weight of the particles. The particles are no longer in contact with each other and are freely suspended in the fluid. This idealized state of statistical homogeneous fluidized suspension with constant fluidizing velocity is known as the state of uniform fluidization and possesses many desirable heat and mass transfer characteristics with applications to numerous processes of industrial significance.

The stability of this state of uniform fluidization has been the subject of several stability analysis based on "isothermal" continuum models (Anderson and Jackson [1], Drew and Segel [8], Garg and Pritchett [9], Homsy, El-Kaissy and Didwania [12], Didwania and Homsy [7], Batchelor [3]). By isothermal we mean these models neglect spatio-temporal variation of particle velocity fluctuations also known as particle phase "Temperature" as defined later in the text. The fluid is generally taken to be a gas or liquid of density ρ^f, viscosity μ^f and particles are assumed to be rigid of density ρ^s and diameter d_p. These stability analysis have provided considerable theoretical insight into the bifurcation structure of these equations with the possible speculation that the origin of particle-free regions or "bubbles" may be related to a particular sequence of bifurcations (Didwania and Homsy [6], Didwania [5], Homsy

W. Ehlers (ed.), IUTAM Symposium on Theoretical and Numerical Methods in Continuum Mechanics of Porous Materials, 75–80.

[11]). In the present work we revisit the earlier stability analsis by employing "non-isothermal" disturbance with a new expression for particle phase pressure and a particle phase mechanical energy conservation equation.

Equations governing conservation of mass and momentum for a dispersed two phase system have the general form

Continuity:

$$Particle: \frac{\partial \phi}{\partial t} + \frac{\partial \phi v_i}{\partial x_i} = 0 \tag{1}$$

$$Fluid: \frac{\partial (1-\phi)}{\partial t} + \frac{\partial (1-\phi) u_i}{\partial x_i} = 0 \tag{2}$$

Momentum:

$$Particle: \rho^s \phi[\frac{\partial v_i}{\partial t} + v_j \frac{\partial v_i}{\partial x_j}] = \frac{\partial \sigma_{ij}^s}{\partial x_j} + \phi \frac{\partial \sigma_{ij}^f}{\partial x_j} + F_i + \phi \rho^s g_i \tag{3}$$

$$Fluid: \rho^f (1-\phi)[\frac{\partial u_i}{\partial t} + u_j \frac{\partial u_i}{\partial x_j}] = (1-\phi) \frac{\partial \sigma_{ij}^f}{\partial x_j} - F_i + (1-\phi)\rho^f g_i \tag{4}$$

where v_i, u_i and ϕ are respectively particle and fluid phase velocities and particle concentration. The second term on the right hand side of (3) reflects the "effective stress" principle (Didwania and de Boer [4]).

The dispersed solids phase satisfies an additional conservation equation for the mechanical energy given as

$$\frac{3}{2}\rho^s \phi(\frac{\partial T}{\partial t} + v_j \frac{\partial T}{\partial x_j}) = \sigma_{ij}^s \frac{\partial v_i}{\partial x_j} + \frac{\partial}{\partial x_i}(\lambda \frac{\partial T}{\partial x_i}) + \Gamma_s - \Gamma_d \tag{5}$$

The particle "Temperature" is defined as $\frac{3}{2}T = \frac{1}{2} < (v_{ip} - v_i)(v_{ip} - v_i) >$, where v_{ip} is individual particle velocity and $< .. >$ is an appropriate average. It is a measure of the kinetic energy associated with the random particle motion or "psuedo-thermal" energy (i.e. the fluctuational energy of the particles) and unlike Batchelor ([3]) we take it to scale with $|w_i|^2 = |u_{oi} - u_{mfi}|^2$, where u_{oi} is the fluidization velocity. This reflects the fact that the particle temperature results from the fluid kinetic energy that exceeds the amount required to support the weight of the particles. The first term on the right hand side in (5) represents the generation of psuedo-thermal energy by the working of particle phase stress, the second term its accumulation by conduction. The term Γ_s accounts for the generation of particle fluctuation as a result of relative motion between the phases and Γ_d, the dissipation. The source term Γ_s has the form (Koch [13]) $\Gamma_s = ((\rho^s - \rho^f)\phi \overline{v_i' u_i'}/\tau_1)$ where primes indicate fluctuating velocities and the overbar denotes the average. A plausible choice for the relaxation time τ_1 is (u_o/g) which expressed in terms of the drag coefficient is $\tau_1 = ((\rho^s - \rho^f)\phi/\beta)$ Extending heuristically Koch's expression for $\overline{u_i'v_i'}$ we choose $\Gamma_s = \frac{4d_p\beta^2|w_i - v_i|^2}{(\rho^s - \rho^f)\sqrt{\pi T}}$. The form of the dissipation term is presented at the end of this section after discussing the assumed constitutive behaviour of the two phases.

The interphase interaction force F_i in (3)-(4) is taken to consist of drag and added mass contributions and is of the form

$$F_i = F_i^d + F_i^a = (1-\phi)\beta(u_i - v_i) + \phi C\rho_f \frac{d(u_i - v_i)}{dt} \tag{6}$$

where C is the added mass coefficient and the drag coefficient is $\beta = \frac{\phi(\rho^s - \rho^f)g}{u_t}(1-\phi)^{1-n}$ with n being the Richardson-Zaki exponent and $u_t = u_o(1-\phi)^{-n}$, the single particle terminal velocity. The form of the relative accleration term is discussed in details in Homsy, El-Kaissy and Didwania ([12]). Both phases are assumed to be newtonian with the respective stresses given as

$$\sigma_{ij}^s = -p^s(\phi, T)\delta_{ij} + \zeta^s(\phi, T)\frac{\partial v_k}{\partial x_k}\delta_{ij} + 2\eta^s(\phi, T)S_{ij} \tag{7}$$

$$\sigma_{ij}^f = -p^f\delta_{ij} + \zeta^f\frac{\partial u_k}{\partial x_k}\delta_{ij} + 2\eta^f D_{ij} \tag{8}$$

where ζ and η are the bulk and shear viscosities, $S_{ij} = \frac{1}{2}(\frac{\partial v_i}{\partial x_j} + \frac{\partial v_j}{\partial x_i} - \frac{1}{3}\frac{\partial v_k}{\partial x_k})$ and $D_{ij} = \frac{1}{2}(\frac{\partial u_i}{\partial x_j} + \frac{\partial u_j}{\partial x_i} - \frac{1}{3}\frac{\partial u_k}{\partial x_k})$. Dimensional considerations suggest the particle phase pressure to have the form

$$p^s = p_o^s(\phi) + \mu^f d_p^{-1} f_1(\phi)\, T^{1/2} + \rho^s f_2(\phi, \frac{\rho^f}{\rho^s})T, \tag{9}$$

Based on a recent work of Savage ([14]), we take $\eta^s = p^s A/\epsilon, \zeta^s = p^s B/\epsilon, \Gamma_d = p^s d_p \epsilon$, where the root-mean square strain fluctuations ϵ has the form $\epsilon = (\xi T^{1/2})/d_p$. The constants A, B, D depend on the angle of friction between the particles and are treated as constants. Since A, B, D, ξ are all order unity (Savage [14]), we set them equal to 1 in the rest of the paper. The ratio of the third to the second term of the right hand side of (9), $(\rho^s d_p T^{1/2}/\mu^f)(f_2/f_1)$, defines a Stokes number based on particle velocity fluctuations. In the present work, we are concerned with regions close to packed-fluid bed transition where this Stokes number is exceedingly small. The term $p_o^s(\phi)$ in (9) is the usual voidage dependent particle pressure included in other theoretical works (Homsy et. al. [12], Anderson et. al. [2], Glasser et. al. [10]). We choose $f_1 = [1 - (\phi/\phi_m)]^{-2} - 1$ such that $\eta^s = \dfrac{p^s d_p A}{T^{1/2}}$ approaches the viscosity of a concentrated suspension at nonzero larger values of T. We obtain $\Gamma_d = \dfrac{p^s T^{1/2}}{d_p}, \zeta^s = \eta^s, \lambda = s_1 \eta^s$ where s_1 is a constant. We take $\eta^f \approx \zeta^f \approx \mu^f$.

Non-dimensionalizing the material constants and parameters appearing in the above equations with d_p, u_o and other appropriate variables $R = \dfrac{\rho^f}{\rho^s}, T = \dfrac{T}{u_o^2}, p^f = \dfrac{p^f}{\rho^f u_o^2}, \sigma_{ij}^s = \dfrac{\sigma_{ij}^s}{\rho^s u_o^2}, \beta = \dfrac{\beta d_p^2}{\eta^f}, Re = \dfrac{\rho^s d_p u_o}{\mu^f}$,

$F = \dfrac{u_o^2}{g d_p}$.

2. "Isothermal" homogeneous fluidization

The state of 'Homogeneous' fluidization, whose temporal linear stability has been widely investigated is thus represented by

$$u_i = \delta_{i3}, \quad v_i = 0, \quad \phi = \phi_o, \quad T = T_o \; p^f = P_f, \quad p^s = P_s \tag{10}$$

The conservation equations (1)-(5) reduce to

$$\frac{\partial P_f}{\partial x_i} + (\frac{R\beta_o}{Re} + \frac{1}{F})\delta_{i3} = 0 \qquad \frac{R\beta_o}{Re} = \frac{\phi_o(1-R)}{F} \tag{11}$$
$$T_o^{3/2} + \frac{Rep_o^s}{f_1} T_o - \frac{4\beta_o^2 w_o^2}{Re(1-R)\sqrt{\pi} f_1} = 0$$

3. Linear stability analysis

We analyse the temporal linear stability of this base state (12) to a general three dimensional disturbance by introducing the linearized perturbed variables, $\psi' = (\phi'(x_i,t), u_i'(x_i,t), p^{f'}(x_i,t), v_i'(x_i,t), T'(x_i,t))$. We assume these perturbed variables to be of the form $\psi'(x_i,t) = \psi e^{\iota k_j x_j + \sigma t}$ where k_j is the wavenumber and the real and imaginary part of σ are the growth constant and frequency of the perturbations respectively. The governing equations (1)-(5) are linearized about the base state and after substituting for the perturbed variables yield a linear system of equations of the form $L\psi = \sigma N \psi$. The resulting eigenvalue problem $det(L - \sigma N) = 0$ is solved to obtain the dispersion relation describing the growth and propagation characteristics of the disturbances. Here we analyze the dispersion relation for a class of disturbance which depend only on z and t. A complete analysis of general three dimensional disturbance will be presented elsewhere. For this special class of disturbance $\psi(z,t)$, the dispersion relation reduces to

$$(\sigma^2 + a\sigma + b)\phi + cT = 0 \tag{12}$$

$$T = (\frac{l_1\sigma + l_2}{l_3\sigma + l_4})\phi \tag{13}$$

where

$$\begin{aligned} a &= \frac{(1-R)}{A_1(1-\phi_o)F} + \frac{(\zeta^s + (4/3)\eta^s)k^2}{A_1\phi_o Re} + \\ &+ \iota\frac{2kR}{A_1}\frac{\phi_o}{(1-\phi_o}(1 + \frac{C}{(1-\phi_o)}) \end{aligned} \tag{14}$$

$$b = k^2(\frac{p^s_\phi}{A_1} - \frac{R}{A_1}\frac{\phi_o}{(1-\phi_o)}(1 + \frac{C}{(1-\phi_o)})) + \iota k\frac{n(1-R)}{A_1(1-\phi_o)F} \tag{15}$$

$$c = k^2\frac{p^s_T}{A_1} \quad l_1 = \frac{p^s_o}{\phi_o} - \iota\frac{\Gamma_{su}}{k\phi_o(1-\phi_o)} \tag{16}$$

$$l_2 = \Gamma_{s\phi} - \Gamma_{d\phi} + \frac{\Gamma_{su}}{1-\phi_o} \quad l_3 = \frac{3}{2}\phi, \quad l_4 = \lambda^s k^2 + \Gamma_{dT} - \Gamma_{sT} \tag{17}$$

where $A_1 = 1 + R\frac{\phi_o}{(1-\phi_o)}(1 + \frac{C}{\phi_o(1-\phi_o)})$ and a derivative with respect to a variable is denoted by the variable appearing as the subscript. For isothermal model the last term in (12) is absent and the stability criteria and propagation characteristics of disturbances are same as those of earlier investigators. Accounting for T leads to appearance of an additional mode and results in a considerably more complex dependence of growth constant on the disturbance wave number. The detailed quantitative results will be presented elsewhere.

Acknowledgements

The present work is supported by NSF Grants INT-9605036, CTS-9510121 and NASA Grant NAG 3-1888. The author is grateful to Prof. J. D. Goddard for many valuable discussions during the course of this work.

References

1. Anderson, T. B. and Jackson, R. Fluid mechanical description of fluidized beds-stability of the state of uniform fluidization. *Indust. Engng. Chem. Fundam.*, **7**, 12–21, 1968.
2. Anderson, K., Sundaresan, S. and Jackson, R. Instabilities and the formation of bubbles in fluidized beds. *J. Fluid Mech.*, **303**, 327–366, 1995.
3. Batchelor, G. K. A new theory of the instability of a uniform fluidized bed. *J. Fluid Mech.*, **193**, 75–110, 1988.
4. Didwania, A. K. and de Boer, R. Saturated Compressible and Incompressible Porous Solids: Macro- and Micromechanical Approaches. *Transport in Porous Media*, **34**, 101–115, 1999.
5. Didwania, A. K. New Beltrami type solutions to continuum equations for fluidization and their stability. *Physica D*, **84**, 532–544, 1995.
6. Didwania, A. K. and Homsy, G. M. Flow regimes and flow transitions in liquid fluidized beds. *Intl. J. Multiphase Flow*, **7**, 563–580, 1981.
7. Didwania, A. K. and Homsy, G. M. Resonant side-band instabilities in wave propagation in fluidized beds. *J. Fluid Mech.*, **122**, 433–438, 1982.
8. Drew, D. A. and Segel, L. A. Averaged equations for two-phase flows. *Studies in Appl. Math.*, **50**, 205–231, 1971.
9. Garg, S. K. and Pritchett, J. W. Dynamics of Gas-fluidized beds. *J. Appl. Phys.*, **46**, 4493–4500, 1975.
10. Glasser, B. J., Kevrekidis, I. G. and Sundaresan, S. One and two dimensional travelling wave solutions in gas-fluidized beds. *J. Fluid Mech.*, **306**, 183–221, 1996.
11. Homsy, G. M. Nonlinear Waves and the Origin of Bubbles in Fluidized Beds. *Appl. Sci. Res.*, **58**, 251–274, 1998.
12. Homsy, G. M., El-Kaissy, M. M. and Didwania, A. K. Instability waves and the origin of bubbles in fluidized beds. Part II. Comparison with theory. *Intl. J. Multiphase Flow*, **6**, 305–318, 1980.
13. Koch, D. L. Kinetic theory for a monodisperse gas-solids suspension. *Phys. Fluids A* , **2**, 1711–1723, 1990.
14. Savage, S. B. Analyses of slow high-concentration flows of granular materials. *J. Fluid Mech.*, **377**, 1–26, 1998.

h-Adaptive Strategies Applied to Multi-Phase Models

W. Ehlers, P. Ellsiepen and M. Ammann
Institute of Applied Mechanics (Civil Engineering),
University of Stuttgart, D-70550 Stuttgart, Germany
e-mail: (Ehlers, Ellsiepen, Ammann)@mechbau.uni-stuttgart.de

Abstract. Based on the error estimator of Zienkiewicz and Zhu, a new error estimator is presented which is especially designed for multi-phase problems. Furthermore, efficient h-adaptive strategies concerning the generation of a new mesh and data transfer between different meshes are pointed out. The efficiency of these tools is demonstrated using a shear banding problem as a numerical example.

Keywords: h-adaptivity, multi-phase model, error estimation, density function

1. Introduction

In the h-adaptive finite element method, the construction of a new mesh is based on *a posteriori* estimates of the discretization error. Zienkiewicz and Zhu [15] proposed a gradient-based error indicator for linear elasticity which compares the finite element (FE) stress field with a higher order stress field generated by some smoothening technique, e. g. the well-known superconvergent patch recovery (SPR) technique [16]. In order to be able to apply h-adaptive strategies to multi-phase computations, apart from the stress field, additional variables have to be taken into account.

Here, a new error indicator is derived taking as an example a two-phase model consisting of an elastic-viscoplastic solid skeleton and a viscous pore-fluid, cf. Diebels *et al.* [5]. This two-phase model is established based on the Theory of Porous Media (cf. e. g. Bowen, [1], [2] and Ehlers, [7], [8], [9]).

After the error estimation, a mesh density function is computed which takes the local error information and provides information where and to which degree the finite element mesh has to be adapted. The next step in the generic h-adaptive procedure is the generation of the mesh. In general, this can be realized by two different methods: a hierarchical strategy which is here based on bisection of the finite elements by adding new edges (cf. Kossaczký, [12]; Ellsiepen, [10]) and a remeshing strategy where a new, independent mesh is generated. In this paper, the remeshing strategy is used.

In order to obtain good numerical results when computing time dependent problems, the data transfer from the old to the new mesh is

W. Ehlers (ed.), IUTAM Symposium on Theoretical and Numerical Methods in Continuum Mechanics of Porous Materials, 81–86.

crucial. Using the h-adaptive strategies presented in this paper together with an efficient time adaptive scheme (cf. Ellsiepen, [10]), it is shown in the numerical example that accurate results can be obtained.

2. Error estimation and density function

In linear elasticity, the exact discretization error in the stresses, $e := ||\boldsymbol{\sigma} - \boldsymbol{\sigma}^h||_2$, can be approximated by using the SPR technique of Zienkiewicz and Zhu [16]

$$e \approx \eta := ||\boldsymbol{\sigma}^* - \boldsymbol{\sigma}^h||_2 := \left(\int_\Omega (\boldsymbol{\sigma}^* - \boldsymbol{\sigma}^h) \cdot (\boldsymbol{\sigma}^* - \boldsymbol{\sigma}^h) \, dv \right)^{1/2}, \quad (1)$$

where $\boldsymbol{\sigma}$ are the exact stresses (usually not known) and $\boldsymbol{\sigma}^h$, $\boldsymbol{\sigma}^*$ are the calculated and the improved stresses, respectively. $|| \cdot ||_2$ represents the L_2-norm as defined in equation (1).

The global error indicator η is computed in an element-wise fashion by integrating over the element areas Ω_e. Thus, element-wise error indicators $\eta^{(e)}$ can be defined which, in the sum of all elements E, yield the global error η:

$$\begin{aligned} \eta^{(e)} &:= ||\boldsymbol{\sigma}^* - \boldsymbol{\sigma}^h||_{2,e} := \left(\int_{\Omega_e} (\boldsymbol{\sigma}^* - \boldsymbol{\sigma}^h) \cdot (\boldsymbol{\sigma}^* - \boldsymbol{\sigma}^h) \, dv \right)^{1/2} \\ \eta &= \left(\sum_{e=1}^{E} \left(\eta^{(e)} \right)^2 \right)^{1/2}. \end{aligned} \quad (2)$$

In case of the considered two-phase model which consists of an elastic-viscoplastic solid skeleton and a viscous pore fluid (cf. Diebels and Ehlers, [4]), the error indicator of Zienkiewicz and Zhu is extended in such a way that all driving quantities of this model are included:

$$\begin{aligned} \eta_1^{(e)} &:= ||\mathbf{T}_E^{S*} - \mathbf{T}_E^{Sh}||_{2,e}, & W_1 &:= ||\mathbf{T}_E^{Sh}||_2, \\ \eta_2^{(e)} &:= ||\boldsymbol{\varepsilon}_{Sp}^* - \boldsymbol{\varepsilon}_{Sp}^h||_{2,e}, & W_2 &:= ||\boldsymbol{\varepsilon}_{Sp}^h||_2, \\ \eta_3^{(e)} &:= ||\mathbf{w}_F^* - \mathbf{w}_F^h||_{2,e}, & W_3 &:= ||\mathbf{w}_F^h||_2. \end{aligned} \quad (3)$$

Therein, $\mathbf{T}_E^S$ is the extra stress tensor of the solid resulting from Hooke's law, $\boldsymbol{\varepsilon}_{Sp}$ is the plastic strain tensor of the solid and $\mathbf{w}_F$ is the so-called seepage velocity (i. e. a natural variable to describe the fluid motion relative to the deforming skeleton). Again, $(\cdot)^*$ are the smoothened values computed on the basis of the FE quantities $(\cdot)^h$ by the SPR technique.

For practical computations, the absolute errors $\eta_i^{(e)}$ are transferred into dimensionless, relative errors by dividing through W_i. Consequently, tolerance-weighted error measures $\xi_{e,i}$ can be defined based on user-defined relative and absolute tolerances ϵ_r and $\epsilon_{a,i}$:

$$\xi_{e,i} = \frac{\eta_i^{(e)}}{\epsilon_r \cdot W_i + \epsilon_{a,i}}, \qquad i = 1, 2, 3\,. \tag{4}$$

Finally, the sum of the $\xi_{e,i}$ weighted by user-specified weighting factors $\alpha_i \geq 0$ provides the error measure ξ_e per element and the global error measure ξ:

$$\xi_e := \left(\sum_{i=1}^{3} \alpha_i \cdot \xi_{e,i}^2 \right)^{1/2} \quad \text{with} \quad \sum_{i=1}^{3} \alpha_i = 1, \qquad \xi := \left(\sum_{e=1}^{E} \xi_e^2 \right)^{1/2}. \tag{5}$$

The factors α_i can be used to assign variable weight to the $\xi_{e,i}$. For example, calculating purely elastic problems, the error indicator $\xi_{e,2}$ for the plastic strains can be discarded by choosing $\alpha_1 = \alpha_3 = 1/2,\ \alpha_2 = 0$.

The solution on the current mesh is accepted if $\xi \leq 1$ and rejected otherwise. In order to refine ($\xi > 1$) or to coarsen ($\xi < 1$) the mesh, a new characteristic element length h_{new} can be computed with the density function

$$h_{\text{new}} = \xi_e^{-1/(p+1)} \left(\sum_{e=1}^{E} \xi_e^{2/(p+1)} \right)^{-1/2p} h_{\text{old}}\,, \tag{6}$$

where p is the convergence order of the FE discretization. The above density function has been derived by Gallimard *et al.* [11] by minimizing the number of elements in the new mesh in order to reach the user-specified tolerances (i. e. $\xi = 1$).

3. Mesh generation and data transfer

With the information given by the density function, a new mesh can be generated. Here, this is done by a complete remeshing of the whole domain. Therefore, the density function is evaluated in such a way that the new nominal areas for each element of the old mesh are written into a file which are taken into account by the mesh generator during the generation of the new mesh. For this procedure, a modified version of the triangular mesh generator `Triangle` by Shewchuk [14] is used here.

Calculating time-dependent problems, the complete data of the current mesh has to be transferred to the new mesh in order to avoid a restart of the computation. When transferring data between FE meshes, two different data types have to be considered: data at nodal points and data at integration points. Considering the transfer of nodal data, the first task is to find the element in the old mesh wherein a given nodal point P_{new} of the new mesh is located. This element search procedure in a FE mesh can be carried out efficiently by use of a quadtree data structure, cf. Krause and Rank [13]. After the proper element has been found, the local coordinates of P_{new} in the detected element are computed [3] and, finally, the FE shape functions are evaluated to yield the transferred data value at the nodal point P_{new} of the new mesh. When transferring data at integration points, in a first step, the element-wise data has to be projected to the nodal points, for example by a local L_2 projection (SPR). Afterwards, the same strategy as for the transfer of nodal data can be applied.

4. Example

In the example, a numerical simulation of the biaxial experiment is investigated, cf. Diebels *et al.* [6]. Thereby, a sand specimen is loaded displacement driven by a rigid load plate ($\dot{u}_2 = 0.0002$ m/s). The two sides of the specimen are stabilized by a constant stress $t_1 = 1000$ kN/m^2 (Figure 1, left). Due to the imperfection, a shear band is initiated at the bottom of the sample.

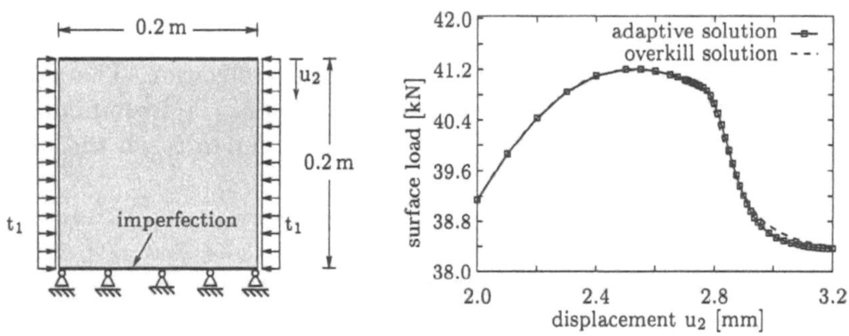

Figure 1. Initial boundary-value problem (left) and load-deflection curves (right)

The shear band is of great interest for practical calculations. Using h-adaptive methods for the computation of this initial boundary value problem, the shear band can be resolved with a substantially lower computational effort than in a non-adaptive FE procedure based on a uniform mesh. This is due to the high gradients of the elastic stresses $\mathbf{T}_E^S$ and – especially – the plastic strains $\boldsymbol{\varepsilon}_{Sp}$ at the edges of the shear band. Consequently, the mesh with 18451 degrees of freedom (d. o. f.) resulting from the adaptive calculation is refined in these areas, cf. Figure 2 (right). The quality of the adaptive calculation is verified by a comparison with an overkill solution which was carried out with a uniformly refined mesh with 57682 d. o. f. The load-deflection curves of the two computations are shown in Figure 1 (right), where both solutions are equivalent to a great extent, whereas the computing times of these two calculations differ significantly. On an SGI Power Challenge R10000/195, the effort for computing the overkill solution was 177 h and for the adaptive solution 2 h.

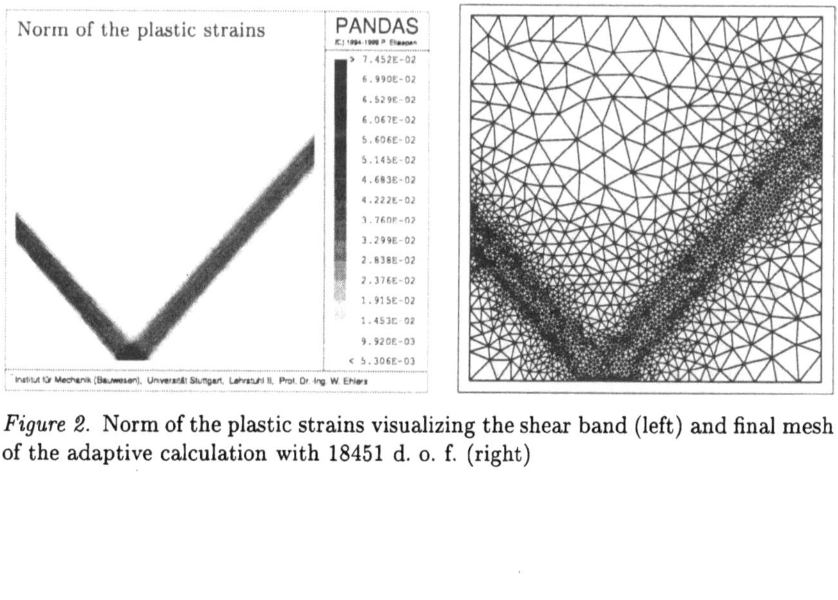

Figure 2. Norm of the plastic strains visualizing the shear band (left) and final mesh of the adaptive calculation with 18451 d. o. f. (right)

5. Conclusions

In this paper, a new error estimator for a two phase-model consisting of an elastic-viscoplastic solid skeleton and a viscous pore fluid has been presented. The reliability and efficiency of this error estimator in combination with the presented h-adaptive strategy and a time adaptive scheme [10] was verified in the numerical example, where a good correspondence of the adaptive and the overkill solution could be shown.

References

1. Bowen, R. M. Incompressible porous media models by use of the theory of mixtures. *Int. J. Engng. Sci.*, **18**, 1129–1148, 1980.
2. Bowen, R. M. Compressible porous media models by use of the theory of mixtures. *Int. J. Engng. Sci.*, **20**, 697–735, 1982.
3. Crawford, R. H. , Anderson, D. C. and Waggenspack, W. N. Mesh rezoning of 2D isoparametric elements by inversion. *Int. J. Numer. Methods Engng.*, **28**, 523–531, 1989.
4. Diebels, S. and Ehlers, W. Dynamic analysis of a fully saturated porous medium accounting for geometrical and non-linearities. *Int. J. Numer. Methods Engng.*, **39**, 81–97, 1996.
5. Diebels, S., Ellsiepen, P. and Ehlers, W. A Two-Phase Model for Viscoplastic Geomaterials. In D. Besdo and R. Bogacz, editors, *Dynamics of Continua*, pages 103 - 112. Shaker Verlag, Aachen, 1998.
6. Diebels, S., Ellsiepen, P. and Ehlers, W. Error-controlled Runge-Kutta Time Integration of a Viscoplastic Hybrid Two-phase Model. *Technische Mechanik*, **19**, 19–27, 1999.
7. Ehlers, W. *Poröse Medien - ein kontinuumsmechanisches Modell auf der Basis der Mischungstheorie.* Habilitationsschrift, Forschungsberichte aus dem Fachbereich Bauwesen, Nr. 47. Universität-GH-Essen, 1989.
8. Ehlers, W. Constitutive equations for granular materials in geomechanical context. In K. Hutter, editor, *Continuum Mechanics in Environmental Sciences and Geophysics*, CISM Courses and Lectures No. 337, 313–402. Springer-Verlag, Wien New York, 1993.
9. Ehlers, W. Grundlegende Konzepte in der Theorie Poröser Medien. *Technische Mechanik*, **16**, 63–76, 1996.
10. Ellsiepen, P. *Zeit- und ortsadaptive Verfahren angewandt auf Mehrphasenprobleme poröser Medien.* Dissertation, Berichte aus dem Institut für Mechanik (Bauwesen), Nr. II-3. Universität Stuttgart, 1999.
11. Gallimard, L., Ladevèze, P. and Pelle, J. P. Error estimation and adaptivity in elastoplasticity. *Int. J. Numer. Methods Engng.*, **39**, 189–217, 1996.
12. Kossaczký, I. A recursive approach to local mesh refinement in two and three dimensions. *J. Comp. Appl. Math.*, **55**, 275–288, 1994.
13. Krause, R. and Rank, E. A fast algorithm for point-location in a finite element mesh. *Computing*, **57**, 49–62, 1996.
14. Shewchuk, J. R. Triangle: A Two-Dimensional Quality Mesh Generator and Delaunay Triangulator. School of Computer Science, Carnegie Mellon University, Pittsburgh, Pennsylvania, 1996. `http://www.cs.cmu.edu/ quake/triangle.html`
15. Zienkiewicz, O. C. and Zhu, J. Z. A simple error estimator and adaptive procedure for practical engineering analysis. *Int. J. Numer. Methods Engng.*, **24**, 337–357, 1987.
16. Zienkiewicz, O. C. and Zhu, J. Z. The superconvergent patch recovery and a posteriori error estimates. Part 1: The recovery technique. *Int. J. Numer. Methods Engng.*, **33**, 1331–1364, 1992.
17. Zienkiewicz, O. C. and Zhu, J. Z. The superconvergent patch recovery and a posteriori error estimates. Part 2: Error estimates and adaptivity. *Int. J. Numer. Methods Engng.*, **33**, 1365–1382, 1992.

A Viscoelastic Two-Phase Model for Cartilage Tissues

W. Ehlers and B. Markert
Institute of Applied Mechanics (Civil Engineering),
University of Stuttgart, D-70550 Stuttgart, Germany
e-mail: (Ehlers, Markert)@mechbau.uni-stuttgart.de

Abstract. Hydrated soft tissues such as cartilage can be considered as porous materials consisting of an organic solid matrix (collagen and proteoglycans) ventilated by a pore-fluid (water and elektrolytes). These types of biological tissues significantly demonstrate viscoelastic phenomena under various configurations of deformation. Two distinct mechanisms exist for this dissipative behaviour, the frictional drag associated with the interstitial fluid flow through the porous solid matrix and the flow-independent viscoelastic properties of the organic solid matrix itself.

The goal of this paper is to describe this complex behaviour by a biphasic viscoelastic model accounting for both the flow-dependent fluid-solid interaction and the intrinsic skeleton viscoelasticity. Therefore, the macroscopic Theory of Porous Media (TPM) is applied, where the effective skeleton stress is determined by an appropriate viscoelasticity law. The fundamental approach of the viscoelasticity law is based on the thermodynamics with internal state variables, where a hyperelastic material formulation is used. Since cartilage tissues exhibit deformation dependent permeability effects, it is assumed that the permeability is a function of the actual porosity of the solid skeleton. Futhermore, both constituents are modelled materially incompressible; thermal effects and mass exchanges are excluded.

The viscoelastic two-phase model is implemented into PANDAS[1], an adaptive finite element code for multi-phase problems. To show the capability, the model is applied to articular cartilage, where some representative problems are computed. On this occasion, the influence of the intrinsic viscoelasticity of the solid skeleton is studied. In addition, the problem of separating the flow-independent dissipative behaviour from the flow-dependent consolidation process is discussed.

Keywords: Finite viscoelasticity, Theory of Porous Media (TPM), articular cartilage, deformation dependent permeability

1. Introduction

Since it is well known that articular cartilage exhibits flow-dependent and flow-independent viscoelastic properties ([9], [12], [16]), a lot of work has been done to consider both dissipative effects within constitutive models ([11], [10], [7]). Due to the complexity of the porous extracellular structure, macroscopic continuum theories are used to describe the multiphasic character of connective tissues.

In the present contribution, the Theory of Porous Media (TPM) is applied to describe the strongly coupled solid-fluid problem. The

[1] Porous media **A**daptive **N**onlinear finite element solver based on **D**ifferential **A**lgebraic **S**ystems

W. Ehlers (ed.), IUTAM Symposium on Theoretical and Numerical Methods in Continuum Mechanics of Porous Materials, 87–92.

Theory of Porous Media is a macroscopic theory based on the theory of mixtures extended by the concept of volume fractions. The reader who is interested in the fundamentals of the TPM approach is referred to Bowen ([2], [3]), de Boer and Ehlers ([4]) and Ehlers ([5], [6]). The energy absorbing behaviour of the extracellular matrix is described by a finite viscoelasticity law with discrete realxation spectrum, where a hyperelastic material formulation is used. Based on thermodynamics with internal state variables, this differential approach significantly saves CPU time compared to integral-type viscoelasticity models with continuous relaxation spectrum [14]. Under physiological levels of pressure, the interstitial fluid and the organic solid matrix are materially incompressible [1]. Therefore, the so-called point of compaction exists, defined as that deformation state, where all pores are closed and no further volume deformation is possible due to the skeleton incompressibility. Furthermore, the permeability of the solid matrix depends on the deformation state and is related to the actual porosity via a power function [8].

2. Governing equations

In the framework of the Theory of Porous Media, a fluid-saturated porous medium can be treated as an immiscible mixture of constituents φ^α ($\alpha = S$: solid skeleton; $\alpha = F$: pore-fluid). Based on the assumption of superimposed continua, the constituents are averaged over a representative elementary volume occupied by the whole mixture. In the biphasic macro model, the local structure of the mixture is represented by scalar variables, the volume fractions n^α. Thus, to avoid any vacant space, the saturation constraint $n^S + n^F = 1$ is assumed.

For the numerical treatment of the problem within the finite element method (FEM), weak forms of the governing field equations, i. e. the mixture balance of momentum and the mixture volume balance, are required. Therefore, after eliminating the seepage velocity by use of the Darcy filter law, the balance relations weighted by independent test functions and integrated over the spatial domain Ω with surface $\partial\Omega$ result in the respective weak formulations $\mathcal{G}_{\text{MM}}$ and $\mathcal{G}_{\text{VM}}$:

$$\begin{aligned} \mathcal{G}_{\text{MM}} \equiv & \int_\Omega \operatorname{grad} \delta\mathbf{u}_S \cdot (\mathbf{T}_E^S - p\,\mathbf{I})\, \mathrm{d}v \\ & - \int_\Omega \delta\mathbf{u}_S \cdot (\rho^S + \rho^F)\, \mathbf{b}\, \mathrm{d}v - \int_{\partial\Omega} \delta\mathbf{u}_S \cdot \bar{\mathbf{t}}\, \mathrm{d}a \;=\; 0\,, \end{aligned} \tag{1}$$

$$
\begin{aligned}
\mathcal{G}_{\mathrm{VM}} \equiv & \int_{\Omega} \delta p \operatorname{div}(\mathbf{u}_S)'_S \, dv + \int_{\Omega} \operatorname{grad} \delta p \cdot \frac{k^F}{\gamma^{FR}} \operatorname{grad} p \, dv \\
& - \int_{\Omega} \operatorname{grad} \delta p \cdot \frac{k^F}{\gamma^{FR}} \rho^{FR} \mathbf{b} \, dv + \int_{\partial\Omega} \delta p \, \overline{q} \, da = 0 \,.
\end{aligned} \tag{2}
$$

Herein, $\delta\mathbf{u}_S$ and δp are the test functions corresponding to the solid displacement $\mathbf{u}_S$ and the pore pressure p, $\overline{\mathbf{t}} = (\mathbf{T}_E^S - p\,\mathbf{I})\,\mathbf{n}$ is the external load vector acting on both constituents with the solid Cauchy extra stress $\mathbf{T}_E^S$, the identity tensor $\mathbf{I}$, and the outward oriented unit surface normal $\mathbf{n}$. Furthermore, $\overline{q} = n^F\,\mathbf{w}_F \cdot \mathbf{n}$ denotes the filter volume flux of the fluid draining through the surface $\partial\Omega$, k^F is the macroscopic Darcy permeability coefficient, $\mathbf{b}$ is the body force density, ρ^{FR} is the effective (material) fluid density, and γ^{FR} is the specific weight of the pore-fluid. Note that in the framework of a macroscopic theory, the friction stress of the fluid-phase effects no contribution and is neglected a priori.

In general, the permeability parameter k^F is deformation dependent, i. e. the value of k^F decreases when the pore volume decreases. Thus, it is assumed that the permeability depends on the porosity n^F of the solid skeleton which is described by a power function

$$
k^F(n^F) = k_0^F \left(\frac{n^F}{n_{0S}^F} \right)^{\kappa} , \tag{3}
$$

where k_0^F is the initial permeability and the exponent κ governs the deformation dependency [8].

3. Finite viscoelasticity law

To describe the intrinsic dissipative phenomenon of the extracellular skeleton structure, an adequate viscoelastic material formulation is required allowing macroscopic compressibility due to porosity changes. In the framework of a finite deformation theory of materials with elastic and inelastic (here: viscous) behaviour, we proceed from a multiplicative split of the solid deformation gradient [15]

$$
\mathbf{F}_S = \mathbf{F}_{Se}\,\mathbf{F}_{Si} \tag{4}
$$

into an elastic part $\mathbf{F}_{Se}$ and an inelastic part $\mathbf{F}_{Si}$. From elastoplasticity, it is commonly known that the concept of the multiplicative decomposition of deformation gradients is connected with the suggestion of a stress-free, geometrically incompatible intermediate configuration (Fig. 1), where the purely inelastic state of deformation is frozen into the memory of the material. Furthermore, from thermodynamics

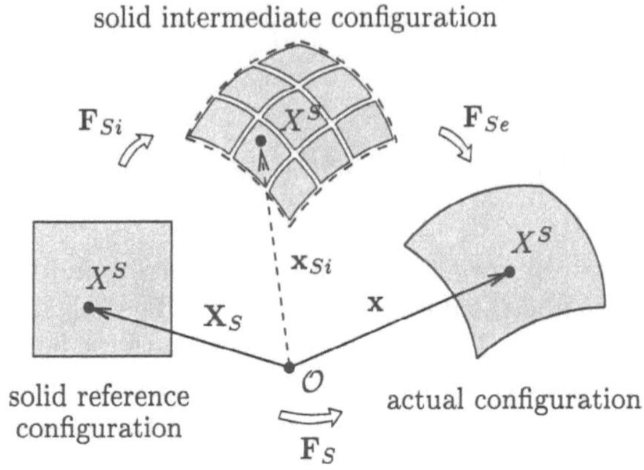

Figure 1. Solid configurations.

with internal state variables and rheological considerations [13], one obtains the ansatz of a decomposed solid Helmholtz free energy density and a decomposed solid extra stress

$$\psi^S(\mathbf{F}_S, \mathbf{F}_{Se}) = \psi^S_{EQ}(\mathbf{F}_S) + \psi^S_{NEQ}(\mathbf{F}_{Se})\,, \tag{5}$$

$$\boldsymbol{\tau}^S_E(\mathbf{F}_S, \mathbf{F}_{Se}) = \boldsymbol{\tau}^S_{EQ}(\mathbf{F}_S) + \boldsymbol{\tau}^S_{NEQ}(\mathbf{F}_{Se}) \tag{6}$$

into equilibrium parts (Index *EQ*) and non-equilibrium parts (Index *NEQ*) vanishing in the thermodynamic equilibrium. Note that the Kirchhoff extra stress is a weighted Cauchy stress $\boldsymbol{\tau}^S_E = J_S\,\mathbf{T}^S_E$, where $J_S = \det \mathbf{F}_S$ is the Jacobian determinant.

The solid extra stresses are determined from hyperelastic strain energy functions of the Neo-Hookean type which imply the point of compaction [8]. Thus,

$$\boldsymbol{\tau}^S_{EQ} = \mu^S_0\,(\mathbf{B}_S - \mathbf{I}) + \lambda^S_0\,(1 - n^S_{0S})^2 \left(\frac{J_S}{1 - n^S_{0S}} - \frac{J_S}{J_S - n^S_{0S}}\right) \mathbf{I}\,, \tag{7}$$

$$\boldsymbol{\tau}^S_{NEQ} = \mu^S_1\,(\mathbf{B}_{Se} - \mathbf{I}) + \lambda^S_1\,(1 - n^S_i)^2 \left(\frac{J_{Se}}{1 - n^S_i} - \frac{J_{Se}}{J_{Se} - n^S_i}\right) \mathbf{I}\,. \tag{8}$$

Therein, $J_{Se} = \det \mathbf{F}_{Se}$ is an elastic Jacobian determinant, $\mathbf{B}_S = \mathbf{F}_S\,\mathbf{F}_S^T$ and $\mathbf{B}_{Se} = \mathbf{F}_{Se}\,\mathbf{F}_{Se}^T$ are left Cauchy-Green deformation tensors, n^S_{0S} is the initial solidity and $n^S_i = n^S_{0S} \det \mathbf{F}_{Si}^{-1}$ is the inelastic solid volume fraction with respect to the intermediate configuration. The Lamé constants μ^S_0, λ^S_0 and μ^S_1, λ^S_1 are macroscopic parameters of the solid skeleton structure and not of the microscopic skeleton material itself. The inelastic strain as internal variable is obtained from an evolution equation which has to fulfill the residual dissipation inequality:

$$\overset{\triangle}{\mathbf{A}}_{Si} = \overset{4}{\mathbf{D}}{}^{-1}\,\boldsymbol{\tau}^S_{NEQ} \quad \text{with} \quad \overset{4}{\mathbf{D}}{}^{-1} = \frac{1}{2\eta^S_1}\,[\,\overset{4}{\mathbf{I}} - \tfrac{1}{3}\,(\mathbf{I}\otimes\mathbf{I})\,] + \frac{1}{9\zeta^S_1}\,(\mathbf{I}\otimes\mathbf{I})\,. \tag{9}$$

Therein, $\overset{\triangle}{\mathbf{A}}_{Si}$ is the upper Oldroyd derivative of the Almansian strain [6] and $\overset{4}{\mathbf{D}}^{-1}$ is a positive definite, isotropic fourth order viscosity tensor with the macroscopic viscosity parameters η_1^S and ζ_1^S.

4. Examples

Two canonical problems have been computed in order to show the applicability of the presented model. The first example is a torsion test which is used to determine the intrinsic viscoelastic behaviour of cartilage under shear [16]. In this test, a direct parameter identification is possible because the intrinsic dissipation is decoupled since no viscous fluid flow is induced. The computed results (Fig. 2) show the typical fast stress relaxation of cartilage under shear. It would also be desirable to determine the volumetric viscoelastic response of the solid skeleton directly from a hydrostatic compression test. But the problem occurs that volumetric deformation of a fluid-saturated porous medium always causes viscous fluid motion. Thus, in order to separate the flow-independent volumetric dissipation of the matrix material, two different computations are performed. One with a viscoelastic solid skeleton as substitute for the experimental data and one with an elastic solid skeleton using the same elastic parameters as for the viscoelastic computation. The difference between both results (Fig. 2) is an indicator of the flow-independent volumetric viscoelastic material behaviour, since the properties of the fluid are the same in both computations. An occuring problem is the large time period to consolidate the specimen in order to determine the elastic compression modulus of the matrix material. Due to the very low permeability of cartilage, the computed consolidation process has taken about several decades of model time.

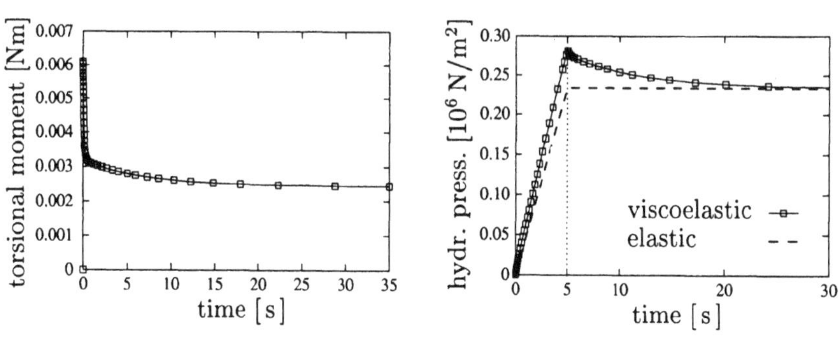

Figure 2. Progression of the torsional moment (left) and the hydrostatic pressure (right) of the computed torsion test and the hydrostatic compression test.

Thus, the direct determination of the intrinsic volumetric viscoelastic properties through a quasi-static experiment is not possible.

References

1. Bachrach, N. M., Mow, V. C. and Guilak, F. Incompressibility of the solid matrix of articular cartilage under high hydrostatic pressures. *J. Biomechanics*, **31**, 445–451, 1998.
2. Bowen, R. M. Theory of Mixtures. In A. C. Eringen, editor, *Continuum Physics, Vol. III*, pages 1–127. Academic Press, New York, 1976.
3. Bowen, R. M. Incompressible porous media models by use of the theory of mixtures. *Int. J. Engng. Sci.*, **18**, 1129–1148, 1980.
4. de Boer, R. and Ehlers, W. Theorie der Mehrkomponentenkontinua mit Anwendung auf bodenmechanische Probleme. Forschungsberichte aus dem Fachbereich Bauwesen 40, Universität Essen, Essen, 1986.
5. Ehlers, W. Poröse Medien – ein kontinuumsmechanisches Modell auf der Basis der Mischungstheorie. Forschungsberichte aus dem Fachbereich Bauwesen 47, Universität Essen, Essen, 1989.
6. Ehlers, W. Constitutive equations for granular materials in geomechanical context. In K. Hutter, editor, *Continuum Mechanics in Environmental Sciences*, CISM Courses and Lectures 337, pages 313–402. Springer-Verlag, Wien, 1993.
7. Ehlers, W. and Markert, B. A linear viscoelastic biphasic model for soft tissues based on the Theory of Porous Media. Berichte aus dem Institut für Mechanik (Bauwesen), Nr. 99-II-3, Universität Stuttgart, Stuttgart, 1999.
8. Eipper, G. Theorie und Numerik finiter elastischer Deformationen in fluidgesättigten porösen Festkörpern. Dissertation, Bericht Nr. II-1 des Instituts für Mechanik (Bauwesen), Universität Stuttgart, Stuttgart, 1998.
9. Hayes, W. C. and Bodine, A. J. Flow-independent viscoelastic properties of articular cartilage matrix. *J. Biomechanics*, **11**, 407–419, 1978.
10. Lai, W. M., Mow, V. C. and Zhu, W. Constitutive modeling of articular cartilage and biomacromolecular solutions. *J. biomech. Engng*, **115**, 474–480, 1993.
11. Mak, A. F. The apparent viscoelastic behaviour of articular cartilage - the contribution from the intrinsic matrix viscoelasticity and interstitial fluid flows. *J. biomech. Engng*, **108**, 123–130, 1986.
12. Mow, V. C., Kuei, S. C., Lai, W. M. and Armstrong, C. G. Biphasic creep and stress relaxation of articular cartilage in compression: theory and experiments. *J. biomech. Engng*, **102**, 73–84, 1980.
13. Reese, S. and Govindjee, S. A theory of finite viscoelasticity and numerical aspects. *Int. J. Solids Structures*, **35**, 3455–3482, 1998.
14. Suh, J.-K. and Bai, S. Finite element formulation of biphasic poroviscoelastic model of articular cartilage. *J. biomech. Engng*, **120**, 195–201, 1998.
15. Sidoroff, F. Un modèle viscoélastique non linéaire avec configuration intermédiaire. *Journal de Mécanique*, **13**, 679–713, 1974.
16. Zhu, W. B., Lai, W. M. and Mow, V. C. Intrinsic quasi-linear viscoelastic behavior of the extracellular matrix of cartilage. *Trans. Orthop. Res. Soc.*, **11**, 407, 1986.

A Poroelastic Material with a Scale Independent Pressure-Volume Relation

D. Elata
Technion - Israel Institute of Technology, Haifa 32000, Israel

Abstract. A poroelastic material with arbitrary pore space geometry is considered. The porous material is made of a single phase solid constituent. The pressure of the non-porous solid constituent is considered to be a unique function of relative volume, and the average solid pressure is assumed to be a unique function of solid relative volume. The assumption that this function has the same form as the pressure-volume relation of the non-porous solid is examined. It is found that this assumption is correct only for a special type of elastic solid described in this work.

Keywords: Nonlinear poroelasticity, constitutive equations

1. Introduction

The mechanical response of poroelastic materials is more complex than the mechanical response of a non-porous material made of the same solid constituents. The question of interest in this work is: to what extent can the mechanical response of the porous material be analytically related to the mechanical response of its (nonporous) solid constituent, without reference to the pore space geometry? This question is a matter of scientific curiosity and may provide some insight to the constitutive modeling of poroelastic materials. This study considers large deformations, nonlinear response of the solid constituents, and *arbitrary* pore space geometry. The discussion is limited to equilibrium states in a poroelastic material made of a single phase solid constituent, and it is assumed that the solid skeleton and the pore space are each a well-connected network.

In many problems concerning the mechanical response of poroelastic materials, the length scale of interest is much larger than a characteristic length of the pore space. In the larger macroscopic scale the material responds as a homogeneous continuum, even though the stress and deformation at the smaller microscopic pore scale are inhomogeneous.

Let R_0 be a macroscopic material point of a porous material in the stress-free reference configuration, and let R_{s0} and R_{p0} be the parts of R_0 that contain the solid constituent and the pore space, respectively ($R_0 = R_{s0} + R_{p0}$). Also, let R, R_s, and R_p, be the projections of R_0, R_{s0}, and R_{p0}, in the current (i.e., stressed) configuration. The volume of R_0, R_{s0}, and R_{p0}, is V, V_s, and V_p, respectively, and the volume of

W. Ehlers (ed.), IUTAM Symposium on Theoretical and Numerical Methods in Continuum Mechanics of Porous Materials, 93–98.

R, R_s, and R_p, is v, v_s, and v_p, so that

$$V = V_s + V_p \,, \qquad v = v_s + v_p \,. \tag{1}$$

The reference porosity Φ and its current value ϕ are then defined by

$$\Phi = \frac{V_p}{V} \,, \qquad \phi = \frac{v_p}{v} \,. \tag{2}$$

The macroscopic relative volume J, and the macroscopic solid relative volume J_s, are given by

$$J = \frac{v}{V} \,, \qquad J_s = \frac{v_s}{V_s} \,, \tag{3}$$

and it follows that

$$J_s = J \frac{1-\phi}{1-\Phi} \,. \tag{4}$$

The macroscopic Cauchy stress $\mathbf{T}$ (i.e., confining stress) in a porous material is given by [1]

$$\mathbf{T} = (1-\phi)\mathbf{T}_s - \phi p_p \mathbf{I} \,, \tag{5}$$

where $\mathbf{T}_s$ is the average stress in the solid constituent and p_p is the average fluid pressure in the pore space. These are defined by

$$\mathbf{T}_s = \frac{1}{v_s}\int_{R_s} {}^{\mu}\mathbf{T}_s dv_s \,, \qquad p_p = \frac{1}{v_p}\int_{R_p} {}^{\mu}p_p dv_p \,, \tag{6}$$

where ${}^{\mu}\mathbf{T}_s$ is the microscopic stress in the solid and ${}^{\mu}p_p$ is the microscopic pressure in the solid. Equation (5) is independent of material response of the solid and fluid constituents. Decomposing the stress into its spherical and deviatoric parts Eqn. (5) may be rewritten in the form

$$\mathbf{T}' = (1-\phi)\mathbf{T}'_s \,, \qquad p = (1-\phi)p_s + \phi p_p \,, \tag{7}$$

where the average deviatoric stress in the solid $\mathbf{T}'_s$ and the average pressure in the solid p_s are given by

$$\mathbf{T}'_{,s} = \frac{1}{v_s}\int_{R_s} {}^{\mu}\mathbf{T}'_s dv_s \,, \qquad p_s = \frac{1}{v_s}\int_{R_s} {}^{\mu}p_s dv_s \,. \tag{8}$$

Here ${}^{\mu}\mathbf{T}'_s$ is the deviatoric part of the microscopic stress in the solid and ${}^{\mu}p_s$ is the microscopic pressure in the solid.

Let ${}^{\mu}\mathbf{F}_s$ be the microscopic deformation gradient tensor. Following [3], this tensor may be decomposed into its volumetric part ${}^{\mu}J_s$ and its unimodular part ${}^{\mu}\mathbf{F}^{\circ}_s$

$${}^{\mu}\mathbf{F}^{\circ}_s = {}^{\mu}J_s^{-1/3}\, {}^{\mu}\mathbf{F}_s \,, \qquad {}^{\mu}J_s = \frac{dv_s}{dV_s} = det({}^{\mu}\mathbf{F}_s) \,, \tag{9}$$

so that ${}^{\mu}\mathbf{F}_s^{\circ}$ is independent of volumetric deformations and $det({}^{\mu}\mathbf{F}_s^{\circ}) = 1$. Similarly, the macroscopic deformation gradient $\mathbf{F}$ may be decomposed into its volumetric part J and its unimodular part $\mathbf{F}^{\circ}$.

2. Analysis

The uniform pore pressure may be controlled independently of the confining stress, and both loads affect the macroscopic deformation and the porosity. In general, the confining stress may be written as a function of macroscopic deformation $\mathbf{F}$ and pore pressure p_p

$$\mathbf{T} = \tilde{\mathbf{T}}(\mathbf{F}, p_p) , \tag{10}$$

or with reference to Eqn. (5) in the alternative form

$$\mathbf{T}_s = \tilde{\mathbf{T}}_s(\mathbf{F}, p_p) , \qquad \phi = \tilde{\phi}(\mathbf{F}, p_p) . \tag{11}$$

The functional forms of $\tilde{\phi}$ and $\tilde{\mathbf{T}}_s$ can only be determined by experiments. However, for modeling purposes it is occasionally assumed that $\tilde{\mathbf{T}}'_s$ (i.e., the deviatoric part of $\tilde{\mathbf{T}}_s$) is only a function of $\mathbf{F}^{\circ}$, that it has the same functional form as the microscopic stress

$$\mathbf{T}'_s = {}^{\mu}\tilde{\mathbf{T}}'_s(\mathbf{F}_s^{\circ}) , \qquad where \qquad {}^{\mu}\mathbf{T}'_s = {}^{\mu}\tilde{\mathbf{T}}'_s({}^{\mu}\mathbf{F}_s^{\circ}) , \tag{12}$$

and that the relation between $\mathbf{F}^{\circ}$ and ${}^{\mu}\mathbf{F}_s^{\circ}$ is known (e.g., [6]; [5]). This assumption greatly simplifies the task of developing constitutive models of poroelastic solids. It can be shown however, that the unimodular part of the macroscopic deformation $\mathbf{F}^{\circ}$ is not uniquely related to the unimodular part of the average of the microscopic deformation gradient ${}^{\mu}\mathbf{F}_s^{\circ}$. Therefore, Eqn. (12) has no physical basis.

In contrast to this, the macroscopic solid relative volume J_s [Eqn. (3)] is identical to the average of the microscopic solid relative volume ${}^{\mu}J_s$

$$J_s = \frac{1}{V_s}\int_{R_{s0}} {}^{\mu}J_s dV_s = \frac{v_s}{V_s} . \tag{13}$$

Motivated by this, the discussion is limited to the pressure response of the porous material and the pressure response of solid constituent, where both responses are assumed to be independent of distortion

$$\tilde{p}(J) = \left(1 - \tilde{\phi}(J)\right)\tilde{p}_s(J_s) , \qquad {}^{\mu}p_s = {}^{\mu}\tilde{p}_s({}^{\mu}J_s) . \tag{14}$$

For simplicity, only zero pore pressure is considered[1]. In this model of the pressure response of a porous solid, the functional form of $\tilde{p}_s$ can

[1] For the special material derived in the following, the affect of uniform pore pressure can be analytically predicted [2]

be determined from experiments by continuously measuring p, J, and ϕ, in a compaction test, and determining J_s with the help of Eqn. (4).

A further significant simplification may be achieved by assuming that the functional form of the solid pressure-volume relations is the same in both macroscopic and microscopic scales (e.g., [6])

$$\tilde{p}_s(J_s) = f(J_s) , \qquad {}^{\mu}\tilde{p}_s({}^{\mu}J_s) = f({}^{\mu}J_s) . \tag{15}$$

In this case, the pressure-volume relation of the porous material take the form

$$\tilde{p}(J) = \left(1 - \tilde{\phi}(J)\right) f(J_s) . \tag{16}$$

The functional form ϕ can be deduced from macroscopic pressure and volume measurements by solving Eqns. (16) and (4). This in turn determines the pore space volume in deformed states. An independent measurement of the pore space volume in deformed states can be obtained from compaction experiments of jacketed, semi-saturated samples of the porous material [4]. The validity of the assumption (15) can be determined by comparing the measured and calculated pore space volumes.

Equation (15) is analytically correct only if

$$^{\mu}\tilde{p}_s(J_s) = \frac{1}{v_s} \int_{R_s} {}^{\mu}\tilde{p}_s({}^{\mu}J_s) dv_s . \tag{17}$$

It can be shown that the only solution of Eqn. (17) that satisfies ${}^{\mu}\tilde{p}_s(1) = 0$ is

$$\begin{aligned} {}^{\mu}\tilde{p}_s({}^{\mu}J_s) &= k_0 \left(\frac{1}{{}^{\mu}J_s} - 1\right) , \\ \tilde{p}_s(J_s) &= k_0 \left(\frac{1}{J_s} - 1\right) , \end{aligned} \tag{18}$$

where k_0 is a constant. The bulk modulus k_0 of the solid constituent is defined by

$$^{\mu}k_0 = -{}^{\mu}J_s \frac{\partial^{\mu} p_s}{\partial^{\mu} J_s} = \frac{k_0}{{}^{\mu}J_s} , \tag{19}$$

and the material constant k_0 is seen to be the bulk modulus of the solid constituent at the stress free state $J = 1$. The nonlinear character of the microscopic pressure response is significant only if the microscopic relative volume is sufficiently different than one.

It is emphasized that the identity in the functional form of solid pressure in the macroscopic and microscopic scales [Eqns. (18)], is independent of pore space geometry and therefore independent of any inhomogeneities in ${}^{\mu}J_s$ within the domain R_{s0}. For any functional form of the microscopic solid pressure that is different than Eqn. (18), the functional form of the macroscopic solid pressure will include a term that depends on the inhomogeneities of ${}^{\mu}J_s$. Therefore, any functional form of the microscopic solid pressure that is different than Eqn. (18), will not be correct for arbitrary pore space geometry.

If the solid constituent described above is hyperelastic, then its strain energy ${}^{\mu}\psi_s$ has the functional form

$$ {}^{\mu}\psi_s = {}^{\mu}\hat{\psi}_{Is}({}^{\mu}J_s) + {}^{\mu}\hat{\psi}_{IIs}({}^{\mu}\mathbf{C}_s^{\circ}) \,, \tag{20} $$

where

$$ {}^{\mu}\rho_{s0}{}^{\mu}\hat{\psi}_{Is}({}^{\mu}J_s) = k_0 \left({}^{\mu}J_s - 1 - ln({}^{\mu}J_s)\right) \,, \tag{21} $$

${}^{\mu}\rho_{s0}$ is the microscopic solid density in the reference configuration, and ${}^{\mu}\mathbf{C}_s^{\circ}$ is the unimodular part of the Cauchy deformation tensor (${}^{\mu}\mathbf{C}_s^{\circ} = {}^{\mu}\mathbf{F}_s^{\circ\mathrm{T}\mu}\mathbf{F}_s^{\circ}$). Both ${}^{\mu}\hat{\psi}_{Is}$ and ${}^{\mu}\hat{\psi}_{IIs}$ are non-negative functions of their arguments with ${}^{\mu}\hat{\psi}_{Is}(1) = 0$ and ${}^{\mu}\hat{\psi}_{IIs}(\mathbf{I}) = 0$. Otherwise, the functional form of ${}^{\mu}\hat{\psi}_{IIs}({}^{\mu}\mathbf{C}_s^{\circ})$ remains unrestricted so that Eqn. (20) defines a family of nonlinear hyperelastic solids. Applying a uniform pore pressure to a prestressed porous solid made of this material, induces a microscopically uniform compaction [2].

The macroscopic strain energy term associated with ${}^{\mu}\hat{\psi}_{Is}({}^{\mu}J_s)$ [Eqn. (21)] is given by the volume average

$$ \psi_{Is} = \frac{1}{V_s}\int_{R_{s0}} {}^{\mu}\hat{\psi}_{Is}({}^{\mu}J_s)dV_s = {}^{\mu}\hat{\psi}_{Is}(J_s) + \tilde{\psi}_{Is} \,, \tag{22} $$

$$ \rho_{s0}\tilde{\psi}_{Is} = k_0 ln(J_s) - \frac{k_0}{V_s}\int_{R_{s0}} ln({}^{\mu}J_s)dV_s, \tag{23} $$

and ρ_{s0} is the macroscopic solid density in the reference configuration ($\rho_{s0} ={}^{\mu} \rho_{s0}$). The first term in the right hand side of Eqn. (22) is the same function that appears in Eqn. (21), only that the argument here is J_s rather than ${}^{\mu}J_s$. By analogy to the relation between the arithmetic and geometric averages of a series of positive numbers, it can be shown that $\tilde{\psi}_{Is} \geq 0$ where $\tilde{\psi}_{Is} = 0$ only if ${}^{\mu}J_s$ is uniform. This means that the term $\tilde{\psi}_{Is}$ measures the free energy stored in the volumetric deformation inhomogeneities, and therefore depends on the pore space geometry. Consequently, the free energy term that is associated with pressure cannot have the same functional form in the microscopic and macroscopic scales.

3. Summary

This work presents a model of the volumetric response of a theoretical nonlinear poroelastic material. In this model, the pressure-volume relation of the solid constituent in the macroscopic and microscopic scales have the same functional form. In this model, the porosity that is deduced from the values of macroscopic pressure and volume, is necessarily identical to the porosity that is calculated by solving the related boundary value problem at the microscale. The properties of the model are independent of the pore space geometry and are also independent of the mechanical response of the solid constituent to distortion.

Acknowledgements

This research was supported by the *Technion V.P.R. Fund*, the *Argentinian Research Fund*, and the *Leah and Donald Lewis* Academic Lectureship.

References

1. Carroll, M. M. Mechanical response of fluid-saturated porous materials. In F. P. J. Rimrott and B. Tabarrok, editors, *Theoretical and Applied Mechanics*. North Holland, Amsterdam, 251–262, 1980.
2. Elata, D. Pure volumetric compaction of a prestressed nonlinear hyperelastic solid with reference to poroelastic materials. *Mech. Mat.*, **31**, 141–147, 1999.
3. Flory, P. Thermodynamic relations for high elastic materials. *Trans. Faraday Soc.*, **57**, 829–838, 1961.
4. Heard, H. C., Bonner, B. P., Duba, A. G., Shock, R. N., Stephens, R. N. and Stephens, D. R. High Pressure Mechanical Properties Of Mt. Helen, Nevada, Tuff. Report No. UCID-16261, Lawrence Livermore National Laboratory, Livermore, CA., 1973.
5. Morland, L. W. A simple constitutive theory for a fluid-saturated porous solid. *J. Geophys. Res.*, **77**, 890–900, 1972.
6. Rubin, M. B. and Elata, D. Modeling added compressibility of porosity and the thermomechanical response of wet porous rock with application to Mt. Helen tuff. *Int. J. Solids Struct.*, **33**, 761–793, 1996.

Numerical modelling of Cartilage as a Deformable Porous Medium

A. J. H. Frijns and E. F. Kaasschieter
Eindhoven University of Technology,
Department of Mathematics and Computing Science,
P.O. Box 513, NL-5600 MB Eindhoven, the Netherlands

J. M. Huyghe
Eindhoven University of Technology,
Department of Mechanical Engineering,
P.O. Box 513, NL-5600 MB Eindhoven, the Netherlands

Abstract. Soft biological tissues, like cartilage and intervertebral disc tissue, exhibit swelling and shrinking behaviour due to mechanical and chemical loadings. A mixture theory is used to simulate this behaviour.

First, the biphasic mixture theory is investigated. In this theory, the mechanical behaviour is described by mass and momentum balances, and constitutive equations. As a result a system of coupled, time-dependent, non-linear equations is obtained. These equations are discretised in space by a mixed-hybrid finite element method, and in time by a suitable implicit time integrator. Unlike the u-p formulation (the displacements and the fluid pressure are the unknowns), the mixed-hybrid finite element method yields a conservative velocity-field of the fluid, which is a sound basis for accurate computations of, for example, diffusion of particles inside the fluid.

Then, the biphasic mixture theory is extended to a four components mixture theory in order to model chemical and electrical phenomena inside the material.

1. Introduction

Porous media, such as soils, rocks and biological tissues, can be idealised as deformable two-phase media [3, 4]. These materials consist of a deformable solid skeleton saturated with a fluid. The solid skeleton models the grains, the rock with pores or the biological fibres.

The study of deformable porous materials was started by Terzaghi [12]. He expressed the main phenomena in a consistent manner for a one-dimensional case. Later, Biot developed a successful three-dimensional theory for soil consolidation [1]. His theory was mainly used to analyse and design structures and buildings on a porous soil. It was also used in the field of oil industry. Later, it was adapted in order to describe the mechanics of biological tissues, for example articular cartilage [8]. In order to describe the influence of particles inside the

W. Ehlers (ed.), IUTAM Symposium on Theoretical and Numerical Methods in Continuum Mechanics of Porous Materials, 99–104.

material, the theory was extended to a triphasic theory [7, 10] and later to a four components mixture theory [5].

For some 'simple' problems, analytic solutions were derived [1, 9, 12]. But in general the geometry and the boundary conditions are too complex to derive an analytic solution. Therefore, finite element approximations are used to solve the problems.

Until now, conforming finite elements are used [2, 10, 11]. Since cartilage is an inhomogeneous material and the permeability depends strongly on the deformation, these methods can give inaccurate solutions for the fluid fluxes [6]. Since the fluid flow and the particle flow influence each other severely, we need accurate solutions for these flows.

The goal of this research is to find an appropriate finite element description of the mechanical behaviour of soft biological tissues.

In this paper, a short description of the biphasic mixture theory is given. A mixed-hybrid finite element approximation for the biphasic mixture theory is derived and the existence and uniqueness of the solution is investigated. At the end, the biphasic mixture theory is extended to the four components mixture theory.

2. Physical model

Soft biological tissues, like articular cartilage and the cartilaginous intervertebral disc tissue, exhibit swelling and shrinking behaviour due to mechanical loadings. A biphasic mixture theory is used to simulate this behaviour. In this theory, the tissue is represented as a deformable saturated porous medium. The collagen fibres and the proteoglycans network in the tissue are represented by the solid matrix. It is assumed that there are no chemical reactions between the components. Further, it is assumed that the fluid and the solid are intrinsically incompressible. However, the tissue can swell or shrink due to absorption or expulsion of fluid.

The material behaviour is described by a set of coupled equations. The material has to fulfil the momentum equation

$$\nabla \cdot \boldsymbol{\sigma} = \mathbf{f}, \tag{1}$$

where the tensor $\boldsymbol{\sigma}$ describes the internal stresses. In this momentum equation, the inertial terms and the body forces are represented by $\mathbf{f}$. We can split the stress tensor up into two parts: the effective stress tensor $\boldsymbol{\sigma}^e$ and the hydrodynamic fluid pressure p: $\boldsymbol{\sigma} = \boldsymbol{\sigma}^e - p\mathbf{I}$.

So, the momentum equation can be rewritten as

$$\nabla \cdot \boldsymbol{\sigma}^e - \nabla p = \mathbf{f}. \tag{2}$$

The material also has to fulfil the mass balance for each component:

$$\frac{\partial(n^\alpha \rho^\alpha)}{\partial t} + \nabla \cdot (n^\alpha \rho^\alpha \mathbf{v}^\alpha) = 0, \qquad \alpha = s, f, \tag{3}$$

where s stands for the solid components and f for the fluid component. In this equation n^α, ρ^α and $\mathbf{v}^\alpha$ are respectively the volume fraction, the density and the velocity of component α. The first term in the equation is the accumulation term. This term accounts for the influence of the deformation of the solid, through which the porosity n^f changes, and it accounts for the deformation of each component, through which the density ρ^α changes.

Further, we need constitutive equations. The first one describes the solid behaviour. We assume that the solid is an elastic material. Then, the mechanical behaviour is described by Hooke's law:

$$\boldsymbol{\sigma}^e = \mu \, \boldsymbol{\epsilon}(\mathbf{u}) + \lambda \, (\nabla \cdot \mathbf{u})\mathbf{I}, \tag{4}$$

where $\boldsymbol{\epsilon}(\mathbf{u}) = \frac{1}{2}((\nabla\mathbf{u}) + (\nabla\mathbf{u})^T)$ is the strain tensor, $\mathbf{u}$ is the solid displacement and μ and λ are the Lamé constants.

The second constitutive equation describes the fluid flow. The average pore size in intervertebral disc tissue is about 3.5 nm. We assume that this pore size is large enough to neglect boundary effects. Then, on a macroscopic scale, this fluid flow is described by Darcy's law:

$$n^f(\mathbf{v}^f - \mathbf{v}^s) = -\mathbf{K}\nabla p, \tag{5}$$

where $\mathbf{K}$ is the permeability tensor, that can be strain-dependent. Darcy's law states that the fluid flow is proportional to the pressure gradient.

After substitution of the constitutive equations into the balance equations, summation of the mass balances of the fluid and the solid (equations (3)) and assuming the densities ρ^α to be constant, we end up with a set of coupled equations:

$$-\nabla \cdot (2\mu \, \boldsymbol{\epsilon}(\mathbf{u}) + \lambda \, (\nabla \cdot \mathbf{u}) \, \mathbf{I}) + \nabla p = \mathbf{f}, \tag{6}$$

$$\mathbf{K}^{-1}\mathbf{v} + \nabla p = 0, \tag{7}$$

$$\nabla \cdot \frac{\partial}{\partial t}\mathbf{u} + \nabla \cdot \mathbf{v} = 0, \tag{8}$$

with adequate boundary conditions. Here $\mathbf{v} = n^f(\mathbf{v}^f - \mathbf{v}^s)$ is the fluid flux relatively to the solid matrix.

This problem can be written in a variational form :

$$\begin{array}{lllll} a(\mathbf{u},\mathbf{w}) & & + \; b(\mathbf{w},p) & = & \langle \mathbf{g}_N, \mathbf{w}\rangle + (\mathbf{f},\mathbf{w}), \\ & c(\mathbf{v},\mathbf{s}) & + \; d(\mathbf{s},p) & = & 0, \\ b(\frac{\partial}{\partial t}\mathbf{u},q) & + \; d(\mathbf{v},q) & & = & 0, \end{array} \tag{9}$$

where

$$\begin{aligned} a(\mathbf{u},\mathbf{w}) &= \int_\Omega 2\mu\, \boldsymbol{\epsilon}(\mathbf{u}) : \boldsymbol{\epsilon}(\mathbf{w}) + \lambda\, \boldsymbol{\nabla}\cdot\mathbf{u}\, \boldsymbol{\nabla}\cdot\mathbf{w}\, dV, \\ b(\mathbf{w},p) &= -\int_\Omega \boldsymbol{\nabla}\cdot\mathbf{w}\, p\, dV, \\ c(\mathbf{v},\mathbf{s}) &= \int_\Omega (\mathbf{K}^{-1}\mathbf{v})\cdot\mathbf{s}\, dV, \\ d(\mathbf{s},q) &= -\int_\Omega \boldsymbol{\nabla}\cdot\mathbf{s}\, q\, dV, \\ (\mathbf{f},\mathbf{w}) &= \int_\Omega \mathbf{f}\cdot\mathbf{w}\, dV, \\ \langle \mathbf{g}_N,\mathbf{w}\rangle &= -\int_{\Gamma_N} \mathbf{w}\cdot\mathbf{g}_N\, dS. \end{aligned}$$

The problem has a unique solution, if the bilinear forms $a(.,.)$ and $c(.,.)$ are coercive and the bilinear forms $b(.,.)$ and $d(.,.)$ satisfy the *inf-sup* condition. These conditions are satisfied if $\mathbf{u} \in H_{1,0}(\Omega)^d$, $\mathbf{v} \in H_0(\mathrm{div},\Omega)$, $p \in L_2(\Omega)$ and the domain is a Lipschitz domain.

3. Finite element model

In the finite element model we approximate the infinite-dimensional spaces by finite-dimensional spaces, for example by using mixed finite elements. In the simplest mixed element that fulfils the unicity conditions, the displacements $\mathbf{u}_h$ are spanned by continuous linear polynomials, the relative velocities $\mathbf{v}_h$ by the lowest order Raviart-Thomas space and the fluid pressure p_h by piecewise constant functions. An advantage of this formulation, compared to a formulation with just the displacements and pressures as unknowns (the *u-p* formulation: equation (7) is substituted into equation (8)), is that the continuity of the fluid flow is enforced.

A method to reduce the resulting large matrix system, is the introduction of an extra variable: the Lagrange multiplier. The Lagrange multipliers enforce the continuity of the normal component of the fluxes $\mathbf{v}$ across the inter element boundaries. In a similar way as described by Kaasschieter and Huiben [6], the mixed-hybrid finite element description can be derived. The resulting system has some nice properties: some submatrices are irreducible M-matrices and have element-wise positive inverses. Others are diagonal matrices and can also be inverted element-wise. By using these properties and by applying an implicit time discretisation, we end up with the saddle-point problem of the form

$$\begin{bmatrix} A & B \\ B^T & -\Delta t\, C \end{bmatrix} \begin{bmatrix} U \\ \Lambda \end{bmatrix} = \begin{bmatrix} F_1(U,\Lambda,t) \\ F_2(U,\Lambda,t) \end{bmatrix}, \qquad (10)$$

where A and C are positive-definite matrices and U and Λ are vectors with respectively the displacements and Lagrange multipliers. The fluxes and the fluid pressures can be computed a posteriori.

4. Extending to the four components mixture model

Now, the biphasic mixture theory can be extended to a four components mixture theory [2, 5]. In this theory, two extra components are introduced: cations and anions. Also, the porous solid can be charged. We assume that the average pore size is large enough to neglect some boundary effects, such as the Knudsen diffusion and the Klinkenberg effect [3]. Therefore, extra equations are needed: mass balances for the cations and the anions, electro-neutrality, an extended Darcy's law and an extended Fick's law for the diffusion of the ions inside the fluid:

$$\nabla \mu^{\beta} + \sum_{\alpha=f,+,-} \frac{\mathbf{B}^{\beta\alpha}}{n^{\beta} n^{\alpha}} n^{\alpha} (\mathbf{v}^{\alpha} - \mathbf{v}^{s}) = \mathbf{0}, \qquad \beta = f, +, -, \tag{11}$$

where μ^{β} is the electrochemical potential defined by

$$\mu^{\beta} = \mu_0^{\beta}(T) + p + \frac{RT}{\overline{V}^{\beta}} ln\left(\frac{f^{\beta} c^{\beta}}{\sum_{\alpha=f,+,-} c^{\alpha}}\right) + \frac{z^{\beta}}{\overline{V}^{\beta}} F \xi. \tag{12}$$

Here, R the universal gas constant, T the absolute temperature, $\overline{V}^{\beta}$ the molar volume of component β, f^{β} the activity coefficient of component β, c^{β} the concentration per fluid volume of component β, z^{β} the valence of component β, F Faraday's constant and ξ the electrical potential. The terms $\mathbf{B}^{\beta\alpha}$ are frictional tensors and depend on the the permeability and the diffusion tensors. These tensors can also depend on the tissue deformation.

The resulting variational formulation of the system is similar to that of the biphasic mixture theory, where, instead of one fluid pressure, three electrochemical potentials are used, and, instead of one fluid flux, three fluxes for fluid, cations and anions are used. Similar to the biphasic mixture theory, three Lagrange multipliers are introduced in the four components mixture theory. After reduction of the variables, we end up with a matrix system of the form

$$\begin{bmatrix} A & B & C & D \\ B^T & E & F & G \\ C^T & F^T & H & L \\ D^T & G^T & L^T & M \end{bmatrix} \begin{bmatrix} U \\ \Lambda^f \\ \Lambda^+ \\ \Lambda^- \end{bmatrix} = \begin{bmatrix} F_1(U, \Lambda^f, \Lambda^+, \Lambda^-, t) \\ F_2(U, \Lambda^f, \Lambda^+, \Lambda^-, t) \\ F_3(U, \Lambda^f, \Lambda^+, \Lambda^-, t) \\ F_4(U, \Lambda^f, \Lambda^+, \Lambda^-, t) \end{bmatrix}. \tag{13}$$

This matrix system can be solved in a similar way as the biphasic problem.

Acknowledgements

This research is a joined project between the Scientific Computing Group of the Eindhoven University of Technology and the department of Movement Sciences of the University of Maastricht. The research of J.M. Huyghe has been made possible through a fellowship of the Royal Netherlands Academy of Arts and Sciences.

References

1. Biot, M. A. General theory of three-dimensional consolidation. *Journal of Applied Physics*, **12**, 155–164, 1941.
2. Frijns, A. J. H., Huyghe, J. M. and Janssen, J. D. A validation of the quadriphasic mixture theory for intervertebral disc issue. *International Journal of Engineering Science*, **35**, 1419–1429, 1997.
3. Helmig, R. Multiphase flow and transport processes in the subsurface: a contribution to the modelling of hydrosystems.. Berlin: Springer, 1997.
4. Huyghe J. M. Non-linear finite element models of the beating left ventricle and the intramyocardial coronary circulation. Ph.D. thesis, Eindhoven University of Technology, Eindhoven, 1986.
5. Huyghe J. M. and Janssen, J. D. Quadraphasic mechanics of swelling incompressible porous media. *International Journal of Engineering Science*, **35**, 793–802, 1997.
6. Kaasschieter, E. F. and Huijben, A. J. M. Mixed-hybrid elements and streamline computation for the potential flow problem. *Numerical Methods for Partial Differential Equations*, **8**, 221–266, 1992.
7. Lai, W. M., Hou, J. S. and Mow, V. C. A triphasic theory for the swelling and deformation behaviors of articular cartilage. *ASME Journal of Biomechanical Engineering*, **113**, 245–258, 1991.
8. Mow, V. C., Kuei, S. C., Lai, W. M. and Armstrong, C. G. Biphasic creep and stress relaxation of articular cartilage: theory and experiments. *ASME Journal of Biomechanical Engineering*, **102**, 73–84, 1980.
9. Simon, B. R., Zienkiewicz, O. C. and Paul, D. K. An analytical solution for the transient response of saturated porous elastic solids. *International Journal for Numerical and Analytical Methods in Geomechanics*, **8**, 381–398, 1984.
10. Snijders, H., Huyghe, J. M. and Janssen, J. D. Triphasic finite element model for swelling porous media. *International Journal for Numerical Methods in Fluids*, **20**, 1039–1046, 1995.
11. Spilker, R. L. and Suh, J.-K. Formulation and evaluation of a finite element model for the biphasic model of hydrated soft tissues. *Computers and Structures*, **35**, 425–439, 1990.
12. Terzaghi, K. Erdbaumechanik auf bodenphysikalischer Grundlage. Wien: Deuticke, 1925.

Numerical Description of Elastic-Plastic Behavior of Saturated Porous Media

J. Skolnik
Institute of Mechanics, FB-10,
University of Essen, D-45117 Essen, Germany

1. Introduction

In this paper, the numerical description of the elasto-plastic behavior of saturated porous materials, consisting of a solid skeleton with liquid filled pores, will be discussed. In particular, frictional materials will be described which show small elastic but mostly plastic deformations.

The elastic-plastic deformable solid skeleton and the liquid are assumed to be incompressible, where the liquid does not exhibit viscous properties.

The description of the stress state is done within the framework of the geometrically-linear theory, because small deformations are assumed. The elastic strain state shall be expressed by Hooke's law and the plastic strain state with the help of a single surface yield criterion as well as a non-associated flowrule.

The numerical description of the initial- and boundary-value problems is done by the finite element method within the framework of the standard Galerkin procedure.

2. Basis

The porous media theory – mixture theory restricted by the concept of volume fractions – is the basis for the numerical description of the elasto-plastic behavior of saturated frictional porous materials, see de Boer [1] and Ehlers [2]. The description of the frictional materials will be done with a binary model, where the constituent φ^S represents a solid skeleton and the second constituent φ^F a liquid in the saturated pores. Both constituents are immiscible and no mass exchange occurs between them.

Assuming both phases to be incompressible, the real density ρ^{SR} of the solid skeleton and ρ^{FR} of the fluid are constant. With the concept of volume fractions, the partial density of the constituents are obtained

$$\rho^\alpha = n^\alpha \rho^{\alpha R} \quad \alpha = S, F\ , \tag{1}$$

W. Ehlers (ed.), IUTAM Symposium on Theoretical and Numerical Methods in Continuum Mechanics of Porous Materials, 105–110.

where n^α is the volume fraction of the constituent φ^α. For the volume fractions n^α, the following saturation condition holds:

$$n^S + n^F = 1\,. \tag{2}$$

In the case of isothermal quasi-statical processes, and assuming that no mass exchange takes place between the constituents, the balance of mass and the balance of momentum of the constituents in the local statement have the following form:

$$\begin{aligned}(\rho^\alpha)'_\alpha + \rho^\alpha \operatorname{div} \mathbf{x}'_\alpha &= 0\,,\\ \operatorname{div} \mathbf{T}^\alpha + \rho^\alpha \mathbf{b} + \hat{\mathbf{p}}^\alpha &= \mathbf{0}\,,\end{aligned} \tag{3}$$

where $\mathbf{T}^\alpha$ represents the partial Cauchy stress tensor, $\mathbf{b}$ the external acceleration,and $\hat{\mathbf{p}}^\alpha$ the interaction force of the constituents φ^α. In the balance of mass, Equation (3), $\mathbf{x}'_\alpha$ is the velocity of the constituents φ^α. For the partial Cauchy stress tensor, the following characteristic is obtained from the balance of moment of momentum:

$$\mathbf{T}^\alpha = (\mathbf{T}^\alpha)^T\,. \tag{4}$$

As the sum of the interaction forces must vanish for a mixture, the following relation holds for the binary model

$$\hat{\mathbf{p}}^F = -\,\hat{\mathbf{p}}^S\,. \tag{5}$$

In order to close the system of equations, constitutive relationsare required forthe partial Cauchy stress tensor and the interaction forces. The partial stress tensor $\mathbf{T}^\alpha$ is additively decomposed into two terms in the following manner.

$$\mathbf{T}^\alpha = -\,n^\alpha p \mathbf{I} + \mathbf{T}^\alpha_E\,. \tag{6}$$

where $\mathbf{T}^\alpha_E$ denotes the partial effective stress tensor and p the pore-water pressure. For the fluid, it is assumed that the effective stress tensor $\mathbf{T}^F_E$ can be neglected. The interaction forces are determined by a constitutive equation as follows:

$$\hat{\mathbf{p}}^F = p \operatorname{grad} n^F - \frac{(n^F)^2 \gamma^{FR}}{k^F} \mathbf{w}_F = -\,\hat{\mathbf{p}}^S\,, \tag{7}$$

with γ^{FR} being the real specific weight of the fluid and k^F the Darcy permeability parameter.The quantitity $\mathbf{w}_F$ represents the seepage velocity of the fluid, which is described as follows:

$$\mathbf{w}_F = \mathbf{x}'_F - \mathbf{x}'_S \, . \tag{8}$$

In order to determine the partial effective stress tensor $\mathbf{T}_E^S$ of the solid, showing elasto-plastic behavior, additional constitutive relationships are required. The linear elastic behavior of the solid is expressed by Hooke's law which, in the geometrically-linear theory (where the Cauchy stress tensor is approximately equal to the second Piola Kirchhoff stress tensor), is expressed as:

$$\mathbf{T}_E^S \approx \mathbf{S}_E^S = 2\mu^S \mathbf{E}_{Se} + \lambda^S (\mathbf{E}_{Se} \cdot \mathbf{I})\mathbf{I} \, , \tag{9}$$

where $\mathbf{E}_{Se}$ is the elastic part of the Greens strain tensor of the solid skeletonand μ^S, λ^S are the Lamé constants of the solid phase.

In order to describe the plastic material behavior, a yield condition, determining the beginning of the ideal-plastic flow, and a flowrule must be introduced. The yield condition for ideal-plastic frictional materials of de Boer[3] is defined through the following function:

$$F = \sqrt{\phi}(1+\gamma\vartheta)^{1/3} + \beta \mathrm{I}_{\boldsymbol{T}_E^S} + \varepsilon \mathrm{I}^2_{\boldsymbol{T}_E^S} - \kappa = 0 \tag{10}$$

$$\phi = \mathrm{II}_{\boldsymbol{T}_E^{SD}} + 0.5\alpha^2 \mathrm{I}^2_{\boldsymbol{T}_E^S} \, , \quad \vartheta = \frac{\mathrm{III}_{\mathbf{T}_E^{SD}}}{\sqrt{\phi^3}} \, . \tag{11}$$

The quantities $\alpha, \beta, \gamma, \epsilon, \kappa$ are material dependent constants and $\mathrm{I}_{\boldsymbol{T}_E^S}, \mathrm{II}_{\boldsymbol{T}_E^{SD}}$ and $\mathrm{III}_{\mathbf{T}_E^{SD}}$ being the first invariant of the effective stress tensor $\mathbf{T}_E^S$ and the second, third invariants of the deviatoric part of the effective stress tensor $\mathbf{T}_E^{SD}$, respectively.

The increase of the plactic strain is determined through the flow rule of de Boer[4]:

$$\Delta \mathbf{E}_{Sp} = \Delta\lambda[\mathbf{T}_E^{SD} + \eta(\mathbf{T}_E^S \cdot \mathbf{I})\mathbf{I}] \, , \tag{12}$$

where λ denotes the plastic multiplicator and η is a function depending on $\mathrm{I}_{\boldsymbol{T}_E^S}, \mathrm{II}_{\boldsymbol{T}_E^{SD}}$ and $\mathrm{III}_{\mathbf{T}_E^{SD}}$, having the following form:

$$\eta = \eta(\mathrm{I}_{\boldsymbol{T}_E^S}, \mathrm{II}_{\boldsymbol{T}_E^{SD}}, \mathrm{III}_{\boldsymbol{T}_E^{SD}}) = c\frac{a_1}{a_2 \mathrm{I}_{\boldsymbol{T}_E^S}} \, , \tag{13}$$

with

$$a_1 = \frac{\partial F}{\partial \mathrm{I}_{\boldsymbol{T}_E^S}}, \tag{14}$$

$$a_2 = \frac{\partial F}{\partial \mathrm{II}_{\boldsymbol{T}_E^{SD}}}. \tag{15}$$

The response parameter c, can be a function of the first invariant of the partial effective stress tensor or a constant. In case of $c = 1$, the non-associative flow rule turns into an associated flow rule.

3. Numerical solution

The treatment of initial- and boundary-value problems within the domain of the porous media theory has proven itself to be efficient, evaluating the process forms of the balance of momentum of the mixture, the balance of momentum of fluid and the volume balance of the mixture, which are written here in their local statements as:

$$\mathrm{div}\,(\mathbf{T}_E^S - p\mathbf{I}) + (\rho^S + \rho^F)\mathbf{b} = \mathbf{0}\,, \tag{16}$$

$$\mathrm{grad}\, p - \rho^{FR}\mathbf{b} + \frac{n^F\gamma^{FR}}{k^F}\mathbf{w}_F = \mathbf{0}\,, \tag{17}$$

$$\mathrm{div}\,(\mathbf{x}'_S + n^F\mathbf{w}_F) = 0\,. \tag{18}$$

The evaluation of these field equations takes place numerically with the help of the finite element method. For this method, one requires the weak formulation of the above-mentioned field equations, which can be formulated whithin the framework of the standard Galerkin procedure. Whithin the framework of this procedure, the balance of momentum of the mixture is to be multiplied with the weighting function $\delta\mathbf{u}_S$, which is to be understood as a variation of the solid displacement, leading to the equation

$$\int_{B_S} [(\mathbf{T}_E^S - p\mathbf{I}) \cdot \mathrm{grad}\,\delta\mathbf{u}_S - (\rho^S + \rho^F)\mathbf{b}\cdot\delta\mathbf{u}_S]\,\mathrm{dv} = \int_{\partial B_S} \mathbf{t}\cdot\rho\mathbf{u}_S \mathrm{da}\,, \tag{19}$$

$$\mathbf{t} = (\mathbf{T}^F + \mathbf{T}^S)\mathbf{n}\,, \tag{20}$$

where $\mathbf{t}$ is the stress vector of the mixture. Furthermore, the balance of momentum for the fluid will be multiplied with the weighting function $\delta\mathbf{w}_F$, the variation of the seepage velocity; written here in the evaluated form as:

$$\int_{B_S} (-p \operatorname{div} \delta\mathbf{w}_F + \frac{n^F \gamma^{FR}}{k^F} \mathbf{w}_F \cdot \delta\mathbf{w}_F - \rho^{FR}\mathbf{b} \cdot \rho\mathbf{w}_F)\,\mathrm{dv} = \tag{21}$$

$$- \int_{\partial B_S} p\delta\mathbf{w}_F \cdot \mathbf{n}\mathrm{da}\,. \tag{22}$$

The weak form of the volume balance equation of the mixture is obtained by multiplying this equation with the weighting function δp, the variation of the porewater pressure. This is transformed with the help of the local statement of the balance of momentum of the fluid, in order to achieve an association of the porewater pressure with the variation of the porewater pressure so that numerical problems are avoided. With this, the weak form of the volume balance can be written as follows:

$$\int_{B_S} [\mathrm{div}\mathbf{x}'_S \delta p - \frac{k^F \rho^{FR}}{\gamma^{FR}} (\mathbf{b} - \frac{1}{\rho^{FR}} \operatorname{grad} p) \cdot \operatorname{grad} \delta p]\,\mathrm{dv} \tag{23}$$

$$= \int_{\partial B_S} n^F \mathbf{w}_F \cdot \mathbf{n} \delta p \,\mathrm{da}\,. \tag{24}$$

Solving the given system of equations consisting of the weak form of the balance of momentum of the mixture, the balance of momentum of the fluid, and the volume balance of the mixture, the unknown fields $\mathbf{u}_S$, $\mathbf{w}_F$, and p can now be determined.

4. Example

The numerical example describes the failure of a vertical slope consisting of a fluid saturated soil, as a consequence of an applied surface load, without consideration of its own weight, see Figure 1. The failure state is reached when a shear band occurs. The surface load q(t) is increased with passing time till the failure state arises.

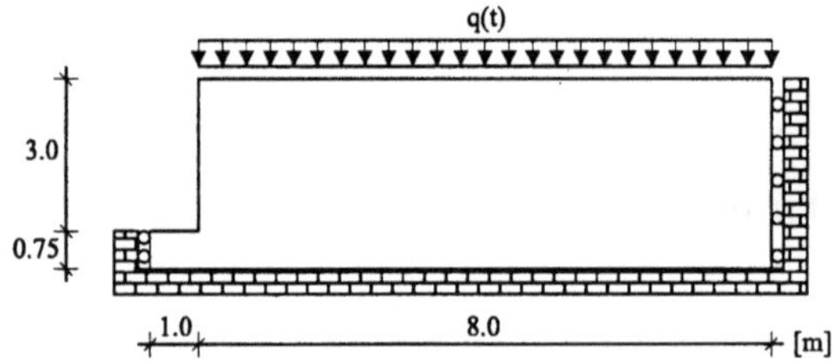

$\alpha = 0.0001$	$\beta = 0.14$
$\varepsilon = 2.0\ 10^{-4}\ m^2/kN$	$\gamma = 1.299$
$\kappa = 15\ kN/m^2$	$k^F = 10^{-3}\ m/s$
$n^F = 0.33$	$n^S = 0.67$
$E^S = 14515\ kN/m^2$	$\nu = 0.3$

Figure 1. Vertical slope under a surface load and its material parameters

Figure 2 shows the equivalent stress for the failure state. Clearly a shear band can be seen in the front part of the slope, which is achieved by a surface load of $q(t) = 47,25kN/m^2$.

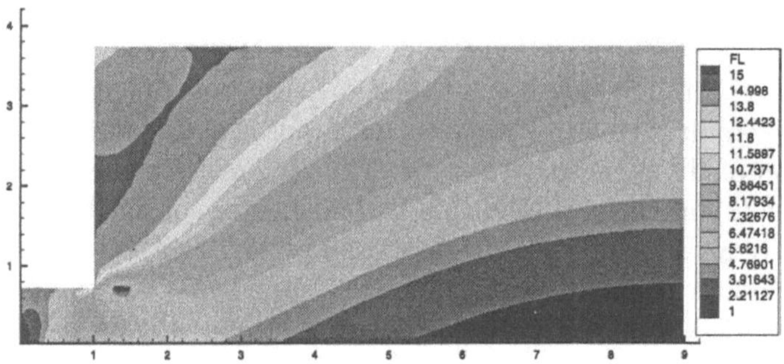

Figure 2. Vertical slope under a surface load and its material parameters

Besides the above mentioned example there are various other problems from soil mechanics, and geotechnical engineering, respectively, which can be numerically simulated whith the help of the process of the theory of elastic-plastic porous media presented here, as for example base failure problems.

References

1. de Boer, R. Highlights in the historical development of the porous media theory: towards a consistent macroscopc theory. *Appl. Mech.Rev.*, 1996.
2. Ehlers, W. Poröse Medien - ein kontinuumsmechanisches Modell auf der Basis der Mischungstheorie. Forschungsbericht aus dem Fachbereich Bauwesen 47. Universität-GH Essen, 1989.
3. de Boer, R. and Lade, P. Towards a general plasticity theory for empty and saturated porous solids. Forschungsbericht aus dem Fachbereich Bauwesen 55. Universität-GH Essen, 1991.
4. de Boer, R. On plastic deformation in soil mechanics. *Int. J. Plasticity*, 1987/1988.

Effective Physical Properties of Sandstones

J. Widjajakusuma[1] and R. Hilfer [1,2]

[1]*ICA-1, Universität Stuttgart, Pfaffenwaldring 27, 70569 Stuttgart*
[2]*Institut für Physik, Universität Mainz, 55099 Mainz, Germany*

Abstract. In this paper we continue the investigation of the effective transport parameters of a digitized sample of Fontainebleau sandstone and three reconstruction models discussed previously in Biswal et. al., Physica A **273**, 452 (1999). The effective transport parameters are computed directly by solving the disordered Laplace equation via a finite-volume method. We find that the transport properties of two stochastic models differ significantly from the real sandstone. Moreover, the effective transport parameters are predicted by employing local porosity theory and various traditional mixing-laws (such as effective medium approximation or Maxwell-Garnet theory). The prediction of local porosity theory is in good agreement with the exact result.

Keywords: Porous materials, effective material parameters, self-consistent method, microstructure

1. Introduction

It is well-known that the overall mechanical and transport properties of porous materials depend strongly on the microstructure [1, 7–12, 19–22]. Because in general the exact microstructures of porous media are not known in detail, one often uses models for calculating the effective macroscopic properties. For detail see [1–3, 8–12] and references therein.

The objective of the present article is to continue the study of transport properties of digitized realistic porous media. Simultaneously, we are testing the validity of local porosity theory (LPT) [9, 11, 13]. Local porosity theory was employed successfully to distinguish the microstructures of various porous media [4, 13], and the mixing-law based on LPT was used to determine the effective transport parameters of porous media [12, 20, 22].

2. Effective transport parameters

For homogeneous and isotropic random media the effective parameter $\overline{C}$ is defined through an ensemble average of the local constitutive

W. Ehlers (ed.), *IUTAM Symposium on Theoretical and Numerical Methods in Continuum Mechanics of Porous Materials*, 113–118.

equation

$$\langle \mathbf{J}(\mathbf{r}) \rangle = -\overline{C}\, \langle \nabla U(\mathbf{r}) \rangle. \tag{1}$$

where $\mathbf{J}(\mathbf{r})$ is the local current and $\nabla U(\mathbf{r})$ is the local potential gradient. If one knows the exact microstructure of a porous medium, one can calculate $\mathbf{J}(\mathbf{r})$ and $\nabla U(\mathbf{r})$ numerically by solving the local continuity equation

$$\nabla \cdot \mathbf{J}(\mathbf{r}) = 0 \qquad \mathbf{r} \in \mathbb{G} \quad (\mathbb{G} = \mathbb{P}, \mathbb{M}) \tag{2}$$

combined with the local constitutive equation

$$\mathbf{J}(\mathbf{r}) = -C(\mathbf{r})\, \nabla U(\mathbf{r}) \tag{3}$$

and appropriate internal and external boundary conditions. In eq. (2) is $C(\mathbf{r}) = C_{\mathbb{P}}\, \chi_{\mathbb{P}}(\mathbf{r}) + C_{\mathbb{M}}\, \chi_{\mathbb{M}}(\mathbf{r})$, where $C_{\mathbb{P}}$ and $C_{\mathbb{M}}$ are material constants of pore space and matrix space, respectively. Eqs. (2) and (3) have been solved recently for a sample of Fountainebleau sandstone via finite-volume method [20, 22, 12]. After taking the average of $\mathbf{J}(\mathbf{r})$ and $\nabla U(\mathbf{r})$, and inserting the two averaged values in eq. (1) the effective conductivity $\overline{C}$ can be directly obtained.

3. Local porosity theory

In practice, the exact microstructure of porous media is usually not known in detail. Therefore, only approximate effective transport parameters can be computed based on partial microgeometric knowledge, such as porosity, specific internal surface, connectedness or correlation lengths [9–13, 16, 20–22]. The partial microgeometric knowledge included into LPT is information about porosity and connectivity fluctuations of porous media in terms of *local porosity distribution* $\mu(\phi, L)$ and *local percolation probability* $\lambda(\phi, L)$. $\mu(\phi, L)$ is the probability density to find a local porosity ϕ within the cubic measurement cell of sidelength L and $\lambda(\phi, L)$ gives the fraction of percolating cells with prescribed local porosity ϕ.

The mixing-law based on local porosity theory reads [9, 11, 22]

$$\int_0^1 \mu(\phi, L) \quad [\, \lambda(\phi, L)\, \frac{\overline{C}_{\mathbb{P}}(\phi) - \overline{C}_{\mathrm{LPT}}}{\overline{C}_{\mathbb{P}}(\phi) + 2\,\overline{C}_{\mathrm{LPT}}} + [1 - \lambda(\phi, L)]\, \frac{\overline{C}_{\mathbb{M}}(\phi) - \overline{C}_{\mathrm{LPT}}}{\overline{C}_{\mathbb{M}}(\phi) + 2\,\overline{C}_{\mathrm{LPT}}}\,]\, d\phi = 0, \tag{4}$$

where

$$\overline{C}_{\mathrm{P}} = C_{\mathrm{P}} \left(\frac{3C_{\mathrm{M}} + 2\overline{\phi}(C_{\mathrm{P}} - C_{\mathrm{M}})}{3C_{\mathrm{P}} - \overline{\phi}(C_{\mathrm{P}} - C_{\mathrm{M}})} \right) \tag{5}$$

$$\overline{C}_{\mathrm{M}} = C_{\mathrm{M}} \left(\frac{2C_{\mathrm{M}} + C_{\mathrm{P}} + 2\overline{\phi}(C_{\mathrm{P}} - C_{\mathrm{M}})}{2C_{\mathrm{M}} + C_{\mathrm{P}} - \overline{\phi}(C_{\mathrm{P}} - C_{\mathrm{M}})} \right). \tag{6}$$

The mixing-law based on LPT can be viewed as a generalization of the classical self-consistent approximation. In the limit $L \to 0$, Eq. (4) reduces to the classical effective medium approximation (EMA) [6]

$$\overline{\phi}\, \frac{C_{\mathrm{P}} - C_{\mathrm{EMA}}}{C_{\mathrm{P}} + 2\, C_{\mathrm{EMA}}} + (1 - \overline{\phi})\, \frac{C_{\mathrm{M}} - C_{\mathrm{EMA}}}{C_{\mathrm{M}} + 2\, C_{\mathrm{EMA}}} = 0\,. \tag{7}$$

For $\mu(\phi, L) = \delta(\phi - \overline{\phi})$ and $\lambda(\phi, L) = 1$ we obtain the Maxwell-Garnett approximation with $\mathbb{P}$ as the background phase (MGP)

$$\overline{C}_{\mathrm{MGP}} = \overline{C}_{\mathrm{P}}\,. \tag{8}$$

For $\mu(\phi, L) = \delta(\phi - \overline{\phi})$ and $\lambda(\phi, L) = 0$ Eq. (4) recovers the Maxwell-Garnett approximation with $\mathbb{M}$ as background phase (MGM)

$$\overline{C}_{\mathrm{MGM}} = \overline{C}_{\mathrm{M}}\,. \tag{9}$$

For more detail see [11, 19].

4. Results and discussion

We apply now LPT to analyze quantitatively four different samples. The first sample is a digitized real Fountainebleau sandstone (EX), which is obtained by microtomographic imaging [3, 4]. The second sample is a diagenesis model (DM), which is obtained by imitating the natural sandstone-forming processes. As in the natural processes, the numerical modelling of DM is performed in three main steps: grain sedimentation, compaction and diagenesis described in detail in [2, 17]. The Gaussian field reconstruction model (GF) is generated in such a way that the two-point correlation function of this model is identical to a given reference correlation function by filtering Gaussian random variables. Here the reference correlation function is the correlation function of the real Fountainebleau sandstone G_{EX}. For further information see [1, 18]. The last model, denoted as SA, is reconstructed by employing a simulated annealing technique. This technique produces a configuration $\mathbb{S}_{SA}$ by minimizing the deviations between its correlation function G_{SA}

Table I. Overview of geometric properties of the Fontainebleau sandstone and its reconstruction models and their effective conductivity $\overline{C}$ and standard deviation

	Samples			
	$\mathbb{S}_{EX}$	$\mathbb{S}_{DM}$	$\mathbb{S}_{GF}$	$\mathbb{S}_{SA}$
$\overline{\phi}$	0.1355	0.1356	0.1421	0.1354
$\overline{C}$	0.01880	0.01959	0.00234	0.00119
stand. dev.	±0.00852	±0.00942	±0.00230	±0.00234

and the reference correlation function G_{EX}. Besides the correlation function G another statistical property, which has to match with the reference statistical property, is porosity ϕ. The detailed description of the simulated annealing method can be found in [14, 23]. Note that we can choose other statistical properties, which should be equal to the prescribed reference properties, such as linear or spherical contact distributions [15].

We choose that the matrix space as nonconducting, i.e. $C_{\mathbb{M}} = 0$, and pore space as conducting, i.e. $C_{\mathbb{P}} = 1.0$ (dimensionless). The boundary conditions are chosen so that potential values were prescribed at two parallel faces of the cubic sample $\mathbb{S}$ (Dirichlet boundary condition), and the flux across the four remaining faces of $\mathbb{S}$ is set to zero (Neumann boundary condition).

The disorder average of fields $\langle \mathbf{J}(\mathbf{r}) \rangle$ and $\langle \nabla U(\mathbf{r}) \rangle$ are known to fluctuate strongly from one sample to another. To improve the statistics all of the samples were cut into eight pieces of dimension $128 \times 128 \times 128$. For each piece three values of $\overline{C}_i$ were obtained from the exact solution corresponding to the application of the potential gradient in the x-, y- and z-direction. Then, the values of $\overline{C}$ of samples are obtained by taking the arithmetic average of $\overline{C}_i$. The results are displayed in Table I. The standard deviations in Table I show that the fluctuations in $\overline{C}$ are indeed rather strong. For ergodic geometries $\overline{C}$ can be calculated directly from the exact solution for the full sample. For sample EX the exact transport coefficient for the full sample is $\overline{C}_x = 0.02046$ in the x-direction, $\overline{C}_y = 0.02193$ in the y-direction, and $\overline{C}_z = 0.01850$ in the z-direction [22]. All of these are seen to fall within one standard deviation of $\overline{C}$. The effective conductivity of the DM model matches quite closely to that of EX. The effective transport parameters $\overline{C}$ of GF and SA differ strongly from $\overline{C}$ of EX. This was already predicted in [4] on a pure connectivity analysis based on λ from LPT.

For LPT calculations we have to choose the sidelength L of the measurement cell $\mathbb{K}$. In the case of small L the local geometries become strongly correlated, and this is at variance with the basic assumption of weak or no correlations. On the other hand, for large L the assumption that the local geometry is sufficiently simple becomes invalid [11]. Hence, one expects that formula (4) will yield good results only for intermediate L. We use the so called percolation length L_p, which is defined through the condition

$$\left.\frac{d^2p}{dL^2}\right|_{L=L_p} = 0 \tag{10}$$

assuming that it is unique. The idea behind this definition is that at the inflection point the function $p_3(L)$ changes most rapidly from its trivial value $p(0) = \overline{\phi}$ at small L to its equally trivial value $p(\infty) = 1$ at large L (assuming that the pore space percolates). For another choice of L see [11, 12, 20, 22]. At $L = L_p$ we find for the effective conductivity of EX the value $\overline{C}_{\text{LPT}} = 0.025115$ in good agreement with the exact result.

In contrast one has $C_{\text{EMA}} = 0$ for the effective medium approximation because the porosity $\overline{\phi}$ is below the percolation threshold $\overline{\phi}_c = 1/3$. Similarly, the result $\overline{C}_{\text{MGM}} = 0$ obtained by MGM-approximation underestimates the effective parameter $\overline{C}$ of EX. On the other hand, $\overline{C}_{\text{MGP}} = 0.094606$ obtained from the MGP-approximation overestimates the exact result. These results reflect the nature of the approximations involved, because in both cases the inclusion phase is always dispersed in the background phase without having a connecting path. The functions μ and λ, which are used in LPT, provide more information about the underlying microgeometry than a single parameter $\overline{\phi}$, which are employed in the traditional mixing-laws (EMA, MGM, MGP). Therefore, the estimate of $\overline{C}_{\text{LPT}}$ obtained by LPT seems to be better than those obtained by traditional mixing-laws.

Acknowledgements

We are deeply indebted to Dr. B. Biswal and C. Manwart for their close cooperation on many aspects of this work. We thank the Deutsche Forschungsgemeinschaft and the Graduiertenkolleg "Modellierung und Diskretisierungsmethoden für Kontinua und Strömungen" for financial support.

References

1. Adler, P. *Porous Media*. Boston: Butterworth-Heinemann, 1992.

2. Bakke, S. and Øren, P. 3-d pore-scale modeling of sandstones and flow simulations in pore networks. *SPE Journal*, **2**, 136, 1997.
3. Biswal, B., Manwart, C. and Hilfer, R. Three-dimensional local porosity analysis of porous media. *Physica A*, **225**, 221, 1998.
4. Biswal, B., Manwart, C., Hilfer, R., Bakke, S. and Øren, P. Quantitative analysis of experimental and synthetic microstructures for sedimentary rock. *Physica A*, **273**, 452, 1999.
5. Boger, F., Feder, J., Jøssang, T. and Hilfer, R. Microstructural sensitivity of local porosity distributions. *Physica A*, **187**, 55, 1992.
6. Bruggeman, D. A. G. Berechnung verschiedener physikalischer Konstanten von heterogenen Substanzen. I. Dielektrizitätskonstanten und Leitfähigkeiten der Mischkörper aus isotropen Substanzen. *Ann. Phys.*, **24**, 636, 1935.
7. Ehlers, W. Poröse Medien - ein kontinuummechanisches Modell auf der Basis der Mischungstheorie. Habilitationsschrift. Forschungsberichte aus dem Fachbereich Bauwesen der Universität-GH-Essen, Vol. 47, Essen, 1989.
8. Ehlers, W. Grundlegende Konzepte in der Theorie poröser Medien. *Technische Mechanik*, **16**, 63–76, 1996.
9. Hilfer, R. Geometric and dielectric characterization of porous media. *Phys. Rev. B*, **44**, 60, 1991.
10. Hilfer, R. Local porosity theory for electrical and hydrodynamical transport through porous media. *Physica A*, **194**, 406, 1993.
11. Hilfer, R. Transport and relaxation phenomena in porous media. *Adv. in Chem. Phys.*, **XCII**, 299, 1996.
12. Hilfer, R., Widjajakusuma, J. and Biswal, B. Dielectric relaxation in water saturated sedimentary rocks. *Granular Matter*, in print, 1999.
13. Hilfer, R. Local porosity theory and stochastic reconstruction for porous media. *in preparation*, 2000.
14. Manwart, C. and Hilfer, R. Reconstruction of random media using Monte Carlo methods. *Phys. Rev. E*, **59**, 5596, 1999.
15. Manwart, C., Torquato, S. and Hilfer, R. Stochastic reconstruction of sandstones. *preprint*, 1999.
16. Manwart, C., Widjajakusuma, J., Hilfer, R. and Ohser, J. Local specific surface distribution of sandstones. *in preparation*, 2000.
17. Øren,P., Bakke, S. and Arntzen, O. Extending predictive capabilities to network models. *SPE Journal*, p. SPE 38880, 1998.
18. Quiblier, J. A new three dimensional modeling technique for studying porous media. *J. Colloid Interface Sci.*, **98**, 84, 1984.
19. Sheng, P. *Introduction to Wave Scattering, Localization, and Mesoscopic Phenomena.* San Diego: Academic Press, 1995.
20. Widjajakusuma, J. Biswal, B. and Hilfer, R. Quantitative prediction of effective material properties of heterogeneous media. *Computational Materials Science*, **16**, 70, 1999.
21. Widjajakusuma, J., Manwart, C., Biswal, B. and Hilfer, R. Exact and approximate calculations for the Conductivity of Sandstones. *Physica A*, **270**, 325, 1999.
22. Widjajakusuma, J., Biswal, B. and Hilfer, R. Predicting transport parameters of heterogeneous media. *Preprint*, 1999.
23. Yeong, C. and Torquato, S. Reconstructing random media. *Phys. Rev. E*, **57**, 495, 1998.

Perspective of Computational Micro–Macro–Transition for the Postcritical Analysis of Localized Failure

C. Miehe and M. Lambrecht
Institut für Mechanik (Bauwesen)
Universität Stuttgart, 70550 Stuttgart
e-mail: cm@mechbau.uni-stuttgart.de

Abstract. The simulation of localized failure in elastic–plastic solids, e.g. in the form of shear bands, yields the typical mesh–dependent postcritical results within standard finite element formulations. We here propose a new relaxation technique based on a micromechanically motivated approach which overcomes this problem. The key idea is the introduction of a micro–structure at a typical Gauss–point of the finite element mesh which bifurcates in the form of an assumed fluctuation field when macroscopic localization occurs. The method can be understood as an approximated quasi–convexification of a non–convex incremental stress potential function.

1. Introduction

The simulation of localized failure in standard rate–independent plasticity formulations for strain softening materials yields the typical mesh–dependent postcritical results within standard finite element formulations. The cause is the loss of ellipticity of the linearized boundary value problem. The literature contains a broad spectrum of methods to overcome this problem, which may be subdevided into two classes: (i) higher order continuum–based models and (ii) approaches based on the modelling of discontinuities. This paper presents, in line with the second class of methods, a new micromechanically motivated relaxation technique which models locally at a point of the continuum (regularized) discontinuities based on a two–scale homogenization technique. The key idea is the introduction of a micro–structure at a typical integration point of the finite element mesh which bifurcates when macroscopic localization occurs. The fluctuation field is assumed to have the form of a regularized discontinuity with given band width (length scale parameter) aligned to a critical direction obtained from the localization analysis. The intensity of the fluctuation is determined based on the assumed equilibrium state of the micro–structure. The proposed method provides an approximated quasi–convexification of an incremental stress potential function which looses rank–1 convexity when localization occurs, see for example Dacorogna [1] for general definitions of weak convexity. The proposed numerical implementation of the approximated quasi–convexification method is inspired by recently

W. Ehlers (ed.), *IUTAM Symposium on Theoretical and Numerical Methods in Continuum Mechanics of Porous Materials*, 119–124.

developed computational homogenization concepts of heterogeneous materials with micro–structures at finite strains as outlined in Miehe, Schotte, Schröder [3].

2. Postcritical Relaxation Analysis

We focus our considerations on the postcritical analysis. Assume that at a local point $\bar{\mathbf{X}} \in \bar{\mathcal{B}}$ of a macro–continuum $\bar{\mathcal{B}}$ the deformation field $\bar{\mathbf{x}} : \bar{\mathcal{B}} \to \bar{\mathcal{S}}$ is critical in the sense that the incremental formulation of a strain–softening constitutive model looses rank–one convexity (ellipticity) in a critical direction θ (for plane problems), for example detected by an accompanying check of the determinant of the acoustic tensor as outlined in Miehe and Schröder [4]. We then introduce at this point a micro–structure and perform a relaxation analysis as follows.

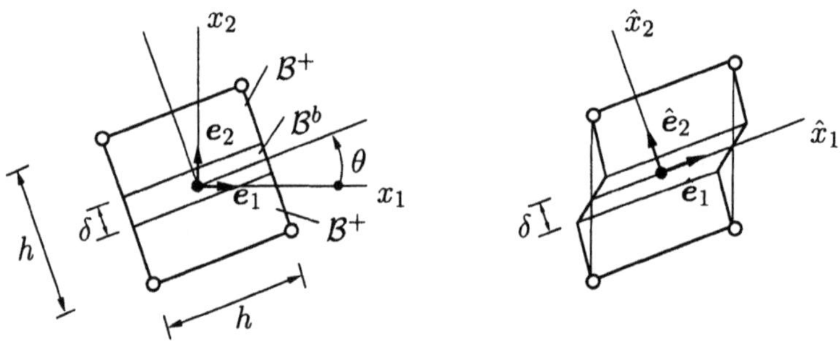

Figure 1. Definition and deformation of the micro–structure $\mathcal{B}$.

2.1. Definition of a Micro–Structure

The considered micro–structure $\mathcal{B}$ is depicted in Figure 1 and consists of two parts $\mathcal{B} = \mathcal{B}^b \cup \mathcal{B}^+$. It constitutes essentially fractions of two layers of the material where different homogeneous deformations may occur. The part $\mathcal{B}^b$ is viewed as a localized band with width δ aligned to the critical localization direction θ. The key idea of the proposed numerical relaxation concept is to relate the volume of the micro–structure to the volume $\bar{V}_g$ associated with a typical integration point of the macroscopic finite element mesh, i.e.

$$V \equiv \bar{V}_g \quad \Rightarrow \quad h = \sqrt{\bar{V}_g} \tag{1}$$

where $V := h^2$ denotes the volume of the micro–structure $\mathcal{B}$. The band width δ is considered as an intrinsic length scale parameter independent of the finite element discretization. $V^b = \delta h$ and $V^+ = (h - \delta)h$ are the volumes associated with the constituents of the micro–structure.

2.2. Definition of the Fine–Scale Fluctuations

Let $\mathbf{F}$ be the deformation gradient at a local point of the micro–structure. This gradient is assumed to consist of the constant macroscopic part $\bar{\mathbf{F}} := \nabla \bar{\mathbf{x}}$ and a superimposed fluctuation part $\tilde{\mathbf{F}}$, i.e.

$$\mathbf{F} = \bar{\mathbf{F}} + \tilde{\mathbf{F}} \ . \tag{2}$$

For plane problems, the fluctuation part is assumed to be the gradient of a regularized discontinuity fluctuation field as visualized in Figure 1. We write

$$\tilde{\mathbf{F}}^b = \frac{1}{\delta}\mathbf{G}\mathbf{d} \quad \text{and} \quad \tilde{\mathbf{F}}^+ = -\frac{1}{h-\delta}\mathbf{G}\mathbf{d} \tag{3}$$

in terms of the geometric matrix

$$\mathbf{G} := \begin{bmatrix} -\sin\theta\cos\theta & \sin^2\theta \\ \sin\theta\cos\theta & \cos^2\theta \\ \cos^2\theta & -\sin\theta\cos\theta \\ -\sin^2\theta & -\sin\theta\cos\theta \end{bmatrix} \tag{4}$$

and the discrete fluctuation vector

$$\mathbf{d} := [d_{||}, d_{\perp}]^T \tag{5}$$

of the micro–structure. The discrete fluctuations $d_{||}$ and $d_{\perp}$ govern superimposed mode–II and mode–I discontinuity modes, which are regularized by the parameter δ.

Note that the fluctuations $\tilde{\mathbf{F}}^b$ and $\tilde{\mathbf{F}}^+$ are constant within the two parts $\mathcal{B}^b$ and $\mathcal{B}^+$ of the micro–structure. The associated first Piola stresses in these parts are then determined by function evaluations

$$\mathbf{P}^b = \hat{\mathbf{P}}(\bar{\mathbf{F}} + \tilde{\mathbf{F}}^b) \quad \text{and} \quad \mathbf{P}^+ = \hat{\mathbf{P}}(\bar{\mathbf{F}} + \tilde{\mathbf{F}}^+) \ , \tag{6}$$

where $\hat{\mathbf{P}}$ represents an incremental stress update algorithm associated with a standard model of strain–softening elastoplasticity, see for example Miehe [2]. For inelasticity models with normal structure of the evolution equations, the function $\hat{\mathbf{P}}$ can be considered as the derivative of an incremental work expression $\hat{\Psi}$, i.e.

$$\hat{\mathbf{P}}(\mathbf{F}) = \partial_{\mathbf{F}} \hat{\Psi}(\mathbf{F}) \ . \tag{7}$$

The incremental stress potential function $\hat{\Psi}$ looses rank–one convexity in the postcritical range under consideration.

2.3. Determination of the Fine–Scale Fluctuations

The assumed deformation–fluctuations $\tilde{\mathbf{F}}^b$ and $\tilde{\mathbf{F}}^+$ are governed by the two scalar values $d_{\|}$ and $d_{\perp}$, which develop when the micro–structure bifurcates. These degrees are determined by an equlibrium condition for the micro–stress field which can be written in the weak form

$$\frac{1}{|\mathcal{B}|}\int_{\mathcal{B}} \delta\tilde{\mathbf{F}} : \mathbf{P}\, dV = 0 \ , \tag{8}$$

where $\delta\tilde{\mathbf{F}}$ denotes a virtual fluctuation–deformation. The condition can be understood as the necessary condition associated with the quasi–convexification problem of the incremental stress potential function $\hat{\Psi}$

$$Q^h\hat{\Psi}(\bar{\mathbf{F}}) = \inf_{\mathbf{d}} \frac{1}{|\mathcal{B}|}\int_{\mathcal{B}} \hat{\Psi}(\bar{\mathbf{F}} + \tilde{\mathbf{F}})dV \ . \tag{9}$$

Insertion of (3) into (8) and linearization yields the algebraic system

$$\mathbf{K}\Delta\mathbf{d} = -[\,\mathbf{L}\Delta\bar{\mathbf{F}} + \mathbf{r}\,] \tag{10}$$

in terms of the two fluctuation stiffness matrices

$$\mathbf{K} = \mathbf{G}^T[\,\mathbf{C}^b/\delta + \mathbf{C}^+/(h-\delta)\,]\,h\,\mathbf{G} \quad , \quad \mathbf{L} = \mathbf{G}^T[\,\mathbf{C}^b - \mathbf{C}^+\,]\,h \tag{11}$$

and the residual vector

$$\mathbf{r} = \mathbf{G}^T[\,\mathbf{P}^b - \mathbf{P}^+\,]\,h \tag{12}$$

of the micro–structure. Here, $\mathbf{C}^b$ and $\mathbf{C}^+$ denote the consistent tangent moduli associated with the stress update algorithm (6) within the two parts of the micro–structure. The fluctuation field is then determined by a local Newton iteration at frozen macroscopic deformation, i.e. for $\Delta\bar{\mathbf{F}} = \mathbf{0}$, by updates $\mathbf{d} \Leftarrow \mathbf{d} - \mathbf{K}^{-1}\mathbf{r}$ until convergence is obtained.

2.4. Homogenized Macroscopic Stresses and Moduli

We define the macroscopic stresses at $\bar{\mathbf{X}} \in \bar{\mathcal{B}}$ in the postcritical range by the volume average $\bar{\mathbf{P}} = \frac{1}{V}\int_{\mathcal{B}} \mathbf{P} dV$ of the stresses $\mathbf{P}$ defined on the micro–structure $\mathcal{B}$, yielding the simple representation

$$\bar{\mathbf{P}} = \frac{\delta}{h}\mathbf{P}^b + \frac{h-\delta}{h}\mathbf{P}^+ \ . \tag{13}$$

The macroscopic consistent tangent moduli appear in the form

$$\bar{\mathbf{C}} = \frac{\delta}{h}\mathbf{C}^b + \frac{h-\delta}{h}\mathbf{C}^+ - \frac{1}{h^2}\mathbf{L}^T\mathbf{K}^{-1}\mathbf{L} \ . \tag{14}$$

Observe that these moduli consist of the volume average of the microscopic consistent tangent moduli and a softening part which is the consequence of the flexibility of the micro–structure.

3. Example: Plane Strain Indentation Problem

We apply the proposed computational relaxation procedure to a computational model of large–strain von Mises–type elastoplasticity with linear strain–softening, see Miehe [2] for details of the formulation and numerical implementation.

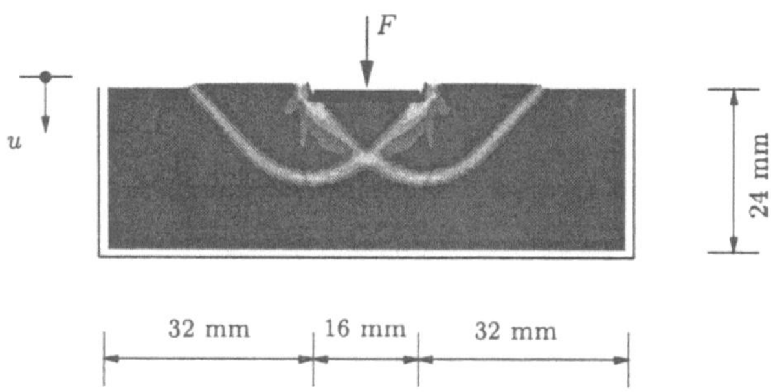

Figure 2. System and intensity of developed mode–II fluctuations.

As a typical numerical benchmark we investigate the indentation test at plane strain conditions, where a curved shear band develops in the postcritical range. The initial geometry of the specimen is depicted in Figure 2. A rigid bloc indents within a strain–controlled process into a material up to a final displacement $u = 1.8\,mm$. The half specimen has been discretized by different meshes of enhanced strain finite elements. The intrinsic length scale parameter has been set to $\delta = 0.25\,mm$.

Figure 2 points out the intensity of the mode–II fluctuations $d_{||}$ which have been developed in the postcritical analysis. Figure 3 depicts the deformed 45×27 element mesh, where some developed micro–structures within the shear band have been zoomed out. Observe the alignment of the deformed micro–structures to the global orientation of the shear band.

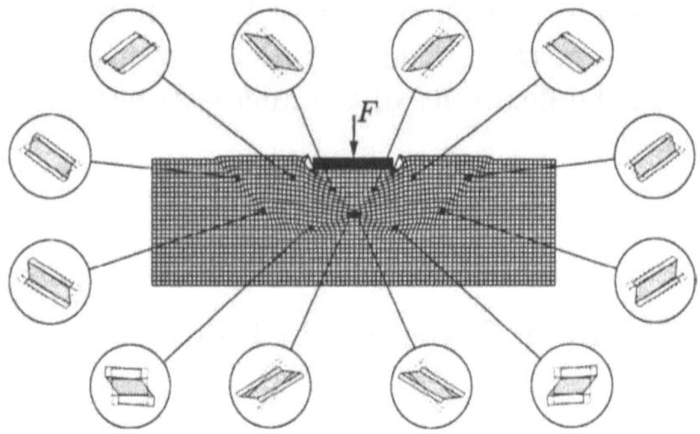

Figure 3. Deformed mesh with zoomed-out micro-structures.

The performance of the proposed relaxation technique is demonstrated by a comparison of load-deflection curves for different mesh sizes. Figure 4a reports mesh-dependent load-displacement paths for the standard formulation. Figure 4b depicts the mesh-invariant curves which are obtained by the proposed relaxation technique.

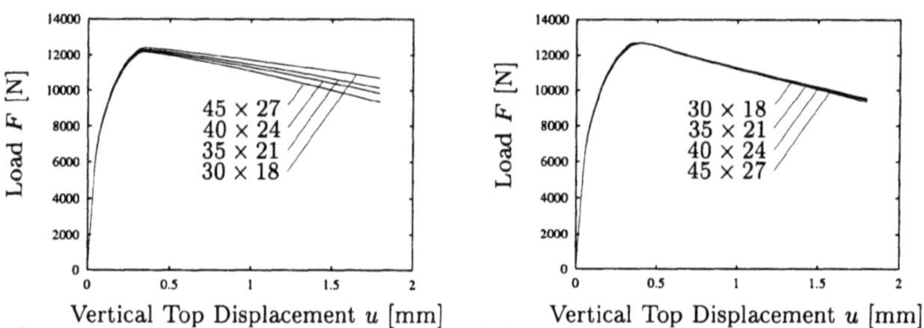

a) b)

Figure 4. Load displacement curves for different mesh sizes with a) standard formulation and b) proposed relaxation technique.

References

1. Dacorogna, B. Direct Methods in the Calculus of Variations. *Springer-Verlag*, Berlin Heidelberg, 1989.
2. Miehe, C. A Formulation of Finite Elastoplasticity Based on Dual Co- and Contra-Variant Eigenvector Triads Normalized with Respect to a Plastic Metric. *Comp. Meth. Appl. Mech. Engng.*, **159**, 223–260, 1998.
3. Miehe, C., Schotte, J. and Schröder, J. Computational Micro-Macro Transitions and Overall Moduli in the Analysis of Polycrystals at Large Strains. *Computational Materials Science*, **16**, 372–382, 1999.
4. Miehe, C. and Schröder, J. Post-Critical Discontinuous Localization Analysis of Small-Strain Softening Elastoplastic Solids. *Archive of Applied Mechanics*, **64**, 267–285, 1994.

Micromechanics of Unsaturated Porous Media

X. Chateau
LMSGC, (UMR113), CNRS-LCPC, 2 Allée Kepler, 77420 Champs sur Marne, France
e-mail: chateau@lcpc.fr

L. Dormieux
CERMMO, ENPC, 6-8 Avenue Blaise Pascal, 77455 Marne la Vallée, France
e-mail: dormieux@enpc.fr

Abstract. The macroscopic mechanical behaviour of unsaturated porous media under isothermal conditions is studied within the framework of upscaling techniques. First the main features of the homogenization method are recalled from ([4], [5]). Then, the macroscopic state equation including the capillary effects is derived. Finally, a morphological model is used in order to clarify the link between the sorbtion-desorbtion hysteretic phenomena and the macroscopic mechanical behaviour.

1. Statics of unsaturated porous media

The Representative Elementary Volume Ω is made up of a solid phase, a liquid phase and a gazeous phase which respectively occupy the domains Ω^s, Ω^ℓ and Ω^g. $\omega^{\alpha\beta}$ denotes the interface between the α- and $\beta-$phases. At the microscopic scale, the internal forces are described by a Cauchy stress tensor field $\boldsymbol{\sigma}$ in the domains Ω^s, Ω^g and Ω^ℓ. The cauchy stress in the fluid phases is defined by the fluid pressure p^α ($\alpha = \ell,\ g$), that is $\boldsymbol{\sigma} = -p^\alpha \boldsymbol{\delta}$, where $\boldsymbol{\delta}$ denotes the unit tensor of second order. The capillary effects introduce internal forces of the membrane type located in the interfaces between phases. The latter are represented by a tensor field of surface tension $\gamma^{\alpha\beta}\boldsymbol{\delta}_{T_\omega}$ in the surface $\omega^{\alpha\beta}$, where $\boldsymbol{\delta}_{T_\omega}$ denotes the unit tensor to the tangent plane to surface ω and $\gamma^{\alpha\beta}$ the surface tension in the $\alpha\beta$ interface, with $(\alpha\beta) = (s\ell)$, (sg) and (ℓg). As usual within the framework of homogenization approach to the behaviour of the porous media skeleton ([1], [2]), this internal forces must comply with no body force equilibrium conditions. Thus, to be statically admissible, the microscopic stress field has to comply with the momentum balance equation $\mathrm{div}\boldsymbol{\sigma} = 0$. In particular, this implies that p^ℓ and p^g are uniform in Ω^g and Ω^ℓ respectively. Taking the capillary effects into account, the classical condition of continuity of the stress vector at the interfaces $\omega^{s\alpha}$ is replaced by the following

W. Ehlers (ed.), IUTAM Symposium on Theoretical and Numerical Methods in Continuum Mechanics of Porous Materials, 125–130.

conditions:

$$\boldsymbol{\sigma} \cdot \underline{n} = -p^{\alpha}\underline{n} - \gamma^{s\alpha}(\boldsymbol{\delta}_{T_\omega} : \boldsymbol{b})\underline{n} \quad (\forall\, \underline{x} \in \omega^{s\alpha})\ (\alpha = g, \ell) \tag{1}$$

where $\underline{n}$ denotes the outer unit normal vector to Ω^s and $\boldsymbol{b}$ is the tensor of curvature. $\boldsymbol{\delta}_{T_\omega} : \boldsymbol{b}$ thus represents the mean curvature of $\omega^{s\alpha}$. The corresponding condition at the interface $\omega^{g\ell}$ is classically referred to as Laplace law:

$$(\forall\, \underline{x} \in \omega^{\ell g}) \qquad \gamma^{\ell g}\boldsymbol{\delta}_{T_\omega} : \boldsymbol{b} = p^{\ell} - p^{g} = -p^{c} \tag{2}$$

p^c will be referred to as the capillary pressure. Hill's lemma provides the link between the macroscopic stress tensor $\boldsymbol{\Sigma}$ and the microscopic stress field ([4]):

$$\boldsymbol{\Sigma} = (1-f) < \boldsymbol{\sigma} >_s - f^{\ell}p^{\ell}\boldsymbol{\delta} - f^{g}p^{g}\boldsymbol{\delta} + \frac{1}{|\Omega|}\int_{\omega} \gamma\,(\underline{x})\, \boldsymbol{\delta}_{T_\omega} dS \tag{3}$$

where f denotes the porosity, $< \cdot >_\alpha$ the volume average in Ω^α and f^α the volume fraction of the α-phase.

In the sequel, we examine the situation of an homogeneous linear elastic solid matrix. The microscopic state equation for the solid phase thus reads $\boldsymbol{\sigma} = \mathbb{C}^s : \boldsymbol{\varepsilon}$ or $\boldsymbol{\varepsilon} = \mathbb{S}^s : \boldsymbol{\sigma}$.

2. Simlified model

First, it is assumed that the surface tension $\gamma^{\alpha\beta}$ can be neglected in equations (1) and (3): the capillary effects are only taken into account through the difference between the fluid and gas pressures. For the sake of simplicity, it is assumed that the boundary of the elementary representative volume $\partial\Omega$ is located within the solid phase.

We have to express that the solid is a linear elastic material and that the stress is of the form $-p^\alpha\boldsymbol{\delta}$ in the fluid phases. The idea is to deal with each of the 3 phases in the same way. This can be achieved in writting the microscopic stress tensor $\boldsymbol{\sigma}$ as follows:

$$\boldsymbol{\sigma} = \mathbb{C}(\underline{x}) : \boldsymbol{\varepsilon} + \boldsymbol{\sigma}_p(\underline{x}) \tag{4}$$

In (4), the tensor of elastic moduli $\mathbb{C}(\underline{x})$ is $\mathbb{C}^s$ in the solid whereas $\mathbb{C}(\underline{x}) = 0$ in the fluid phases. The pre-stress $\boldsymbol{\sigma}_p$ is 0 in the solid whereas $\boldsymbol{\sigma}_p^\alpha = -p^\alpha\boldsymbol{\delta}$ in the fluid phases. The r.e.v. thus appears as an heterogeneous structure as regards its elastic properties as well as the pre-stress field $\boldsymbol{\sigma}_p(\underline{x})$. The macroscopic mechanical loading to which

Ω is subjected is defined by the macroscopic strain tensor $\boldsymbol{E}$ according to Hashin boundary conditions on the displacement $\underline{\xi}$:

$$(\forall \underline{x} \in \partial\Omega) \quad \underline{\xi} = \boldsymbol{E}.\underline{x} \tag{5}$$

A classical result of micromechanics often referred to as Levin theorem then allows to state that the macroscopic state equation is of the same form as the microscopic one ([9]), namely:

$$\boldsymbol{\Sigma} = \mathbb{C}^{hom} : \boldsymbol{E} + \boldsymbol{\Sigma}_p \tag{6}$$

with $\mathbb{C}^{hom} = f^s\mathbb{C}^s :< \mathbb{A} >_s$ and $\boldsymbol{\Sigma}_p = -f^\ell p^\ell < \mathbb{A} >_\ell - f^g p^g < \mathbb{A} >_g$. $\mathbb{A}(\underline{x})$ denotes the strain concentration tensor at point $\underline{x}$. It relates the macroscopic strain tensor $\boldsymbol{E}$ to the microscopic strain field $\boldsymbol{\varepsilon}$ according to $\boldsymbol{\varepsilon}(\underline{x}) = \mathbb{A}(\underline{x}).\boldsymbol{E}$ in drained condition ($p^\ell = p^g = 0$).

We now introduce the assumption that there is no morphological difference between the domains Ω^g and Ω^ℓ. This implies that $< \mathbb{A} >_g = < \mathbb{A} >_\ell$. $\mathbb{I}$ denoting the unit tensor of the fourth order, the condition $< \mathbb{A} > = \mathbb{I}$ allows to write the macroscopic state equation in the following form

$$\boldsymbol{\Sigma} + (p^g - S_r p^c)\boldsymbol{B} = \mathbb{C}^{hom} : \boldsymbol{E} \quad \text{with} \quad \boldsymbol{B} = \boldsymbol{\delta} : \left(\mathbb{I} - \mathbb{S}^s : \mathbb{C}^{hom}\right) \tag{7}$$

where S_r denotes the degree of saturation. We note that (7) introduces an effective stress of the Bishop type with $\chi = S_r$ ([3]). In the particular case of an incompressible solid matrix ($\boldsymbol{\delta} : \mathbb{S}^s \to 0$, $\boldsymbol{B} \to \boldsymbol{\delta}$), the above reasoning provides a micromechanical interpretation to the effective stress formulation which has been proposed by several authors ([8]). The importance of the assumptions that Ω^g and Ω^ℓ have the same morphology and that the mechanical effects of surface tension are only taken into account through the capillary pressure is emphasized.

3. Surface tension effect

For the sake of simplicity, it is assumed in the sequel that the liquid perfectly wets the solid and that the surface tension in the solid-liquid interface is naught. Then we have $\gamma^{\ell g} = \gamma^{sg} = \gamma$ and $\gamma^{s\ell} = 0$.

In order to introduce the capillary effects in a more rigorous way than in section 2, the capillary terms in (1) and (3) which have been neglected in the simplified model must be taken into account. This requires to specify the geometry of the interfaces between phases. Let us assume that the porous space is made up of N subsets of spherical pores, each subset being characterized by the pore radius R_β and the volume fraction f_β ($\beta = 1, \dots, N$). The distribution of liquid and gas in the pore network is described by the set $\mathcal{I}_g$ according to the following rule: if $\beta \in \mathcal{I}_g$, all the pores of the subset β are filled with gas whereas they are filled with liquid if $\beta \notin \mathcal{I}_g$. Since the solid-gas interface is made up of spheres, (1) now writes

$$\beta \in \mathcal{I}_g : \quad (\forall \underline{x} \in \omega^{s\beta})\, \boldsymbol{\sigma} \cdot \underline{n} = -(p^g - p^\beta)\underline{n} \quad \text{with} \quad p^\beta = \frac{2\gamma}{R_\beta} \tag{8}$$

Furthermore, it can be shown that the contribution of the solid-gas interface associated with a gas-filled pore to the macroscopic stress tensor $\boldsymbol{\Sigma}$ is identical to that of a negative pressure $-p^\beta$ in the volume of the pore. The capillary effects in (1) and (3) can therefore be taken into account in replacing the "true" pressure p^g in a gas-filled pore of radius R^β by the corrected pressure $p^g - p^\beta$.

Then, as for the simplified model, it is possible to write the behaviour of each phase of unsaturated porous media in the form (4) with $\boldsymbol{\sigma}_p^\alpha \to -(p^g - p^\alpha)\boldsymbol{\delta}$ and $\mathbb{C}^\alpha \to 0$ for $\alpha = \beta \in \mathcal{I}_g$, $\boldsymbol{\sigma}_p^\beta \to -(p^\ell)\boldsymbol{\delta}$ and $\mathbb{C}^\beta \to 0$ for $\alpha = \beta \notin \mathcal{I}_g$ and $\boldsymbol{\sigma}_p^\beta \to -0$ for $\alpha = s$. The same kind of reasoning as in section 2 yields the macroscopic state equation:

$$\boldsymbol{\Sigma} + (p^g - S_r p^c)\boldsymbol{B} - \boldsymbol{\Sigma}(\mathcal{I}_g) = \boldsymbol{\Sigma}' - \boldsymbol{\Sigma}(\mathcal{I}_g) = \mathbb{C}^{hom} : \boldsymbol{E} \tag{9}$$

with:

$$\boldsymbol{\Sigma}(\mathcal{I}_g) = \sum_{\beta \in \mathcal{I}_g} \frac{f^\beta p^\beta}{f} \left(\mathbb{I} - \mathbb{C}^{hom} : \mathbb{S}^s\right) : \boldsymbol{\delta} \tag{10}$$

With respect to (6), (9) comprises the correcting term $\boldsymbol{\Sigma}(\mathcal{I}_g)$ which takes the surface tension at the solid-gas interface into account.

Although the evolutions of $\mathcal{I}_g$ and p^c are linked, $\mathcal{I}_g$ is not, in general, a one-to-one function of p^c. We now study a simpler situation in which the pore network is made up of spherical pores of decreasing radius R_β, related to one another by capillary necks of decreasing radius r_β as indicated on Figure 1.

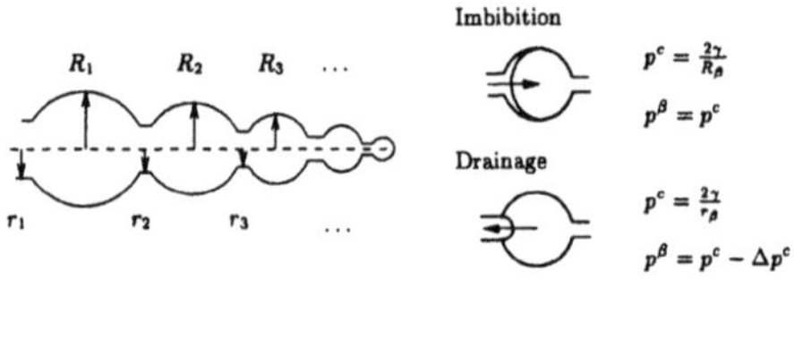

Figure 1. Morphological model

In fact, this model defines a one to one relation between S_r and $\mathcal{I}_g$. For an incremental variation dS_r of the current saturation ratio S_r, a single β-subset is being filled with liquid or with gas ($dS_r = \pm f^\beta/f$). It is then possible to relate the variation of $\boldsymbol{\Sigma}(\mathcal{I}_g)$ to dS_r

$$d\boldsymbol{\Sigma}(\mathcal{I}_g) = -\boldsymbol{B} dS_r p^\beta \tag{11}$$

The hysteresis of the capillary curve is classically explained by the concepts of access radius and pore radius ([7]). The capillary pressure at which a gas-filled pore is being filled with liquid (imbibition) depends on the pore radius R_β whereas the capillary pressure at which a liquid-filled pore is being filled with gas (drainage) is controlled by the neck radius r_β. Taking (11) and the sorbtion-desorbtion mechanism described above into account, the differentiation of (9) yields the differential form of the macroscopic state equation

$$\begin{aligned} &\text{If } dS_r > 0\text{: } p^\beta = p^c \;\rightarrow\; \delta\overline{\boldsymbol{\Sigma}} = \mathbb{C}^{hom} : d\boldsymbol{E} \\ &\text{If } dS_r < 0\text{: } p^\beta \neq p^c \;\rightarrow\; \delta\overline{\boldsymbol{\Sigma}} - \boldsymbol{B}\Delta p^c(S_r) dS_r = \mathbb{C}^{hom} : d\boldsymbol{E} \end{aligned} \tag{12}$$

with

$$\delta\overline{\boldsymbol{\Sigma}} = d\boldsymbol{\Sigma} + (dp^g - S_r dp^c)\,\boldsymbol{B} \tag{13}$$

and where $\Delta p^c(S_r) = p^c(S_r, dS_r < 0) - p^c(S_r, dS_r > 0)$ denotes the value of the hysteresis in the sorbtion desorbtion curve (Figure 2).

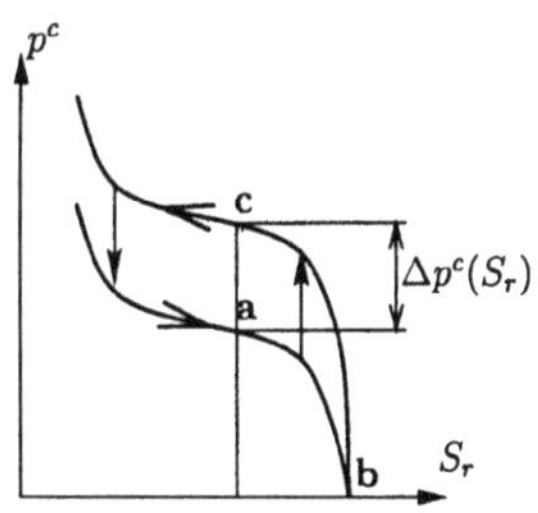

Figure 2. Sorbtion-desorbtion curve

According to the present micromechanical approach, we note that the state equation $\delta\overline{\boldsymbol{\Sigma}} = \mathbb{C}^{hom} : d\boldsymbol{E}$ proposed by Coussy ([6]) requires that the hysteresis in the capillary curve be negligible.

Consider a loading in which the capillary pressure is subjected to a cycle, whereas the macroscopic stress $\boldsymbol{\Sigma}$ and the gas pressure p^g are kept constant. Starting from point **a** on the imbibition curve (Figure 2), the sample is first totally saturated (point **b**) and then dried such as the saturation reaches its initial value (point **c** on the drainage curve). The integration of equation (12) over this loading cycle yields an irreversible strain

$$\Delta \boldsymbol{E} = \mathbb{C}^{hom-1} : \boldsymbol{B} \Delta p^c(S_r) S_r \qquad (14)$$

$\Delta \boldsymbol{E}$ appears as the mechanical counterpart of the hysteresis in the capillary pressure curve. Though the matrix behaves elastically, an inelastic behaviour is observed at the macroscopic scale. The macroscopic constitutive law when the porous space is saturated with liquid, obtained by putting $S_r = 1$ and $\mathcal{I}_g = \emptyset$ in equation (9), writes $\boldsymbol{\Sigma} + p^\ell \boldsymbol{B} = \mathbb{C}^{hom} : \boldsymbol{E}$. The comparison of this equation and (14) shows that the mechanical effect of the capillary hysteresis is the same as the one due to a variation $\Delta p = \Delta p^c(S_r) S_r$ of the pore pressure applied to a saturated sample at constant macroscopic stress.

Within the framework of the pore network of Figure 1, it is emphasized that the capillary pressure curves and the drained mechanical properties allow to characterize the mechanical behaviour in unsaturated conditions.

References

1. Auriault, J. L. and Sanchez-Palencia, E. Etude du comportement d'un milieu poreux saturé déformable. *J. Méc.*, **16**, 575–603, 1977.
2. Auriault, J. L. Nonsaturated deformable porous media: quasistatics. *Transport in Porous Media*, **2**, 45–64, 1987.
3. Bishop, A. W. The principle of effective stress. *Teknisk Ukeblad*, **39**, Oct., 859–863, 1959.
4. Chateau, X. and Dormieu, L. Homogenization of a non-saturated porous medium: Hill's lemma and applications. *C. R. Acad. Sci.*, **320**, II, 627–634, 1995, (in french with abridged english version).
5. Chateau, X. and Dormieux, L. A micromechanical approach to the behaviour of unsaturated porous media, in *Poromechanics*, ed. J. F. Thimus *et al.*, Balkema, 47–52, 1998.
6. Coussy, O. *Mechanics of porous continua*, John Wiley, 1995.
7. Dullien, F. A. L. *Porous media - Fluid transport and pore structure*, Academic Press, 1992.
8. Schrefler, B. A. *Recent advances in numerical modelling of geomaterials.* Key note lecture, III EPMESC CONT. MACAU 1990, Meccanica, 1990.
9. Zaoui, A. Structural morphology and constitutive behaviour of microheterogeneous materials, in *Continuum micromechanics*, ed. P. Suquet, Springer, 1997.

Influence of Porosity on the Response of Fibrous Composites

S. C. Baxter
University of South Carolina, U.S.A
e-mail: baxter@engr.sc.edu

C. T. Herakovich
University of Virginia, U.S.A
e-mail: herak@virginia.edu

A. M. Roerden
Navy Tactical Aircraft Strength, U.S.A
e-mail: roerdenam@navair.navy.mil

Abstract. This paper presents a micromechanical examination of the effect of microscale porosity over three length scales of an intermetallic hybrid fiber composite. The reinforcing fiber is a multifilament tow infiltrated with a porous binder. Predictions of the elastic isotropic properties of the binder are developed using Aboudi's three-dimensional generalized method of cells (GMC), micromechanics model. The second stage of modeling uses the effective properties of the binder in a two dimensional GMC analysis of the elastic properties of the hybrid fiber. Finally, the effective properties of the hybrid fiber and the nonlinear properties of the intermetallic matrix are used to predict the elastic properties and inelastic response of the composite.

1. Introduction

The search for low cost lightweight materials, that perform well at elevated temperatures, has led to an investigation of composites reinforced with multifilament, hybrid ceramic fibers. In this work a hybrid fiber was constructed by infiltrating a ceramic aluminum oxide filament tow with an aluminum oxide binder. Fabrication considerations motivated the use of a slurry to infiltrate the binder among the filaments. Heat-treatment then transformed the penetrating slurry into a porous ceramic binder. While the preprocessing properties of filaments and binder are the same, the effect of the induced porosity creates an anisotropic hybrid fiber. When the hybrid fiber is finally embedded in a matrix material, the effect of the porosity extends across three length scales; first as it influences the property of the binder, second as it affects the properties of the fiber and finally as it contributes to the effective properties of the composite. The multifilament fiber tow used was the Nextel 610 oxide fiber (3M Corporation). The matrix material was the intermetallic nickel aluminide, Ni_3Al.

W. Ehlers (ed.), IUTAM Symposium on Theoretical and Numerical Methods in Continuum Mechanics of Porous Materials, 131–136.

2. The Micromechanical model

The generalized method of cells (GMC), a micromechanics model developed by Paley and Aboudi [7], to predict the response of periodic, reinforced composite materials, was used to model the microstructures in this work. Periodic microstructures in GMC are characterized by a repeating unit cell, which captures the pattern of the microstructure. The repeating unit cell is subdivided into subcells. Each subcell is assigned the properties of one constituent. In using GMC to model a porous material, the pores are considered a second phase and assigned physical properties corresponding to a material with very small elastic stiffnesses, C_{ij}. Within the model, thermal and/or mechanical load histories can be imposed A detailed description of the model is available in Paley and Aboudi [7].

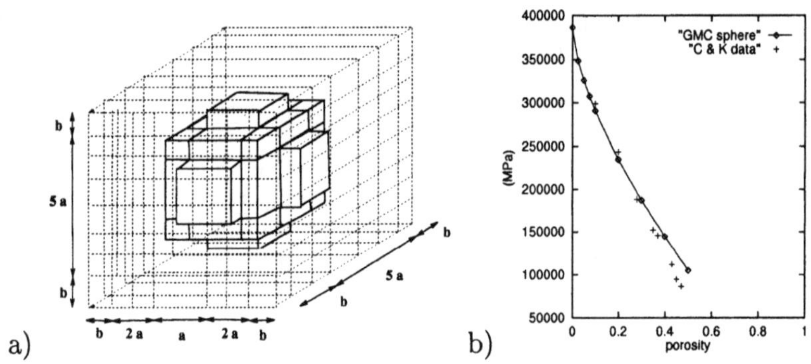

Figure 1. a) Spherical pore. b) Binder properties with various porosity.

A spherical pore was modeled in GMC using 343 subcells as shown in Fig. 1a. (Additional pore shapes were considered in Baxter and Herakovich [6]). The shape of the pore is held constant with varying porosity by defining its proportions in terms of a characteristic dimension a. The pore volume fraction can be calculated for $0 \leq P < .5236$, which is the maximum volume of a true sphere enscribed within a cube.

3. Elastic properties of the binder

The effective elastic properties of the porous alumina are $E = 386.$ GPa, $G = 163.$ GPa, Poisson's ratio, $\nu = 0.19$. The pore was assigned the properties $E = 0.001$ MPa, $G = 0.0004$ MPa, and $\nu = 0.25$. The elastic modulus of the isotropic Al_2O_3 binder, as a function of porosity, is shown in Fig. 1b, where the model predictions are compared to the experimental data of Coble and Kingery [4].

An increase in porosity results in a degradation of the elastic modulus. GMC's predictions for the spherical pore correspond well with the experimental data up to a value of about 40% porosity, beyond this value the agreement is good but not excellent. This agreement between theory and experiment is consistent with suggestions by Ashby and Jones [1], that densification of ceramic powder through sintering results in porosity in the form of small, nearly spherical pores.

3.1. Comparison with empirical/analytic models

Comparisons of GMC's predictions were made to two empirical models and to one analytically derived bound. The first model has an exponential form, the second is a geometric model. Finally, the results are compared to an analytic upper bound.

3.1.1. *Exponential form*

Spriggs [10], suggests an empirical formula for the effective elastic moduli, M^*, of ceramic materials with the following form

$$M^* = M_0 e^{-bP}, \tag{1}$$

where M_0 is the modulus of the non-porous material, P is the porosity, and b is an empirical constant determined by experiment. This relationship has provided good approximations for moduli for $P \leq 0.4$, (Spriggs [10] and Rice [8]). The value of the constant b has been strongly linked to pore shape through the observation that b depends on the processing method by which the porosity is produced. The values of b have been found to range from 4.08-4.35 for hot-pressing, 3.44-3.55 for cold-pressing and sintering and approximately 2.73 for slipcasting and sintering. The predictions made by the GMC model were fit to an exponential form using a modified least squares method. The resulting value, $b \approx 2.57$, is comparable to empirical predictions for the spherical pores generated by slip casting and sintering. Figure 2a shows this model.

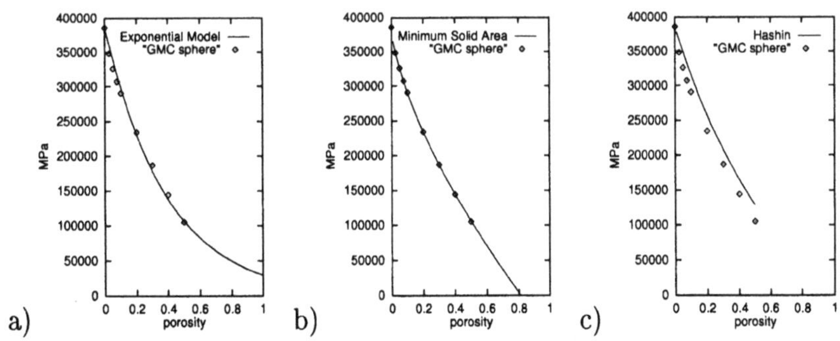

Figure 2. GMC versus a) Exponential model b) MSA model c) Upper bound.

3.1.2. *Minimum solid area model*
Rice [8], associated the effective properties of porous media with pore shape and packing arrangement, by noting that these geometric characteristics are related to the minimum area that transfers load. For one-dimensional loading this is the minimum solid area normal to the stress. Under this assumption, the ratio of the modulus of the porous material to its non-porous base is equal to the minimum solid area (MSA)

$$\frac{M^*}{M_0} = MSA. \tag{2}$$

Since GMC's spherical pore is explicitly defined, it is possible to derive the following relationship

$$\mathrm{MSA}_{sphere} \approx 1 - 1.15P^{2/3}. \tag{3}$$

GMC's predictions are in excellent agreement with the MSA model, shown in Fig. 2b.

3.1.3. *Analytic composite sphere model*
A fundamental work on the properties of multiphase materials is the composite sphere model developed by Hashin [5]. In his work, a representative sub-volume of a two phase material is modeled by a sphere of one material surrounded by a sheath of the second. A collection of these layered spheres, of different sizes, describes the continuum material. The relative proportions of the constituent materials are preserved. Hydrostatic pressure is imposed under the assumption that each composite sphere experiences the same pressure. In the limiting case, when one phase is a void, the model produces an upper bound on the elastic modulus as

$$\frac{E^*}{E_0} \leq \frac{2(7-5\nu)(P-1)}{-14+10\nu-13P+2\nu P+15\nu^2 P} \tag{4}$$

GMC's predictions, lie below this upper bound. The comparison for the axial modulus can be seen in Fig. 2c.

4. Elastic properties of the hybrid fiber

The hybrid fiber was modeled, using a two-dimensional analysis, as a composite fiber with a 55% volume fraction of the Al_2O_3 filaments infiltrated by the porous binder. The filaments were assigned the properties of the non-porous alumina.

The effective properties of the binder were used as the second phase of the hybrid fiber. Figure 3 illustrates the expected decay of both the axial and transverse modulus of the hybrid fiber with varying porosity. Additionally the figure highlights a loss of isotropy in the fiber, as demonstrated by the growing difference between moduli in the axial and transverse directions, with increasing porosity.

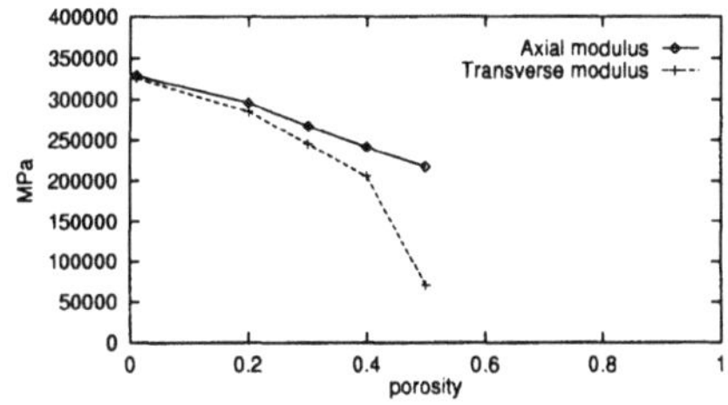

Figure 3. Axial and Transverse Elastic Modulus of Hybrid Fiber

5. Response of the composite material

The axial and transverse tensile responses of the composite material for various binder porosities are shown in Figs. 4a and 4b. The bulk properties of the Ni_3Al matrix are $E = 236.$ GPa, $G = 95.$ GPa, with Poisson's ratio $\nu = 0.27$. The inelastic response was modeled using Bodner-Partom viscoplasticity theory, Bodner and Partom [3].

The axial response of the composite is bilinear. The initial Young's modulus, yield stress and the secondary (strain-hardening) modulus vary with binder porosity. Assuming that axial tensile failure is controlled by the strain to failure of the fiber, the results in Fig. 4a indicate that the ultimate tensile stress in the axial direction is a strong function of binder porosity.

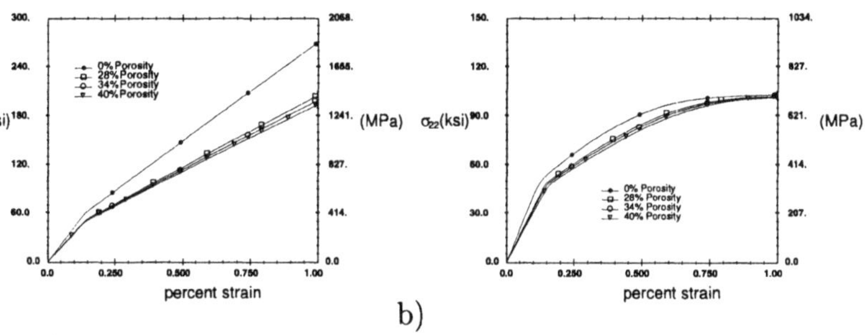

Figure 4. a) Axial response of composite b) Transverse response of composite

The transverse behavior of the composite, Fig. 4b, exhibits elastic strain-hardening behavior with a gradual decrease in stiffness in the plastic region. The effect of the binder porosity on the transverse elastic response is similar to that of the axial response. The final flow stress, however, appears independent of binder porosity. This is a reflection of the fact that the final flow stress for transverse loading of a continuous fiber composite is a matrix dominated property; influenced primarily by the flow stress of the matrix. Composite response, based on this model, has been extended to predictions of the stresses and plastic strains due to processing in Baxter and Pindera [2], and to an examination of temperature dependent effects for elastic and inelastic responses of the hybrid fiber composite in Roerden and Herakovich [9].

6. Conclusions

A micromechanical examination of the effect of microscale porosity over three length scales of an intermetallic hybrid fiber composite is presented. The generalized method of cells (GMC) is used to model the microstructure at all three length scales.

Predictions of the properties of a porous aluminum oxide binder with spherical pores correspond well to experimental results and empirical models. The elastic properties of the porous binder degrade rapidly with increasing porosity under the assumption of constant pore shape. Using a multi-level modeling approach, the loss of isotropy in the hybrid fiber, due to the reinforcing architecture and the porosity of the binder, is demonstrated.

The response of the hybrid fiber composite is largely dominated by the properties of the matrix material. The final flow stress in the transverse direction is independent of the binder porosity. The composite axial response however, does show some influence due to the percent binder porosity. If it is assumed that the axial strain to failure is controlled by the failure of the fiber, then the effect of binder porosity is significant in predictions of the ultimate tensile stress.

References

1. Ashby, M. F. and Jones, D. R. H. Engineering Materials 2, An Introduction to Microstructures, Processing and Design, Pergamon Press, Oxford, 179, 1986.
2. Baxter, S. C. and Pindera, M-J. Stress and Plastic Strain Fields During Unconstrained and Constrained Fabrication Cool Down of Fiber Reinforced Intermetallic Matrix Composites, *J. Composite Materials*, **33**, 351-375, 1999.
3. Bodner, S. R. and Partom, Y. Constitutive Equations for Elastic Viscoplastic Strain-Hardening Materials, *J. Applied Mechanics*, **42**, 385-389, 1975.
4. Coble, R. L., and Kingery, W. D. Effect of Porosity on Physical Properties of Sintered Alumina, *J. Amer. Ceram. Soc.*, **39**, 377, 1956.
5. Hashin, Z. The Elastic Moduli of Heterogeneous Materials, *J. Applied Mechanics*, **29**, 143-150, 1962.
6. Herakovich, C. T. and Baxter, S. C. Influence of Pore Geometry on the Effective Response of Porous Media", *J. Materials Science*, **34**, 1595-1609, 1998.
7. Paley, M. and Aboudi, J. Micromechanical analysis of composites by the generalized cells model, *Mechanics of Materials*, **14**, 127-139, 1992.
8. Rice R. W. Comparison of physical property-porosity behavior with minimum solid area models, *Journal of Materials Science*, **31**, 1509-1528, 1996.
9. Roerden, A. M., and Herakovich, C. T. Inelastic Response of Porous, Hybrid Fiber Composites, presented 1998 ASME IMECE Symposium in honor of Nick Pagano's 65th birthday, to be published *Composites Science and Technology*, 1999.
10. Spriggs, R. M. Expressions for the Effect of Porosity on Elastic Modulus of Polycrystalline Refractory Materials, Particularly Aluminum Oxide *J. Amer. Ceram. Soc.*, **44**, 628, 1961.

Session B2: Biot's Theory

Chairman: L. Dormieux

Non Linear Thermomechanical Couplings in Unsaturated Clay Barriers

P. Dangla and O. Coussy
Laboratoire Central de Ponts et Chaussées 58 bd Lefèbvre 75732 Paris, France

E. Olchitzky and C. Imbert
Commissariat à l'Énergie Atomique CEN Saclay, 91191 Gif-sur-Yvette, France

Abstract. The constitutive equations of clay barriers are investigated within the framework of continuum thermodynamics. This energy approach gives rise to a non linear poroelasticity theory ranging in the Biot's theory. A model for unsaturated clay barrier is proposed based on the knowledge of the saturation curve. The predictions of this very constraint model are confirmed by the results of a sorption isotherm experiment performed at different temperatures.

1. Introduction

In concepts which consider the placement of nuclear waste in deep geological formation, compacted clays are currently adopted as appropriate engineered barriers. Once installed between the saturated host rock and the waste containers, these barriers hydrate, heat, swell, undergo stresses thereby experiencing thermo-hydro-mechanical couplings. The understanding of these couplings is, of course, a key step in the analysis and development of design concepts of nuclear waste disposal. To this end a reliable framework to formulate the constitutive equations of porous materials is provided by a thermodynamic approach which have led to the Biot theory of poroelasticity. In light of the Biot theory the goal of this paper is to extend the poroelasticity to non linear and non saturated behavior. A model is proposed for clay barriers. The results of experiments performed on FoCa clay are reported and compared to the predictions given by the model.

2. Energy approach

Submitted to external actions the porous material not only deforms but can gain or lose fluid mass. Therefore, referring the deformation to that of the solid skeleton, a representative volume element of porous material can be considered as an open thermodynamic system. Two non miscible fluids, saturating the porous space, are distinguished in

W. Ehlers (ed.), IUTAM Symposium on Theoretical and Numerical Methods in Continuum Mechanics of Porous Materials, 139–144.

this paper: the liquid water (subscript l) and the gas air (subscript g). The energy balance of this open thermodynamic system can be written as [1] (summation convention on repeated subscript $\alpha = l, g$ is adopted from here)

$$dE = \sigma_{ij} d\epsilon_{ij} + \mu_\alpha dm_\alpha + T dS \tag{1}$$

In Eq. 1 σ_{ij} and ϵ_{ij} are the stress and strain tensors, T is the temperature and μ_α the chemical potential per mass unit of the fluid phase α. The quantities $Ed\Omega$, $Sd\Omega$ and $m_\alpha d\Omega$ represent the energy, the entropy and the fluid mass content of fluid phase α contained in the volume element of porous material $d\Omega$. At the microscopic scale there are generally mechanical and chemical interactions between the fluids and solid materials. When only mechanical interactions occur, or in other words when chemical or distant interactions between the solid and fluid particles can be neglected at the microscopic scale, it can be stated that firstly the energy of the system reduces to the sum of its constitutive materials part, secondly the chemical potentials are given by specific free enthalpies. Therefore it entails that $E = E_s + e_\alpha m_\alpha$, $S = S_s + s_\alpha m_\alpha$, $\mu_\alpha = (e + \frac{p}{\rho} - Ts)_\alpha$ where e_α is the specific internal energy, s_α the entropy, p_α the pressure and ρ_α the density of the fluid α. Besides E_s and S_s represent the internal energy and the entropy of the *skeleton* system composed not only of the solid constituent but of the phase interfaces as swell since the latter possess their own thermomechanical properties such as their surface tensions. From the last definitions and Eq. 1 it can be shown that [2]

$$dE_s = \sigma_{ij} d\epsilon_{ij} + p_\alpha d\phi_\alpha + T dS_s \tag{2}$$

where $\phi_\alpha d\Omega$ represents the volume occupied by the fluid α in $d\Omega$, thereby $m_\alpha = (\rho\phi)_\alpha$. Eq. 2 implies that E_s is a potential function of the state variables $(\epsilon_{ij}, \phi_l, \phi_g, S_s)$ since dE_s turns out to be the exact differential of a function $E_s(\epsilon_{ij}, \phi_l, \phi_g, S_s)$. As a result, the conjugate state variables $(\sigma_{ij}, p_l, p_g, T)$ are given by the partial derivatives of E_s. Therefore the determination of the constitutive equations of the porous skeleton reduces to the identification of the potential function $E_s(\epsilon_{ij}, \phi_l, \phi_g, S_s)$.

In granular materials as considered in this paper, and at constant temperature, the grains generally undergo no significant volume change as compared to that of the total porosity ϕ. The isothermal volumetric strain ϵ_{ii} is then only due to the variation of the porosity

$$\epsilon_{ii} = \phi - \phi_0 \tag{3}$$

Since $\phi = \phi_l + \phi_g$ Eq. 3 implies that only two among the set of variables $(\epsilon_{ii}, \phi_l, \phi_g)$ are independent. Let us choose (ϵ_{ii}, ϕ_l). Thus at constant temperature the free energy $\Psi_s = E_s - TS_s$ depends only on (ϵ_{ii}, ϕ_l) since Eq. 2 and Eq. 3 entail

$$d\Psi_s = (\sigma_{ij} + p_g\delta_{ij})d\epsilon_{ij} - p_c d\phi_l \tag{4}$$

where $p_c = p_g - p_l$ is the capillary pressure. Eq. 4 states that Ψ_s is a potential function of the state variables (ϵ_{ij}, ϕ_l) the partial derivatives of which give $(\sigma_{ij}+p_g\delta_{ij}, p_c)$. The use of the state variables (ϵ_{ij}, p_c) instead of (ϵ_{ij}, ϕ_l) being more convenient, let us introduced the Legendre-Fenchel transform of Ψ_s with respect to ϕ_l, namely $G_s = E_s + p_c\phi_l$. From Eq. 4 it is easily shown that

$$dG_s = (\sigma_{ij} + p_g\delta_{ij})d\epsilon_{ij} + \phi_l dp_c \tag{5}$$

As a result the formal isothermal constitutive equations that are discussed in this paper are then given by

$$\sigma_{ij} + p_g\delta_{ij} = \frac{\partial G_s}{\partial \epsilon_{ij}}(\epsilon_{ij}, p_c) \tag{6}$$

$$\phi_l = \frac{\partial G_s}{\partial p_c}(\epsilon_{ij}, p_c) \tag{7}$$

In most situations the potential G_s cannot be approximated by quadratic functions of their arguments since experimental evidence shows that the constitutive equations 6, 7 do not rely linearly the conjugate set of variables for usual materials. That means that the second order derivatives of the potential function G_s, involved in the differentiation of the above equations are not constant and do depend on the state variables (ϵ_{ij}, p_c). However an usual assumption consists in considering that the deviatoric behavior is linear and not coupled with the volume change behavior. In other words $G_s(\epsilon_{ij}, p_c) = W(\epsilon, p_c) + Ge_{ij}e_{ij}$ with $\epsilon = \epsilon_{ii}$ and $e_{ij} = \epsilon_{ij} - \frac{1}{3}\epsilon\delta_{ij}$. Differentiating Eq. 6 and Eq. 7 yields

$$d(\sigma + p_g) = K(\epsilon, p_c)d\epsilon + b(\epsilon, p_c)dp_c \tag{8}$$

$$d\phi_l = b(\epsilon, p_c)d\epsilon + N(\epsilon, p_c)dp_c \tag{9}$$

where $\sigma = \frac{1}{3}\sigma_{ii}$ is the overall mean stress. The function $K(\epsilon, p_c)$ is the tangent bulk modulus, $b(\epsilon, p_c)$ is identified as the tangent Biot coefficient while $N(\epsilon, p_c)$ is the inverse of a modulus. The complete identification of the model presented here requires the evaluation of the functions $K(\epsilon, p_c)$, $b(\epsilon, p_c)$ and $N(\epsilon, p_c)$. This is the goal of the next section with a special attention to clay material.

3. A non linear poroelastic model for clay

A model based on the incremental constitutive equations 8 and 9 is applied to unsaturated clays. To specify the expression of the functions $K(\epsilon, p_c)$, $b(\epsilon, p_c)$ and $N(\epsilon, p_c)$, the following assumptions are considered:

Assumption 1 *At constant temperature, the liquid saturation is a function of the capillary pressure only, reading :* $\frac{\phi_l}{\phi} = S_l(p_c)$.

Assumption 2 *By reaching the saturation, the effective stress principle applies and the mechanical behavior of the clay satisfies a logarithmic relationship between the void ratio and the effective stress :* $de = -\kappa \frac{d(\sigma+p_l)}{\sigma+p_l}$.

The first assumption is generally well accepted in the litterature as soon as the hysteretic phenomena can be neglected during drainage and drying cycles [3]. The saturation function $S_l(p_c)$ is a monotonic decreasing function, starting from unity for a non negative capillary pressure p_b. In the following it will be convenient to extend the definition of $S_l(p_c)$ for any arbitrary value of the capillary pressure such as $S_l(p_c) = 1$ on $]-\infty, p_b]$. In the next section such a function obtained from experiments performed on FoCa clay is reported. The second assumption is consistent with the behavior of saturated clays as shown by some experimental results also reported in the next section.

Let us draw the conclusions of these two assumptions in the framework of the formal model described above. A comparison of the first assumption reading $\phi_l = (\phi_0+\epsilon)S_l(p_c)$ and Eq. 9 shows that $N(\epsilon, p_c) = (\phi_0+\epsilon)\frac{dS_l}{dp_c}$ and $b(\epsilon, p_c) = S_l(p_c)$. From the Maxwell's symmetry relation which imposes that $\frac{\partial K}{\partial p_c} = \frac{\partial b}{\partial \epsilon}$ it results that the bulk modulus doesn't depend on the capillary pressure. Therefore Eq. 8 takes the form

$$d(\sigma + \pi) = K(\epsilon)d\epsilon \quad (10)$$

$$\pi(p_l, p_g) = p_g - \int_0^{p_c} S_l(x)dx \quad (11)$$

In Eq. 11 the function $\pi(p_l, p_g)$ was obtained by integration of the differential form $dp_g - S_l(p_c)dp_c$ the constant of which was adjusted so as to have $\pi = p_l$ when $S_l = 1$. Therefore the stress $\sigma + \pi$ recovers the effective stress when the saturation reaches unity. Furthermore Eq. 10 stipulates that there is a one to one relation between $\sigma + \pi$ and the volumetric strain ϵ or equivalently the void ratio e, since $\epsilon = \frac{e-e_0}{1+e_0}$. Hence the bulk modulus may be expressed in terms of either ϵ or e or $\sigma+\pi$. Then by comparing Eq. 10 for $S_l = 1$, namely $d(\sigma+p_l) = K(\epsilon)d\epsilon$, and the logarithmic stress-strain relationship assumed above it follows that the bulk modulus is necessarily proportional to $\sigma + \pi$. Therefore Eq. 10 necessarily reads

$$de = -\kappa \frac{d(\sigma + \pi)}{\sigma + \pi} \tag{12}$$

The material parameter κ is the same as that mentionned in the assumption 2 and thus can be obtained from tests performed at saturation. The form of the constitutive equation 12 obtained from thermodynamical considerations is extremely constraint and thus can be validated through experiments involving unsaturated clays. This is the goal of the next section.

The previous model can be extended to non isothermal conditions consistent with a thermodynamic potentiel of the form $G_s(\epsilon_{ij}, p_c, T)$. It can be shown that Eq. 12 is replaced by $de = -\kappa \frac{d(\sigma+\pi)}{\sigma+\pi} + \alpha dT$ where α stands for the dilatation of the grains and π depends on the temperature through Eq. 11 where $S_l(p_c)$ is replaced by $S_l(p_c, T)$.

4. Sorption isotherm experiment

During a sorption isotherm experiment a piece of unsaturated clay is wetted by controlling the relative humidity of the air h. Salt solution technique is used to impose the relative humidity. Once equilibrium is reached, the weight and volume of the sample are measured and the liquid saturation is computed. The clay sample was compacted at 50 MPa. The dry density is 1.9 and the initial water content is 12.5%. The determination of the capillary pressure inside the sample is carried out through the Kelvin's law : $p_c = -\rho_l \frac{RT}{M_v} \ln h$, where R is the gas constant and M_v the water vapor molecular weight. The figure 1a represents the liquid saturation obtained experimentally in terms of the capillary pressure for $20°C$ and $80°C$. As shown in the same figure, these experimental points can be fitted accurately by the Brutsaert's model : $S_l(p_c) = (1 + \frac{p_c}{A})^{-B}$ where B is a constant and A is temperature dependent. During such an experiment, the sample swells. Since $\sigma = p_g = 0$ (the atmospheric pressure can be neglected), the model predicts a linear relationship between the void ratio and $\ln \pi$: $e - e_0 = -\kappa \ln \frac{\pi}{\pi_0}$ with $\pi = -\int_0^{p_c} S_l(x)dx$. To test this prediction, the void ratio measured experimentally is plotted in the figure 1c in terms of π which can be computed from the capillary pressure using the saturation curve identified previously. This plot confirms the prediction of the model and the value 0.1 is found for the parameter κ which is independent of the temperature as predicted by the model. Finally, as shown in the figure 1b which represents a classical consolidation test for the saturated clay, the value $\kappa = 0.1$ is found again for the slope of the elastic response between the void ratio and the log of the effective mean stress.

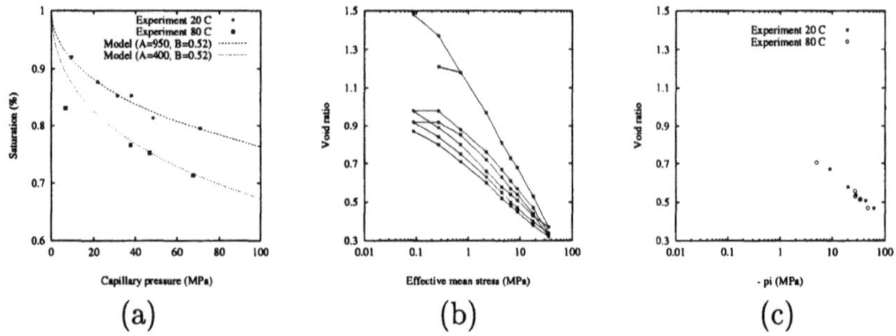

Figure 1. a) Saturation curves; b) Consolidation of the saturated clay; c) Void ratio vs $-\pi$.

5. Conclusion

1. In the framework of the thermodynamics, the internal energy of the solid grains plus that of the phase interfaces has been identified as the appropriate potential function from which the constitutive equations of unsaturated porous materials have to be derived. This approach leads to a Biot's theory of poroelasticity involving two internal pressures of non miscible fluids.

2. Based on the previous general framework, it is shown that the well-known logarithmic stress-strain relationship that the saturated clay exhibits during classical consolidation tests and the $p_c - S_l$ relationship completely determine the mechanical behavior of the clay in the whole range of saturation.

3. The predictions of the proposed model are compared to the results of sorption isotherm experiments perfomed on FoCa clay at two different temperatures. The theory is accurately confirmed.

References

1. Coussy, O. Mechanics of porous continua *J. Wiley & Sons*, 1995.
2. Dangla, P. and Coussy, P. Non linear poroelasticity for unsaturated porous materials: an energy approach. *Poromechanics – A Tribute to Maurice Biot.* Thimus *et al. Louvain-La-Neuve*, 14–16 Sept. 1998.
3. Fredlund, D. G. and Rahardjo, H. Soil mechanics for unsaturated soils. *J. Wiley & Sons*, 1993.

Porothermoelasticity in Transversely Isotropic Porous Materials

Y. Abousleiman
Rock Mechanics Institute, The University of Oklahoma, Norman, USA (On leave from the Lebanese American University, Byblos, Lebanon)

S. Ekbote
Rock Mechanics Institute, The University of Oklahoma, Norman, USA

1. Introduction

The past few decades have shown an increased interest in the mechanics of saturated porous media. Research in this field has resulted in the theory of poroelasticity, formulated and derived by Biot [1], in a form that can be applied to problems commonly encountered in the geotechnical, petroleum and mining engineering fields [2]. Biot's isotropic formulation has subsequently been extended to the general anisotropic case [3, 4] and the associated material parameters have been linked to engineering constants [5]. Accordingly, analytical solutions of fundamental problems, such as Mandel's problem [6], borehole problem and cylinder problem [7], have been extended for the transversely isotropic case. Analysis of the transversely isotropic poroelastic problems have shown unpredicted results when compared to their elastic counterparts [7]. It has been found that in addition to the time-dependency of the stress and pore pressure, the anisotropic material coefficients play an important role in calculating the in-plane stress fields. These new developments were motivated by the simple natural deposition of sedimentary rocks exhibiting transverse isotropy.

In deep drilling, where temperature gradients are sufficiently large, or in nuclear waste storage facilities, where the applied boundary temperatures are of great concern, modelling a fully coupled porothermoelastic response in a transversely isotropic rock formation becomes a must. In this paper the solution to the inclined borehole problem in a transversely isotropic poroelastic formation is extended to include thermal effects [7, 8]. In low-permeability formations heat convection caused by pore fluid flow is insignificant and can be neglected for all practical purposes. Analytical solutions from transversely isotropic poroelasticity for the borehole and cylinder problems can then be extended to incorporate the thermal effects.

W. Ehlers (ed.), IUTAM Symposium on Theoretical and Numerical Methods in Continuum Mechanics of Porous Materials, 145–152.

2. Governing equations

Using tension as positive convention, the constitutive equations for linear anisotropic porothermoelasticity are given as:

$$\sigma_{ij} = D_{ijkl}\epsilon_{kl} - \alpha_{ij}p - \beta_{ij}T \tag{1}$$

$$p = M[\zeta - \alpha_{ij}\epsilon_{ij} + \beta_m T] \tag{2}$$

The above equations relate the response of the dynamic variables, σ_{ij} (total stress tensor), p (pore pressure), and T (temperature), to the kinematic quantities, ϵ_{ij} (solid strain tensor) and ζ (variation of fluid content). The connection between the dynamic and kinematic quantities is characterized by the material constants, D_{ijkl} (drained elastic modulus tensor), α_{ij} (Biot's effective stress coefficient tensor), M (Biot's modulus), β_{ij} (thermic coefficient tensor related to the solid skeleton), and β_m (thermic coefficient related to the pore fluid). For the most general anisotropic case the response behavior is described using 35 constants (21 D_{ijkl}'s, 6 α_{ij}'s, 1 M, 6 β_{ij}'s and 1 β_m). In addition, the material parameters resulting from the balance and conduction laws need to be accounted for. Table 1 gives a comparison of the constants required for a complete solution for various forms of anisotropy in elasticity, poroelasticity, and porothermoelasticity.

Table I. Material constants required in various constitutive formulations.

Material Type	**Elastic**	**Poroelastic**	**Porothermoelastic**
Isotropic	2	4 + 1 (fluid flow)	4 + 3 (thermic) + 1 (fluid flow) + 1 (heat flow)
Transversely Isotropic	5	8 + 2 (fluid flow)	8 + 4 (thermic) + 2 (fluid flow) + 2 (heat flow)
Orthotropic	9	13 + 3 (fluid flow)	13 + 5 (thermic) + 3 (fluid flow) + 3 (heat flow)

Experimental studies have been conducted to measure constants of anisotropic elasticity. For the transversely isotropic poroelasticity only one set of parameters measured in the laboratory has been reported in the literature [9]. For porothermoelasticity the parameters for isotropic materials have been measured in the field as well as in the lab [10, 11]. However, we are unaware of any attempts made to measure constants of transversely isotropic porothermoelasticity.

For the transversely isotropic material it is assumed that the z-axis coincides with the axis of elastic symmetry. The constitutive relations can be expressed in terms of five drained elastic constants, E, E', ν, ν', G', three poroelastic constants α, α', M, and three thermic constants, β, β', β_m, in which the unprimed variables are material coefficients in the isotropic plane and the primed variables are material coefficients in the direction perpendicular to the isotropic plane (z-direction). Though stress-strain relation given in equation (1) are in terms of D_{ijkl}'s, the coefficients therein can easily be related to the drained elastic constants [6]. In fact for engineering applications it is more convenient to use the drained Young's moduli, drained Poisson's ratio, and the shear modulus. For the transversely isotropic material the constitutive equations for porothermoelasticity are given as follows:

$$\begin{Bmatrix} \sigma_{xx} \\ \sigma_{yy} \\ \sigma_{zz} \\ \tau_{xy} \\ \tau_{yz} \\ \tau_{zx} \end{Bmatrix} = \begin{bmatrix} D_{11} & D_{12} & D_{13} & 0 & 0 & 0 \\ D_{12} & D_{11} & D_{13} & 0 & 0 & 0 \\ D_{13} & D_{13} & D_{33} & 0 & 0 & 0 \\ 0 & 0 & 0 & D_{44} & 0 & 0 \\ 0 & 0 & 0 & 0 & D_{55} & 0 \\ 0 & 0 & 0 & 0 & 0 & D_{55} \end{bmatrix} \begin{Bmatrix} \epsilon_{xx} \\ \epsilon_{yy} \\ \epsilon_{zz} \\ \gamma_{xy} \\ \gamma_{yz} \\ \gamma_{zx} \end{Bmatrix} - \begin{bmatrix} \alpha \\ \alpha \\ \alpha' \\ 0 \\ 0 \\ 0 \end{bmatrix} p - \begin{bmatrix} \beta \\ \beta \\ \beta' \\ 0 \\ 0 \\ 0 \end{bmatrix} T \tag{3}$$

$$p = M\left[\zeta - \alpha(\epsilon_{xx} + \epsilon_{yy}) - \alpha'\epsilon_{zz} + \beta_m T\right] \tag{4}$$

in which

$$D_{11} = \frac{E(E' - E\nu'^2)}{(1+\nu)(E' - E'\nu - 2E\nu'^2)}; \; D_{12} = \frac{E(E'\nu + E\nu'^2)}{(1+\nu)(E' - E'\nu - 2E\nu'^2)}; \tag{5}$$

$$D_{13} = \frac{EE'\nu'}{(E' - E'\nu - 2E\nu'^2)}; \qquad D_{33} = \frac{E'^2(1-\nu)}{(E' - E'\nu - 2E\nu'^2)}; \tag{6}$$

$$D_{44} = G = \frac{D_{11} - D_{12}}{2} = \frac{E}{2(1+\nu)}; \qquad D_{55} = G'; \tag{7}$$

The constitutive relations are combined with the balance and conduction laws to give field equations. The field equations for the transversely isotropic porothermoelastic material are given as follows:

$$\frac{1}{2}(D_{11}-D_{12})\nabla^2 u_i+\frac{1}{2}(D_{11}+D_{12})u_{k,ki}=\alpha p_{,i}+\beta T_{,i} \tag{8}$$

$$\frac{\partial T}{\partial t}-c_0\nabla^2 T=0; \qquad c_o=\frac{\lambda}{C_h} \tag{9}$$

$$\frac{\partial p}{\partial t}-\kappa M\nabla^2 p=-\alpha M\frac{\partial \epsilon}{\partial t}+\beta_m M\frac{\partial T}{\partial t} \tag{10}$$

$$\frac{\partial \zeta}{\partial t}-c_T\left[\nabla^2\zeta+c_{hT}\nabla^2 T\right]=0 \tag{11}$$

$$c_T=\frac{\kappa M D_{11}}{(D_{11}+\alpha^2 M)}; \quad c_{hT}=\frac{\alpha\beta-D_{11}\beta_m}{D_{11}} \tag{12}$$

in which equation (8) is the Navier-type equation, equation (9) is the heat diffusion equation, equation (10) is the pore pressure diffusion equation, and equation (11) is the diffusion equation in terms of the variation of fluid content ζ. In the above, u_i is the displacement vector, c_o is the heat diffusivity, λ is the thermal conductivity in the isotropic plane, C_h is the heat capacity of the bulk material, c_T is the fluid diffusivity, κ is the permeability in the isotropic plane, and c_{hT} is a coupling constant given by equation (12). It should be noted here that equation (9) has been written ignoring heat transfer resulting from convection and equation (11) has been written assuming an irrotational vector displacement condition.

3. Inclined borehole problem

It is assumed that an infinitely long borehole is drilled perpendicular to the isotropic plane of a transversely isotropic poroelastic formation. The borehole is inclined and its axis deviated from the in-situ stress orientation. A schematic of the inclined borehole is shown in Figure 1(a). The formation, described using a Cartesian coordinate system $x'y'z'$, is characterized by in-situ stresses $S_{x'}$, $S_{y'}$, and $S_{z'}$, virgin pore pressure p_o, and formation temperature T_o. The borehole deviation is measured by two angles φ_z and φ_y, which are the inclination and azimuth angles respectively. A local coordinate system is chosen to represent the borehole in which the z-axis is assumed to coincide with the borehole axis. The far-field in-situ stresses in the $x'y'z'$ coordinate system are transformed to the local xyz coordinate system via a transform matrix [12]. In the local coordinate system, the borehole is subject to normal as well as shear components of stress given as S_x, S_y, S_z, S_{xy}, S_{xz}, and S_{yz} as shown in Figure 1(b).

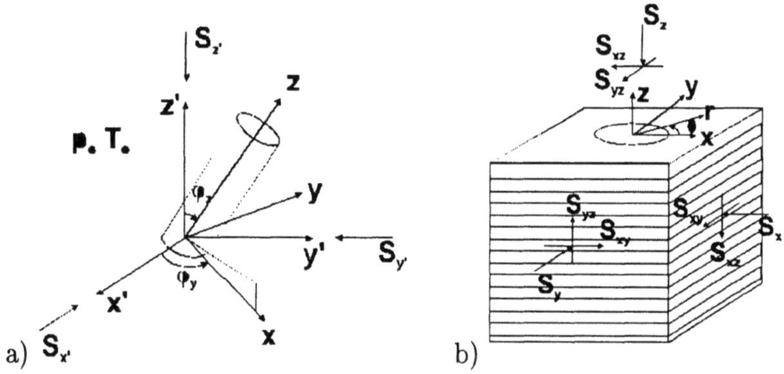

Figure 1. (a) Schematic of an inclined borehole; (b) Far-field stresses in the xyz coordinate system

The solution for the inclined borehole problem is obtained as a superposition of three sub-problems [12]. Of these, the first problem, which is a modified plane strain problem, accounts for the in-plane normal and shear stresses and also the pore pressure and temperature perturbations. This problem shows full coupling of the fluid and heat diffusion processes with the deformation. The other two problems which are described as the uniaxial problem and the anti-plane problem [12] are purely elastic since they do not trigger fluid or heat diffusion. The complete solution for the inclined borehole problem is given as:

$$\sigma_{rr} = -P_o + S_o \cos 2(\theta - \theta_r) + \sigma_{rr}^{(1)} + \sigma_{rr}^{(2)} + \sigma_{rr}^{(3)} \tag{13}$$

$$\sigma_{\theta\theta} = -P_o - S_o \cos 2(\theta - \theta_r) + \sigma_{\theta\theta}^{(1)} + \sigma_{\theta\theta}^{(2)} + \sigma_{\theta\theta}^{(3)} \tag{14}$$

$$p = p_o + p^{(2)} + p^{(3)} \tag{15}$$

$$T = T_o + T^{(2)} \tag{16}$$

$$\begin{aligned}\sigma_{zz} = {} & \nu'(\sigma_{rr} + \sigma_{\theta\theta}) - (\alpha' - 2\nu'\alpha)p - (\beta' - 2\nu'\beta)T \\ & -S_z + [\nu'(S_x + S_y) + (\alpha' - 2\nu'\alpha)p_o + (\beta' - 2\nu'\beta)T_o]\end{aligned} \tag{17}$$

$$\sigma_{r\theta} = -P_o + S_o \sin 2(\theta - \theta_r) + \sigma_{r\theta}^{(3)} \tag{18}$$

$$\sigma_{rz} = -(S_{xz}\cos\theta + S_{yz}\sin\theta)\left[1 - (R^2/r^2)\right] \tag{19}$$

$$\sigma_{\theta z} = (S_{xz}\sin\theta + S_{yz}\cos\theta)\left[1 + (R^2/r^2)\right] \tag{20}$$

where $\sigma_{rr}^{(1)}$, $\sigma_{rr}^{(2)}$, $\sigma_{rr}^{(3)}$, $\sigma_{\theta\theta}^{(1)}$, $\sigma_{\theta\theta}^{(2)}$, $\sigma_{\theta\theta}^{(3)}$, $\sigma_{r\theta}^{(3)}$, $p^{(2)}$, $p^{(3)}$, and $T^{(2)}$ are solutions for the loading modes of a plane strain problem. In the above, θ_r, P_o, and S_o are given by:

$$\theta_r = \frac{1}{2}\tan^{-1}\left[2S_{xy}/(S_x+S_y)\right] \tag{21}$$
$$P_o = (S_x+S_y)/2 \tag{22}$$
$$S_o = 0.5\sqrt{(S_x-S_y)^2+4S_{xy}^2} \tag{23}$$

4. Plane strain problem

Solution of the plane strain problem is obtained by a decomposition of the boundary conditions into three contributing loading modes [12]. Of these, mode 1 accounts for the hydrostatic part of the boundary stresses, mode 2 accounts for both the pore pressure and temperature perturbations, and mode 3 takes into account the deviatoric part of the boundary stresses. Only modes 2 and 3 are time-dependent in which mode 2 is characterized by coupling between the pore fluid and heat diffusion processes. Although mode 3 shows characteristics of full poroelastic coupling it is still not affected by temperature perturbations and can be directly adopted from [7]. The boundary conditions at the borehole wall are as follows:

$$\text{Mode 1}: \quad \sigma_{rr}^{(1)} = P_o - p_w; \quad \sigma_{r\theta}^{(1)} = 0; \quad p^{(1)} = 0; \quad T^{(1)} = 0; \tag{24}$$
$$\text{Mode 2}: \quad \sigma_{rr}^{(2)} = 0; \quad \sigma_{r\theta}^{(2)} = 0; \quad p^{(2)} = (p_w - p_o); \quad T^{(2)} = (T_w - T_o); \tag{25}$$
$$\text{Mode 3}: \quad \sigma_{rr}^{(3)} = -S_o\cos 2(\theta-\theta_r); \; \sigma_{r\theta}^{(3)} = S_o\sin 2(\theta-\theta_r); \quad p^{(3)} = 0; \; T^{(3)} = 0; \tag{26}$$

In the above p_w is the wellbore fluid pressure, and T_w is the wellbore fluid temperature. Solutions for mode 2 in the Laplace domain are as follows:

$$s\tilde{T}^{(2)} = (T_w - T_o)\Phi(\omega) \tag{27}$$
$$s\tilde{p}^{(2)} = \left[(p_w-p_o) - \chi(T_w-T_o)\right]\Phi(\xi) + \chi(T_w-T_o)\Phi(\omega) \tag{28}$$
$$s\tilde{\sigma}_{rr}^{(2)} = \alpha\left(1-\frac{D_{12}}{D_{11}}\right)\left\{\left[(p_w-p_o)-\chi(T_w-T_o)\right]\Psi(\xi) + \chi(T_w-T_o)\Psi(\omega)\right\} + \beta\left(1-\frac{D_{12}}{D_{11}}\right)\left\{(T_w-T_o)\Psi(\omega)\right\} \tag{29}$$

$$s\tilde{\sigma}_{\theta\theta}^{(2)} = -\alpha(1-\frac{D_{12}}{D_{11}})\{[(p_w - p_o) - \chi(T_w - T_o)]\,\Omega(\xi) + \chi(T_w - T_o)\Omega(\omega)\} - \beta(1-\frac{D_{12}}{D_{11}})\{(T_w - T_o)\Omega(\omega)\} \quad (30)$$

in which

$$\Phi(x) = \left[\frac{\mathrm{K}_0(xr)}{\mathrm{K}_0(xR)}\right] \quad (31)$$

$$\Psi(x) = \left[\frac{\mathrm{K}_1(xr)}{xr\mathrm{K}_0(xR)} - \frac{R\mathrm{K}_1(xR)}{xr^2\mathrm{K}_0(xR)}\right] \quad (32)$$

$$\Omega(x) = \left[\frac{\mathrm{K}_1(xr)}{xr\mathrm{K}_0(xR)} - \frac{R\mathrm{K}_1(xR)}{xr^2\mathrm{K}_0(xR)} + \frac{\mathrm{K}_0(xr)}{\mathrm{K}_0(xR)}\right] \quad (33)$$

$$\chi = \frac{c^{fT}}{(1 - c_T/c_o)};\ c^{fT} = \frac{c_T}{\kappa}\left[\beta_m - \frac{\alpha\beta}{D_{11}}\right]; \quad (34)$$

$$\omega = \sqrt{\frac{s}{c_o}};\ \ \xi = \sqrt{\frac{s}{c_T}}; \quad (35)$$

where K_n is the modified Bessel function of the second kind of order n.

5. Conclusions

A porothermoelastic theory for saturated porous elastic materials has been presented. Constitutive equations for the most general case were presented and specializations for orthotropy, transverse isotropy, and isotropy were carried out to identify the material constants required to define the system. Although theoretical developments have accomplished development of governing equations to describe behaviors of anisotropic poroelastic materials under non-isothermal conditions, the experimental research is still lacking in measurement of these fundamental constants. The solution for an inclined borehole in a transversely isotropic poroelastic formation subject to a temperature gradient via difference in borehole fluid and formation temperature has been presented. Although the solution is limited for the case where the isotropic plane is perpendicular to the borehole axis, it is still a unique solution in transversely isotropic porothermoelasticity and can serve as a benchmark for validating numerical codes.

References

1. Biot, M. A. General theory of three-dimensional consolidation, *J. Appl. Phys.*, **12**, 155-164, 1941.

2. Rice, J. R. and Cleary, M. P. Some basic stress diffusion solutions for fluid-saturated elastic porous media with compressible constituents, *Reviews of Geophysics and Space Physics*, **14**, 227-241, 1976.
3. Biot, M. A. Theory of Elasticity and consolidation of a porous anisotropic solid, *J. Appl. Phys.*, **26**, 182-185, 1955.
4. Thompson, M. and Willis, J. R. A reformulation of the equations of anisotropic poroelasticity, *J. Appl. Mech., ASME*, **58**, 612-616, 1991.
5. Cheng, A. H-D. Material coefficients of anisotropic poroelasticity, *Int. J. Rock Mech. Min. Sci.*, **34**, 199-205, 1997.
6. Abousleiman, Y. Cheng, A. H-D., Cui, L., Detournay, E. and Roegiers, J.-C. Mandel's problem revisited, *Geotechnique*, **46**, 187-195, 1996.
7. Abousleiman, Y. and Cui, L. Poroelastic solutions in transversely isotropic media for wellbore and cylinder, *Int. J. Solids Structures*, **35**, 4905-4929, 1998.
8. Coussy. O. Mechanics of Porous Continua, John Wiley and Sons, New York, 1995.
9. Aoki, T., Tan, C. P., and Bamford, W. E. Effects of elastic and strength anisotropy on borehole failures in saturated rocks, *Int. J. Rock Mech. Min. Sci.*, **30**, 1031-1034, 1993.
10. Charlez, P. A., and Heugas, O. Measurement of thermoporoelastic properties of rocks: theory and applications, *ISRM Symp: Eurock '92*, Rock Characterization, ed. J. A. Hudson, 42-46, 1992.
11. Berchenko, I., Detournay, E., Chandler, N., Martino, J., and Kozak, E. In-situ measurement of some thermoporoelastic parameters of a granite, In *Poromechanics, A Tribute to Maurice A. Biot, Proc. Biot Conf. on Poromechanics, Louvain-La-Neuve,* Balkema, Rotterdam, 545-550, 1998.
12. Cui, L., Cheng, A. H.-D., and Abousleiman, Y. Poroelastic solution of an inclined borehole, *J. Appl. Mech., ASME*, **64**, 32-38, 1997.

Constitutive Description of Fluid-Porous Solid Immiscible Mixtures. Derivation of the Effective Stress-Strain Relation

J. Kubik and M. Cieszko
Department of Environmental Mechanics, Pedagogical University of Bydgoszcz, 85-064 Bydgoszcz, Chodkiewicza 30
e-mail: kubik@rose.man.poznan.pl

Abstract. The effective stress concept for elastic porous solid filled with barotropic fluid is discussed. The saturated porous solid is considered as the immiscible mixture consisting of two physically identifiable components. The derived constitutive stress equations (in general non-linear) for each component are used to formulate the effective stress expression. In particular, the linear relation between incremental stresses and the porous solid strains in the vicinity of an initial stress state, the effective stress-strain relation and expression for the effective stress coefficient are derived.

Keywords: Saturated porous solids, effective stress, constitutive relations

1. Introduction

The principle of effective stresses for porous media filled with fluid plays an important role especially in soil mechanics, rock and concrete mechanics as well as in mechanics of materials such as ceramics or powder metals. The effective stress is the stress that controls the strain, volume change and strength behaviour of a given porous medium, independent of the magnitude of the pore pressure (see e.g. Lade and de Boer [6]). The commonly discussed expression for the effective stress $\boldsymbol{\sigma}^e$ is proposed as the difference between the total stress $\boldsymbol{\sigma}$ and the fraction of the pore water pressure p^f, i.e.

$$\boldsymbol{\sigma}^e = \boldsymbol{\sigma} - \eta p^f \mathbf{I} \tag{1}$$

where η expresses the fraction of the pressure that should be employed to make equation (1) to satisfy the effective stress definition. The formulation of the concept of effective stress is most often attributed to Terzaghi (see e.g. Skempton [8]). His original effective stress expression is of the form (1) for $\eta = 1$.

The review of the historical development of the effective stress concept and discussion concerning pioneering Terzaghi's and Fillunger's contributions to the subject of effective stresses is given by de Boer and Ehlers [2].

W. Ehlers (ed.), IUTAM Symposium on Theoretical and Numerical Methods in Continuum Mechanics of Porous Materials, 153–160.

Although Terzaghi's expression (1) (for $\eta = 1$) works well for most geotechnical applications, it is not adequate for some porous media such as concrete and rock and for porous materials under high pressure conditions. Another form of η involving two compressibilities of porous skeleton, different from that proposed by Terzaghi, is proposed by various authors e.g. Lade and de Boer [6], Geertsma [5], Biot and Willis [1], Skempton [9], Šuklje [10], Nur and Byerlee [7] and the results are summarised in the paper Lade and de Boer [6]. At the same time the analytical development and laboratory experiments are presented by Lade and de Boer [6] to obtain the η-expression satisfying the concept of effective stress for porous media in more general cases by the introducing four skeleton compressibilities to analysis.

The main purpose of this work is to formulate within the continuum description of fluid saturated porous media a stress-strain relation satisfying the effective stress concept in which the stress is determined by strain measures of porous solid only. In this approach the saturated porous solid is considered as the immiscible mixture consisting of physically identifiable components: porous solid and pore fluid preserving their own individual, physical properties during deformation process. In the considerations the components are assumed to be elastic and in the reference state pore structure has isotropic and homogeneous properties in the macroscopic sense. This enables one to define the internal energy for each component independently by the field quantities describing its own state of deformation. Then, the constitutive stress-strain relations (in general non-linear) for each component derived from the balance equation for the internal energy of the whole composition are basic relations used to formulate the effective stresses. In such a case considering the linear relation between incremental stresses and the porous solid strains in the vicinity of an initial stress state the effective stress is derived and the expression for η-parameter is established.

It is shown that parameter η is defined by the volume porosity and two compressibilities of the porous sample that can be measured in jacketed test. It is also shown that coefficients appearing in stress-strain relations are functions of the initial stresses in porous medium.

2. Constitutive relations for the elastic fluid-porous solid immiscible mixture

The starting point for the considerations is the macroscopic non-linear constitutive description of an elastic porous skeleton filled with barotropic fluid. Applying the method of derivation of the constitutive relations analogous to the approach used for the hyperelastic medium we propose the fluid internal energy e^f and the internal energy for porous elastic solid e^s in the following form, [3],

$$e^f = \hat{e}^f(\varrho^f) \; ; \qquad e^s = \hat{e}^s(\mathbf{C}, \varrho^s) \tag{2}$$

that should satisfy the balance equation for the internal energy of the whole system. Therefore we have

$$\begin{aligned} &tr[2\overline{\varrho}^s \mathbf{F} \frac{\partial \hat{e}^s}{\partial \mathbf{C}} \mathbf{F}^T - (1-f_v)p^f \mathbf{I} - \mathbf{T}^s]\mathbf{L}^s - [p^f f_v \mathbf{I} + \mathbf{T}^f]\mathbf{L}^f + \\ &\left(-\frac{p^f}{(\varrho^s)^2} + \frac{\partial \hat{e}^s}{\partial \varrho^s}\right) \overline{\varrho}^s \frac{D^s}{Dt} \varrho^s + [p^f grad\,(f_v) + \boldsymbol{\pi}] \cdot (\mathbf{v}^s - \mathbf{v}^f) = 0 \, , \end{aligned} \tag{3}$$

where

$$\overline{\varrho}^s = \varrho^s(1-f_v) \, , \quad p^f = (\varrho^f)^2 \frac{d\hat{e}^f}{d\varrho^f} \, .$$

In the above relations ϱ^f and ϱ^s stand for the effective density of fluid and porous skeleton, respectively, f_v is the volume porosity, $\mathbf{C} = \mathbf{F}^T\mathbf{F}$ is the right Cauchy-Green deformation tensor, $\mathbf{F}$ is the deformation gradient of porous solid and p^f denotes the effective pore pressure. Tensors $\mathbf{T}^f$ and $\mathbf{T}^s$ stand for the partial Cauchy stress of fluid and porous solid, respectively, and are assumed to be symmetric. Vector $\boldsymbol{\pi}$ is the internal interaction force between solid and fluid. Quantities $\mathbf{L}^f = grad(\mathbf{v}^f)$ and $\mathbf{L}^s = grad(\mathbf{v}^s)$ are the gradients of phase velocities $\mathbf{v}^f$ and $\mathbf{v}^s$ of the corresponding constituents.

From equation (3) that is the linear function of independent quantities $\mathbf{L}^f$, $\mathbf{L}^s$, $D^s\varrho^s/Dt$ and $(\mathbf{v}^s - \mathbf{v}^f)$ one can obtain the constitutive relations for partial stresses of the elastic porous skeleton and the barotropic pore fluid, respectively,

$$\mathbf{T}^s = -(1-f_v)p^f\mathbf{I} + 2\overline{\varrho}^s\mathbf{F}\frac{\partial \hat{e}^s}{\partial \mathbf{C}}\mathbf{F}^T \, , \tag{4}$$

$$\mathbf{T}^f = -f_v p^f \mathbf{I} \, , \tag{5}$$

and the condition for internal mechanical equilibrium between porous solid and fluid

$$p^f = (\varrho^s)^2 \frac{d\hat{e}^s}{d\varrho^s} \, . \tag{6}$$

relating the pore pressure p^f with the deformation state of porous skeleton. Equations (4)-(6) form an appropriate basis to derive and discuss the effective stress concept.

3. The effective stress when material of porous skeleton is incompressible

From theoretical point of view and for many practical problems (e.g. in soil mechanics) it is reasonable to consider the mechanical behaviour of saturated porous media when the solid material of skeleton is incompressible. The incompressibility condition has following form

$$\varrho^s = \varrho_o^s$$

where ϱ_o^s stands for the initial effective mass density of porous skeleton. Thus the internal energy of porous solid depends only on tensor $\mathbf{C}$, i.e.

$$e^s = \hat{e}^s(\mathbf{C})$$

In such a case the condition (6) does not appear in the set of constitutive relations. When we combine together (4) and (5) we obtain the non-linear stress-strain relation

$$\mathbf{T}^s + \mathbf{T}^f + p^f\mathbf{I} = 2\overline{\varrho}^s\mathbf{F}\frac{\partial\hat{e}^s}{\partial\mathbf{C}}\mathbf{F}^T , \tag{7}$$

that corresponds to the effective stress principle. The expression on the LHS of (7) is the effective stress and has the form

$$\mathbf{T}^s + \mathbf{T}^f + p^f\mathbf{I} = \mathbf{T} + p^f\mathbf{I} \tag{8}$$

which is identical to that proposed by Terzaghi.

4. Linear stress-strain relation. The effective stress expression

To find the expression for the effective stress when both constituents of fluid - porous solid mixture are compressible we consider small incremental stresses and strains in the vicinity of a given initial stress state.

Assuming that the porous medium in the reference configuration is homogeneous and isotropic and additionally that the initial stress state is isotropic, the initial state of the porous medium, can be characterised by the following set of quantities:

$$\mathbf{T}_o^s = -(1 - f_v^o)p_o^f \mathbf{I}, \quad \varrho_o^s, \quad f_v^o, \quad p_o^f, \quad \varrho_o^f .$$

By the linearization procedure applied to egs. $(3)_3$, (4) and (6) we obtain the corresponding set of the linear constitutive relations in the following incremental form, [4],

$$\begin{aligned} \Delta\mathbf{T}^s + (1 - f_v^o)\Delta p^f \mathbf{I} = 2[\mu - (1 - f_v^o)(p_o^s - p_o^f)]\, \mathbf{E} + \\ +\{[\lambda + (1 - f_v^o)p_s^o]\, tr(\mathbf{E}) + [K + (1 - f_v^o)p_o^f]\, \frac{\Delta\varrho^s}{\varrho_o^s}\}\mathbf{I}\, , \end{aligned} \tag{9}$$

$$(1 - f_v^o)\Delta p^f = K\, tr(\mathbf{E}) + [K_c + 2(1 - f_v^o)p_o^f]\, \frac{\Delta\varrho^s}{\varrho_o^s}\, , \tag{10}$$

$$\Delta p^f = K^f\, \frac{\Delta\varrho^f}{\varrho_o^f}\, , \tag{11}$$

where $\mathbf{E}$ is the infinitesimal strain tensor of the skeleton and μ, λ, K, K^c and K^f are independent material constants the first four of which characterise elastic properties of porous skeleton and the last one describes the mechanical properties of pore fluid.

Combining egs. (9) and (10) and making use of the linear form of the skeleton continuity equation, we obtain the expression relating the increments of total stresses and pore pressure to small deformations of porous skeleton

$$\begin{aligned} \Delta\mathbf{T} + [1 - (1 - f_v^o)\alpha^I]\Delta p^f \mathbf{I} = 2[\mu + (1 - f_v^o)(p_o^s - p_o^f)]\, \mathbf{E} + \\ +[\lambda - (1 - f_v^o)(p_f^o - p_o^s) - K\alpha^I]\, tr(\mathbf{E})\, \mathbf{I} \end{aligned} \tag{12}$$

where

$$\alpha^I = \frac{K}{K_c + 2(1 - f_v^o)p_o^f}\, .$$

For the hydrostatic case, equation (12) reduces to

$$\begin{gathered}\Delta\sigma + [1 - (1 - f_v^o)\alpha^I]\Delta p^f = \\ = [\frac{2\mu + 3\lambda}{3} - \frac{1}{3}(1 - f_v^o)(p_o^f - p_o^f) - K\alpha^I]\ tr(\mathbf{E})\end{gathered} \tag{13}$$

where $\Delta\sigma$ is the increment of the spherical part of the total stress, i.e. $\Delta\sigma = tr(\Delta\mathbf{T})/3$.

From equation (13) (as well as from (12)) one can find that the expression on LHS of this equation is the effective stress

$$\Delta\sigma^e = \Delta\sigma + [1 - (1 - f_v^o)\alpha^I]\Delta p^f \tag{14}$$

for the isotropic, elastic fluid-porous solid mixture and the η-parameter (see formula (1)) gets the form

$$\eta = 1 - (1 - f_v^o)\alpha^I . \tag{15}$$

It is worth to note that, in general, stress-strain relation (12) or (13) is influenced by the initial pressure conditions and is characterised by four material constants. It can be also seen that in the case of very low compressibility of skeleton material ($K^c \to \infty$), $\eta = 1$, and the effective stress (14) takes form that of Terzaghi. The same effect is observed for the very high value of initial pore pressure .

To provide the determination of the effective stress and an measurement of the coefficient α^I one can consider jacketed and unjacketed compressibility tests (at simplified initial condition: $p_o^s = p_o^f$).

JACKETED COMPRESSIBILITY TEST

The stress conditions are: $\Delta\sigma = -\Delta p_c^f$, $\Delta p^f = 0$ where Δp_c^f is the increment of confining pressure. In such a case from (10) and (13) one can find

$$\alpha^I = \frac{K}{K_c + 2(1 - f_v^o)p_o^f} = \frac{\varepsilon_{ms}}{\varepsilon}|_J , \tag{16}$$

$$\frac{\Delta p_c^f}{\varepsilon}|_J = \frac{2\mu + 3\lambda}{3} - K\alpha^I , \tag{17}$$

where ε_{ms} and ε stand for dilatations of matrix material and porous skeleton, respectively.

The expression (16) provides an interpretation of the α^I-quantity that is the ratio of relative volume change of matrix material to the relative volume change of porous sample in jacketed test. Another form of (16) can be written in terms of the compressibilities of matrix material and porous sample. We have

$$\alpha^I = \frac{\varepsilon_{ms}/\Delta p_c^f}{\varepsilon/\Delta p_c^f}|_J = \frac{C_{ms}}{C}|_J \leq 1 . \tag{18}$$

The relation (17) determines the elastic coefficient of porous sample in the jacketed test that can be also represented by

$$K_s = 1/C|_J = 1/\kappa \tag{19}$$

where κ is the coefficient of jacketed compressibility introduced by Biot and Willis [1].

UNJACKETED COMPRESSIBILITY TEST

The stress conditions are: $\Delta\sigma = -\Delta p_c^f$, $\Delta p^f = \Delta p_c^f$.
With such conditions, from (13), we can determine the unjacketed compressibility coefficient δ analogous to that defined by Biot and Willis [1],

$$\delta = -\frac{\varepsilon_{ms}}{\Delta p_c^f}|_{UN} = C|_J(1 - f_v^o)\alpha^I = \kappa(1 - f_v^o)\alpha^I . \tag{20}$$

Combaining (15) and (20) we find the expression

$$\eta = 1 - (1 - f_v^o)\alpha^I = 1 - \frac{\delta}{\kappa} \tag{21}$$

that proves the derived here effective stress coefficient is equivalent to that proposed by Biot and Willis [1].

5. Conclusions

1. From the foregoing discussion it is seen that the derived results for the effective stresses allows to obtain the expressions corresponding to some limiting cases;

-material of porous matrix is incompressible

$$C_{ms} = 0 \quad \Rightarrow \quad \alpha^I = 0 \quad \Rightarrow \quad \eta = 1 , \tag{22}$$

which is the result analogous to that of Terzaghi;

-deformation of porous sample at the constant porosity

$$f_v = const \quad \Rightarrow \quad \alpha^I = 1 \quad \Rightarrow \quad \eta = f_v \, ; \tag{23}$$

-deformation at very high pore pressure

$$p_o^f \uparrow \; \Rightarrow \quad \alpha^I \to 0 \quad \Rightarrow \quad \eta = 1 \, ; \tag{24}$$

(compressibility of matrix material becomes exhausted);
-very low volume porosity

$$f_v \downarrow 0 \quad \Rightarrow \quad \alpha^I \to 1 \quad \Rightarrow \quad \eta = 0 \, . \tag{25}$$

2. The effective stresses deduced in this work from constitutive analysis, explicitly depending on initial stress conditions and volume porosity parameter, on some identified conditions contain the results for effective stress formulations proposed by Biot and Willis [1], Šuklje [10], Nur and Byerlee [7].

3. From (15), (16) and (20) one can conclude that the volume change of porous sample in unjacketed compressibility test is equal to the volume change of matrix material in jacketed compressibility test. This result is very useful in experimental works.

References

1. Biot, M. A. and Willis, D. G. The Elastic Coefficients of the Theory of Consolidation. *J. Appl. Mech.*, **24**, 594–601, 1957.
2. de Boer, R. and Ehlers, W. The Development of the Concept of Effective Stresses. *Acta Mechanica*, **83**, 77–92, 1990.
3. Cieszko, M. and Kubik, J. Constitutive Relations and Internal Equilibrium Condition for Fluid-Satureted Porous Solid. Non-Linear Theory. *Arch. Mech.*, **48**, 893–910, 1996.
4. Cieszko, M. and Kubik, J. Constitutive Relations and Internal Equilibrium Condition for Fluid-Satureted Porous Solid. Linear description. *Arch. Mech.*, **48**, 911–923, 1996.
5. Geertsma, J. The Effect of Fluid Pressure Decline on Volumetric Changes of Porous Rocks. *Transactions AIME*, **210**, 331–340, 1957.
6. Lade, P. V. and de Boer, R. The Concept of Effective Stress for Soil, Concrete and Rock. *Geotechnique*, **47**, 61–78, 1997.
7. Nur, A. and Byerlee, J. D. An Exact Effective Stress Law for Elastic Deformation of Rock with Fluids. *J. Geophys. Res.*, **76**, 6414–6419, 1971.
8. Skempton, A. W. Significance of Terzaghi's Concept of Effective Stress (Terzaghi's Discovery of Effective Stress). In: From Theory to Practice in Soil Mechanics (L. Bjerrum, A. Casagrande,R.B. Peck, A.W. Skempton, eds). New York - London: John Wiley and Sons, 1960.
9. Skempton, A. W. Effective Stress in Soils, Concrete and Rock. Conf. on Pore Pressure and Suction in Soils, Butterworths, 4–16, 1960.
10. Šuklje, L. Rheological Aspects of Soil Mechanics. Wiley Interscience, New York, 1969.

Two-Phase Flow Modelling of Flood Defense Structures

M. Paul, R. Hinkelmann, H. Sheta and R. Helmig
Institut für ComputerAnwendungen im Bauingenieurwesen
Technische Universität Braunschweig,
Pockelsstr. 3, 38106 Braunschweig
e-mail: (m.paul, r.hinkelmann, h.sheta, r.helmig)@tu-bs.de

Abstract. In this paper a two-phase flow model concept for the flow processes in flood defense structures is presented. The governing equations as well as the discretization and solver techniques, which are included in the numerical simulator MUFTE-UG, are explained briefly. Results of the simulation of water overtopping a dike are shown.

1. Introduction

In the field of coastal engineering, a problem of major importance results from the development that the height and frequency of storm surge levels will increase in the near future. Therefore, the probability of an overtopping of flood defense structures such as dikes, dunes, or breakwaters by waves will increase, too.

Overtopping evokes very complex multiphase flow fields inside the structure, which, together with the forces from the outside, enhance the stress on these structures and have a significant influence on the stability of such systems. Because of the periodically varying distribution of the pressure and flow fields, hysteresis effects may occur, which lead to a further strain on the soil structure. For the numerical simulation of seepage problems and flows with an almost linear surface, the commonly used approaches for single-phase flow are very useful. In order, however, to allow for more complex forms of the free surface called forth by the simultaneous infiltration from the side and from the top, and in order to consider the interaction between air and water, a two-phase approach should be preferred.

Fundamental work dealing with fluid-structure-interaction based on fully saturated soils, has been accomplished by the research group of Ehlers *et al.* [1]. Soil mechanical experiments in this area have been carried out by Richwien and Weißmann [7], who investigated representative soils of dikes both in situ and in the laboratory.

In this paper, the numerical simulation of two-phase flow processes in an idealized dike is presented. The numerical algorithms, including the two-phase flow equations, constitutive relationships and discretization techniques applied are explained and some results regarding the

W. Ehlers (ed.), IUTAM Symposium on Theoretical and Numerical Methods in Continuum Mechanics of Porous Materials, 163–168.

infiltration of water in the dike due to overtopping water are shown. The paper closes with final remarks.

2. Numerical algorithm

2.1. Two-phase flow equations

The balance equations for the flow of two immiscible fluid phases in porous media are given by the conservation of mass (eq. (1)) and the extended Darcy-law (eq. (2)). For the two-phase flow of water and air, water is considered the *wetting* and air the *nonwetting* phase. In the following the different phases will be indicated by the index w and n, respectively.

$$\frac{\partial(S_\alpha\,\phi\,\varrho_\alpha)}{\partial t} + \operatorname{div}(\mathbf{v}_\alpha\,\varrho_\alpha) - \varrho_\alpha\,q_\alpha = 0 \tag{1}$$

$$\mathbf{v}_\alpha = -\frac{k_{r\alpha}}{\mu_\alpha}\,\mathbf{K}\,(\operatorname{grad} p_\alpha - \varrho_\alpha\,\mathbf{g}) \qquad \alpha \in \{w,n\} \tag{2}$$

In these equations S denotes unknown saturation, ϕ porosity, ϱ density, t time, $\mathbf{v}$ velocity, q a source or sink term, k_r relative permeability, $\mathbf{K}$ intrinsic permeability tensor, μ dynamic viscosity, p unknown pressure, and $\mathbf{g}$ vector of gravitational acceleration. Inserting eq. (2) into eq. (1) yields the two-phase flow equation

$$\begin{aligned} L_\alpha(p_\alpha, S_\alpha) :=\; & \frac{\partial(S_\alpha\,\phi\,\varrho_\alpha)}{\partial t} - \operatorname{div}\left(\varrho_\alpha\lambda_\alpha\,\mathbf{K}\,(\operatorname{grad} p_\alpha - \varrho_\alpha\,\mathbf{g})\right) \\ & - \varrho_\alpha\,q_\alpha = 0 \qquad \alpha \in \{w,n\} \end{aligned} \tag{3}$$

with the mobility $\lambda_\alpha = k_{r\alpha}/\mu_\alpha$.

This coupled dynamic system of partial differential equations is completed by two algebraic relations:
The void space in the porous medium is completely filled by the different fluid phases (eq. (4)) and the difference of the pressure at every point is a function of the capillary pressure p_c (eq. (5)):

$$S_w + S_n = 1 \tag{4}$$

$$p_c = p_n - p_w \tag{5}$$

With the use of these relations two of the four unknowns in eq. (3) (p_w, p_n, S_w, S_n) can be eliminated. For the present case this leads to a pressure-saturation-formulation (for further details see [2]).

For the unknowns being the pressure of the wetting phase, p_w, and the saturation of the nonwetting phase, S_n, the equations are given by:

$$\begin{aligned} L_\alpha(p_w, S_n) \quad := \quad & (-1)^{\delta_{\alpha w}} \phi \frac{\partial(\varrho_\alpha S_n)}{\partial t} \\ & - \operatorname{div}\{\lambda_\alpha \varrho_\alpha \mathbf{K} (\operatorname{grad} p_w + \delta_{\alpha w} \operatorname{grad} p_c - \varrho_\alpha \mathbf{g})\} \\ & - \varrho_\alpha q_\alpha = 0 \qquad\qquad \alpha \in \{w, n\} \end{aligned} \tag{6}$$

$\delta_{\alpha w}$: Kronecker-symbol

2.2. Constitutive relationships

Constitutive relationships account for material properties, which influence the flow of fluids in porous media. Many authors have developed various approaches for describing these relationships. For the present case of gas-water-flow the formulations after van Genuchten were chosen.

2.2.1. *Capillary pressure*

The relationship between the capillary pressure and the water saturation is stated as follows:

$$p_c(S_w) = \frac{1}{\alpha}\left(S_e^{-1/m} - 1\right)^{1/n} \qquad ; \quad p_c > 0 \tag{7}$$

with $1/\alpha$ representing the necessary entry pressure for an infiltration into a pore and n describing the distribution of the different pore sizes. m is derived from n by $m = 1-1/n$. S_e is called the effective saturation of the wetting phase and is given by

$$S_e = \frac{S_w - S_{wr}}{1 - S_{wr}} \qquad ; \quad S_{wr} \le S_w \le 1 \tag{8}$$

Here S_{wr} represents the residual saturation of the wetting phase.

2.2.2. *Relative permeability*

The hydraulic conductivity of a porous medium is composed of the soil parameter $\mathbf{K}$ (absolute permeability tensor) and the relative permeability $k_{r\alpha}$, which depends on the phase α and its saturation. $k_{r\alpha}$ accouts for the fact that two or more phases in the porous medium influence the flow of each other. Van Genuchten describes this by the following equations:

$$k_{rw} = S_e^{\frac{1}{2}} \left[1 - \left(1 - S_e^{1/m}\right)^m\right]^2 \tag{9}$$

$$k_{rn} = (1 - S_e)^{\frac{1}{2}} \left[1 - S_e^{1/m}\right]^{2m} \tag{10}$$

For more details on the constitutive relationships see [2].

2.3. Discretization

For numerical solution, the $p_w - S_n$-formulation (eq. (3)) is discretized in space by a finite volume based method, a so-called box scheme. The time is discretized using an implicit Euler scheme. This leads to the following system of nonlinear algebraic equations for each time step:

$$\begin{aligned}
&(-1)^{\delta_{\alpha w}}\, \phi\, \{[\hat{S}_n\, \varrho_\alpha]_i^{n+1} - [\hat{S}_n\, \varrho_\alpha]_i^n\}\, \frac{B_i}{\Delta t} \\
&- \sum_{j \in \eta_i} (\lambda_{\alpha ij}\, \varrho_{\alpha ij})_{ups}^{n+1} \oint_\Gamma \mathbf{K}\, \mathrm{grad}\, N_j\, \mathbf{n}\, d\Gamma\, (\psi_{\alpha j} - \psi_{\alpha i})^{n+1} \\
&- B_i\, [\varrho_\alpha\, \hat{q}_\alpha]_i^{n+1} = 0 \qquad \alpha \in \{w, n\} \qquad (11)
\end{aligned}$$

In these equations the symbol ^ indicates, that the discrete value of a variable at node i is used. n represents the current time step of the iteration. B_i stands for the volume of a box that is assigned to node i. η_i is the set of neighbour nodes of i whose boxes share an interface with B_i. *ups* indicates that an upwind technique is applied for the flux terms. The basic function is referred to as N_j, and $\mathbf{n}$ denotes the unit vector perpendicular to boundary Γ of box B_i. $(\psi_{\alpha j} - \psi_{\alpha i})$ is the direction of the discrete flow of phase α with $\psi_{\alpha i} = p_{wi} + \delta_{\alpha n}\, p_{ci} - \varrho_{\alpha i}\, g_i$.

Eq. (11) is solved by combining an outer Newton iteration with an inner BiCGSTAB solver. As a preconditioner a multigrid cycle is used [3].

3. Application

For the case of overtopping flow processes were simulated in an idealized dike consisting of homogeneous sand. The system was chosen in accordance to experiments carried out on a wooden dike by H. Oumeraci and H. Schüttrumpf [4]. The aim of these experiments was to measure the pressure and velocity fields on the surface of the dike as well as the amount of the overtopping water. These data were used as boundary conditions for the numerical simulation. In future experiments measurements will also be taken inside the dike to make comparison between experimental and numerical results possible. The thickness of the overflowing water layer was averaged and ranges from 2 cm to 3 mm. For further initial and boundary conditions as well as the geometry see Figure 1.

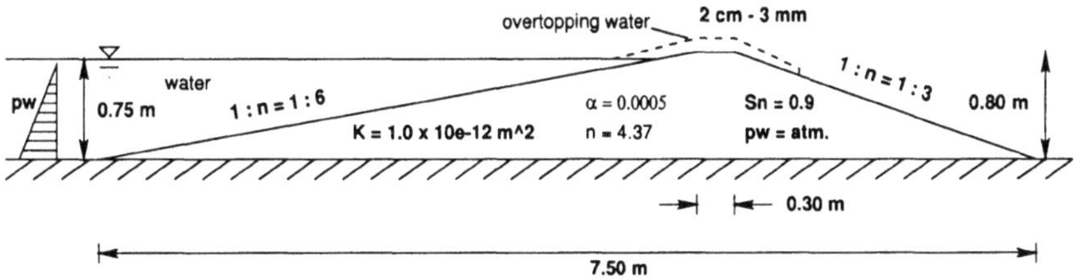

Figure 1. System geometry, initial and boundary conditions

Figure 2 shows the water saturation after an infiltration time of 180 minutes. In the middle of the domain the zone where the air will be entrapped becomes visible. The applied two-phase flow model approximates this fact well.

Figures 3 and 4 show the velocity of water and air, respectively. The vertical infiltration of water and the horizontal retreating of air can be seen clearly.

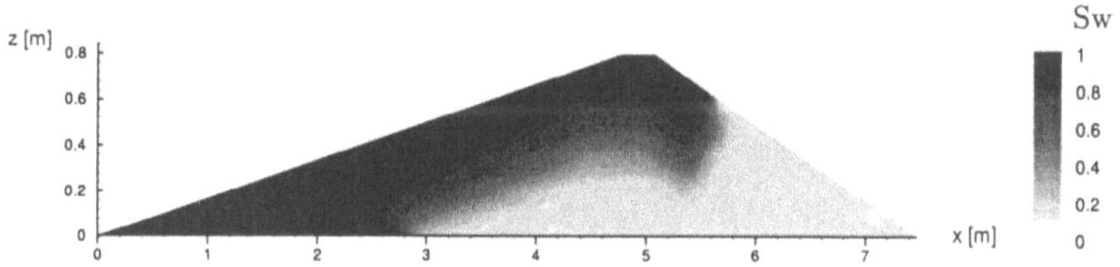

Figure 2. Water saturation, t = 180 min

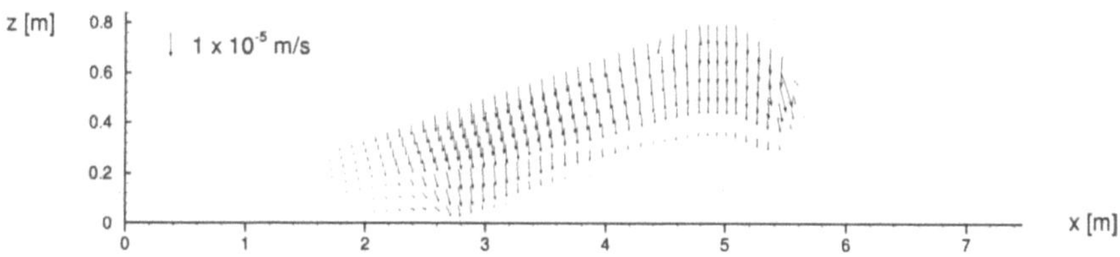

Figure 3. Velocity of water, t = 180 min

More detailed results and parameter studies for constitutive relationships are presented in [6].

4. Final remarks

The two-phase flow formulation used by the numerical program MUFTE-UG is capable of simulating the flow processes in dikes for the case of overtopping water. As there will be investigations carried out considering a permeable dike [5] it will in the near future be possible to compare numerical and experimental data and therefore validate the simulated results. Future steps for developing powerful tools for the

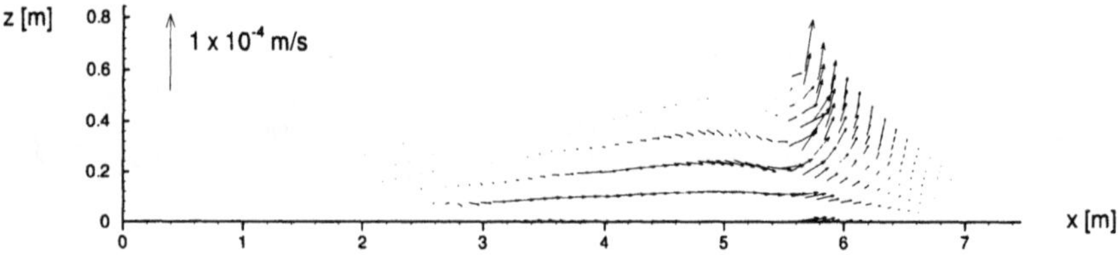

Figure 4. velocity of air, t = 180 min

simulation of breaching flood defense structures could include a consideration of the multilayered inhomogeneous structure of real dikes with imperfections (holes), the pressure increase resulting from breaking waves and the interaction of soil and fluid.

Acknowledgements

The work presented here was carried out with kind support of Prof. H. Oumeraci and Dipl.-Ing. H. Schüttrumpf. The project is funded by the DFG-Graduiertenkolleg "Wechselwirkung von Struktur und Fluid".

References

1. Ehlers, W., Ellsiepen, P., Blome, P., Mahnkopf, D. and Markert, B. Theoretische und numerische Studien zur Lösung von Rand- und Anfangswertproblemen in der Theorie Poröser Medien. Bericht Nr. 99-II-1 aus dem Institut für Mechanik, Universität Stuttgart, 1998.
2. Helmig, R. Multiphase Flow and Transport Processes in the Subsurface. *Springer-Verlag*, Heidelberg, 1997.
3. Helmig, R., Braun, C. and Emmert, M. Architecture of the Modular Program System MUFTE-UG for Simulating Multiphase Flow and Transport Processes in Heterogeneous Porous Media. *Mathematische Geologie*, Band 2, 1998.
4. Oumeraci, H. and Schüttrumpf, H. DFG-Projekt OU 1/2-1: Hydrodynamische Belastung der Binnenböschung von Seedeichen durch Wellenüberlauf. TU Braunschweig, 1997.
5. Oumeraci, H. and Richwien, W. Forschungsantrag „Belastung der Binnenböschung von Seedeichen durch Wellenüberlauf". Braunschweig/Essen, 1999
6. Paul, M. Numerische Simulation der Zweiphasenströmungsprozesse in Deichen mit Wellenüberlauf. Diplomarbeit, TU Braunschweig, 1999.
7. Richwien, W. and Weißmann, R. Zur Standsicherheit von Deichbinnenböschungen bei Wellenüberlauf. Essen, 1995.

A Sand Erosion Problem in Axial Flow Conditions on the Example of Contact Erosion due to Horizontal Groundwater Flow

A. Scheuermann
University of Karlsruhe, Institute of Soil and Rock Mechanics, Engler-Bunte-Ring, 76131 Karlsruhe, Germany

I. Vardoulakis
National Technical University of Athens, Department of Mechanics, GR-15773, Zougraphou, Greece

P. Papanastasiou
Schlumberger Cambridge Research, High Cross, Madingley Road, Cambridge CB3 0EL, U.K.

M. Stavropoulou
O.T.M. Consulting Engineering Company S.A., 6-8 Koumarianou & Plapouta Str., GR-11473, Athens, Greece

1. Introduction

Below water retaining structures, such as embankment dams, and during groundwater draw downs horizontal groundwater seepage prevails. If the subsoil is layered - as often encountered - contact erosion along the contact faces can occur. Subsoils commonly subject to this phenomena consist of fluviatile sediments with rather horizontal sand-gravel layers. Between these adjoining layers abrupt changes of grain size can occur. As a result the adjoining layers are not filter-proof to each other. In the case of an increased horizontal groundwater flow, the finer-grained soil will be eroded and transported in the direction of the seepage through the void structure of the coarser soil. As a consequence of the material loss, cavities will develop in the subsoil, which can induce large-scaled soil-deformations.

Only few considerations and experimental investigations were performed to examine and understand the physical connections of this kind of erosion. Brauns [1] found in large scale tests that for great differences in the grain size of the concerning gravel (filter) and sand (base) layers, the erosion process is governed by hydraulic influences only and practically free of geometrical compulsion. Such a combination of adjoining filter and base layers is characerized by a nearly constant Froude-number and it can be considered as a pure hydraulic problem.

W. Ehlers (ed.), IUTAM Symposium on Theoretical and Numerical Methods in Continuum Mechanics of Porous Materials, 169–175.

Small scale tests by van der Meulen [3] have shown that dislodged base particles remain near the filter interface, transport processes take place within a distance of one grain size of the filter material and in spite of increasing dispersion at higher flow velocities, the base particles remain within a distance of one filter particle from the interface. The velocity of the dislodged particles, which was determined experimentally by van der Meulen, is roughly half the flow velocity (factor β in equation 1).

2. Mathematical modelling

In the present study we consider the hydro-mechanical aspect of contact erosion due to horizontal groundwater flow for great differences in the grain size of the concerning filter and base material. For such a condition it is justified to consider the problem of contact erosion as a surface erosion as a first approximation. For this kind of erosion the basic definitions, mass balances and constitutive laws governing the erosion process were presented in a previous publication [6], from which we adopt the same notation. In the considered problem of axial flow (figure 1a and b) convection is assumed to be equal to zero and hydrodynamical dispersion dominates. Thus both the dispersion and the mass-generation term must be specified through appropriate evolution equations as functions of state variables.

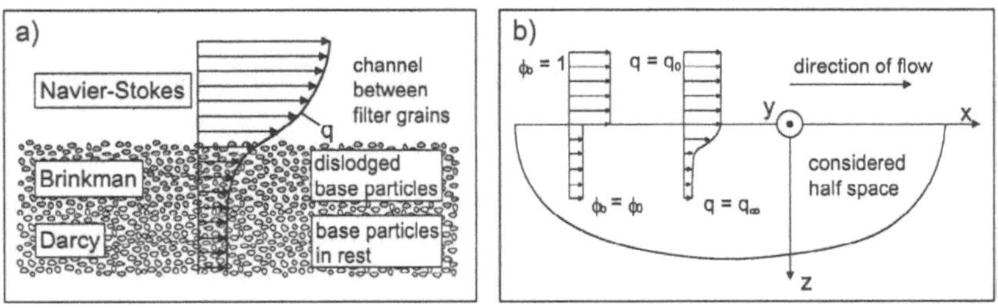

Figure 1. a) Conditions of flow b) Considered half space

The mass-generation term was given in [5] by a constitutive equation attributed to Einstein and Irmay and Sakthivadivel [4] where λ is the spatial frequency of erosion starter points. The porosity of filter ϕ_{fl} takes into account the fact that only between the filter grains the erosion process on the base material can occur. As it can be seen from equation (1), the rate of eroded mass $\dot{m}$ is proportional to the discharge of the fluidized particles $\beta c \|q_i\|$. As long as the transport concentration

c is nonzero, the erosion process modelled by such a law will continue until all base mass is eroded away.

$$\dot{m} = \rho_S \lambda \left(1 - \phi_b\right) \phi_{fl} \beta c \left\| q_i \right\| \tag{1}$$

As for the dispersion coefficient D we adopt here a proposition of Bear in which the dispersion is assumed to be a function of flow rate and a characteristic lenght l of the medium as well as permeability of the base layer k and thus of porosity . Moreover, it is assumed that the dispersion also depends on the tortuosity T and its scaling factor α of the concerning base material as a function of the actual and minimum porosity (2). Here, we may adopt the Carman-Kozeny equation for the permeability coefficient κ.

$$D = D_0 \alpha T \left(q - q_\infty\right) \sqrt{k} \; ; \; k = l^2 \kappa\left(\phi_b\right) \; ; \; \kappa\left(\phi_b\right) = \frac{\phi_b^3}{\left(1 - \phi_b\right)^2} \tag{2}$$

Considering the mass-balance equation for the solid skeleton a simple expression that links porosity changes to the produced eroded mass can be written

$$\frac{\dot{m}}{\rho_S} = \frac{\partial \phi_b}{\partial t} \tag{3}$$

Since for axial flow surface-parallel erosion inside the porous medium leads to strong variations of porosity and permeability perpendicular to the direction of flow, the validity of Darcy's law is questionable, and a modified form has to be adopted. One possibility is to use Brinkman's [2] generalization of Darcy's law, which allows for energy dissipation not only due to friction of fluid flowing around grains, but also due to the viscous shearing stress, which is acting within the fluid phase itself. This generalization was done by adding the viscous-stress term to Darcy's law that typically appears in the Navier-Stokes equation. In terms of pore-fluid preassure p_f and fluid-discharge $\|q_i\|$ Brinkman's generalization of Darcy's law can be written in the following form:

$$i = -\frac{\partial p_f}{\partial x_i} = \rho_f \left(\frac{\eta_k}{k} q_i - \eta_k \nabla^2 q_i \right) \tag{4}$$

In the preceding equation k is the permeability of the porous medium, ρ_f and η_k are the density and kinematic viscosity of the fluid, respectively. It should be noted that Brinkman's extension of Darcy's law could be seen as an interpolation between the Navier-Stokes equation, covering the flow in the "channel" between the coarse filter grains, and the Darcian flow at some distance from the erosion region. We assume here the existence of a (possibly) thin boundary layer that leads assymptotically from one condition to the other. With an empirical relationship it is possible to calculate the width of the boundary layer of these perturbed particles (5). Essentially, this equation means that the boundary layer stops where the flow rate is one percent higher than that of Darcy's flow.

$$\delta := q(\delta) = 1.01 q_\infty \tag{5}$$

The whole derivation of the system, the initial and boundary conditions as well as the numerical solution of the governing partial differential equations are represented in [5].

3. Summary of governing equations and conditions

It is useful to normalise the system of governing equations by its boundary values. The normalisation procedure is given in [5]. For easy reference the normalised governing equations and conditions of the considered problem of contact erosion in axial flow are summarised below.

$$\frac{\partial \phi_b}{\partial \tau} = \hat{f} \;\; ; \;\; \hat{f} = \hat{\lambda}\,(1-\phi_b)\,\phi_{fl}\beta c\,\hat{q} \tag{6}$$

$$\frac{\partial \theta}{\partial \tau} = \frac{\partial}{\partial \xi}\left[\hat{D}\,\frac{\partial \theta}{\partial \xi}\right] + \hat{f} \;\; ; \theta = c\phi_b,\; \hat{D} = D_0 \alpha T \left(\hat{q} - \hat{q}_\infty\right)\sqrt{\kappa(\phi_b)} \tag{7}$$

$$-\frac{d^2 \hat{q}}{d\xi^2} + \frac{1}{\kappa(\phi_b)}\,\hat{q} = \hat{q}_0 \;\; ; \;\; \hat{q}_0 = \frac{1}{q_0}\frac{il^2}{\rho_f \eta_k},\;\; \hat{q}_\infty = \hat{q}_0\,\kappa_0,\; \kappa_0 = \kappa(\phi_0) \tag{8}$$

$$I.C. \quad \tau = 0: \quad \phi_b = \phi_0 \; ; c = c_0 \,(=0)$$

$$\hat{q} = \hat{q}_\infty + \left(1 - \hat{q}_\infty\right)\exp\left(-\tfrac{\xi}{\sqrt{\kappa_0}}\right)$$

$$B.C. \quad \xi = 0: \quad \hat{q}\,(0,\tau) = 1 \; ; \; \theta\,(0,\tau) = \theta_0$$

$$\xi = \infty: \quad \hat{q}_\infty\,(\infty,\tau) = \hat{q}_\infty \; ; \; \theta\,(\infty,\tau) = 0$$

4. Discussion of numerical results

Selected computational results will be presented for the problem of contact erosion under axial flow conditions with the following set of parameters [2]:

flow rate in the filter q_0 [m/s]	0.046
flow rate in the base layer q_∞ [m/s]	$1.96{\cdot}10^{-5}$
erosion starter points λ [1/m]	100
char. length of the base layer l [m]	$2.4{\cdot}10^{-4}$
dispersion coefficient D_0 [/]	$1{\cdot}10^{-4}$
initial transport concentration c_0 [/]	$1{\cdot}10^{-5}$
initial porosity of the base layer ϕ_b [/]	0.4
factor of tortuosity α [/]	10

This system of parameters is based on experimental investigations performed by Brauns [1]. A parametric study on these experiments is given in [5].

For better orientation figure 2a schematically depicts the geometric and hydraulic situation for the numerical analysis. Figure 2b shows the variation of the normalised flow rate in space with time. We verify that Brinkman's law accounts for a smooth transition between channel-flow at the free surface and Darcian flow at some depth. Moreover, Brinkman's law is justified here, since it is capable of concentrating high flow rates in the near-surface region where erosion will be dominant.

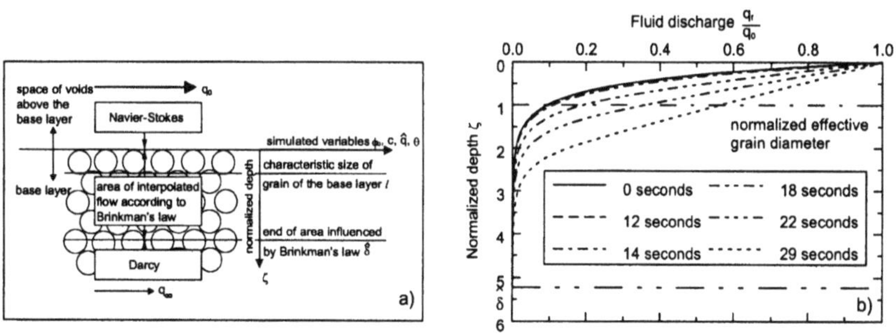

Figure 2. a) Introduction into the numerical result b) Evolution of fluid discharge

The evolution of the normalised fluidized-particles density is shown in figure 3a. It is clear that $\theta = c\phi$, takes a maximum value at the free surface as a direct result of the assumed Neumann-type boundary condition.

The evolution of transport concentration is represented in figure 3b. In concordance with the basic constitutive equations for erosion kinetics, this figure shows that erosion progresses in time as a „front" of high transport concentration. Although the governing equation (6) is diffusion-like, this result is justified by the highly non-linear character of the dominating erosion source term (7). The local peak of c at some depth from the free surface indicates the region where the scouring effect of the fluidized base particles is maximum.

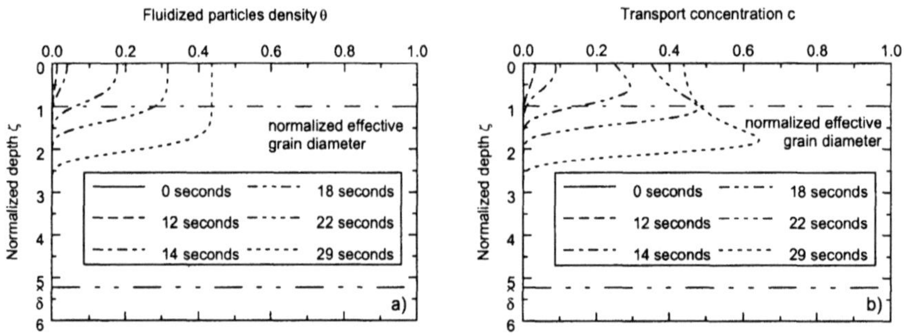

Figure 3. Evolution of a) fluidized particles density and b) transport concentration

Figure 4a shows the evolution of porosity in space with time. It is clear that the erosion process, in terms of porosity, is maximum at the free surface. The evolution of porosity at the free surface with time is represented in figure 4b. We emphasise here that a base particle will be eroded when the porosity reaches a critical value. This critical value can be used as a boundary condition necessary for the definition of the moving eroded boundary.

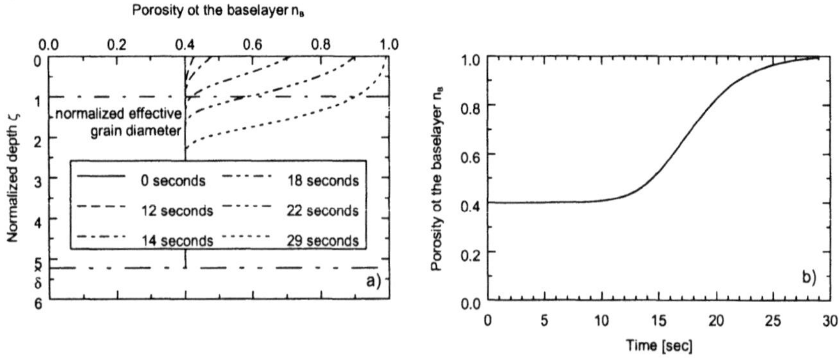

Figure 4. Evolution of porosity a) with depth and b) in time for $\xi = 0$

5. Concluding remarks

The presented system of equations allows the qualitative interpretation of time dependent changes caused by a hydraulic force parallel to an interface of a base/filter combination. Thus it can be used to guide the design of experiments for studying the contact erosion and the sand erosion problem, respectively. The numerical results for the erosion process of a thin perturbed layer consisting of only few moving particles is justified by the observations of van der Meulen [3]. The transfer of the theoretical model to a real problem consideration is not possible as long as at least the order of magnitude of the parameters can't be estimated. The factor for the erosion starter points λ, for example, should be determinable by small scale tests with different materials. If these parameters are estimated first, it is essential to define a suitable criterion for the interpretation of the simulated results. As soon as a practical criterion is formulated and the parameters are specifically applicable, it is possible to calculate the loss of material and hence the subsidence without further ado.

References

1. Brauns, J. Erosionsverhalten geschichteten Bodens bei horizontaler Durchströmung. *Wasserwirtschaft 75*, 448–453, 1985.
2. Brinkman, H. C. A calculation of the viscous force exerted by a flowing fluid on a dense swarm of particles. *Appl. Sci. Res. Al.*, 27–34, 1947.
3. den Adel, H. and Bakker, K. J. A transport modell for filtration: Control of erosion processes. *proc. Geo-Filters 92 in Karlsruhe, Balkema, Rotterdam* , 189–196, 1993.
4. Sakthivadivel, R. and Irmay, S. A review of filtration thesis *HEL 15-4, University of California, Berkley*, 1966
5. Scheuermann, A. Schichterosion bei horizontaler Durchströmung. diploma thesis, not published, 1998.
6. Vardoulakis, I., Stavropoulou, M. and Papanastasiou, P. Hydro-Mechanical Aspects of the Sand Production Problem. *Transport in Porous Media 22*, 225–144, 1996.

5. Concluding remarks

Rayleigh Waves in Porous Media Saturated with Liquid

A. A. Gubaidullin and O. Y. Kuchugurina
Tyumen Institute of Mechanics of Multiphase Systems SB RAS
Taymirskaya Str.74, Tyumen, 625000 Russia
e-mail: timms@sbtx.tmn.ru

1. Introduction

Surface Rayleigh waves in porous media have been the subject of a number of investigations before (Dobrin, [1]; Press, Healy, [2]; Macdonald, [3]; Deresiewicz, Rice, [4]; Takeuchi, Saito, [5]; White, [6]; Liu Philip, Wen Jiangang, [7]). In the present paper the features of the Rayleigh waves propagating along the free surface of a porous medium saturated with liquid are investigated by means of two-velocity, two-stress tensor model. The dispersion relation is obtained. The frequency dependence of phase velocity, linear decrement of attenuation and coefficients exhibiting damping of amplitude with depth for monochromatic waves are studied. The pattern of deformation of the near-surface layer during Rayleigh wave propagation is considered.

2. Equations of motion

The linearized equations of motion of a granular or porous medium are given in the form (Nigmatulin, [8]):

$$\frac{\partial \rho_s}{\partial t} + \rho_{s0}\nabla^k v_s^k = 0, \quad \frac{\partial \rho_l}{\partial t} + \rho_{l0}\nabla^k v_l^k = 0, \tag{1}$$

$$\rho_{s0}\frac{\partial v_s^k}{\partial t} = -\alpha_{s0}\nabla^k p_l + \nabla^n \sigma_{s*}^{nk} + F^k, \quad \rho_{l0}\frac{\partial v_l^k}{\partial t} = -\alpha_{l0}\nabla^k p_l - F^k$$

$$\begin{aligned} \sigma_{s*}^{kn\prime} &= \alpha_{s0}(\lambda_*\delta^{kn}\varepsilon_s^{mm} + 2\mu_*\varepsilon_s^{kn} + \frac{K_*}{K_s}\delta^{kn}p_l'), \\ K_* &= \lambda_* + 2\mu_*/3, \\ \frac{\partial \varepsilon_s^{kn}}{\partial t} &= \frac{1}{2}(\nabla^k v_s^n + \nabla^n v_s^k) \end{aligned} \tag{2}$$

W. Ehlers (ed.), IUTAM Symposium on Theoretical and Numerical Methods in Continuum Mechanics of Porous Materials, 179–186.

$$p'_s = K_s\rho_s^{o\prime}/\rho_{s0}^o, \quad p'_l = K_l\rho_l^{o\prime}/\rho_{l0}^o \tag{3}$$
$$p'_{s*} = \alpha_{s0}(p'_s - p'_l) + \alpha'_s(p_{s0} - p_{l0}), \quad p_{s*} = -\sigma_{s*}^{mm}/3$$
$$\rho'_l = \alpha'_l\rho_{l0}^o + \alpha_{l0}\rho_l^{o\prime}, \quad \rho'_s = \alpha'_s\rho_{s0}^o + \alpha_{s0}\rho_s^{o\prime}$$
$$\alpha'_l + \alpha'_s = 0$$

Here $\alpha_j, \rho_j, \rho_j^o, v_j, p_j$ are the volume fraction, effective and true densities, velocity and pressure of j -phase ($j = s$ for solid phase and $j = l$ for liquid), σ_{s*} is the effective stress of skeleton; F is the force of interaction between the liquid and skeleton; $\alpha_s\lambda_*, \alpha_s\mu_*$ are the elastic moduli of skeleton, K_s, K_l are the elastic bulk moduli for the solid phase material and liquid correspondingly. The total stress σ of the porous medium is expressed as $\sigma^{ik} = -\delta^{ik}p_l + \sigma_{s*}^{ik}$. The superscripts are coordinate indices, summation is implied over the repetitive superscripts; subscript 0 corresponds to the equilibrium value of quantity, primed symbol means the deviation of quantity from its equilibrium value ($w' = w - w_0$).

The above equations of motion are conceptually identical to those of the Biot theory.

The expression for the interaction force F for harmonic oscillations of frequency ω is written as follows

$$\begin{aligned} F &= F_m + F_\mu + F_B, \\ F_m &= 0.5\eta_m\alpha_{l0}\alpha_{s0}\rho_{l0}^o i\omega(v_l - v_s), \\ F_\mu &= \eta_\mu a_*^{-2}\alpha_{l0}\alpha_{s0}\mu_l(v_l - v_s), \\ F_B &= \eta_B a_*^{-1}\alpha_{l0}\alpha_{s0}\sqrt{2\rho_{l0}^o\mu_l\omega}(1+i)(v_l - v_s) \end{aligned} \tag{4}$$

where F_m is the associated mass force, F_μ is the Stokes force of viscous friction, F_B is the analogue of the Basse force due to the nonstationary viscous layer near the interphase boundary; i is an imaginary unit; μ_l is the liquid viscosity, a_* is the characteristic size of pores; η_m, η_μ, η_B are factors taking into account the structure of the medium.

To investigate the surface Rayleigh waves that are the sum of heterogeneous longitudinal (LW) and transverse waves (TW), the dispersion dependences for LW and TW are needed. The dispersion relations determining the dependences of complex wave numbers of longitudinal $k_{(l)}$ and transverse $k_{(t)}$ waves on frequency ω have been obtained by Gubaidullin, Kuchugurina [9]. Dispersion relation for LW is quadratic in $k_{(l)}^2$ while that for TW is linear in $k_{(t)}^2$, therefore, there are two types of LW and one type of TW in a saturated porous medium. On solving the dispersion equations, the phase velocity and linear attenuation decrement for each wave are calculated as

$$C_{(l1,2)} = \omega/\mathrm{Re}k_{(l1,2)}, \quad \delta_{(l1,2)} = -\mathrm{Im}k_{(l1,2)},$$
$$C_{(t)} = \omega/\mathrm{Re}k_{(t)}, \quad \delta_{(t)} = -\mathrm{Im}k_{(t)}$$

3. Surface rayleigh waves

Consider liquid-saturated porous medium which occupies half-space $z > 0$ and is bounded with a plane free surface $z = 0$. The Rayleigh wave is an elastic wave traveling along the free surface and decreasing highly with depth. We consider the two-dimensional motion of the medium corresponding to Rayleigh wave propagation along the x-axis, that is, all values depend on the spatial coordinates x, z only. To study a monochromatic surface Rayleigh wave of frequency ω as a combination of heterogeneous LW and TW, we use scalar and vector potentials of displacements u_s, u_l in the form $(j = s, l)$ (Viktorov, [10]):

$$\vec{v}_j = \frac{\partial \vec{u}_j}{\partial t}, \quad \vec{u}_j = grad\Phi_j + rot\vec{\Psi}_j, \quad \vec{\Psi}_j = \{0, \Psi_j, 0\}, \tag{5}$$
$$\Phi_j = A_j e^{-\alpha z} \exp(i\omega t - ik_{(R)}x), \quad \Psi_j = B_j e^{-\beta z} \exp(i\omega t - ik_{(R)}x)$$

where A_j, B_j are the amplitudes, $k_{(R)}$ is complex wave number, α and β are coefficients showing decreasing of the wave with depth. It should be remembered that wave numbers $k_{(l)}, k_{(t)}, k_{(R)}$ and coefficients α, β are related as follows:

$$k_{(l)}^2 = k_{(R)}^2 - \alpha^2, \quad k_{(t)}^2 = k_{(R)}^2 - \beta^2 \tag{6}$$

At a free surface $z = 0$ the deviations of normal and shear components of total stress are equal to zero:

$$\sigma^{xz\prime} \mid_{z=0} = 0, \quad \sigma^{zz\prime} \mid_{z=0} = 0 \tag{7}$$

Using Eqs.1 - 7, we obtain the dispersion relation which allows to determine the wave number $k_{(R)}$ as a function of frequency ω:

$$\left(k_{(R)}^2 - \frac{1}{2}\frac{E_*}{\mu_*}k_{(l)}^2\right)\left(k_{(R)}^2 - \frac{1}{2}k_{(t)}^2\right) - k_{(R)}^2\sqrt{k_{(R)}^2 - k_{(l)}^2}\sqrt{k_{(R)}^2 - k_{(t)}^2} = 0 \tag{8}$$

and $\alpha = \sqrt{k^2_{(R)} - k^2_{(l)}}$, $\beta = \sqrt{k^2_{(R)} - k^2_{(t)}}$,
here E_* is expressed via elastic moduli of the skeleton and liquid.

As it was mentioned above, there are two types of longitudinal waves in a porous medium. That is why one can assume that there are two types of Rayleigh waves as well, that is, as a sum of heterogeneous longitudinal wave of each type and heterogeneous transverse wave. It is required of $k_{(R)}$, α, β that the following inequalities hold:

$$\mathrm{Re}k_{(R)} > \mathrm{Re}k_{(l)}, \quad \mathrm{Re}k_{(R)} > \mathrm{Re}k_{(t)}, \quad \mathrm{Re}\alpha > 0, \quad \mathrm{Re}\beta > 0 \qquad (9)$$

The first and second inequalities imply that the phase velocity of a Rayleigh wave is less than that of homogeneous longitudinal and homogeneous transverse waves. The two latter inequalities mean that the amplitudes of Rayleigh wave decrease with depth.

The complex roots of Eq.8 are found when $k_{(l)} = k_{(l1)}$ corresponding to the fast *LW* and at $k_{(l)} = k_{(l2)}$ corresponding to the slow one. Dispersion Eq.8 is easily transformed into a cubic equation in $k^2_{(R)}$. On finding the roots of the cubic equation with the parabola method, it is checked if they are also the roots of Eq.8 and satisfy to Ineq.9. It is found that when $k_{(l)} = k_{(l1)}$ there is one root of Eq.8 which satisfies Ineq.9, that is, there is the surface Rayleigh wave as a sum of heterogeneous fast *LW* and *TW*. On the other hand, at $k_{(l)} = k_{(l2)}$ none of the roots of Eq.8 satisfies to Ineq.9 at all real range of parameters of the porous medium. It implies that there is no Rayleigh wave being a sum of heterogeneous slow *LW* and *TW*. Hence only one type of Rayleigh wave which is a sum of heterogeneous fast *LW* and *TW* may propagate along the free surface of saturated porous medium.

4. Calculation results

The velocity and damping coefficients of Rayleigh wave are calculated as $C_{(R)} = \omega/\mathrm{Re}k_{(R)}$, $\delta_{(R)} = -\mathrm{Im}k_{(R)}$, the comparison with that of *LW* and *TW* is made; the deformation of the near-surface layer at monochromatic Rayleigh wave propagation was considered in detail. In calculations the parameters of porous medium of quartz saturated with water are the following: $p_0 = 1 \cdot 10^5$ Pa, $\rho^o_{s0} = 2500$ kg/m^3, $\rho^o_{l0} = 1000$ kg/m^3, $K_s = 1{\cdot}10^{10}$ Pa, $K_l = 2.25{\cdot}10^9$ Pa, $\mu_l = 1{\cdot}10^{-3}$ Pa·s, $\alpha_{l0} = 0.33$, $a_* = 0.25$ mm, $\lambda_* = 5 \cdot 10^9$ Pa, $\mu_* = 7 \cdot 10^9$ Pa, $\eta_m = 1$, $\eta_\mu = 100$, $\eta_B = 1.5$; the characteristic time value $t_\mu = \rho^o_{l0} a^2_*/(\eta_\mu \mu_l) = 0.625{\cdot}10^{-3}$ s.

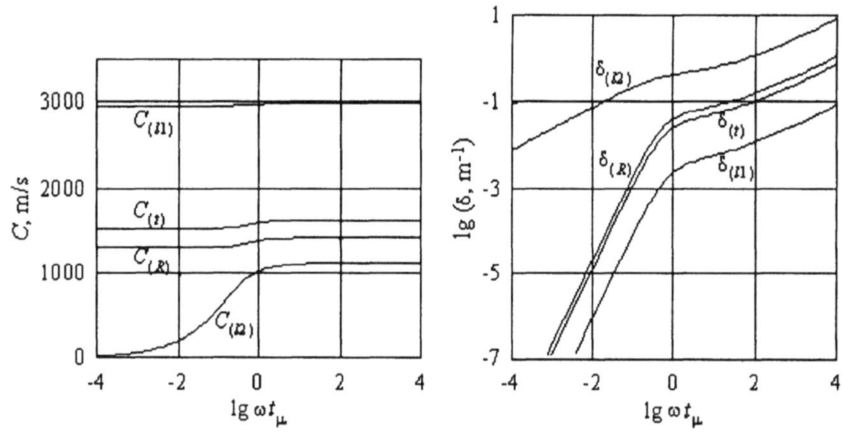

Figure 1. Phase velocity and linear attenuation decrement of longitudinal, transverse and Rayleigh waves vs frequency

In Fig.1 phase velocities of two longitudinal ($C_{(l1)}, C_{(l2)}$), transverse ($C_{(t)}$) and the Rayleigh ($C_{(R)}$) waves and their linear attenuation decrements $\delta_{(l1)}, \delta_{(l2)}, \delta_{(t)}, \delta_{(R)}$ are presented. The phase velocity of the surface wave $C_{(R)}$ is somewhat less than that of the transverse one $C_{(t)}$. The similar relation is also true for a usual elastic medium (Viktorov, 1982). The attenuation decrements $\delta_{(l1)}, \delta_{(l2)}, \delta_{(t)}, \delta_{(R)}$ characterize damping of each wave along its direction. Here, unlike a usual elastic medium, there is non-zero attenuation decrement of the Rayleigh wave. The curve for $\delta_{(R)}$ is similar to the curves $\delta_{(l1)}, \delta_{(t)}$; the Rayleigh wave damps somewhat stronger than the TW but less than the slow LW.

Fig.2 presents the coefficients α, β showing decreasing of Rayleigh wave with depth as functions of frequency ω. The real parts of complex values α, β are directly proportional to ω, whereas their imaginary parts are negative and are negligibly small when compared with real ones. It should be pointed out that in usual elastic medium α and β are real, positive and directly proportional to frequency ω.

To calculate the deformation of the near-surface layer at monochromatic Rayleigh wave propagation, the initial amplitude was chosen in order that $\sigma^{xx'}|_{x,z=0} = -p_0 exp(i\omega t)$

Fig.3 shows the momentary (at $t = 0$) pattern of deformations of the near-surface layer due to the Rayleigh wave propagation. The wave of frequency $\omega = 1000$ s^{-1} propagates along the x-axis. Here wave length $\lambda = \lambda_{(R)} = 9.1$ m, solid lines correspond to grids of the solid phase and dashed lines to grids of liquid, the displacements u_s, u_l are increased

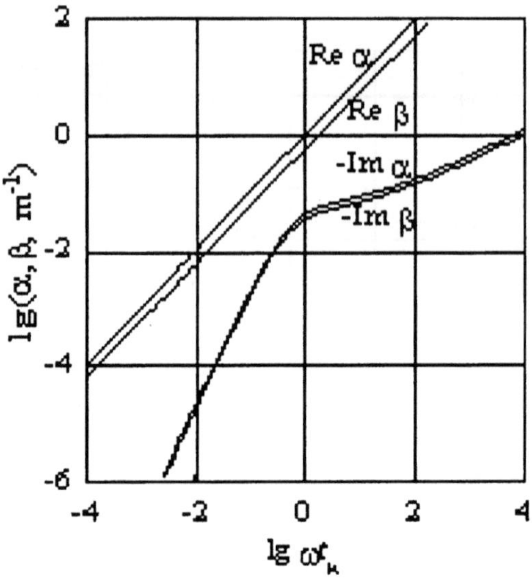

Figure 2. Coefficients showing decreasing of Rayleigh wave with depth vs frequency

15 000 times. ¿From this figure we notice that the amplitudes of displacements of liquid and solid phases decrease with depth. Near the surface the displacements of solid and liquid are just different: motion of liquid relative to skeleton is out of phase by $\pi/2$ approximately. This takes place because liquid is involved in motion by the solid skeleton due to forces of interaction. The calculations show that at the lower frequencies displacements of skeleton and liquid are nearly equal, so that they are displaced in phase. With increasing frequency, difference in their displacements increases.

5. Conclusions

The main results and conclusions are the following:

1. The features of Rayleigh waves propagating along the free surface of a liquid-saturated porous medium are investigated in the scheme of two-velocity model.

2. The dispersion relation for Rayleigh waves in a liquid-saturated porous medium is derived. It is found that for the wide range of parameters of porous medium being considered there is only one type of Rayleigh waves which is the combination of transverse and fast longitudinal waves

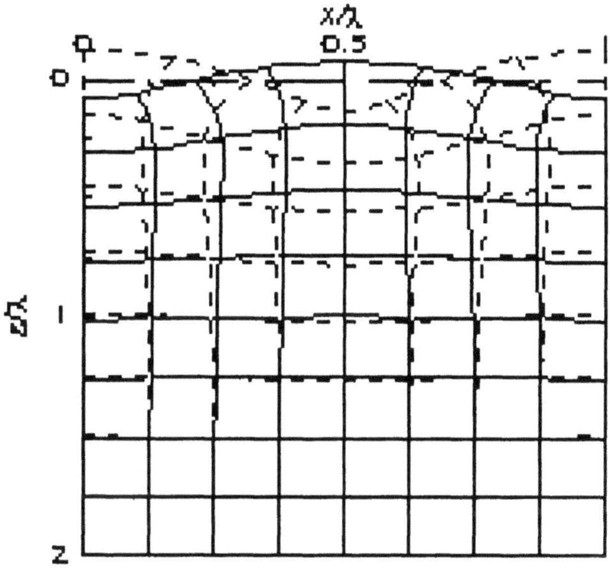

Figure 3. Deformation of near-surface area of saturated porous medium due to Rayleigh wave propagation

3. The dependences of the phase velocity, linear attenuation decrement of a monochromatic wave and coefficients showing wave decreasing with depth on frequency are studied. The pattern of deformation of the near-surface area during Rayleigh wave propagation is considered in detail.

Acknowledgements

The work was supported by Ministry of Education of Russia (Grant No.97-0-4.2-130) and Russian Foundation for Basic Research (Grant No.98-01-00831).

References

1. Dobrin, M. B. Rayleigh waves from small explosions, *Trans. Am. Geophys. Union*, **32**, 822–832, 1951.
2. Press, F. and Healy, J. Absorption of Rayleigh waves in mellow-loss media, *J. Appl. Phys.*, **28**, 1323–1325, 1957.
3. Macdonald, J. R. Rayleigh wave dissipation functions in lowloss media, *Geophys. J. R. Astron. Soc.*, **2**, 132–135, 1959.
4. Deresiewicz, H. and Rice, I. T. The effect of boundaries on wave propagation in a liquid-filled porous solid; 3. Reflection of plane waves at a free boundary (general case), *Bull. Seism. Soc. America*, **52**, 595–625, 1962.
5. Takeuchi, H. and Saito, M. Seismic surface waves, *In: B. A. Bolt (Editor) Methods in Computational Physics*, **11**, Academic Press, New-York, 217–295, 1972.
6. White, J. E. *Underground sound. Application of seismic waves*, Elsevier, Amsterdam-Oxford-New York, 1983.
7. Liu Philip, L.-F. and Wen Jiangang Nonlinear diffusive surface waves in porous media, *J. Fluid Mech.* **347**, 119–139, 1997.
8. Nigmatulin, R. I. *Dynamics of Multiphase Media*, Hemisphere Publ. Corp., New York, 1991.
9. Gubaidullin, A. A. and Kuchugurina, O. Y. One-dimensional longitudinal and transverse linear waves in saturated porous media, *Festschrift zum 60. Geburstag von Prof. Dr.-Ing. Reint de Boer. Beitrage zur Mechanik. Universität-Essen, Germany*, **66**, 135–144, 1995.
10. Viktorov, I. A. *Acoustic surface waves in solids*, Nauka, Moscow (in Russian), 1982.

Geometry Effects on Sound in Porous Media

A. Cortis and D. M. J. Smeulders
Delft University of Technology, PO Box 5028, 2600 GA Delft, The Netherlands

D. Lafarge
LAUM, UMR 6613, Av. O. Messiaen, 72017, Le Mans, France

M. Firdaouss and J. L.Guermond
LIMSI, UPR 3251 (CNRS), BP 133, 91403 Orsay, France

Abstract. The problem of sound propagation in rigid porous media is investigated. Two so-called scaling functions are introduced to describe the dynamic viscous and thermal interaction of the pore fluid and the porous structure. These scaling functions are characterized by the viscous and thermal permeabilities k_0 and k'_0, the viscous and thermal tortuosities α_∞, α_0, and α'_0, and the characteristic length scales Λ and Λ'. These parameters can be numerically evaluated from steady-state descriptions. For a pore geometry consisting of an arrangement of cylinders, the characteristic parameters are presented. The full microscopic dynamic flow and heat problems for this configuration were solved, averaged, and compared with the scaling functions. We found that for this configuration the scaling functions gave an accurate description of the oscillatory flow and heat phenomena.

1. Introduction

Sound propagation in porous media is of importance in many fields of engineering science. In air-filled sound absorbing materials, the frequency dependence of the compressibility varies from isothermal at low frequencies to adiabatic in the high-frequency regime. Similarly, the frequency dependence of the gas density, which can be described in terms of dynamic flow permeability or in terms of dynamic frame tortuosity, varies from viscosity-dominated at low frequencies to inertia-dominated in the high-frequency regime. The understanding of such behaviour in equally important for the oil industry, where acoustic borehole logging is commonly practiced. A borehole is drilled in a potential hydrocarbon reservoir and probed with an acoustic tool. The inversion process comprises the delineation of the reservoir properties from the acoustic signals, and is complicated because of the inherent inhomogeneity of the reservoir with its multiple inclusions and microcracks on the microscale. On the macroscale, where the measurements are performed, these inhomogeneities affect the viscous and thermal behaviour of the porous fluid-solid system. This work presents a numerical study on the macroscopic dynamic properties of a schematic rigid porous medium, starting from the microscopic geometry of the pore space.

W. Ehlers (ed.), IUTAM Symposium on Theoretical and Numerical Methods in Continuum Mechanics of Porous Materials, 187–192.

2. Theory

The process of sound propagation in gas-filled rigid porous media can be characterized by a complex-valued tortuosity and compressibility. These two so-called scaling functions are frequency-dependent and governed by the microgeometry of the pores. The linear (i.e., small-amplitude) response of the pore fluid to a macroscopic pressure gradient $\nabla\langle\hat{p}\rangle e^{i\omega t}$ is usually described in terms of the macroscopic fluid velocity $\langle\hat{\mathbf{v}}\rangle e^{i\omega t}$ and the dynamic tortuosity $\alpha(\omega)$ [4]:

$$i\omega\alpha(\omega)\rho_f\langle\hat{\mathbf{v}}\rangle = -\nabla\langle\hat{p}\rangle, \tag{1}$$

where the dynamic tortuosity takes into account the viscous and inertial interaction of air with the porous frame. The combination $\alpha(\omega)\rho_f$ is sometimes called the dynamic gas density. The symbol $\langle\rangle$ denotes an intrinsic air-phase average. An alternative formulation is based on a dynamic extension of Darcy's law [6, 7]:

$$\frac{\eta\phi}{k(\omega)}\langle\hat{\mathbf{v}}\rangle = -\nabla\langle\hat{p}\rangle, \tag{2}$$

where $k(\omega)$ is the dynamic permeability, ϕ is the porosity, and η the dynamic viscosity. This means that $\alpha(\omega)$ and $k(\omega)$ are not independent: $\alpha(\omega) = -i\alpha_\infty k_0\omega_c/\omega k(\omega)$, where we have introduced a characteristic frequency $\omega_c = \eta\phi/\alpha_\infty k_0\rho_f$ with k_0 the stationary Darcy permeability. Locally, $\hat{\mathbf{v}}$ and $\hat{p}$ have to satisfy the unsteady Stokes equation for incompressible media:

$$i\omega\rho_f\hat{\mathbf{v}} = -\nabla\hat{p} + \eta\nabla^2\hat{\mathbf{v}}. \tag{3}$$

On the basis of a microstructural approach, Auriault [7] and Johnson [4] proved that for high frequencies

$$\lim_{\omega\to\infty}\alpha(\omega) = \alpha_\infty\left[1 + (1-i)\frac{\delta(\omega)}{\Lambda}\right], \tag{4}$$

where α_∞ is the so-called tortuosity, $\delta(\omega)$ is the viscous boundary layer thickness $\sqrt{2\eta/\rho_f\omega}$, and Λ is a viscous length scale. For a porous material consisting of an ensemble of parallel identical tubes, for example, Λ equals the radius of those tubes.

For low frequencies, we simply find that $k(\omega) \to k_0$. A straightforward analytical scaling function was proposed by Johnson [4] which satisfies both high- and low-frequency limits:

$$\alpha(\omega) = \alpha_\infty[\,1 - i\frac{\omega_c}{\omega}F(\omega)\,], \tag{5}$$

where $F(\omega) = \sqrt{1 + \frac{1}{2} i M \frac{\omega}{\omega_c}}$, with M the so-called similarity parameter $8k_0\alpha_\infty/\phi\Lambda^2$.

Following Champoux & Allard [1], we write the mass continuity law of a perfect gas to address the thermal dissipation problem:

$$i\omega \frac{\beta(\omega)}{\gamma\langle p\rangle}\langle\hat{p}\rangle = -\nabla\cdot\langle\hat{\mathbf{v}}\rangle, \tag{6}$$

where $\beta(\omega)$ is the reduced dynamic incompressibility $\gamma\langle p\rangle/K_f(\omega)$, with γ the specific heat ratio, and $K_f(\omega)$ the dynamic compressibility. Locally, the temperature equation for a perfect gas has to be satisfied:

$$i\omega\rho_f c_p \hat{T} = i\omega\hat{p} + \lambda_g \nabla^2 \hat{T}. \tag{7}$$

Here, the specific heat at constant pressure is denoted by c_p, λ_g is the thermal conductivity, and $\hat{T}$ is the excess temperature. Introducing the characteristic thermal frequency $\omega_c' = a_g\phi/k_0'$, with k_0' the stationary thermal permeability and a_g the thermal diffusivity $\lambda_g/\rho_f c_p$, a straightforward analytical scaling function was proposed by Champoux & Allard [1] to satisfy both the high-frequency adiabatic limit and the low-frequency isothermal limit:

$$\beta(\omega) = \gamma - (\gamma - 1)[\, 1 - i\frac{\omega_c'}{\omega}F'(\omega)\,]^{-1}, \tag{8}$$

where $F'(\omega) = \sqrt{1 + \frac{1}{2} i M' \frac{\omega}{\omega_c'}}$, with M' the thermal similarity parameter $8k_0'/\phi\Lambda'^2$. A thermal lengthscale Λ' is introduced here. The viscous and thermal behaviour can elegantly be described by means of the relaxation functions $\chi(\omega) = 1 - \alpha_\infty/\alpha(\omega)$ and $\chi'(\omega) = (\beta(\omega)-1)/(\gamma-1)$. They show similar behaviour over the entire frequency regime and they represent the so-called Johnson-Allard (JA) model. We will also consider the low-frequency exension of this JA-model, suggested by Pride [5]. Here the functions $F(\omega)$ and $F'(\omega)$ are defined as:

$$F(\omega) = 1 - p + p\sqrt{1 + \frac{1}{2} i \frac{M}{p^2}\frac{\omega}{\omega_c}}, \tag{9}$$

$$F'(\omega) = 1 - p' + p'\sqrt{1 + \frac{1}{2} i \frac{M'}{p'^2}\frac{\omega}{\omega_c'}}, \tag{10}$$

where $p = M/4(\alpha_0/\alpha_\infty - 1)$ and $p' = M'/4(\alpha_0' - 1)$, with α_0 and α_0' the low-frequency viscous and thermal equivalents of the tortuosity. This model will be referred to as the Pride model.

3. Numerical computations

Numerical computations were performed on a 2-D configuration of solid cylinders surrounded by gas drawn in the upper left corners of Figs. 1 and 2. The unit cell used in the computations has identical width and height $B = H = 2L$. The radius of the cylinders is $r = 0.7136L$, corresponding to a cell porosity $\phi = 0.60$. Using the SEPRAN finite element package (Cuvelier [2]), we computed α_∞, α_0, α_0', Λ, Λ', k_0', and k_0. The results are presented in Table I. It was shown by Johnson [4] that $\alpha_\infty = \langle|\mathbf{v}_\infty|^2\rangle/|\langle\mathbf{v}_\infty\rangle|^2$, and that $2/\Lambda = \int|\mathbf{v}_\infty|^2 dS/\int|\mathbf{v}_\infty|^2 dV$, where the latter integral describes a velocity-weighted surface-to-volume ratio. The velocity field $\mathbf{v}_\infty = \nabla\psi$, limiting the oscillatory viscous flow for the high–frequencies regime, follows from the potential problem $\nabla^2\psi = 0$, with Neumann boundary conditions on the fluid–solid interface and periodicity on the inlet-outlet surfaces, i.e., the sides of the unit cell in this case. Furthermore, we computed k_0 by solving the Stokes problem $\eta\nabla^2\mathbf{v}_0 - \nabla p + \mathbf{e} = 0$, and $\nabla\cdot\mathbf{v}_0 = 0$, the quantity $\mathbf{e}$ being a unit force vector. No–slip boundary conditions at the pore walls, and periodicity of $\mathbf{v}_0$ and p were prescribed.

The thermal parameters can be expressed as follows: $\alpha_0' = \langle T_0^2\rangle/\langle T_0\rangle^2$, and $2/\Lambda' = \int dS/\int dV$. This means that both α_0' and k_0' can be computed from the problem $\nabla^2 T_0 + e = 0$, where e is the thermal scalar equivalent of the unit force vector in the flow problem. Dirichlet boundary conditions on the fluid–solid surface, and periodicity on the inlet–outlet boundaries were used for T_0.

The full dynamic flow problem (3) was solved using a finite element method developed by Guermond (Firdaouss *et al.* 1999). The full dynamic heat problem (7) was solved using the SEPRAN package. Results are presented in Figs. 1 and 2, where we plotted the two relaxation functions $\chi(\omega)$ and $\chi'(\omega)$. The shape of both scaling functions is identical. We notice that a perfect agreement between the numerical computations and the Pride model can be found for both the flow and the heat problem. The JA–model gives a reasonable prediction of the numerical results, but shows some deviations in the rollover frequency zone. It was suggested previously that the scaling functions are accurate for a wide range of morphologies [4], but that they break down for more extreme configurations where the pore flow channels contain sharp-edged intrusions [3].

Table I. *Characteristic parameters.*

ϕ	k_0/L^2	α_0	α_∞	Λ/L	k_0'/L^2	α_0'	Λ'/L
0.6000	0.0184	1.7301	1.4105	0.7186	0.0505	1.2761	1.0705

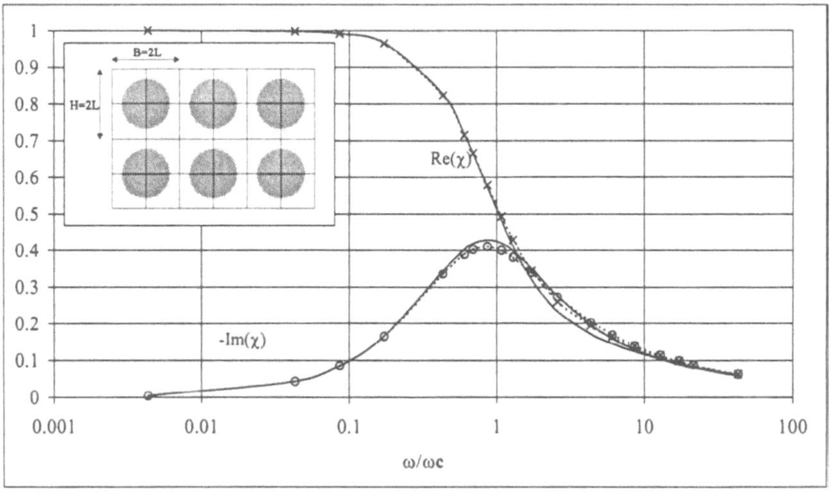

Figure 1. Viscous relaxation function vs dimensionless frequency ω/ω_c for an arrangement of solid cylinders surrounded by fluid. Crosses and circles represent direct numerical simulations. The solid line represents Johnson's model and the dashed line Pride's model.

4. Conclusions

For a cylindrical pore geometry we have numerically computed the characteristic parameters determining the scaling functions defined by Johnson, Allard and Pride *et al.* These scaling functions were compared with a full solution of the microscopic dynamic flow and heat problems. An excellent agreement was found for the Pride model, whereas the Johnson model did show some minor deviations. We remark that the shape of the scaling functions for flow and heat are identical, due to the similarity between the dissipation processes in the viscous and thermal boundary layers.

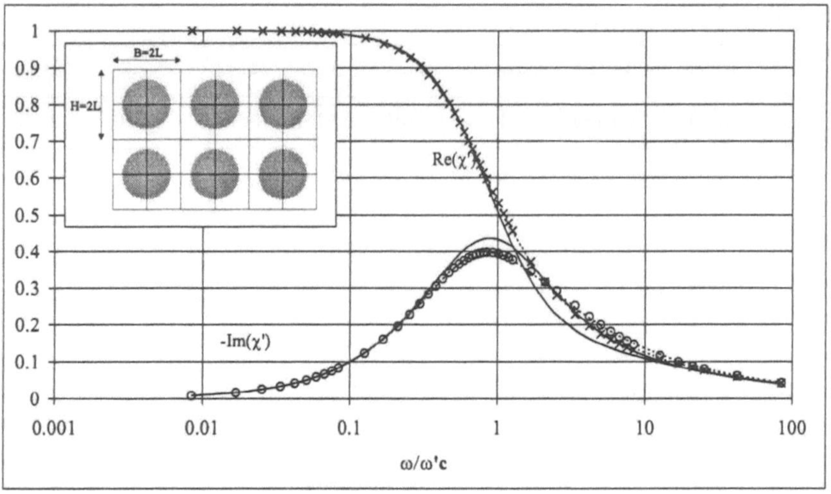

Figure 2. Thermal relaxation function vs dimensionless frequency ω/ω_c' for an arrangement of solid cylinders surrounded by fluid. Crosses and circles represent direct numerical simulations. The solid line represents Johnson's model and the dashed line Pride's model.

References

1. Champoux, Y. and Allard, J.-F. Dynamic tortuosity and bulk modulus in air-saturated porous media. *J. Appl. Phys.*, **70**, 1975–1979, 1991.
2. Cuvelier, C., van Steenhoven, A. A. and Segal, G. Finite Element Methods and Navier-Stokes Equations. Reidel, 1986.
3. Firdaouss, M., Guermond, J. L., Lafarge, D. and Smeulders, D. M. J. Sound propagation in gas–filled rigid framed porous media: effect of cusped pore geometries. *Bolletino di Geofisica*, **40**, 80–81, 1999.
4. Johnson, D. L., Koplik, J. and Dashen, R. Theory of dynamic permeability and tortuosity in fluid-saturated porous media.*J. Fluid Mech.*, **176**, 379–402, 1987.
5. Pride, S. R., Morgan, F. D. and Gangi, A. F. Drag forces of porous-medium acoustics. *Phys. Rev. B*, **47**, 4964–4978, 1993.
6. Lévy, T. Propagation of waves in a fluid–saturated porous elastic solid. *Int. J. Engng Sci.*, **17**, 1005–1014, 1979.
7. Auriault, J. L., Borne, L. and Chambon, R. Dynamics of porous saturated media, checking of the generalized law of Darcy. *J. Acous. Soc. Am.*, **77**, 1641–1650, 1985.

Mechanics of Liquefaction in Saturated Granular Soils

A. Sawicki
Institute of Hydro-Engineering, IBW PAN
ul. Kościerska 7, 80-953 Gdańsk-Oliwa, Poland

Abstract. A simple model describing the pore pressure generation in saturated sands and subsequent liquefaction is briefly outlined. Some applications of this model, dealing with an earthquake induced liquefaction of a soil stratum, propagation of a uni-axial liquefaction wave and liquefaction induced by surface waves are discussed.

1. Introduction

Saturated granular soils can be treated, on a macroscopic scale, like solid materials since they possess shearing resistance. Under certain loading conditions, the shearing resistance of saturated soils gradually decreases and, in an extreme case, practically disappears. Therefore, an initially solid material is transformed into a liquid as regards flow behaviour. Such a process is designated as the soil liquefaction. Liquefaction is preceded by progressive pore pressure build-up and associated reduction of effective stresses in the soil skeleton which, in turn, leads to reduction of the shearing resistance of saturated soil. In the extreme case of liquefaction, effective stresses vanish and the saturated soil behaves macroscopically as a liquid. Pore pressure generation in saturated soils, and subsequent liquefaction, is caused mainly by dynamic and quasi-dynamic (cyclic) loadings as, for example, earthquakes, machine and traffic induced vibrations, water waves, etc. These phenomena have been the cause of much damage to buildings and earth structures, including such effects as excessive settlements, floating, sinking and tilting.

The first attempt to explain the liquefaction phenomenon was made by Casagrande [3] and then, for nearly thirty years, the problem was not of much interest to researchers in geotechnical engineering and applied mechanics. The subsequent two decades were characterised by extensive investigations on soil liquefaction and associated phenomena. Respective references can be found in Bazant and Krizek [1], Finn [5], Finn *et al.* [6], Ishihara and Towhata [8], Martin and Seed [10], Nemat-Nasser and Shokooh [13], Seed and Lee [20], Valanis and Read [21], Zienkiewicz *et al.* [22]. Various theoretical models of liquefaction were proposed, but it seems that none of them has been accepted as a

W. Ehlers (ed.), IUTAM Symposium on Theoretical and Numerical Methods in Continuum Mechanics of Porous Materials, 193–198.

geotechnical standard, see Cakmak [2], Das [4], Haupt and Herrmann [7]. Although we understand more and more about the behaviour of saturated soils under various loading conditions, the problem of soil liquefaction still needs further investigations which should lead to a sound mechanical model of the phenomenon.

The aim of this paper is to outline briefly the low-resolution model of soil liquefaction and to show its various applications. The model is characterised by simple constitutive equations and a minimal set of material parameters, which can be determined from fairly simple experiments.

2. Model of liquefaction

The model, proposed by Morland and Sawicki [11], is based on the analogy between the compaction of a dry granular soil and the pore pressure generation in the same, but saturated soil loaded in undrained conditions, cf. Martin *et al.* [9]:

$$u = \frac{n_0}{\kappa(1-n_0)}\Phi, \tag{1}$$

where: u = pore pressure; n_0 = initial porosity; κ = compressibility of the soil skeleton and Φ is the compaction (irreversible porosity change) defined as follows:

$$\Phi = \frac{n_0 - n}{n_0} - \Delta, \tag{2}$$

where n = current porosity, Δ = reversible (elastic) porosity change.

The compaction is governed by the following evolutionary law:

$$\frac{d\Phi}{d\xi} = D_1 K \exp(-D_2\Phi), \tag{3}$$

where D_1 and D_2 = coefficients which have to be determined experimentally for a given sand, ξ = the loading parameter which is defined as accumulated deviatoric strain, K = the second invariant of deviatoric reversible strain $\mathbf{E}$. These quantities are defined as follows:

$$K = \frac{1}{2}\mathrm{tr}\mathbf{E}^2, \tag{4}$$

$$d\xi = \sqrt{\left|\frac{1}{2}\mathrm{tr}(d\mathbf{E})^2\right|}. \tag{5}$$

A reversible shear response of the saturated sand is described by the hypoelastic deviatoric relation:

$$\frac{d\mathbf{S}}{d\xi} = 2G(p')\frac{d\mathbf{E}}{d\xi}, \tag{6}$$

where $\mathbf{S}$ = stress deviator, G = shear modulus, $p' = p - u$ = mean effective stress, p = mean total stress. For small strains it is convenient to use the following form of the shear modulus:

$$G = G_0 + G_1\sqrt{p'}, \tag{7}$$

where G_0 and G_1 = coefficients which have to be determined experimentally for a given sand. Eq. (7) reflects the degradation of shearing resistance of saturated sand during pore pressure generation, down to its residual value of G_0 in an extreme case of liquefaction.

Morland and Sawicki [12] presented some modifications and generalisations of the above model, and Sawicki [14] proposed an engineering model of liquefaction formulated in terms of the cyclic stress and strain amplitudes, and the number of loading cycles instead of the parameter ξ. However, a general structure of governing equations is similar in all these cases.

3. Earthquake induced liquefaction in a soil stratum

Earthquake induced ground shaking can affect stability of structures founded on the surface as well as those embedded in the soil. It is then important to study the one-dimensional wave propagation problem for a saturated sand layer subjected to a cyclic horizontal acceleration at its base. Such a problem became the classic test for various models of liquefaction, see previously cited papers. The constitutive model of liquefaction, outlined in Section 2, was also applied by Sawicki and Morland [16] to this test problem. It was shown that the shear and bulk waves propagate through the soil stratum, but the effects caused by the bulk wave are much smaller than those due to the shear wave, so the bulk wave can be neglected. Pore pressure histories at various depths, up to the onset of liquefaction, as well as corresponding stresses and strains were determined numerically for typical field data. The results obtained were roughly similar to those calculated using other methods.

4. Uni-axial liquefaction wave

In contrast to the previous example, liquefaction associated with blast-induced ground motions is still not well understood. Therefore, it was interesting to apply the liquefaction model to the analysis of one-dimensional, compressional wave propagation through saturated sand, in undrained conditions, see Sawicki [15]. The governing wave equations were solved with a help of the method of characteristics. Numerical examples show that liquefaction is possible during either loading or unloading, depending on the magnitude of applied stress and mechanical properties of saturated sand. It was shown that the stress wave propagates without change of shapes, but produces regrouping of the intrinsic stresses in the soil matrix and pore fluid, i.e. generating excess pore pressure and reducing mean effective stress, which leads to liquefaction. Numerical results support empirical observations that it is easier to liquefy loose sands than the dense ones.

5. Liquefaction induced by surface waves

The liquefaction model was also applied in order to study the behaviour of water-saturated half-space, the motion of which is induced by the passage of both harmonic in time and transient Rayleigh waves, see Sawicki and Staroszczyk [19]. Numerical computations, performed by means of the Finite Element Method, illustrate the history of pore pressure generation, development of liquefaction zones and changes of displacement amplitudes in the subsoil. The method applied allows also for analysing the phenomenon of pore pressure dissipation where the assumption of undrained conditions is not satisfied, see also Sawicki and Świdziński [17].

6. Conclusions

The liquefaction model, outlined in this paper, differs essentially from the theories of saturated granular materials based on the Biot type approach, since it includes a description of the phenomena of pore pressure generation and subsequent reduction of shearing resistance of saturated soil. The model can be applied successfully to a wide range of engineering problems which cannot be analysed using traditional methods. Therefore, it seems that the model of liquefaction can serve as a useful tool in geotechnical engineering. An engineering version of this model has been applied to some other problems, such as the analysis of seabed during storms or settlements of machine foundations, Sawicki and Świdziński [18].

References

1. Bazant, Z. P. and Krizek, R. J. Endochronic constitutive law for liquefaction of sand, *J. Engng. Mech. Div.*, ASCE, EM2, 225–238, 1976.
2. Cakmak, A. S. (Editor) Soil Dynamics and Liquefaction, *Elsevier and Computational Mechanics Publications*, Amsterdam/Oxford/New York/Tokyo, 1987.
3. Casagrande, A. Characteristics of cohesionless soils affecting the stability of slopes and earthfills, *Journal of the Boston Society of Civil Engineers*, **23**, 257–276, 1936.
4. Das, B. M. Fundamentals of Soil Dynamics, *Elsevier*, New York/ /Amsterdam/Oxford, 1984.
5. Finn, W. D. L. Dynamic response analyses of saturated sands, in: *Soil Mechanics - Transient and Cyclic Loads* (eds. G. N. Pande and O. C. Zienkiewicz), John Wiley and Sons, 105, 1982.
6. Finn, W. D. L., Pickering, D. J. and Bransby, P. L. Sand liquefaction in triaxial and simple shear tests, *J. Soil Mechanics Foundation Division*, ASCE, **97**, SM4, 639–659, 1971.
7. Haupt, W. and Herrmann, R. Dynamische Bodenkennwerte, *LGA Publications*, Heft 48, Nürnberg, 1987.
8. Ishihara, K. and Towhata, I. Dynamic response analysis of level ground based on the effective stress method, in: *Soil Mechanics - Transient and Cyclic Loads* (eds. G. N. Pande and O. C. Zienkiewicz), John Wiley and Sons, 133, 1982.
9. Martin, G. R., Finn, W. D. L. and Seed, H. B. Fundamentals of liquefaction under cyclic loading, *J. Geotech. Engng. Div.*, ASCE, **101**, GT5, 423–438, 1975.
10. Martin, P. P. and Seed, H. B. One-dimensional dynamic ground response analyses, *J. Geotech. Engng. Div.*, ASCE, **108**, GT7, 935, 1982.
11. Morland, L. W. and Sawicki, A. A mixture model for the compaction of saturated sand, *Mechanics of Materials*, **2**, 217–231, 1983.
12. Morland, L. W. and Sawicki, A. A model for compaction and shear hysteresis in saturated granular materials, *J. Mech. Physics Solids*, **33**, 1–24, 1985.
13. Nemat-Nasser, S. and Shokooh, A. A unified approach to densification and liquefaction of cohesionless sand in cyclic shearing, *Canadian Geotechnical J.*, **16**, 659–678, 1979.
14. Sawicki, A. An engineering model for compaction of sand under cyclic loading, *Engineering Transactions*, **35**, 4, 677–693, 1987.
15. Sawicki, A. Uni-axial liquefaction wave in saturated sands, *J. Theoretical and Applied Mechanics*, **36**, 681–701, 1998.
16. Sawicki, A. and Morland, L. W. Pore pressure generation in a saturated sand layer subjected to a cyclic horizontal acceleration at its base, *J. Mech. Physics Solids*, **33**, 545–559, 1985.
17. Sawicki, A. and Świdziński, W. Pore pressure generation, dissipation and resolidification in a saturated soil, *Soils and Foundations*, **29**, 1–18, 1989.
18. Sawicki, A. and Świdziński, W. Mechanics of a sandy subsoil subjected to cyclic loadings, *Int. J. Num. Analyt. Methods in Geomechanics*, **13**, 511–529, 1989.
19. Sawicki, A. and Staroszczyk, R. Development of ground liquefaction due to surface waves, *Archives of Mechanics*, **47**, 557–576, 1995.

20. Seed, H. B. and Lee, K. L. Liquefaction of saturated sands during cyclic loadings, *J. Soil Mech. Found. Div.*, ASCE, **92**, SM6, 105-134, 1966.
21. Valanis, K. C. and Read H. E. A new endochronic plasticity model for soils, in: *Soil Mechanics – Transient and Cyclic Loads* (eds. G. N. Pande and O. C. Zienkiewicz), John Wiley and Sons, 375–417, 1982.
22. Zienkiewicz, O. C., Chang, C. T. and Hinton, T. Nonlinear seismic response and liquefaction, *Int. J. Num. Analyt. Methods in Geomechanics*, **2**, 381–404, 1978.

Fluid Mechanics in Minkowski Space. Modelling of Fluid Motion in Porous Materials with Anisotropic Pore Space Structure

M. Cieszko

Department of Environmental Mechanics Pedagogical University of Bydgoszcz, 85-064 Bydgoszcz, Chodkiewicza 30

e-mail: mkk@rose.man.poznan.pl

Abstract. In the paper a fluid motion in a rigid porous medium of anisotropic pore space structure is described. Considerations are based on the new macroscopic model of saturated porous medium (Cieszko [4], [5]) in which the fluid flow through porous skeleton of anisotropic pore structure is described as a motion of the material continuum in the plane anisotropic metric space - Minkowski space - immersed in Euclidean one that is the model of the physical space. This model takes into account the fundamental fact for kinematics of fluid-saturated porous solid that pore space of permeable skeleton forms the real space for a fluid motion and its structure imposes restriction on that motion. In such approach the metric tensor of the Minkowski space is used to characterise the anisotropic structure of the skeleton pore space. It enabled one to determine the measures of any line, surface and volume elements in Minkowski and Euclidean spaces and to define the geometrical parameters characterising pore structure of porous materials: tortuosity, surface and volume porosity.

The mass and linear momentum balance equations for fluid are derived and the equation for wave propagation in barotropic inviscid fluid filling orthotropic space of pores is obtained. It is shown that the velocity of the plane wave in such a medium depends on the direction of wave propagation.

It worth to underline that presented description of fluid motion in the Minkowski space is a good starting point for modelling of mechanics of deformable porous solid saturated with fluid where the concept of deforming anisotropic space (Finsler space) as a model of pore space would have to be used.

Keywords: Saturated porous materials, anisotropic pore structure, anisotropic space, wave propagation

1. Notations and basic definitions

Vectors and tensors. We denote by $\mathcal{V}$ a three-dimensional real vector space and by $\mathcal{V}^*$ the dual vector space of $\mathcal{V}$. For $\mathbf{u} \in \mathcal{V}$ and $\mathbf{v}^* \in \mathcal{V}^*$ the scalar $\mathbf{u} \cdot \mathbf{v}^* = \mathbf{v}^* \cdot \mathbf{u} \in R$ is called the dual product of a vector $\mathbf{u}$ and covector $\mathbf{v}^*$ and by dot we denote the bilinear operation (dual multiplication) defined on $\mathcal{V} \times \mathcal{V}^*$. Tensors are elements of linear spaces formed by the tensor products of vector spaces, e.g., any tensor $\mathbf{A} \in \mathcal{V} \otimes \mathcal{V}^*$ is an endomorphism of $\mathcal{V}$ and $\mathcal{V}^*$; $\mathbf{A} \cdot \mathbf{u} \in \mathcal{V}$ and $\mathbf{v}^* \cdot \mathbf{A} \in \mathcal{V}^*$.

W. Ehlers (ed.), IUTAM Symposium on Theoretical and Numerical Methods in Continuum Mechanics of Porous Materials, 201–208.

Representations of surface and volume elements. Similarly as a vector is the algebraic model of a directed segment, skew-symmetric tensors of order two and three are natural models of surface and volume elements, respectively. These elements spanned by two ($\mathbf{u}$, $\mathbf{v}$) and three ($\mathbf{u}$, $\mathbf{v}$, $\mathbf{w}$) linearly independent vectors can be represented by the exterior products of these vectors; $\mathbf{u} \wedge \mathbf{v}$ and $\mathbf{u} \wedge \mathbf{v} \wedge \mathbf{w}$, [3]. The surface elements $\boldsymbol{\sigma}$ form the three-dimensional space $\wedge\mathcal{V}^2$ of skew-symmetric tensors of order two ($\boldsymbol{\sigma} \in \wedge\mathcal{V}^2 \subset \otimes\mathcal{V}^2$) and the volume elements $\boldsymbol{\vartheta}$ form the one-dimensional space $\wedge\mathcal{V}^3$ of skew-symmetric tensors of order three ($\boldsymbol{\vartheta} \in \wedge\mathcal{V}^3 \subset \otimes\mathcal{V}^3$).

Minkowski and Euclidean point spaces. The pair (P, $\mathcal{V}$) composed of a point P and of a vector space $\mathcal{V}$ will be identified with the affine point space. The point P is called the reference point and $\mathcal{V}$ is the space of position vectors of points of the affine space.

The affine space (P,$\mathcal{V}$) becomes a metric space if in the vector space $\mathcal{V}$ a norm is defined. The norm $u_A = L_A(\mathbf{u})$ define the length of vector $\mathbf{u}$ and in general case can be written in the form, [7],

$$L_A^2(\mathbf{u}) = \mathbf{u} \cdot \mathbf{M}_A(\mathbf{u}) \cdot \mathbf{u} \tag{1}$$

where the second order tensor $\mathbf{M}_A(\mathbf{u}) \in \otimes\mathcal{V}^{*2}$ satisfies condition

$$\mathbf{M}_A(k\mathbf{u}) = k\, \mathbf{M}_A(\mathbf{u}) \quad for \quad k > 0 \tag{2}$$

and is symmetric, non-singular and positive definite. The norm (1) enables one to measure a distance between any two points of space (P,$\mathcal{V}$). The tensor $\mathbf{M}_A(\mathbf{u})$ is called the metric tensor of the affine space, and the space equipped with such a general metric is called the Minkowski space (Rund [7], Thompson [9]). According to condition (2) the metric tensor $\mathbf{M}_A(\mathbf{u})$ depends on direction of a measured vector $\mathbf{u}$ and is independent of its length. This property of the metric $\mathbf{M}_A(\mathbf{u})$ determines the anisotropic properties of the Minkowski space.

The Minkowski space and its generalisation - Finsler space (the curved one; Rund [7], Matsumoto [6]) - are more and more often used in the modelling of various physical, biological and mechanical phenomena (Bejancu [2], Antonelli *et al.* [1], Saczuk [8]).

The Euclidean point space is the special case of the Minkowski space and is obtained when the metric tensor $\mathbf{M}_A(\mathbf{u})$ is independent of $\mathbf{u}$. The Euclidean metric tensor will be denoted by $\mathbf{M}$ and the length of vector $\mathbf{u}$ in this space by u.

The existence of a measure of distance in the Minkowski space enables one to define the metrics for elements of surface $\boldsymbol{\sigma} \in \wedge\mathcal{V}^2$ and volume $\boldsymbol{\vartheta} \in \wedge\mathcal{V}^3$. We obtain, [4],

$$S_A^2(\boldsymbol{\sigma}) = \boldsymbol{\sigma} : \mathbf{S}_A(\mathbf{r}) : \boldsymbol{\sigma}\,, \qquad V_A^2(\boldsymbol{\vartheta}) = \boldsymbol{\vartheta} \,\vdots\, \mathbf{V}_A \,\vdots\, \boldsymbol{\vartheta} \tag{3}$$

where

$$\mathbf{S}_A(\mathbf{r}) = \alpha(\mathbf{r})\frac{1}{2!}\begin{pmatrix}\mathbf{M}_A(\mathbf{r})\\ \mathbf{M}_A(\mathbf{r})\end{pmatrix}, \qquad \mathbf{V}_A = \alpha(\mathbf{r})\frac{1}{3!}\begin{pmatrix}\mathbf{M}_A(\mathbf{r})\\ \mathbf{M}_A(\mathbf{r})\\ \mathbf{M}_A(\mathbf{r})\end{pmatrix} \tag{4}$$

are the metric tensors in the spaces of surface elements $\wedge\mathcal{V}^2$ and volume elements $\wedge\mathcal{V}^3$, respectively. The tensor $\mathbf{S}_A(\mathbf{r})$ is of order four and $\mathbf{V}_A$ is of order six. The vector $\mathbf{r}$ in (4) is orthogonal to the surface element $\boldsymbol{\sigma}$ with respect to the metric $\mathbf{M}_A(\mathbf{r})$ and $\alpha(\mathbf{r}) = f_v/det(\mathbf{M}^{-1}\cdot\mathbf{M}_A)$, where f_v is a positive constant parameter.

The Euclidean measures S and V of a surface element $\boldsymbol{\sigma}$ and a volume element $\boldsymbol{\vartheta}$, respectively, are the special cases of expressions (3) and can be obtained taking in (4) $\mathbf{M}_A(\mathbf{r}) = \mathbf{M}$ and $f_v = 1$.

2. Minkowski space as a model of anisotropic pore space

We apply the notion of the Minkowski space as a model of an anisotropic pore space of permeable porous materials and to define parameters characterising the structure the pore space.

Taking into account the fact that real porous materials are immersed in the physical space modelled within the classical mechanics by the Euclidean point space, the Minkowski space as a model of the anisotropic pore space shall be considered as immersed in the Euclidean space. Such immersion enables one to measure any line, surface and volume elements both in Minkowski and Euclidean spaces and to define parameters characterising pore space structure of a porous medium.

For tortuosity of pores $\delta(\mathbf{u})$, surface porosity $\lambda(\boldsymbol{\sigma})$ and volume porosity f_v we have definitions, [4],

$$\delta^2(\mathbf{u}) = \left(\frac{u_A}{u}\right)^2 = \frac{\mathbf{u}\cdot\mathbf{M}_A(\mathbf{u})\cdot\mathbf{u}}{\mathbf{u}\cdot\mathbf{M}\cdot\mathbf{u}} = \mathbf{n}\cdot\mathbf{M}_A(\mathbf{n})\cdot\mathbf{n} \quad \geq 1\,, \tag{5}$$

$$\lambda^2(\boldsymbol{\sigma}) = \left(\frac{S_A}{S}\right)^2 = \frac{\boldsymbol{\sigma}:\mathbf{S}_A(\mathbf{r}):\boldsymbol{\sigma}}{\boldsymbol{\sigma}:\mathbf{S}:\boldsymbol{\sigma}} = \mathbf{E}_2:\mathbf{S}_A(\mathbf{r}):\mathbf{E}_2 \quad \leq 1\,, \tag{6}$$

$$f_v^2 = \left(\frac{V_A}{V}\right)^2 = \frac{\boldsymbol{\vartheta}\,\vdots\,\mathbf{V}_A\,\vdots\,\boldsymbol{\vartheta}}{\boldsymbol{\vartheta}\,\vdots\,\mathbf{V}\,\vdots\,\boldsymbol{\vartheta}} = \mathbf{E}_3\,\vdots\,\mathbf{V}_A\,\vdots\,\mathbf{E}_3 \quad \leq 1 \tag{7}$$

where $\mathbf{n}$, $\mathbf{E}_2$ and $\mathbf{E}_3$ are the Euclidean unit vector, unit surface and volume element, respectively.

3. Balance of mass

Partial and effective mass densities. We consider a domain D of the physical space occupied by a fluid filling porous medium enclosed by the surface ∂D. For any infinitesimal volume element $\boldsymbol{\vartheta} \in \wedge\mathcal{V}^3$ of the domain D we prescribe the mass of fluid dm given by the function

$$dm = m(\mathbf{x}, t, d\boldsymbol{\vartheta}) \; , \tag{8}$$

that should satisfy the homogeneity condition

$$m(\mathbf{x}, t, k\; d\boldsymbol{\vartheta}) = |k|\; m(\mathbf{x}, t, d\boldsymbol{\vartheta}) \qquad for \quad k \neq 0 \; . \tag{9}$$

This property assure that the mass dm contained in the infinitesimal volume element $d\boldsymbol{\vartheta}$ increases proportionally to the extension of the element $d\boldsymbol{\vartheta}$ and does not depend on its orientation.

(8) can be represented in the form

$$dm = \mathbf{R}(\mathbf{x}, t, d\boldsymbol{\vartheta}) \,\vdots\, d\boldsymbol{\vartheta} \tag{10}$$

where

$$\mathbf{R} = \frac{\partial m}{\partial (d\boldsymbol{\vartheta})} \quad \in \wedge\mathcal{V}^{*3} \quad and \quad \mathbf{R}(\mathbf{x}, t, k\; d\boldsymbol{\vartheta}) = \frac{|k|}{k}\, \mathbf{R}(\mathbf{x}, t, d\boldsymbol{\vartheta})$$

is the skew-symmetric tensor of order three and depends only on the orientation of the element $d\boldsymbol{\vartheta}$. Therefore the tensor $\mathbf{R}$ can be interpreted as the mass density tensor of fluid. Applying representations

$$d\boldsymbol{\vartheta} = dV\mathbf{E}_3 = dV_A \mathbf{E}_3^A \; , \tag{11}$$

from (10) we obtain

$$dm = \bar{\varrho}\; dV = \varrho\; dV_A \tag{12}$$

where

$$\bar{\varrho} = \mathbf{R} \,\vdots\, \mathbf{E}_3 \; , \qquad \varrho = \mathbf{R} \,\vdots\, \mathbf{E}_3^A$$

are the partial and effective fluid mass densities, respectively, and $\mathbf{E}_3^A$ is the Minkowski unit volume element ($\mathbf{E}_3^A \,\vdots\, \mathbf{M}_A \,\vdots\, \mathbf{E}_3^A = 1$). Due to the definition (7) both densities are related by: $\bar{\varrho} = f_v \varrho$.

Mass flux of fluid. We consider an arbitrary infinitesimal surface element $d\boldsymbol{\sigma} \in \wedge\mathcal{V}^2$ of the surface ∂D and assume that the mass flux of fluid $d\varphi$ flowing through the element $d\boldsymbol{\sigma}$ is continuous and homogeneous function of this element. We have

$$d\varphi = \varphi(\mathbf{x}, t, d\boldsymbol{\sigma}) \tag{13}$$

and

$$\varphi(\mathbf{x}, t, k\, d\boldsymbol{\sigma}) = k\, \varphi(\mathbf{x}, t, d\boldsymbol{\sigma}) \qquad for \quad k \neq 0\,. \tag{14}$$

Applying (14) one can prove the tetrahedron theorem. Therefore relation (13) may be written in the form

$$d\varphi = \boldsymbol{\Phi}(\mathbf{x}, t) : d\boldsymbol{\sigma} \tag{15}$$

where $\boldsymbol{\Phi} \in \wedge\mathcal{V}^{*2}$ is the tensor of the mass flux density of fluid. This density tensor is independent of any metric.
Representing the surface element $d\boldsymbol{\sigma}$ by the Euclidean and Minkowski unit vectors $\mathbf{n}$ and $\mathbf{N}$, orthogonal to this element,

$$d\boldsymbol{\sigma} = \mathbf{E}_3 \cdot \mathbf{M} \cdot \mathbf{n}\, dS = \mathbf{E}_3^A \cdot \mathbf{M}_A(\mathbf{N}) \cdot \mathbf{N}\, dS_A\,, \tag{16}$$

from (15) we obtain

$$d\varphi = \bar{\mathbf{q}} \cdot \mathbf{M} \cdot \mathbf{n}\, dS = \mathbf{q} \cdot \mathbf{M}_A(\mathbf{N}) \cdot \mathbf{N}\, dS_A \tag{17}$$

where dS, dS_A are the Euclidean and Minkowski measures of $d\boldsymbol{\sigma}$, respectively, and

$$\bar{\mathbf{q}} = \boldsymbol{\Phi} : \mathbf{E}_3\,, \qquad \mathbf{q} = \boldsymbol{\Phi} : \mathbf{E}_3^A \tag{18}$$

are the partial and effective vectors of mass flux densities of fluid. They represent the mass fluxes flowing through the unit surface of medium and pores, respectively. Due to the relation; $\mathbf{E}_3 = f_v \mathbf{E}_3^A$, we have; $\bar{\mathbf{q}} = f_v \mathbf{q}$. The vectors $\bar{\mathbf{q}}$ and $\mathbf{q}$ of mass flux and densities $\bar{\varrho}$ and ϱ enable one to define the velocity field of fluid mass transport in the pore space

$$\mathbf{v}(\mathbf{x}, t) = \bar{\mathbf{q}}/\bar{\varrho} = \mathbf{q}/\varrho\,. \tag{19}$$

Balance equation for mass of fluid. Applying representations for mass dm and mass flux $d\varphi$, e.g. $(12)_2$ and $(17)_2$, the integral balance equation of mass of fluid for the domain D can be written in the form

$$\frac{d}{dt} \int\limits_D \varrho\, dV_A + \int\limits_{\partial D} \mathbf{q} \cdot \mathbf{M}_A(\mathbf{N}) \cdot \mathbf{N}\, dS_A = 0\,. \tag{20}$$

Due to the Gauss-Ostrogradski theorem its local form is

$$\frac{\partial \varrho}{\partial t} + \boldsymbol{\nabla} \cdot \mathbf{q} = 0 \tag{21}$$

where $\boldsymbol{\nabla}$ denotes the Nabla operator. From (21) results that the vector form of the balance equation for mass of fluid does not depend explicitly on the metric of the space in which the fluid motion takes place.

4. Balance of linear momentum

Linear momentum and flux of liner momentum of fluid. For the linear momentum $d\mathbf{p}$ of fluid contained in an infinitesimal volume element $d\vartheta$ and for the flux of linear momentum $d\psi$ connected with the fluid mass transport through the surface element $d\boldsymbol{\sigma}$ we take

$$d\mathbf{p} = \mathbf{v}(\mathbf{x}, t)\ dm\ , \qquad d\psi = \mathbf{v}(\mathbf{x}, t)\ d\varphi \tag{22}$$

where dm and $d\varphi$ are given by (10) (or by (12)) and by (15) (or by (17)), respectively.

Surface forces in fluid. We assume that the surface force $d\mathbf{f}$ acting in a fluid filling porous medium on the infinitesimal surface element $d\boldsymbol{\sigma}$ is continues and homogeneous function of this element. We have

$$d\mathbf{f} = \mathbf{f}(\mathbf{x}, t, d\boldsymbol{\sigma}) \tag{23}$$

and

$$\mathbf{f}(\mathbf{x}, t, k\ d\boldsymbol{\sigma}) = k\ \mathbf{f}(\mathbf{x}, t, d\boldsymbol{\sigma}) \quad for \;\; k \neq 0\ . \tag{24}$$

Taking into account the tetrahedron theorem, the expression (23) can be written in the form

$$d\mathbf{f} = \mathbf{T}'(\mathbf{x}, t) : d\boldsymbol{\sigma} \tag{25}$$

where $\mathbf{T}'(\mathbf{x}, t) \in \mathcal{V} \otimes \mathcal{V}^{*2}$ is the third order tensor and may be interpreted as the stress tensor in a fluid filling pore space of the skeleton. Relation (25) does not depend explicitly on the metric of space occupied by fluid.

In order to obtain representations for surface forces connected with the Euclidean and the Minkowski measures of the surface element $d\boldsymbol{\sigma}$, we apply relations (16). Then

$$d\mathbf{f} = \bar{\mathbf{t}}\ dS = \mathbf{t}\ dS_A \tag{26}$$

where

$$\bar{\mathbf{t}} = \bar{\mathbf{T}} \cdot \mathbf{M} \cdot \mathbf{n}\ , \qquad \mathbf{t} = \mathbf{T} \cdot \mathbf{M}_A(\mathbf{N}) \cdot \mathbf{N} \tag{27}$$

and

$$\bar{\mathbf{T}} = \mathbf{T}' : \mathbf{E}_3\ , \qquad \mathbf{T} = \mathbf{T}' : \mathbf{E}_3^A \qquad \in \otimes\mathcal{V}^2 \tag{28}$$

Quantities $\bar{\mathbf{t}}$ and $\mathbf{t}$ are the partial and effective stress vectors in fluid per unit surface of porous medium and pores, respectively, whereas $\bar{\mathbf{T}}$ and $\mathbf{T}$ are the partial and the effective stress tensors.
The relations (27) represent the modified form of the Cauchy theorem in which the metric tensors are explicitly present.

From relations (26), (28) and definitions (6), (7) results that

$$\bar{\mathbf{t}} = \lambda(\sigma)\,\mathbf{t}\,, \qquad \bar{\mathbf{T}} = f_v\,\mathbf{T}\,. \tag{29}$$

Balance equation for linear momentum of fluid. Having expressions (22) and (26) for: linear momentum, flux of momentum and surface forces one can formulate the integral balance equation of momentum for an inviscid fluid filling anisotropic pore space of porous medium. One of the alternative forms of this equation is

$$\frac{d}{dt}\int\limits_{D} \mathbf{q}\,dV_A = \int\limits_{\partial D}\left[-\frac{1}{\varrho}(\mathbf{q}\otimes\mathbf{q}) + \mathbf{T}\right]\cdot\mathbf{M}_A(\mathbf{N})\cdot\mathbf{N}\,dS_A\,. \tag{30}$$

After application of the Gauss-Ostrogradski theorem we obtain

$$\frac{\partial\mathbf{q}}{\partial t} = -\nabla\cdot\left[\frac{1}{\varrho}(\mathbf{q}\otimes\mathbf{q})\right] + \nabla\cdot(\mathbf{T})\,. \tag{31}$$

The local balance equation of momentum (31), similarly as the mass balance equation (21), does not depend explicitly on the metric of space in which the fluid motion takes place.

5. Influence of pore structure on wave propagation in fluid

In order to exemplify the influence of the anisotropic pore space structure of the skeleton on mechanical behaviour of fluid filling pores, we will consider a wave propagation in such a medium. We assume that the fluid is inviscid and barotropic and the skeleton pore structure is orthotropic characterised by the constant metric tensor $\mathbf{M}_A$. In that case in each point of the pore space the value of the stress vector $\mathbf{t}$ describing the internal forces in the fluid per unit area of pore surface contained in any plane is equal to the fluid pore pressure. Therefore we have

$$\mathbf{t} = -p(\varrho)\,\mathbf{N}\,. \tag{32}$$

Hence, due to the relation $(27)_2$ and arbitrary choice of $\mathbf{N}$ the constitutive relation for the stress tensor $\mathbf{T}$ can be written in the form

$$\mathbf{T} = -p(\varrho)\,\mathbf{M}_A^{-1}\,. \tag{33}$$

Equations (21) and (31) together with relation (33) describe the dynamical behaviour of the barotropic fluid in the orthotropic pore space of rigid skeleton. In the linear case from (21), (31) and (33) we obtain the following equation for small amplitude waves

$$\frac{\partial^2 p}{\partial t^2} = a_o^2 \, \nabla \cdot \mathbf{M}_A^{-1} \cdot \nabla \, (p) \tag{34}$$

where $a_o = \sqrt{dp/d\varrho|_o}$ is the velocity of wave propagation in the bulk fluid.

Equation (34) written in the Cartesian coordinate system related to the principal directions of the metric tensor $\mathbf{M}_A$ takes the form

$$\frac{\partial^2 p}{\partial t^2} = c_1^2 \, \frac{\partial^2 p}{\partial x_1^2} + c_2^2 \, \frac{\partial^2 p}{\partial x_2^2} + c_3^2 \, \frac{\partial^2 p}{\partial x_3^2} \tag{35}$$

where $c_i = a_o/\delta_i$ $(i = 1, 2, 3)$ is the velocity of a plane wave propagating in the principal directions of the orthotropic pore space and δ_i $(\delta_i \geq 1)$ is the tortuosity parameter of pores in the i-th principal direction.

From the above relation results that the velocity of wave propagation in fluid filling anisotropic pore space depends on the direction of wave propagation and is inversely proportional to the tortuosity of pores for this direction. Taking into account that the tortuosity is no less than one, the wave velocity is always no greater than the velocity in the bulk fluid.

References

1. Antonelli, P. L., Ingarden, R. S. and Matsumoto, M. Theory of Spays and Finsler Spaces with Application in Physics and Biology. *Kluver Academic Publishers*, 1993.
2. Bejancu, A. Finsler Geometry and Applications. *Ellis Horwood*, 1990.
3. Bowen, M. and Wang, C. C. Introduction to Vectors and Tensors. *Plenum Press*, 1976.
4. Cieszko, M. Minkowski Space as a Model of Anisotropic Pore Space of Permeable Materials. Modelling of the Skeleton Pore Structure. *In Proccedings of the International Symposium on Trends in Continuum Physics*, World Scientific, 1998.
5. Cieszko, M. Application of Minkowski Space to Description of Anisotropic Pore Space Structure in Porous Materials. *ZAMM*, in press.
6. Matsumoto, M. Foundations of Finsler Geometry and Spatial Finsler Spaces. *Kaiseisha Press*, 1986.
7. Rund, H. The Differential Geometry of Finsler Spaces. *Springer-Verlag*, 1959.
8. Saczuk, J. Finslerian Foundations of Solid Mechanics. *Reports of Institute of Fluid-Flow Machinery*, 472/1427, 1996.
9. Thompson, A. C. Minkowski Geometry. *Cambridge University Press*, 1996.

Modelling of Soils by Use of the Theory of Porous Media

W. Ehlers and P. Blome
Institute of Applied Mechanics (Civil Engineering),
University of Stuttgart, D-70550 Stuttgart, Germany
e-mail: (Ehlers, Blome)@mechbau.uni-stuttgart.de

Abstract. The continuum mechanical description of cohesive and non-cohesive soil materials is one of the main topics in geomechanical engineering. In the present contribution, saturated and unsaturated soils are investigated within the framework of the Theory of Porous Media using a standard binary model of a materially incompressible solid skeleton and a materially compressible or incompressible pore-fluid. The soil skeleton is assumed as a cohesive-frictional elasto-plastic material, whereas the viscous pore-fluid can be water, air or a water-air mixture. Furthermore, the fluid flow is governed by *Darcy*'s filter law including a deformation dependent permeability. Finally, a numerical example of a strongly coupled solid-fluid-problem is carried out by the finite element method.

Keywords: Cohesive-frictional soils, compressible pore-fluid mixtures, soil elasto-plasticity, pore-fluid viscosity

1. Introduction

In the present contribution, frictional materials, as e. g. soils, are considered within the well-founded framework of the Theory of Porous Media (TPM), compare e. g. the work by Bowen ([2], [3]), de Boer and Ehlers ([1]) and Ehlers ([4], [5]). Embedded in this concept, the model under consideration represents a binary medium consisting of a materially incompressible solid skeleton and either a materially compressible or a materially incompressible pore-fluid. In case of the incompressible pore-fluid, the fluid represents a pure pore-water, whereas, in the materially compressible case, the fluid can be a water-air mixture or a pure pore-gas. In any case, if the pore-fluid is a water-air mixture, the fluid constituents are physically bound and are thus exhibiting the same motion. Concerning the solid skeleton, a general elasto-plasticity approach is presented within the geometrically linear range. In particular, to describe granular soils like, e. g., sand, the physically non-linear soil elasticity law by Ehlers and Müllerschön [8] is taken into consideration, which is appropriate to describe the differences of the unload-reload loops at different stress levels, both in the axial and in the hydrostatic regime. In the plastic domain, use is made of the single-surface yield function by Ehlers ([4]), which is extended towards isotropic hardening materials. Finally, based on the above soil model, the volumetrically strongly coupled consolidation problem is computed

W. Ehlers (ed.), IUTAM Symposium on Theoretical and Numerical Methods in Continuum Mechanics of Porous Materials, 209–214.

by the finite element method, thus exhibiting the different consolidation behaviour in case of materially incompressible and materially compressible pore-fluids.

2. The soil model

The soil model is composed by the phases φ^α. These phases are φ^S, the soil skeleton, and φ^F, the pore-fluid, which is considered as a mixture of φ^L (pore-liquid, here water) and φ^G (pore-gas, here air). Furthermore, the saturation constraint holds. Thus,

$$n^S + n^F = 1 \qquad \text{with} \qquad n^F = n^L + n^G \,, \tag{1}$$

where n^α are the volume fractions of the constituents.

It is assumed that a decomposition of the liquid-gas mixture is excluded, i. e. φ^L and φ^G follow the same motion function and, therefore, exhibit the same velocity $\mathbf{v}_F := \mathbf{v}_L = \mathbf{v}_G$.

The primary variables of the model are the solid displacement $\mathbf{u}_S$ and the excess pore-fluid pressure Δp exceeding the air pressure p_0. For their numerical determination, the sum of the partial linear momentum balances and of the partial mass balances of the constituents are used. The corresponding weak forms are [7]

$$\begin{aligned} &\int_\Omega \operatorname{grad}\delta\mathbf{u}_S \cdot (\mathbf{T}^S + \mathbf{T}^F)\,dv \,- \\ &\qquad - \int_\Omega \delta\mathbf{u}_S \cdot (\rho^S + \rho^F)\,\mathbf{b}\,dv \;=\; \int_\Gamma \delta\mathbf{u}_S \cdot \bar{\mathbf{t}}\,da \,, \end{aligned} \tag{2}$$

$$\begin{aligned} &\int_\Omega \delta(\Delta p)\,\big[\,n^F (\rho^{FR})'_S + \rho^{FR}\operatorname{div}(\mathbf{u}_S)'_S\,\big]\,dv \,- \\ &\qquad - \int_\Omega \operatorname{grad}\delta(\Delta p)\cdot\rho^F\,\mathbf{w}_F\,dv \;=\; -\int_\Gamma \delta(\Delta p)\,\bar{q}\,da \,. \end{aligned} \tag{3}$$

Therein, $\mathbf{T}^\alpha$ is the partial *Cauchy* stress tensor related to each φ^α, whereas $\rho^\alpha = n^\alpha\,\rho^{\alpha R}$ is the partial mass density given by the effective (true) mass density $\rho^{\alpha R}$ weighted by the appropriate volume fraction n^α. The symbol $(\,\cdot\,)'_\alpha$ characterizes the material time derivative following the motion of φ^α. Furthermore, $\mathbf{b}$ denotes the volume force per unit mass (gravity) and $\mathbf{w}_F$ represents the seepage velocity. Finally, $\delta\mathbf{u}_S$ and $\delta(\Delta p)$ are test functions corresponding to the solid displacement and the excess pore pressure, respectively.

On the surface Γ of a considered domain Ω, there may act an external stress vector $\bar{\mathbf{t}} = (\mathbf{T}^S + \mathbf{T}^F)\,\mathbf{n}$ or a fluid mass efflux $\bar{q} = n^F \rho^{FR}\,\mathbf{w}_F \cdot \mathbf{n}$, where $\mathbf{n}$ is the outward oriented unit surface normal on Γ.

According to the effective stress principle, the partial stress tensors in (2) are replaced by

$$\mathbf{T}^S = -\,n^S\,\Delta p\,\mathbf{I} + \mathbf{T}^S_E \qquad \text{and} \qquad \mathbf{T}^F = -\,n^F\,\Delta p\,\mathbf{I}\,, \tag{4}$$

where $\mathbf{T}^S_E$ denotes the solid extra stress. Furthermore, the filter velocity $n^F \mathbf{w}_F$ in (3) is replaced by *Darcy*'s general filter law. The dependence of the filter law on the soil deformation and, as a result, of the induced anisotropy can be additionally taken into consideration [9].

3. The elasto-plastic soil skeleton

The extra part $\mathbf{T}^S_E$ of the solid stress is described in the framework of a geometrically linear theory by a physically non-linear elasticity law that takes into account the complex material behaviour of sand:

$$\mathbf{T}^S_E = 2\,\mu^S(\varepsilon_{vp})\,\boldsymbol{\varepsilon}^D_{Se} + k^S(\varepsilon_{ve};\,\varepsilon_{vp},\,\varepsilon_{v\,\min})\,(\boldsymbol{\varepsilon}_{Se}\cdot\mathbf{I})\,\mathbf{I}\,. \tag{5}$$

Therein, μ^S and k^S are not the usual elastic constants. In particular, k^S is a function of the elastic volume strain ε_{ve} at a given state of plastic deformations. Thus, the plastic volume strain ε_{vp} and the maximal achievable volume contractancy $\varepsilon_{v\,\min}$ must be understood as parameters of the elastic process [8].

For the description of the plastic material behaviour of sand, the model proceeds from a single surface yield criterion, isotropic work-hardening conditions and a non-associated flow rule [4, 8]:

$$F = \sqrt{\mathrm{II}_D\,(1+\gamma\,\mathrm{III}_D/(\mathrm{II}_D)^{3/2})^m + \tfrac{1}{2}\,\alpha\,\mathrm{I}^2 + \delta^2\,\mathrm{I}^4} + \beta\,\mathrm{I} + \epsilon\,\mathrm{I}^2 - \kappa\,, \tag{6}$$

$$G = \sqrt{\psi_1\,\mathrm{II}_D + \tfrac{1}{2}\,\alpha\,\mathrm{I}^2 + \delta^2\,\mathrm{I}^4} + \psi_2\,\mathrm{I} + \epsilon\,\mathrm{I}^2\,, \tag{7}$$

$$(\boldsymbol{\varepsilon}_{Sp})'_S = \Lambda\,\partial G/\partial\mathbf{T}^S_E\,. \tag{8}$$

The yield criterion, $F = 0$, is described by the first invariant I and the second and third deviatoric invariants II^D and III^D of $\mathbf{T}^S_E$ together with a set of seven material parameters. The subset (α, β, δ, ϵ, κ) refers to the shape of the yield curve in the hydrostatic plane of the principal stress space, whereas the subset (γ, m) refers to the deviatoric plane. For the consideration of work-hardening conditions, the parameters are developed as functions of the plastic work.

In the framework of the non-associated plasticity concept, an additional plastic potential (7) is used, where the parameters ψ_1 and ψ_2 govern the plastic dilatation angle. Consequently, the flow rule (8) is given with respect to G, where Λ is the plastic multiplier.

4. The compressible pore-fluid mixture

The pore-fluid is modelled as a fluid mixture composed of incompressible water with dispersed, compressible air. From the material incompressibility of water, it follows that a complete compression of the dispersed air means that there is no further possibility to compress the pore-fluid, i. e. the pore-fluid reaches its so-called "point of compression" and therewith its maximal mass density $\rho^{FR}_{\max}$. The density function ρ^{FR} (Figure 1) of the pore-fluid mixture can be given within the TPM as a function of Δp [6]:

$$\rho^{FR}(\Delta p) = \rho^{LR} \frac{1 + C^F_{0S}}{\rho^{LR}/\rho^{GR} + C^F_{0S}} \, . \tag{9}$$

Therein, C^F_{0S} is an integration constant specifying the composition and thus the characteristics of the pore-fluid for a certain initial state via $C^F_{0S} = n^L_{0S}\, \rho^{LR}/(n^G_{0S}\, \rho^{GR}_0(p_0))$. Furthermore, the effective pore-water density is constant ($\rho^{LR} = \text{const.}$), while the effective pore-gas density ρ^{GR} is governed by *Boyle*'s law (ideal gas law). In the present considerations, the temperature of each constituent of the overall model is assumed to be equal and constant (isothermal conditions).

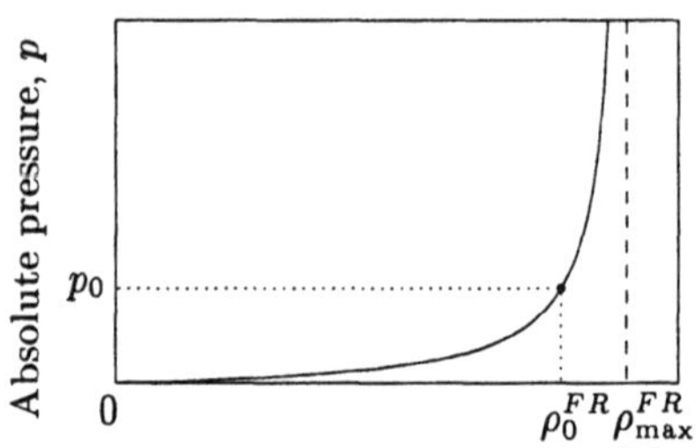

Figure 1. Pressure-density function (9) of the pore-fluid.

The approach (9) automatically includes the limit states of a purely compressible ($n^L = 0 \leadsto C^F_{0S} = 0$) and a purely incompressible ($n^G = 0 \leadsto C^F_{0S} \to \infty$) fluid as well as the limit states of the pressure p:

$$\begin{aligned} \lim_{C^F_{0S}\to 0} \rho^{FR} &= \rho^{GR} \quad \text{and} \quad \lim_{C^F_{0S}\to \infty} \rho^{FR} = \rho^{LR} \, , \\ \lim_{p\to 0} \rho^{FR} &= 0 \quad \text{and} \quad \lim_{p\to \infty} \rho^{FR} = \rho^{FR}_{\max} \, . \end{aligned} \tag{10}$$

Herein, $\rho^{FR}_{\max}$, as the maximal mass density of the pore-fluid mixture, is obtained under an infinite pressure to yield $\rho^{FR}_{\max} := \rho^{LR}\,(1+C^F_{0S})/C^F_{0S}$.

In the limit case of an incompressible pore-fluid ($n^G = 0$), the pore pressure p degenerates to a *Lagrange*an multiplier governed by the boundary conditions of the problem under study [4].

5. Example: the fully coupled consolidation problem

The numerical implementation of the model leads to a differential-algebraic equation (DAE) of first order in time [10]. The solution of this DAE determines the solid deformation $\mathbf{u}_S$, the excess pore-fluid pressure Δp and therewith the seepage velocity $\mathbf{w}_F$.

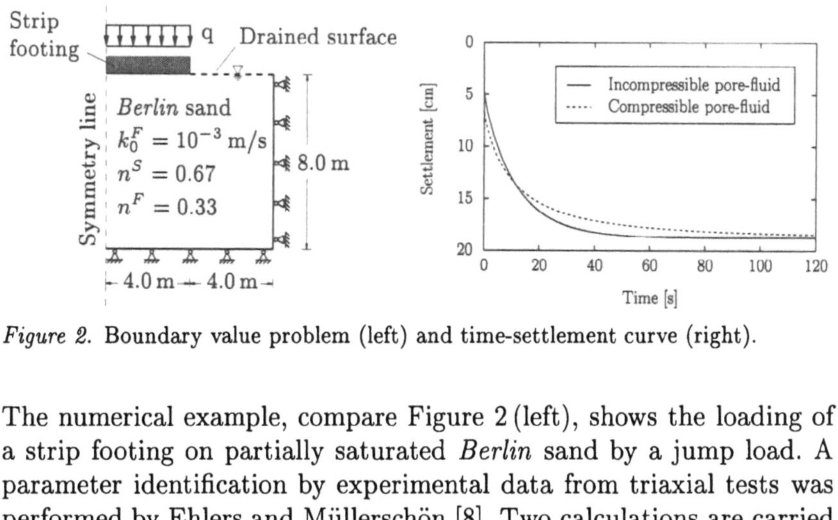

Figure 2. Boundary value problem (left) and time-settlement curve (right).

The numerical example, compare Figure 2 (left), shows the loading of a strip footing on partially saturated *Berlin* sand by a jump load. A parameter identification by experimental data from triaxial tests was performed by Ehlers and Müllerschön [8]. Two calculations are carried out, one for an incompressible and another one for a compressible pore-fluid (air content $n^G_{0S} = 0,05\,n^F_{0S}$). In Figure 2 (right), the time-settlement curves are shown, where the vertical deformation is directly measured at the symmetry line beneath the load. Prior to the beginning of the actual consolidation process, an initial settlement of the footing occurs. In case of the incompressible pore-fluid, the initial settlement is due to the deformation of the underground that maintains a constant volume. In case of the compressible pore-fluid, the instantaneous settlement additionally results from the compressibility of the dispersed air. However, in the example given above, both curves reach the same final settlement value.

6. Conclusion

The above setting of a soil model leads to a realistic description of the soil behaviour under external loads. Especially, the consideration of a compressible fluid-phase is of great importance, since only a very small amount of air in the pore-fluid causes unsaturated (partially saturated) conditions, where the compressiblity of the pore-fluid increases dramatically.

Acknowledgements

The presented paper is related to the research group "Soil-structure Interaction" at the Technical University of Darmstadt, Germany, supported by the Deutsche Forschungsgemeinschaft (DFG).

References

1. de Boer, R. and Ehlers, W. *Theorie der Mehrkomponentenkontinua mit Anwendung auf bodenmechanische Probleme*, Teil I. Forschungsberichte aus dem Fachbereich Bauwesen, Heft 40, Universität-GH-Essen, 1986.
2. Bowen, R. M. Incompressible porous media models by use of the theory of mixtures. *Int. J. Engng. Sci.*, **18**, 1129–1148, 1980.
3. Bowen, R. M. Compressible porous media models by use of the theory of mixtures. *Int. J. Engng. Sci.*, **20**, 697–735, 1982.
4. Ehlers, W. Constitutive Equations for Granular Materials in Geomechanical Context. In K. Hutter, editor, *Continuum Mechanics in Environmental Sciences and Geophysics*, CISM Courses and Lectures No. 337, 313–402, Springer-Verlag, Wien, 1993.
5. Ehlers, W. Grundlegende Konzepte in der Theorie Poröser Medien. *Technische Mechanik*, **16**, 63–76, 1996.
6. Ehlers, W. and Blome, P. On Porous Soil Materials Saturated with a Compressible Pore-fluid Mixture. *ZAMM*, **80**, Suppl.1, 141–144, 2000.
7. Ehlers, W., Ellsiepen, P., Blome, P., Mahnkopf, D. and Markert, B. *Theoretische und numerische Studien zur Lösung von Rand- und Anfangswertproblemen in der Theorie Poröser Medien.* Bericht aus dem Institut für Mechanik (Bauwesen), No. 99-II-1, Universität Stuttgart, 1999.
8. Ehlers, W. and Müllerschön, H. Parameter Identification of a Macroscopic Granular Soil Model Applied to Dense Berlin Sand. *Granular Matter*, **2**, 105–112, 2000.
9. Eipper, G. *Theorie und Numerik finiter elastischer Deformationen in fluidgesättigten porösen Festkörpern.* Dissertation, Bericht aus dem Institut für Mechanik (Bauwesen), No. II-1, 36–39, Universität Stuttgart, 1998.
10. Ellsiepen, P. *Zeit- und ortsadaptive Verfahren angewandt auf Mehrphasenprobleme poröser Medien.* Dissertation, Bericht aus dem Institut für Mechanik (Bauwesen), No. II-3, Universität Stuttgart, 1999.

Integration and Calibration of a Plasticity Model for Granular Materials

L. Jacobsson and K. Runesson
Department of Solid Mechanics, Chalmers University of Technology, SE-412 96 Göteborg, Sweden

Abstract. A new macroscopic plasticity model for non-cohesive granular materials, with the focus on coarse-sized materials (railway ballast), is presented. The corresponding incremental relations are reformulated in the space of stress invariants that is extended with the internal (hardening) variables and the corresponding dissipative stresses. The model is calibrated using data from Conventional Triaxial Compression (CTC) tests. A function evaluation method is used for the optimization. A "multi-vector" strategy for choosing the appropriate start vector is proposed.

1. Introduction

A wealth of literature has been devoted to the development of plasticity-based continuum models for granular materials. Among the relevant literature we note [7], [6], [3], [4].

The model proposed here is formulated within a unifying thermodynamic framework and is calibrated by data from CTC-tests performed on down-scaled Swedish railway ballast. The prototype material is fine-grained crushed granite with a linear gradation of particle size in the range 32–64 mm, which was down-scaled by a factor of eight. By this scaling it was possible to use a standard triaxial cell device. The tests were carried out at the University of Colorado at Boulder in an experimental program, cf. [2]. The model is restricted to monotonic loading in its present form. This paper represents a condensed version of [1].

2. Constitutive relations

2.1. Thermodynamic basis

The thermodynamic state, which is restricted to the isothermal situation, is assumed to be determined by the strain $\boldsymbol{\epsilon}$ and the following internal variables: The plastic strain $\boldsymbol{\epsilon}^{\mathrm{p}}$, the equivalent relative density $\bar{\rho}_{\mathrm{e}}$ (representing volumetric as well as deviatoric hardening) and the frictional hardening variable κ (representing deviatoric hardening).

W. Ehlers (ed.), IUTAM Symposium on Theoretical and Numerical Methods in Continuum Mechanics of Porous Materials, 215–220.

Introducing the elastic strain $\boldsymbol{\epsilon}^{\mathrm{e}} = \boldsymbol{\epsilon} - \boldsymbol{\epsilon}^{\mathrm{p}}$, we propose the following expression of the free energy (per unit bulk volume):

$$\Psi(\boldsymbol{\epsilon}^{\mathrm{e}}, \kappa, \bar{\rho}_{\mathrm{e}}) = c^{\mathrm{e}}(\bar{\rho}_{\mathrm{e}})\,\Psi_0^{\mathrm{e}}(\boldsymbol{\epsilon}^{\mathrm{e}}) + c^{\mathrm{p}}(\bar{\rho}_{\mathrm{e}})\,\Psi_0^{\mathrm{p}}(\kappa) \tag{1}$$

where $\Psi_0^{\mathrm{e}} = \frac{1}{2}H_0\boldsymbol{\epsilon}^{\mathrm{e}} : \boldsymbol{\mathcal{E}}_0^{\mathrm{e}} : \boldsymbol{\epsilon}^{\mathrm{e}}$ and $\Psi_0^{\mathrm{p}} = \frac{1}{2}H_0\kappa^2$ are the elastic and plastic parts of the free energy at the initial (reference) state when $\bar{\rho}_{\mathrm{e}} = \bar{\rho}_0$.

The function $c^{\mathrm{e}}(\bar{\rho}_{\mathrm{e}})$, belonging to the family $c^i(x) = (\frac{1-\bar{\rho}_0}{\bar{\rho}_0})^{m_i}(\frac{x}{1-x})^{m_i}$, $i \in \{\mathrm{e}, \mathrm{p}, \mathrm{v}\}$, determines the change of the elastic response due to hardening, and it thus represents elastic-plastic coupling (which is a non-classical feature). Furthermore, $\boldsymbol{\mathcal{E}}_0^{\mathrm{e}}$ is the initial elastic stiffness tensor and H_0 is the initial value of the friction hardening modulus.

From the Clausius-Duhem-Inequality (CDI), we obtain the constitutive relation for the stress $\boldsymbol{\sigma}$ as follows:

$$\boldsymbol{\sigma} = \frac{\partial \Psi}{\partial \boldsymbol{\epsilon}^{\mathrm{e}}}(\boldsymbol{\epsilon}^{\mathrm{e}}, \bar{\rho}_{\mathrm{e}}) = \boldsymbol{\mathcal{E}}^{\mathrm{e}}(\bar{\rho}_{\mathrm{e}}) : \boldsymbol{\epsilon}^{\mathrm{e}} \quad \text{with} \quad \boldsymbol{\mathcal{E}}^{\mathrm{e}}(\bar{\rho}_{\mathrm{e}}) = c^{\mathrm{e}}(\bar{\rho}_{\mathrm{e}})\boldsymbol{\mathcal{E}}_0^{\mathrm{e}} \tag{2}$$

The (reduced) dissipation inequality then becomes $D = \boldsymbol{\sigma} : \dot{\boldsymbol{\epsilon}}^{\mathrm{p}} + R_{\mathrm{e}}\dot{\bar{\rho}}_{\mathrm{e}} + K\dot{\kappa} \geq 0$, where the dissipative stresses R_{e} and K are given as

$$R_{\mathrm{e}} \stackrel{\mathrm{def}}{=} -\frac{\partial \Psi}{\partial \bar{\rho}_{\mathrm{e}}} = -\left[(c^{\mathrm{e}})'\,\Psi_0^{\mathrm{e}} + (c^{\mathrm{p}})'\,\Psi_0^{\mathrm{p}}\right], \quad K \stackrel{\mathrm{def}}{=} -\frac{\partial \Psi}{\partial \kappa} = -H(\bar{\rho}_{\mathrm{e}})\kappa \tag{3}$$

with the friction hardening modulus defined as $H(\bar{\rho}_{\mathrm{e}}) = c^{\mathrm{p}}(\bar{\rho}_{\mathrm{e}})H_0$.

2.2. Plastic yielding, flow and hardening

Guided by the assessment of the experimental results regarding volumetric strain versus deviatoric strain, we introduce the *equivalent* relative density $\bar{\rho}_{\mathrm{e}}$ via the rate law

$$\dot{\bar{\rho}}_{\mathrm{e}} = \bar{\rho}_{\mathrm{e}}(r_\infty(p)\dot{\epsilon}_{\mathrm{e}}^{\mathrm{p}} - \dot{\epsilon}_{\mathrm{vol}}^{\mathrm{p}}) = \dot{\epsilon}_{\mathrm{e}}^{\mathrm{p}}\bar{\rho}_{\mathrm{e}}(r_\infty(p) - r) \tag{4}$$

where $\dot{\epsilon}_{\mathrm{vol}}^{\mathrm{p}} = \boldsymbol{\delta} : \dot{\boldsymbol{\epsilon}}^{\mathrm{p}}$, $\dot{\epsilon}_{\mathrm{e}}^{\mathrm{p}} = \sqrt{2/3}|\dot{\boldsymbol{\epsilon}}_{\mathrm{dev}}^{\mathrm{p}}|$ and $p = -(\boldsymbol{\sigma} : \boldsymbol{\delta})/3$. The variable r describes the dilatancy and the function $r_\infty(p) = r_{\mathrm{a}} + r_{\mathrm{b}}\ln(p/(p_{\mathrm{c}})_0)$ determines the "stationary" dilatancy rate. Further, $(p_{\mathrm{c}})_0$ is the initial consolidation pressure, cf. below. We note the following: (*i*) $\bar{\rho}_{\mathrm{e}} = \bar{\rho}$ if $\dot{\epsilon}_{\mathrm{e}}^{\mathrm{p}} \equiv 0$, (*ii*) $\dot{\bar{\rho}}_{\mathrm{e}} \to 0$ when $r \to r_\infty$.

The situation (*i*) is pertinent to purely isotropic loading and should be reflected by the flow rule when $q = \sqrt{3/2}|\boldsymbol{\sigma}_{\mathrm{dev}}| = 0$. The situation (*ii*) represents the ultimate behavior (for large deformations), where deformation continues with $r = r_\infty$ without further stress change along the CTC-path.

The rate laws for the plastic strain $\dot{\boldsymbol{\epsilon}}^{\mathrm{p}}$, associated to a plastic potential surface Φ^*, and for the hardening variable κ, are

$$\dot{\boldsymbol{\epsilon}}^{\mathrm{p}} = \dot{\mu}\frac{\partial \Phi^*}{\partial \boldsymbol{\sigma}}, \quad \dot{\kappa} = -\dot{\epsilon}_{\mathrm{e}}^{\mathrm{p}}\left(1 - \frac{K}{K_\infty}\right) \tag{5}$$

where $\dot{\mu}$ is a plastic multiplier and K_∞ is the saturation value of K. The convex set of plastically admissible stresses $\boldsymbol{\sigma}$ for given values of $\bar{\rho}_\mathrm{e}$ and $\bar{K}$ is introduced as $\mathcal{B}(\bar{\rho}_\mathrm{e}, \bar{K}) = \{\boldsymbol{\sigma} | \Phi(\boldsymbol{\sigma}; \bar{\rho}_\mathrm{e}, \bar{K}) \leq 0\}$ where $\Phi(\boldsymbol{\sigma}; \bar{\rho}_\mathrm{e}, \bar{K}) = 0$ is the convex yield surface.

The complete set of constitutive relations are defined by the state equations for $\boldsymbol{\sigma}$ and $\bar{K} = K/G_0$ in (2) and $(3)_2$, by the rate laws for $\boldsymbol{\epsilon}^\mathrm{p}$, $\bar{\rho}_\mathrm{e}$ and κ, and by the loading conditions

$$\dot{\mu} \geq 0 \,, \quad \Phi(\boldsymbol{\sigma}; \bar{\rho}_\mathrm{e}, \bar{K}) \leq 0 \,, \quad \dot{\mu}\Phi(\boldsymbol{\sigma}; \bar{\rho}_\mathrm{e}, \bar{K}) = 0 \tag{6}$$

2.3. Yield surface and plastic potential surface

We shall consider a class of two-invariant yield surfaces, that represent an extension of the Modified Cam-Clay yield surface, cf. Figure 1(a).

$$\Phi(p, q; \bar{\rho}_\mathrm{e}, \bar{K}) = q^2 + b^2 M^2 (p_\mathrm{c})^2 s(x) \quad \text{with} \quad x = p/p_\mathrm{c} - 1/2 \tag{7}$$

where $M(\bar{K}) = M_0 + \bar{K}$ is the mobilized friction at peak stress and $p_\mathrm{c}(\bar{\rho}_\mathrm{e}) = c^\mathrm{v}(\bar{\rho}_\mathrm{e})(p_\mathrm{c})_0$ is equal to the consolidation pressure under isotropic compression.

Note: Restriction to two-invariant representation is made here (without obscuring the generality of the format) since only CTC-test data are available for calibration, cf. below.

The function $s(x)$ is defined by

$$s(x) = (cx + 1)^2 \left(x^2 - 1/4\right) \tag{8}$$

Based on experimental evidence, we choose the plastic potential surface $\Phi^* = 0$, shown in Figure 1(b), from a family of symmetric smooth surfaces as follows:

$$\Phi^*(p, q) = q^2 + (M^*)^2 (p_\mathrm{c}^*)^2 s^*(x) \quad \text{with} \quad x = p/p_\mathrm{c}^* - 1/2 \tag{9}$$

where the function $s^*(x)$ is defined by

$$s^*(x) = (c^* x^2 + 1)^2 \left(x^2 - 1/4\right) \tag{10}$$

3. Integration of constitutive equations

Under strain control, i.e. for a given value ${}^{n+1}\boldsymbol{\epsilon}$ at time ${}^{n+1}t$, we recall from (2), and $(3)_2$ that

$${}^{n+1}\boldsymbol{\sigma} = \boldsymbol{\mathcal{E}}^\mathrm{e}({}^{n+1}\bar{\rho}_\mathrm{e}) : {}^{n+1}\boldsymbol{\epsilon}^\mathrm{e} \quad \text{with} \quad {}^{n+1}\boldsymbol{\epsilon}^\mathrm{e} = {}^{n+1}\boldsymbol{\epsilon} - {}^{n+1}\boldsymbol{\epsilon}^\mathrm{p} \tag{11}$$

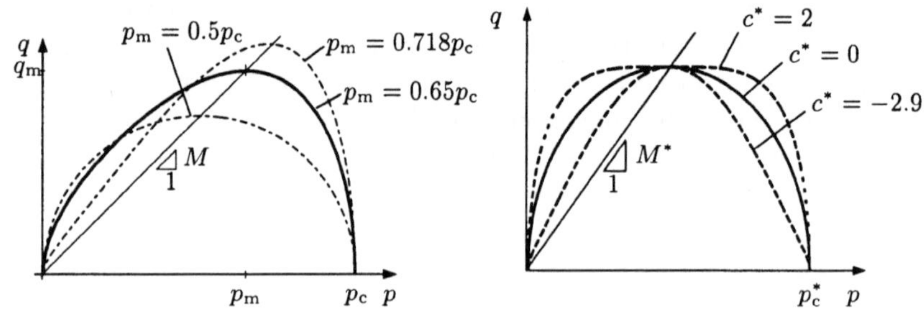

Figure 1. (a) Yield surfaces, $\Phi = 0$, for different values of p_m/p_c, (b) Family of plastic potential surfaces, $\Phi^* = 0$.

$$^{n+1}\bar{K} = -\bar{H}(^{n+1}\bar{\rho}_e)\,^{n+1}\kappa \tag{12}$$

Applying the fully implicit (Backward Euler) rule for integrating the pertinent evolution equations, we obtain the incremental relations.

Remark: When Φ is a scalar function of p and q (for given values of the hardening variables $\bar{\rho}_e$ and $\bar{K}$), it is convenient to obtain the solution in the *generalized invariant stress* space spanned by p, q, $\bar{\rho}_e$ and $\bar{K}$. As soon as the stress has been computed in terms of p and q, the proper transformation to principal and/or Cartesian coordinates can be carried out. □

4. Model calibration using CTC-test data

The calibration of the constitutive model is formulated as an optimization problem, where an objective function E is minimized with respect to the pertinent model parameters $\underline{\beta}$. We may formally state the problem as: Find the optimal solution $\underline{\beta}_*$

$$\underline{\beta}_* = \arg\left[\min_{\underline{\beta}\,\in\,\Omega} E(\underline{\beta})\right] \quad , \quad \Omega = \{\underline{\beta} \in \boldsymbol{R}^n |\ \underline{\beta}^{\min} \le \underline{\beta} \le \underline{\beta}^{\max}\} \tag{13}$$

where the n model parameters are bounded by the "physically reasonable" values $\underline{\beta}^{\min}$ and $\underline{\beta}^{\max}$. The used optimization method is a simplex method described by [5]. We reduce the model parameter vector beforehand by directly estimating following parameters from the CTC-tests: $\bar{\rho}_0 = 0.53$, $G_0 = 50$ kPa, $(K_b)_0 = 232$ kPa.

First, we examine the ability of the model to represent the experimental curves q–ϵ_e and ϵ_{vol}–ϵ_e separately for the four different confining stresses, cf. Figure 2. A single vector $\underline{\beta}_0$ is chosen as the start vector to yield the final parameter vector $\underline{\beta}_*$.

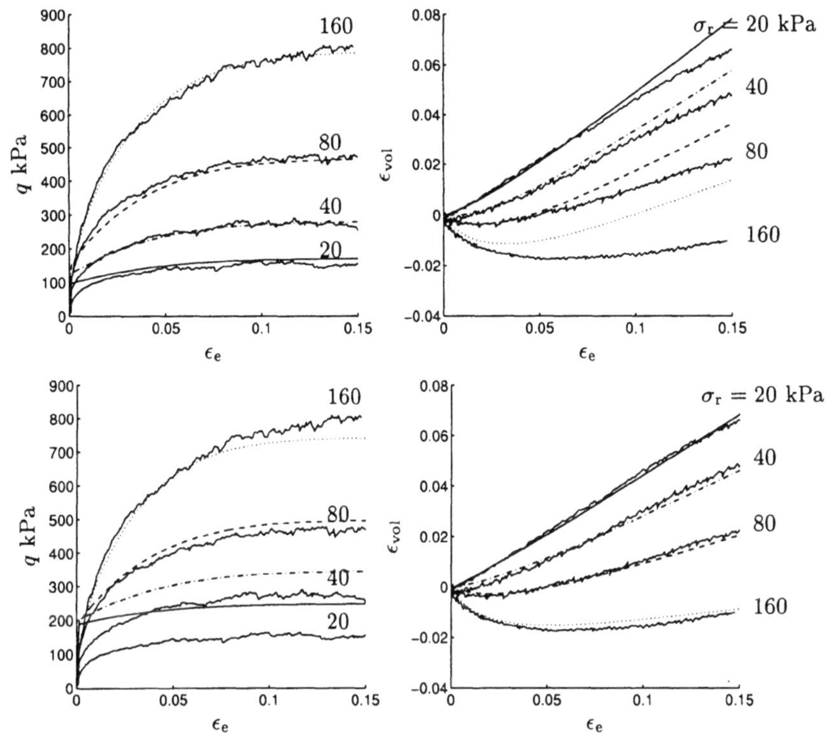

Figure 2. Suboptimization results: The upper figures show curves where q–ϵ_e is calibrated and ϵ_{vol}–ϵ_e is predicted. The lower curves show the opposite case.

It is known that the final solution $\underline{\beta}_*$ is dependent of the choice of start vector $\underline{\beta}_0$ in general. We define the "multi-vector" strategy by first scanning the parameter space Ω in a rectangular grid of equidistantly spaced points between $\underline{\beta}^{\min}$ and $\underline{\beta}^{\max}$ and computing the corresponding value of E in each grid point, cf. Figure 3. Then the corresponding vectors yielding the N lowest values of E define the set $\mathcal{B}_0 = \{\underline{\beta}_0^{(1)}, \underline{\beta}_0^{(1)}, \ldots, \underline{\beta}_0^{(N)}\}$. Optimization for each $\underline{\beta}_0^{(k)}$ gives the final vector $\underline{\beta}_*^{(k)}$, that constitute the set $\mathcal{B}_* = \{\underline{\beta}_*^{(1)}, \underline{\beta}_*^{(1)}, \ldots, \underline{\beta}_*^{(N)}\}$. The chosen

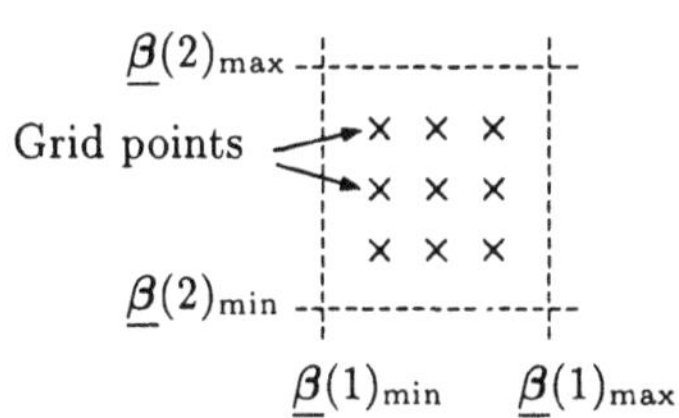

Figure 3. Example of grid for scanning in a two-parameter space.

final vector $\underline{\beta}_*$ is selected as the "best" one in $\mathcal{B}_*$, i.e.

$$\underline{\beta}_* = \arg\left[\min_{\underline{\beta} \in \mathcal{B}_*} E(\underline{\beta})\right] = \underline{\beta}_*^{(K)} \quad \text{for some } K \in (1, N) \tag{14}$$

The corresponding start vector $\underline{\beta}_0$ is defined as $\underline{\beta}_0 = \underline{\beta}_0^{(K)}$.

A calibration using data from the two different types of curves is then performed with the multi-vector strategy, cf. Figure 4. The results in Figure 2 and Figure 4 show that the model can represent the experimental data obtained from a CTC-test for this specific material in an acceptable manner.

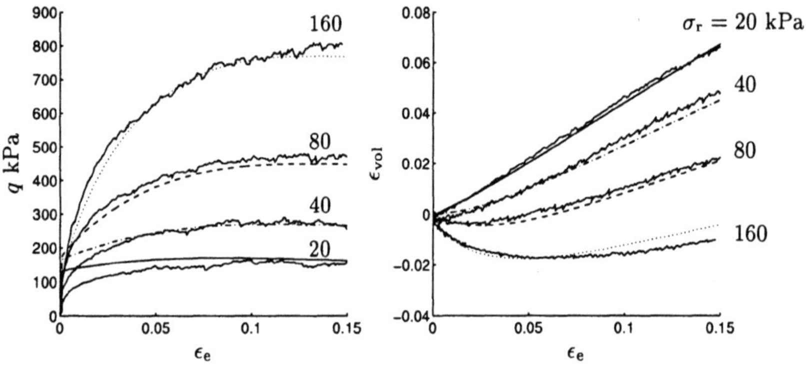

Figure 4. Results of the multi-vector optimization with $N = 50$, using both q–ϵ_e and ϵ_{vol}–ϵ_e data.

References

1. Jacobsson, L. *A plasticity model for cohesionless material with emphasis on railway ballast.* Licentiate Thesis, Department of Solid Mechanics, Chalmers Univ. of Technology, Göteborg, Sweden, 1999.
2. Jernigan, R. *The physical modeling of soils containing oversized particles.* Ph. D. Thesis, Department of CEAE, Univ. of Colorado at Boulder, 1998.
3. Krenk, S. Characteristic state plasticity for granular materials, Part 1: Basic theory, *DCAMM, Danish Center for Applied Mathematics and Mechanics*, Report no. 610, 1999.
4. Lambrecht, M. and Miehe, C. Two non-associated isotropic elastoplastic hardening models for frictional materials, *Acta mechanica*, **135**, 73–90, 1999.
5. Nelder, J. A. and Mead, R. A simplex method for function minimization, *Computing Journal*, **7**, 308–313, 1965.
6. Nova, R. Sinfonietta classica: An exercise on classical soil modelling, *Proc. Int. Workshop on Constitutive equations for granular non-cohesive soils*, Saada. & Bianchini Eds., Balkema, 501–520, 1988.
7. Nova, R. and Wood, D. M. A constitutive model for sand in triaxial compression, *Int. J. Num. Anal. Meth. Geomech.*, **3**, 255–278, 1979.

Mechanical Properties of Modified Wood

S. J. Kowalski
Poznań University of Technology
Pl. Marii Sklodowskiej-Curie 2, 60-695 Poznań, Poland

L. Kyziol
Naval Academy, Institute of Shipboard Machinery Fundamentals ul. Smidowicza 71, 81-919 Gdynia , Poland

1. Introduction

The aim of this paper is to examine the mechanical properties of wood modified with methacrylate of methyl group (MM). Methacrylate is a chemical substance used for wood protection against water soaking. Besides, one states that implantation of the methacrylate into the wood pores changes significantly wood's mechanical properties and increases its strength. The polymerized methacrylate forms a new composite with wood fibers of increased durability and smaller anisotropy than the natural wood. Through this modification, one eliminates not only the natural defects of wood but first of all increases its usable properties. The wood fibers effect the mechanical properties of the polymerized methacrylate better then other synthetic fibers.

Authors' investigations prove that composites obtained on the basis of wood skeleton modified with methacrylate can be applied wherever a protection against magnetism, long working time in connection with elasticity and lightness, stability of dimensions, invariability of shape, ease machining and elements joining, good damping and vibration-isolation properties are required. Besides, the costs of production of such a composite are relatively low, in particular, when such popular and easy growing trees as pines, birches, poplars, asps or alders are use for modification.

Application of wood-polymer composite to special constructions needs exact examination of its strength and other mechanical properties. This paper presents some such results obtained by authors.

W. Ehlers (ed.), IUTAM Symposium on Theoretical and Numerical Methods in Continuum Mechanics of Porous Materials, 221–228.

2. Procedure of wood modification

The modification process of wood with the methacrylate polymer is carried out on the especially prepared wood bars of initial moisture content about 10%, in the autoclave of partial vacuum 0.08 MPa, (see also Aszkenazi and Ganow, [1]; Aszkenazi and Morozow, [2]). First, the air is evacuated from the wood pore space for the period of 1 hour. Next, the stabilized methacrylate is delivered to the autoclave for the purpose of inoculation it in the wood pore spaces for the various time periods: 0.5; 2; 4 and 24 hours, dependent on the required degree of saturation. After saturation, the thermal polymerization process is carried out. To this aim the wood bars are placed in the special polymerization reservoir, which is filled with water solution of sodium nitrate. This solution is first heated to the temperature of 85°C and then kept in that temperature for15 minutes. Next, the temperature is lowered to 75°C and as such held through the period of 4 hours. In order to obtain the complete polymerization of the polymer in wood, one raises the final temperature up to the value of 120°C for 1hour.

After the polymerization, the samples were cut from the composite "wood - metacrylate polymer" for the purpose of mechanical investigations. Depending on the sample form, size and orientation with respect to wood fibers and annual rings, one deals with anisotropy more or less of orthotropic character. Through the thermo-chemical treatment one obtains wood which manifests less anisotropy than natural wood. In spite of this the mechanical properties of modified wood depend strongly on spatial orientation of wood elements cut out from a wood log.

3. Strength of modified wood

The investigations have in view the estimation of influence of methacrylate content in wood on its strength and other mechanical properties. As it was mentioned above, the methacrylate content in wood was estimated for various periods of saturation: t =0.5; 2; 4 and 24 hours. The values are presented in Table 1, Kyziol ([8]).

If the direction of main stresse does not coincide with the direction of wood fibers, the strength of wood decreases. Let α denotes the angle between the wood fibers and the direction of the stress applied to the tested sample. Figure 1 illustrates the influence of anisotropy, represented by angle α, on the yield strength R_m of the modified pine sapwood and pine hardwood by tension, for saturation time t $=$ 0.5 and 24 hours, Kyziol ([7], [8]).

Table I. Polymer content in pine sapwood.

Saturation time [h]	0.5	2.0	4.0	24.0
Methacrylate content in pine sapwood, [%]	35	43	48	56
Methacrylate content in pine hardwood, [%]	4.7	8	13.1	28

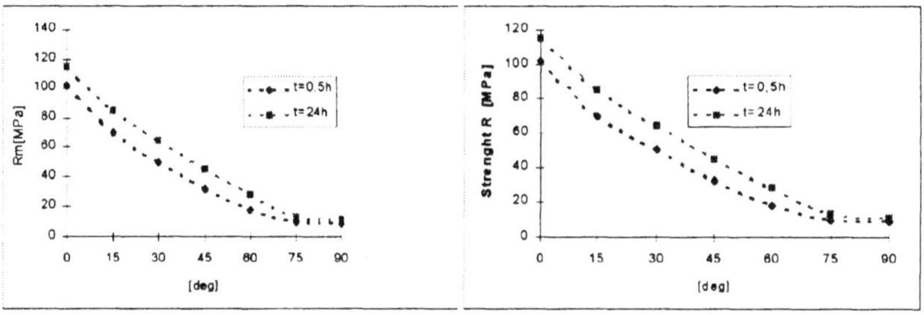

Figure 1. a) pine sapwood, b) pine hardwood.

The additional investigations were concerned with wood strength after wetting in seawater for some time. The test were carried on samples of modified alder (O3), containing 80% of methacrylate, and birch, containing (B1) - 62%, (B2) - 76%, and (B3) - 83% of methacrylate. The samples were placed in a perforated container and immersed in sea water on the depth of 3 m. Four cycles of wetting were carried out: 10, 20, 30 and 60 twenty four hours, Kyziol ([5]). After every cycle, the saturation of samples with seawater was measured and the yield strengths as well as the impact strengths were examined. Results concerning the yield strengths are presented in figure 2.

Each point in figure 2 represents an average value from 5 experimental measurements. It is visible that wood with greater polymer content loses less strength after wetting than that with smaller polymer content.

The impact investigations were carried out on samples of modified pine sapwood of dimensions $10 \times 10 \times 100$ mm. The samples were subjected to impact tests after every cycle of etting at four different temperatures: 293 K, 273 K, 263 K, and 243 K. The given values reflect more or less the average temperatures of the air over the Baltic Sea during the year period. Impact strength was executed through the impact of hammer crosswise to the wood fibers (in tangential direction).

The relation between impact strength U and the time of wetting in seawater can be described by the correlated equation of the form $U = a \log x + b$, where a and b are the regress coefficients, and x is the

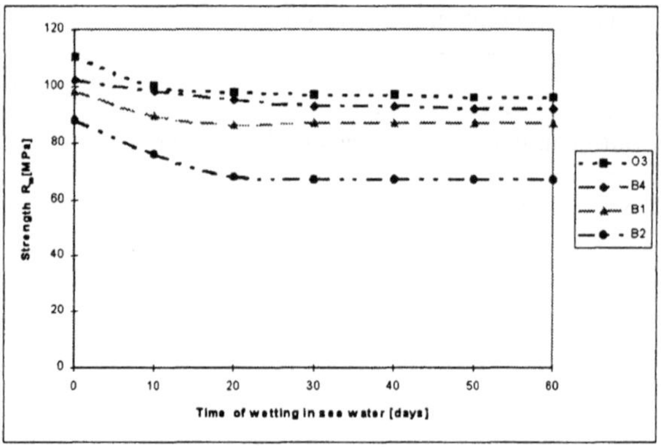

Figure 2. Tensional strength of modified wood after wetting in seawater.

number of twenty-four hours during which the samples were wetted in seawater. The graphical illustration of this equation is given in figure 3.

Summarizing, one states that the greater wood absorbability does not involve much lowering of impact strength, ca. 5 %. Lower temperatures of impact tests, on the other hand, involve an increase of impact strength about 20 %. From earlier investigations follows that absorbability of modified wood after 20 twenty four hours of wetting in seawater is fixed on level of 18-20 %, Kyziol ([8]).

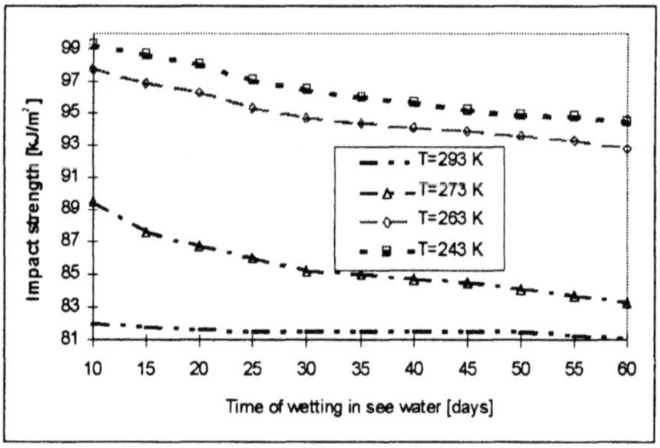

Figure 3. Influence of wetting time in seawater on impact strength for various test temperatures.

It follows from figure 3 that impact strength of modified wood increases with lowering of the test temperature. The highest impact strength manifest the samples of modified wood in the lowest temperature of 243 K and dry state. The lowest impact strength, on the other hand, manifest the samples of highest absorbability and subjected to impact tests in rather high temperatures.

4. Coefficient of elasticity

Coefficients of elasticity for modified wood are the fundamental indicators of usefulness of the modification method. Besides, all mechanical theories contain these fundamental coefficients in constitutive equations (see e.g. Kowalski and Kowal, [3]; Kowal and Kowalski, [4]).

Investigations having in view determination of coefficients of elasticity in this paper were carried out according to the norm PN-75/D-04123 on samples of dimensions 20 × 20 × 300 mm. Samples were cut out in such way to make possible determination of modules of elasticity along wood fibers by static bending. The bending force was acting in a plane crosswise to wood fibers, however, with variable direction with respect to annual rings in each test. In particular, the modules of elasticity E_g along fibers caused by forced applied radial, diagonal and tangential to annual rings directions were examined. The values of E_g for pine sapwood and pine hardwood, modified with polymerized metacrylate, are presented in Kyziol ([8]). The significant variation in mass density ρ of polymerized sapwood as well as the signuficant difference in polymer content S in sapwood is observed. It was not the case in hardwood. Polymer content in pine sapwood after 0.5 h of saturation reached value $S = 35\%$, while in hardwood for the same time only $S = 4.7\%$. The value of elastic modulus for the modified sapwood was lower than for the hardwood for this amount of saturation. It was stated that the saturation of sapwood was $S = 48\%$ after 4 hours of saturation, while that of hardwood only $S = 13,1\%$. The values of elasticity modules for sapwood and hardwood were comparable for the given amount of saturation and equal to $E_g = 1.60 * 104$ MPa. After 24 hours, the saturation of sapwood was $S = 56\%$, while hardwood $S = 28\%$. It appeared that the modulus of elasticity for modified sapwood slightly overcome the modulus for hardwood for this amount of saturation. The greatest value of elasticity modulus was stated for pine sapwood by bending in radial direction. Its value was $E_g = 1.76 * 104$ MPa, Kyziol ([8]).

The investigations were supplemented by estimation of elasticity modules for natural pine sapwood and hardwood. The results are presented in Table 2.

Figure 4 and 5 illustrate the influence of saturation of pine with methacrylate on modules of elasticity along the fibers of pine, when

Table II. Elastic modules of natural pine wood E_g [MPa]

Direction of applied force	Sapwood [MPa]	Hardwood [MPa]
Radial	$1.12x10^4$	$1.25x10^4$
Diagonal	$0.95x10^4$	$1.08x10^4$
Tangential	$0.78x10^4$	$0.95x10^4$

the bending force act radial, diagonal and tangential to the annual rings.

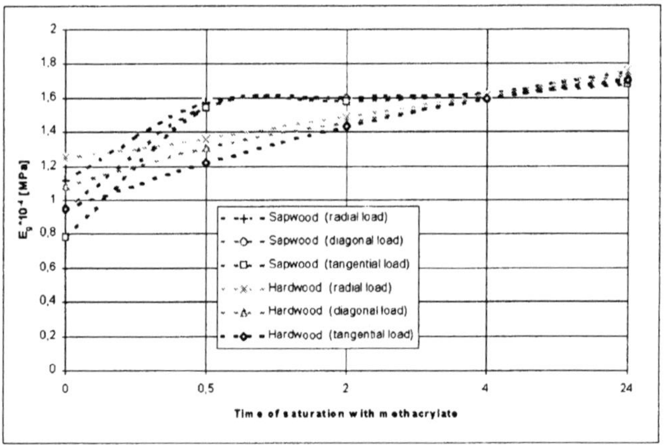

Figure 4. Modules of elasticity along wood fibers for pine modified with methacrylate.

The investigations presented here indicated the significant difference in mechanical properties of modified pine sapwood and pine hardwood. Hardwood soaks methacrylate much slower then sapwood. Essential is the choice of wood part from wood log which is to become subjected to saturation and modification. In the course of wood saturation with polymer the anisotropy of wood properties diminishes, and the values of coefficients of elasticity for considered directions considerably exceed the magnitudes for those of natural wood. For example, the samples of dimensions 20x100x350 cut out from modified pine sapwood reach the yield strength along fibers up to $\dot{R}_m = 114$ MPa, while for natural wood $R_m = 85$ MPa,

Summarizing, one states that modification of both sapwood and hardwood with methacrylate involves increase of the polymer content

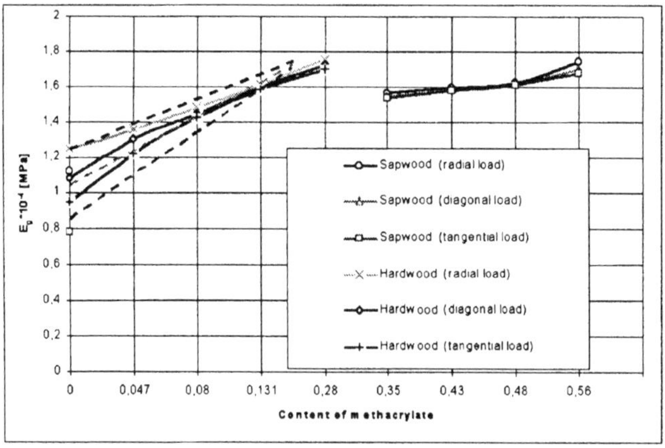

Figure 5. Modules of elasticity for modified pine as a function methacrylate content.

more in sapwood, and increases the modules of elasticity prpoprtionaly to the methacrulate content, Kyziol ([6]).

5. Final remarks

The presented results suggest further investigations having in view initiating of application of modified wood in building of ships. Considerably higher strength properties, lower anisotropy, diminution of absorbability, increase of impact strength, all those suggest a wider application of modified wood in industry at all, and in particular in naval industry. Lowering of production costs can be reached, for example, by surface saturation of wood with polymer only. In relation to basic material, which is wood, one foresees a utilization of fast growing sorts of trees, especially pines and alders, that is, the cheapest trees.

References

1. Aszkenazi, E. K. and Ganow, E. W. Anisotropy of structure materials, Mechanical Engineering, (in Russian), 1980.
2. Aszkenazi, E. K. and Morozow, A. S. Methodology of experimental studies, Mechanical Engineering, (in Russian), 1986.
3. Kowalski, S. J. and Kowal, M. Physical relations for wood at variable humidity, Transport in Porous Media, **31**, 331–346, 1998.
4. Kowal, M. and Kowalski, S. J. Material constants for wet pine sapwood, Studia Geotechnica et Mechanica, **20**, 39–52, 1998.

5. Kyziol, L. Analysis of modified wood properties applied innaval structures, Proceedings of Annual Conference "Polymers and Composites", Kozubnik '95, (in Polish), 1995.
6. Kyziol, L. Influence of wood absorbability on its impact strength at low temperatures, Proceedings of Annual Conference "Polymers and Composites", Ustron'96, (in Polish), 1996a.
7. Kyziol, L. Estimation of possibility of wood application in construction of special naval vessels, Wood Industry, **3**, 1–96, (in Polish), 1996b.
8. Kyziol, L. Influence of pine sapwood saturation with methacrylate on its mechanical properties, Proceedings of Annual Conference "Polymers and Composites", Ustron'98, (in Polish), 1998.

Water Transport in Phase-Changing Snowpacks

S. Sellers
School of Mathematics
UEA
Norwich NR4 7TJ UK
e-mail: shaun_sellers@yahoo.co.uk

Abstract. A one-dimensional model is presented for meltwater transport in a phase-changing snowpack. The model treats the snowpack as a multiphase mixture consisting of a rigid ice matrix along with air and water. Due to applied surface heating or internal temperature gradients, melting and freezing may occur, so that the water and ice phases are interconvertible. The thermodynamics of the phase changes is included by treating the phase fronts as propagating singular surfaces where the equations for the bulk region are supplemented by appropriate interface or jump conditions at the phase fronts, which leads to a generalized Stefan problem. One dimensional numerical illustrations are given for various applied surface heatings in an isothermal snowpack.

Keywords: Multiphase mixtures, phase transitions, porous media, snowpacks, Stefan problem

1. Introduction

Water runoff from melting snowpacks provides a significant proportion of the freshwater supply in northern climates. Predictions of the quantity of water release are also useful for irrigation and flood control. Colbeck ([3, 4, 5]) developed a theory of water transport in snow by treating the snow as a *multiphase porous medium* (cf. Bear, [2]) consisting of a rigid ice matrix, air, and water. A mass balance for the water phase in conjuction with a Darcian constitutive law for the water flux leads to a nonlinear Fokker-Planck equation for the evolution of the water saturation.

Snow however provides an example of a porous medium where the ice matrix and the continuous water phase are, depending upon the local thermodynamic conditions, *interconvertible*. Due to the possible freezing and melting at the atmospheric interface and throughout the bulk, several free boundaries can arise (cf. Crank, [6]): a *moving snow surface*, as well as *percolating melt* and *refreezing fronts* separating cold snow from wet snow (see Fig. 1). The transport of water is thus coupled with the *thermodynamics* of the snowpack.

Here we present a one-dimensional model of water transport in a rigid, phase-changing snowpack, which allows for interconversion of the

W. Ehlers (ed.), IUTAM Symposium on Theoretical and Numerical Methods in Continuum Mechanics of Porous Materials, 229–236.

water phase with the ice matrix due to melting and freezing. It treats the phase fronts as propagating singular surfaces that separate wet snow from dry snow. The model consists of the usual multiphase porous media equations for the bulk region but supplemented with interfacial conditions at the phase fronts. Since the locations of the interfaces vary with time, these locations must also be determined as part of the solution, which leads to a generalized Stefan problem.

This work provides a generalization of the previous isothermal model of Sellers ([10]) to the nonisothermal case. Alternative approaches using volumetric supply terms in the bulk equations have been given by Akan ([1]), Marsh and Woo ([8]), Illangasekare et al. ([7]), Pfeffer et al. ([9]), Tao and Gray ([12]), and Tao ([11]).

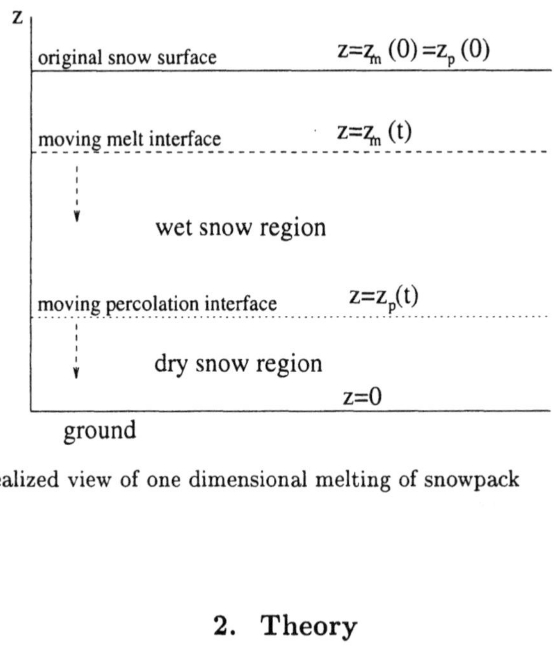

Figure 1. Idealized view of one dimensional melting of snowpack

2. Theory

2.1. Bulk equations

The bulk region is assumed to consist of either *dry snow* (ice matrix and air) or *wet snow* (ice matrix, air, and water). Since the ice matrix is rigid, the dry region is described only by the energy equation for a rigid material. The wet region is described by a mass balance for the water phase, along with constitutive relations for the water flux $\mathbf{u}_w$, the permeability k_w and the capillary pressure P_c. Typically, the relations for k_w and P_c are power laws in terms of the effective saturation S^*. Since the wet region is isothermal, the energy balance is automatically satisfied. The effect of the air is neglected in both regions. The equations are written separately for each region.

– Wet region:

$$0 = \phi_e \frac{\partial S^*}{\partial t} + \operatorname{div} \mathbf{u}_w, \qquad \textit{water mass}$$

$$\mathbf{u}_w = \frac{k_w}{\mu_w}(\operatorname{grad} P_c + \rho_w \mathbf{g}), \qquad \textit{Darcy's law}$$

$$k_w = k_w(S^*), \qquad P_c = P_c(S^*). \qquad \textit{power laws}$$

– Dry region:

$$\operatorname{div}(K \operatorname{grad} T) = (\rho_i \phi_i c_i + \rho_a \phi_a c_a) \frac{\partial T}{\partial t}. \qquad \textit{energy}$$

2.2. Jump relations

The equations for the bulk region must be supplemented with jump conditions at the phase fronts. A given applied heat supply $q(t)$ is assumed specified at the *melt surface* $z_m(t)$ separating the wet snow from the atmosphere. A *percolation front* $z_p(t)$ separates the isothermal wet snow region from the dry snow region, which is subject to an initial temperature gradient. At each phase front, there are jump conditions corresponding to balance of mass for the water, balance of mass for the ice, and balance of energy for the mixture. At the melt front, both water and ice exist only on the plus side of the interface, whereas at the percolation front, water exists only on the minus side of the interface but ice on both sides. Here only two interfaces are considered.

– Melt surface: $z = z_m(t)$

$$-\rho_i \phi_i^+ \dot{z}_m = m_m, \qquad \textit{ice mass}$$

$$-\rho_w(\mathbf{u}_w^+ \cdot \mathbf{n} - \phi_w^+ \dot{z}_m) = m_m, \qquad \textit{water mass}$$

$$q(t) = m_m L. \qquad \textit{energy}$$

These are 3 equations for the 3 unknowns m_m (rate of surface water production), $\mathbf{u}_w^+ \cdot \mathbf{n}$ (volumetric water flux), and $\dot{z}_m$ (interface speed).

– Percolation front: $z = z_p(t)$

$$\rho_i(\phi_i^- - \phi_i^+)\dot{z}_p = m_p, \qquad \textit{ice mass}$$

$$\rho_w(\mathbf{u}_w^- \cdot \mathbf{n} - \phi_w^- \dot{z}_p) = m_p, \qquad \textit{water mass}$$

$$-\mathbf{n} \cdot (K \operatorname{grad} T)^+ = m_p L. \qquad \textit{energy}$$

These are 3 equations for the 3 unknowns m_p (rate of surface water production), $\dot{z}_p$ (interface speed), and ϕ_i^- (ice volume fraction behind moving front).

At the melt front z_m, the applied heating $q(t)$ determines the surface water production m_m due to melting, which then determines the interface speed $\dot{z}_m$ and water flux $\mathbf{u}_w^+ \cdot \mathbf{n}$, thus providing the boundary condition for the bulk equations. At the percolation front z_p, the temperature gradient $(\mathsf{grad}\, T)^+$ in the dry snow determines the surface ice production $-m_p$ due to freezing (m_p is the surface water production), which then determines the interface speed $\dot{z}_p$ and jump $(\phi_i^- - \phi_i^+)$ in ice volume fraction (since the water flux $\mathbf{u}_w^- \cdot \mathbf{n}$ is determined from the bulk equations for the wet region).

The signs are chosen so that q is positive for heat applied at the melt surface z_m, so that m_m is necessarily positive and $\dot{z}_m$ negative (i.e., the surface moves downwards). If the temperature gradient at the percolation front z_p is positive, then the surface water production m_p is negative (i.e., ice is produced), which then slows down the speed $\dot{z}_p$ of the percolation front and increases the ice volume fraction ϕ_i^- behind the front. Thus inhomogeneities can be formed in an initially homogeneous snowpack [7, 13]

3. Examples

Several examples illustrating the model are now given. For simplicity, only an isothermal snowpack is considered, so that the only phase change is the melting that takes place at the snow surface. Fig. 2–4 show the effective saturation S^* as a function of depth for three different applied surface heatings $q(t)$, which are taken to be truncated sinusoids in time. Based on experimental data, the power laws for the constitutive relations were assumed to have an exponent of 3 for the permeability and -2 for the capillary pressure [4]. The nonlinearity of the governing equations leads to front speeds that depend on saturation, hence the coalescence of the meltwaves.

4. Discussion

A one-dimensional model has been presented for the transport of meltwater in a phase-changing snowpack. Such a model can be used to predict water runoff from nonisothermal snowpacks. It treats the snow as a multiphase porous medium where the ice and water phases are interconvertible at the phase fronts, which are assumed to be propagating

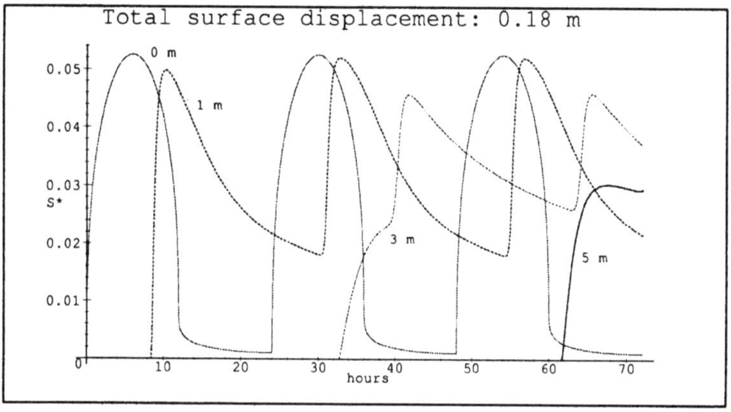

Figure 2. Effective saturation at various depths for an applied periodic heating corresponding to a total surface displacement of 0.18 meters over 3 days.

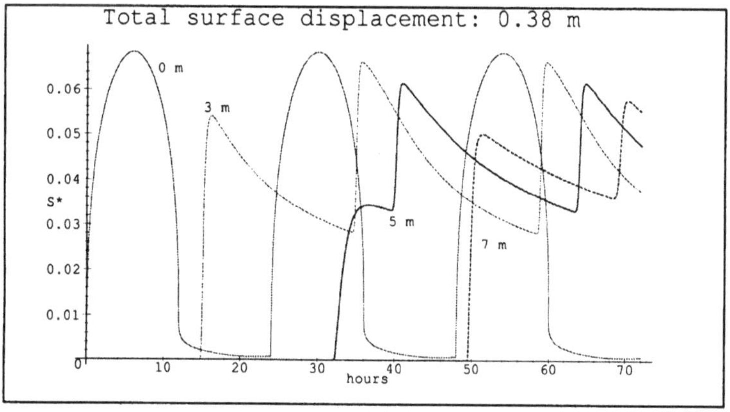

Figure 3. Effective saturation at various depths for an applied periodic heating corresponding to a total surface displacement of 0.38 meters over 3 days.

singular surfaces. The balance equations and constitutive relations for the bulk region are supplemented with jump conditions for the balance laws at the propagating phase fronts. In this model, the ice is rigid in the bulk, but serves as a surface source or sink of water due to melting or freezing at the phase fronts. The jump conditions serve as boundary conditions linking the bulk wet and dry snow regions. The presence of temperature gradients in the snow slows down the propagation of the percolation front due to freezing, which leads to inhomogeneities

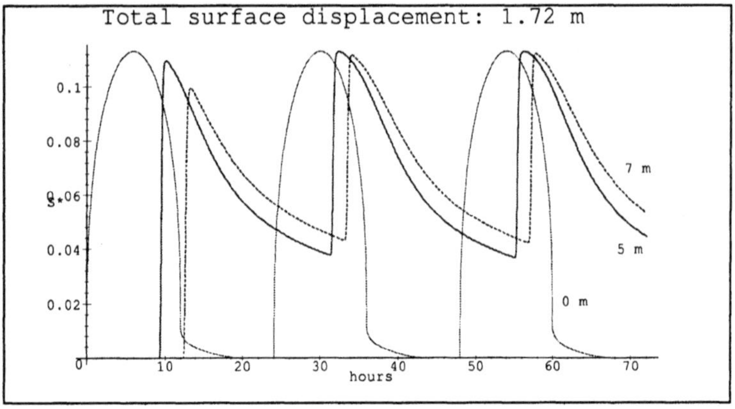

Figure 4. Effective saturations at various depths for an applied periodic heating corresponding to a total surface displacement of 1.72 meters over 3 days.

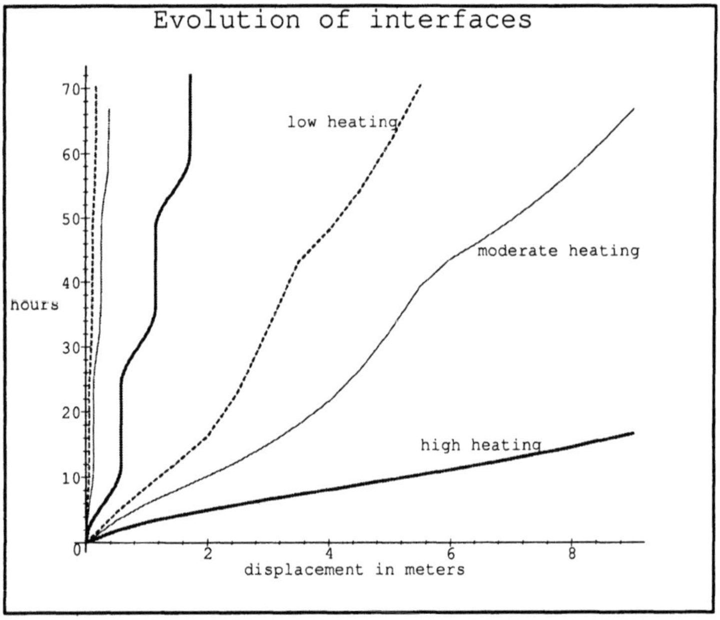

Figure 5. The time evolution of the melt and percolation fronts for the applied surface heatings corresponding to figures 2–4. The three curves on the left are the melt fronts, the three curves on the right are the corresponding percolation fronts.

in the ice volume fraction. For simplicity, the effect of the air has been neglected. This model provides the simplest thermodynamically consistent description of one dimensional water transport in a snowpack that also takes into account freezing and melting.

5. Notation

c_a	heat capacity of air,
c_i	heat capacities of ice,
$\mathbf{g}$	acceleration vector due to gravity,
k_w	permeability to water phase,
K	effective thermal conductivity of the snow,
L	latent heat,
$\mathbf{n}$	interface normal vector,
m_m	rate of surface melting at z_m,
m_p	rate of surface melting at z_p,
P_c	capillary pressure,
$q(t)$	surface heating applied at melt interface at time t,
S^*	effective water saturation,
S_w	water saturation ($= \phi_w/(1-\phi_i)$),
t	time,
T	temperature,
$\mathbf{u}_w$	water volume flux vector,
z	vertical coordinate,
$z_m(t)$	vertical coordinate of melt interface at time t,
$z_p(t)$	vertical coordinate of percolation interface at time t,
ρ_w	true water density,
ρ_i	true ice density,
ϕ_a	volume fraction of air ($= 1 - \phi_w - \phi_i$),
ϕ_e	effective porosity ($= 1 - \phi_i - \phi_{wi}$),
ϕ_i	volume fraction of ice,
ϕ_w	volume fraction of water,
ϕ_{wi}	volume fraction of irreducible or bound water,
μ_w	viscosity of water.

Acknowledgements

This work was supported by the EPSRC (UK).

References

1. Akan, A. O. Simulation of runoff from snow-covered hillslopes, *Water resources Research*, **20**, 707–713, 1984.
2. Bear, J. *Dynamics of Fluids in Porous Media*, Dover, New York, 1988.
3. Colbeck, S. C. A theory of water percolation in snow, *Journal of Glaciology*, **11**, 369–385, 1972.
4. Colbeck, S. C. The capillary effects on water percolation in homogeneous snow, *Journal of Glaciology*, **13**, 85–97, 1974.
5. Colbeck, S. C. An analysis of water flow in dry snow, *Water Resources Research*, **12**, 523–527, 1976.
6. Crank, J. *Free and Moving Boundary Problems*, Clarendon Press, Oxford, 1984.
7. Illangasekare, T. H., Walter, Jr., R. J., Meier, M. F. and Pfeffer, W. T. Modeling of meltwater infiltration in subfreezing snow, *Water Resources Research*, **26**, 1001–1012, 1990.
8. Marsh, P and Woo, M.-K. Wetting front advance and freezing of meltwater within a snow cover. 2. A simulation model, *Water Resources Research*, **20**, 1865–1874, 1984.
9. Pfeffer, W. T., Illangasekare, T. H. and Meier, M. F. Analysis and modeling of melt-water refreezing in dry snow, *Journal of Glaciology*, **36**, 238–246, 1990.
10. Sellers, S. Theory of water transport in melting snow with a moving surface, *Cold Regions Science and Technology*, **31**, 47–57, 2000.
11. Tao, Y.-X. Modeling of melting in heterogeneous snow cover on frozen permeable soil, *Numerical Heat transfer, Part A*, **30**, 143–163, 1996.
12. Tao, Y.-X. and Gray, D. M. Prediction of snowmelt infiltration into frozen soils, *Numerical Heat Transfer, Part A*, **26**, 643–665, 1994.
13. Tseng, P.-H., Illangasekare, T. H. and Meier, M. F..Modeling of snow melting and uniform wetting front migration in a layered subfreezing snowpack, *Water Resources Research*, **30**, 2363–2376, 1994.

Modelling of Localisation at Finite Inelastic Strains in Fluid Saturated Porous Media

L. Sanavia and B. A. Schrefler
Department of Structural and Transportation Engineering, University of Padua, Italy

E. Stein
Institut für Baumechanik und Numerische Mechanik, Universität Hannover, Germany

P. Steinmann
Lehrstuhl für Technische Mechanik, Universität Kaiserslautern, Germany

Abstract. Quasi-static finite strain localisation phenomena in fluid saturated soils are studied. The governing equations at the macroscopic level are derived in a spatial and a material setting. The constituents are assumed to be materially incompressible. The elasto-plastic behaviour of the solid skeleton is described by the multiplicative decomposition of the deformation gradient into an elastic and a plastic part. The effective stress state is limited by the Drucker-Prager yield surface. A particular "apex formulation" is advocated. The fluid is assumed to obey Darcy's law. Numerical examples highlight the developments. Negative water pressures, which are important for strain localisation analysis of undrained water saturated dense sands, are obtained and their values are critically discussed.

Keywords: Saturated porous media, strain localisation, multiplicative elasto-plasticity

1. Introduction

Strain localisation is a strain accumulation in well defined narrow zones of finite amplitude, which is observed in a wide class of engineering materials, including multiphase geomaterials. Experimental investigations of initially water saturated sands have been reported, for instance, in Lade *et al.* ([4]), Vardoulakis ([10]) and Mokni and Desrues ([6]). The analysis of these tests show that the failure in a quasi-static case occurs for large strains. Moreover, they stated also the fundamental role that fluids can have in the strain localisation phenomenon, also as a triggering mechanism.

The aim of this work is the inclusion of these two experimental features in a mathematical and finite element model. Large elasto-plastic strains of the solid skeleton are described by the multiplicative decomposition of the deformation gradient into an elastic and plastic part, using an hyperelastic free energy function. From a micro-mechanical point of view this multiplicative decomposition describes the plastic slip in crystals; for cohesive-frictional soils its validity has been sug-

W. Ehlers (ed.), IUTAM Symposium on Theoretical and Numerical Methods in Continuum Mechanics of Porous Materials, 239–244.

gested by Nemat-Nasser ([7]), where the plastic part of the deformation gradient is viewed as an internal variable related to the amount of slipping, crushing, yielding and, for plate like particles, plastic bending of the granules comprising the soil. The paper presents in the following the balance and the constitutive equations and, finally, the numerical examples.

2. Balance and constitutive equations

The macroscopic governing equations are derived in the spatial setting using averaging theories following Hassanizadeh and Gray ([3]) applied to the microscopic balance equations developed in Lewis and Schrefler ([5]). Incompressible constituents at the microscopic level, isothermal and quasi-static loading conditions are assumed. These balance equations are obtained in the framework of the hybrid mixture theory and are the same of those derived in the classical mixture theory extended by the concept of volume fraction (Diebels and Ehlers [2]) or in the phenomenological Biot's theory (Borja and Alarcon [1]). These governing equations are now summarised.

In the Lagrangian description of the motion the position of each material point in the actual configuration $\mathbf{x}$ is function of its placement in a chosen reference configuration $\mathbf{X}^\pi$ and of the current time t,

$$\mathbf{x} = \chi^\pi(\mathbf{X}^\pi, \mathrm{t}) \ , \tag{1}$$

where $\chi^\pi(\mathbf{X}^\pi, \mathrm{t})$ is a continuous and bijective motion function of each phase. In porous media theory it is common to describe the motion of the fluid phases in terms of velocity relative to the moving solid. This means that the control volume of the porous media is the solid volume (spanned on the total volume by averaging) and that the fluids motion is described with reference to the actual configuration $\mathbf{x}$ occupied by the solid skeleton. When the solid motion is described as function of spatial quantities and the fluid velocity relative to the solid $\mathbf{v}^{ws}(\mathbf{x}, \mathrm{t})$ is introduced, the mass balance equation for the mixture becomes:

$$\mathrm{div}(\mathbf{v}^s + \mathrm{n}\mathbf{v}^{ws}) = 0 \qquad \to \qquad \mathrm{Div}_s(\mathbf{V}^s + \mathrm{N}\mathbf{V}^{ws}) = 0 \ , \tag{2}$$

where $\mathrm{n}(\mathbf{x}, \mathrm{t})$ is the porosity. The term $\mathrm{n}\mathbf{v}^{ws}(\mathbf{x}, \mathrm{t})$ represents the filtration water velocity in terms of spatial co-ordinates, which is assumed to be described by the Darcy's law. The pull back in the reference configuration produces the mass balance equation (2b) with respect the material co-ordinates, where 'Div$_s$' is the divergence operator with respect to material co-ordinates of the solid and $\mathrm{N}\mathbf{V}^{ws}(\mathbf{X}^s, \mathrm{t})$ the material filtration water velocity.

The linear momentum balance equation for the mixture is:

$$\mathrm{div}\boldsymbol{\sigma} + \rho\mathbf{g} = 0 \qquad \to \qquad \mathrm{Div}_s\mathbf{P} + \rho_0\mathbf{g} = 0 \ , \tag{3}$$

where $\rho(\mathbf{x}, t)$ is the spatial density of the binary mixture and $\boldsymbol{\sigma}$ is the total Cauchy stress, which is assumed to be decomposed following the Terzaghi's principle. The pull back of (3a) in the reference configuration gives eq. (3b). $\mathbf{g}$ is the vector of the gravity acceleration in the reference and deformed configuration, $\rho_0(\mathbf{X}^s, t) = \rho_0(\mathbf{X}^s, t_0) + \rho^w(\mathbf{X}^s, t_0)[\mathbf{J}^s(\mathbf{X}^s, t) - 1]$ is the material mass density of the mixture (not constant because dependent on $\mathbf{J}^s$), with ρ^w the intrinsic water density and $\mathbf{J}^s$ the determinant of the solid deformation gradient $\mathbf{F}^s(\mathbf{X}^s, t)$.

As far as the constitutive equations are concerning, the isotropic elasto-plastic behaviour of the solid skeleton at finite strain is based on the multiplicative decomposition of $\mathbf{F}^s$ into the elastic and plastic part $\mathbf{F}^s = \mathbf{F}^{se}\,\mathbf{F}^{sp}$. The effective Kirchhoff stress tensor and the logarithmic principal values of the elastic left Cauchy-Green strain tensor are used (Simo and Hughes [9]).

The weak form of the spatial governing equations presented above has been derived by a standard procedure. Time integration of the weak form of the mass balance equation over a finite time step (t is necessary because of the time dependent term $\mathrm{div}\mathbf{v}^s$ (eq. 2a). The Generalised Trapezoidal Method has been used. The obtained non linear equation system has been linearized for its numerical solution by iterative methods. The linearisation of the weak form of the linear momentum balance equation essentially contains the algorithmic tangent moduli $\mathbf{a}^{ep}$ of the infinitesimal plasticity. For the computation the non-associated Drucker-Prager model with isotropic linear softening behaviour as a phenomenological description of elasto-plastic behaviour of sands has been used and hence the return mapping algorithm and the moduli $\mathbf{a}^{ep}$ have to be computed accordingly. A special treatment of the corner region of the D-P yield surface is necessary to avoid physically meaningless results characterised by the negative norm of the deviatoric stress tensor. Using the concept of multi-surface plasticity, a second yield function is introduced for the corner region and the plastic evolution equations are modified following the Koiter's generalisation. A suitable spatial Finite Element formulation has been derived applying the well known Galerkin procedure using isoparametric finite elements in space, with different shape functions for the solid and the water. A quadratic rate of convergence for the global Newton iteration in each step has been obtained during the computations.

3. Numerical example

The example is concerned with the analysis of a square domain of water saturated frictional porous material loaded by a rigid footing (Fig. 1).

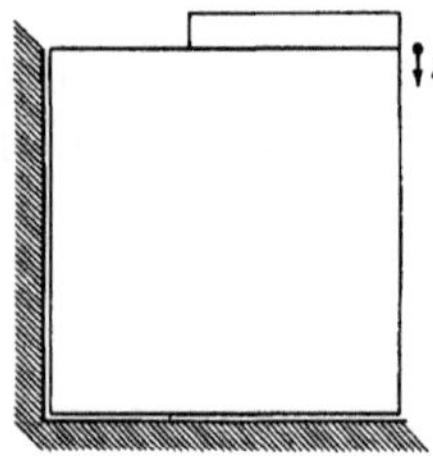

Figure 1. Rigid footing on a square domain of porous elastoplastic material.

The homogeneous soil domain has side length of 10 m, with the rigid footing spans over 6 m at the right part of the top surface. On the boundary, the horizontal and vertical displacements are constrained respectively on the left and bottom surface. Drainage of the water is only allowed through the unloaded part of the top surface of the domain. The loading is applied quasi-statically to the rigid footing by displacement control with constant vertical velocity until the maximum displacement of ∇ = 1 m is obtained after 100 loading steps. The domain has been discretised using 10x10 Q_2Q_1 elements for the solid and the fluid mesh, respectively. Plane strain conditions are assumed. The material parameters used for the computations of the case depicted in figure 2 are: solid bulk modulus K = 8333 kN/m^2, solid shear modulus G = 3486 kN/m^2, linear softening modulus h = -10 kN/m^2, initial apparent cohesion = 100 kN/m^2, isotropic water permeability = 0.0001 m/s, water unit weight = 10 kN/m^3.

Figure 2a shows the distribution of the equivalent plastic strain on the deformed configuration in case of dilatancy angle of 0° and friction angle of 20°. The plastic zone indicates the pronounced accumulation

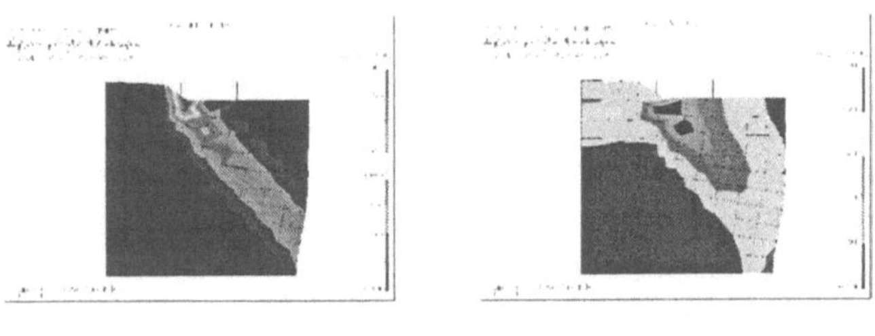

Figure 2. a) Equivalent plastic strain with zero dilatancy angle. b) Excess of water pressure distribution with zero dilatancy angle.

of inelastic strains in a narrow band, while the deformed configuration outlines the classical slip of a part of the domain over the other (no magnification of the displacements has been used in this and the others figures of the paper). Figure 2b shows the excess of the water pressures at the end of the load history. Negative values are observed only in two nodes close to the left corner of the footing and are not significant. The effect of the plastic dilatancy has been analysed in the following, solving the sample with $\phi = 20°$ and $\varphi = 5°$, $\phi = \varphi = 20°$ and $\phi = \varphi = 30°$. Increasing the value of the angle of dilatancy, an increase of the volumetric (plastic) strain and of the horizontal displacement of the right side of the panel was observed. In these cases a diffused and non-localised plastic zone was obtained, even if a negative hardening modulus was used in the computation, and the stiffness seemed to increase with ϕ and φ. This behaviour could be explained introducing the existence of a critical hardening modulus dependent from the solid and fluid material parameters and the yield and potential functions, similarly to the small strain analysis [Schrefler and Zhang]. Numerical computations varying only the value of the softening modulus seem to confirm the previous hypothesis, obtaining a softer deformation pattern with the decrease of the softening modulus. The same behaviour has been obtained increasing only the value of the solid bulk and shear moduli. In particular, a localised plastic zone into a shear band has been described after 200 load step (for a total vertical displacement of 2 m), solving the square panel with the following modified material parameters with respect to those used for Figures 2: h = -100 kPa, K = 250 MPa and G = 100 MPa. The contour of the equivalent plastic strain and of the excess of water pressures are depicted in Figures 3a-b respectively. It is interesting to note that negative value of excess of water pressure is observed only inside the band (Figure 3b), caused by the drop of the water pressure due to the development of the plastic dilatancy. The presence of negative pressure is not surprising. In fact, it was experimentally observed at localisation by Vardoulakis ([10]) and Mokni and Desrues ([6]) during biaxial tests of globally undrained dense sands under imposed displacements. In particular, the value of -91 and -80 kPa was measured by the two authors, respectively. At those pressures, cavitation of the pore fluid was observed, which means the presence of the vapour phase separated from the liquid phase by a meniscus. The second fluid phase hence implies the presence of capillary pressures, which should be modelled by introducing the effect of partial saturation. This improvement will be further pursued. The high values of negative excess of water pressures ($-1.66.10^{+04}$ kPa, Figure 3b) inside the shear band deserve some comments. In fact, they are overestimated because of the use of a saturated model and the D-P model, which do not take into account the influence of the partially saturation in the fluid-solid interaction and in the stiffness of the soil skeleton, and of the use of the Kirchhoff stress tensor.

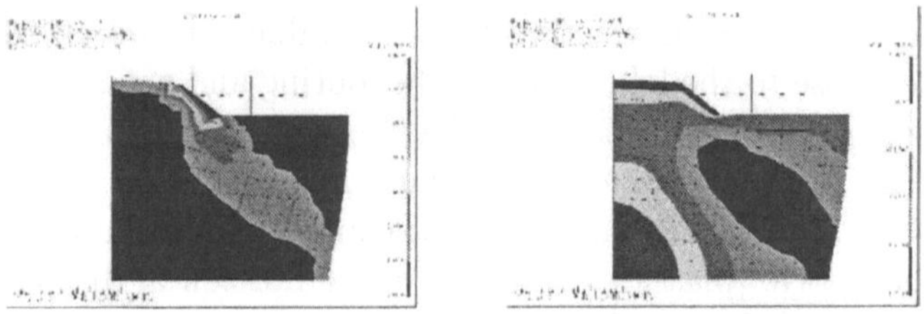

Figure 3. a) Equivalent plastic strain with dilatancy angle of 30°. b) Excess of water pressure distribution with dilatancy angle of 30°.

4. Conclusions

This paper shows early results of a research in progress on finite strain localisation on two phase geomaterials. The governing equations are derived in a material and spatial formulation. The elasto-plastic behaviour of the solid skeleton is based on the multiplicative decomposition of the deformation gradient into an elastic and plastic part. Numerical examples are shown.

References

1. Borja R. I. and Alarcon, E. A mathematical framework for finite strain elastoplastic consolidation. Part 1: Balance laws, variational formulation, and linearization". *Computer Meth. Appl. Mech. Eng.*, **122**, 145–171, 1995.
2. Diebels S. and Ehlers, W. Dynamic analysis of a fully saturated porous medium accounting for geometrical and material non-linearities, *Int. J. Numer. Meth. Eng.*, **39**, 81–97, 1996.
3. Hassanizadeh M. and Gray, W. G. General conservation equations for multiphase system: 1. Averaging technique. *Adv. Water Res.*, **2**, 131–144, 1979.
4. Lade P. V., Bopp P. A. and Peters, J. F. Instability of dilating sand. *Mechanics of Materials*, **16**, 249–264, 1993.
5. Lewis, R. W. and Schrefler, B. A. The Finite Element Method in the Static and Dynamic Deformation and Consolidation of Porous Media. 2a ed., John Wiley, New York, 1998.
6. Mokni M. and Desrues, J. Strain localisation measurements in undrained plane-strain biaxial tests on Hostun RF sand, *Mech. Cohes.-Frict. Mater.*, **4**, 419–441, 1998.
7. Nemat-Nasser S. On finite plastic flow of crystalline solids and geomaterials. *Transactions of the ASME*, **50**, 1114–1126, 1983.
8. Schrefler B. A. and Zhang, H. W. 2-D localisation analysis of saturated porous media. In W. Ehlers (ed.), Proceedings of the IUTAM Symposium on Theoretical and Numerical Methods in Continuum Mechanics of Porous Materials, Stuttgart, Sept. 1999, Kluwer Academic Publishers, 13–19, in this Volume.
9. Simo J. C. and Hughes, T. J. R. Computational Inelasticity, Springer, New York, 1998.
10. Vardoulakis I. Deformation of water-saturated sand: II the effect of pore water flow and shear banding. *Géotechnique*, **46**, 457–471, 1996.

Influence of Density and Pressure on Spontaneous Shear Band Formations in Granular Materials

E. Bauer and W. Huang
Institute of General Mechanics, Technical University Graz, Kopernikusgasse 24, A-8010 Graz, Austria.

Abstract. The aim of the present paper is to investigate the possibility of a spontaneous shear band formation in rate-independent, cohesionless and dry granular materials based on a hypoplastic constitutive model. Since the constitutive equation is incrementally non-linear the question as to whether the velocity gradient on either side of the discontinuity is related to loading, unloading or to a rigid body motion is irrelevant for the bifurcation analysis. By including a pressure dependent relative density the hypoplastic model describes the essential properties of initially dense and initially loose granular materials for a wide range of pressures. Thus, the influence of density and pressure on the stress ratio at the onset of a shear band localisation and the corresponding shear band orientation can be investigated with a single set of constitutive constants.

1. Constitutive model

In order to investigate the influence of pressure and density on shear banding in rate independent, cohesionless and dry granular materials a particular hypoplastic constitutive model proposed by Gudehus ([6]) and Bauer ([1]) is considered. In the following the main properties of the constitutive model are briefly summarised. State changes of the stress are described by an isotropic tensor-valued function depending on the current void ratio e, the Cauchy stress tensor $\mathbf{T}$ and the stretching tensor $\mathbf{D}$. The objective stress rate tensor $\overset{\circ}{\mathbf{T}}$ is represented by the sum of a tensorial function $f_s \, \mathsf{L} : \mathbf{D}$ which is linear in $\mathbf{D}$ and a tensorial function $f_s \, f_d \, \mathbf{N} \, ||\mathbf{D}||$ which is non-linear in $\mathbf{D}$ with respect to $||\mathbf{D}|| = \sqrt{\mathbf{D} : \mathbf{D}}$, i.e.

$$\overset{\circ}{\mathbf{T}} = f_s \, \mathsf{L} : \mathbf{D} + f_s \, f_d \, \mathbf{N} \, ||\mathbf{D}|| \, . \tag{1}$$

Herein the so-called stiffness factor $f_s = f_s(e,p)$ and the density factor $f_d = f_d(e,p)$ are related to the void ratio e and to the mean pressure $p = -\mathrm{tr}\mathbf{T}/3$. The fourth order tensor $\mathsf{L} = \mathsf{L}(\hat{\mathbf{T}})$ and the second order tensor $\mathbf{N} = \mathbf{N}(\hat{\mathbf{T}})$ are functions of the normalised tensor $\hat{\mathbf{T}} = \mathbf{T}/\mathrm{tr}\mathbf{T}$, i.e.

$$\mathsf{L} : \mathbf{D} = \hat{a}^2 \, \mathbf{D} + \hat{\mathbf{T}} \, \mathrm{tr}(\hat{\mathbf{T}}\mathbf{D}) \; , \quad \mathbf{N} = \hat{a} \, (\, \hat{\mathbf{T}} + \hat{\mathbf{T}}^* \,) \, , \tag{2}$$

W. Ehlers (ed.), IUTAM Symposium on Theoretical and Numerical Methods in Continuum Mechanics of Porous Materials, 245–250.

where $\hat{\mathbf{T}}^* = \hat{\mathbf{T}} - (1/3)\,\mathbf{I}$ denotes the deviator of $\hat{\mathbf{T}}$ and $\mathbf{I}$ is the unit tensor. The dimensionless factor $\hat{a}$ is related to stationary states which are included in the constitutive equation for a simultaneous vanishing of the stress rate and the volume strain rate, i.e. for $\mathring{\mathbf{T}} = \mathbf{0}$, $\mathrm{tr}\mathbf{D} = 0$ and $f_d = 1$ the constitutive equation (1) leads to $\hat{\mathbf{T}} = \mathbf{I}/3 - \hat{a}\,\mathbf{D}/||\mathbf{D}||$ and finally to $\hat{a} = ||\hat{\mathbf{T}}^*||$ (Bauer, [3]). Thus, $\hat{a}$ defines the stress limit condition for stationary states. In the present model factor $\hat{a}$ has been adapted to the limit condition given by Matsuoka and Nakai ([11]), i.e.

$$\hat{a} = \frac{\sin\varphi}{3+\sin\varphi}\left[||\hat{\mathbf{T}}^*|| + \sqrt{\frac{(8/3) - 3||\hat{\mathbf{T}}^*||^2 + (\sqrt{6}/2)\,||\hat{\mathbf{T}}^*||^3\,\cos(3\,\theta)}{1 + \sqrt{3/2}\,||\hat{\mathbf{T}}^*||\,\cos(3\,\theta)}}\,\right], \quad (3)$$

where θ represents the Lode-angle, which is defined as

$$\cos(3\,\theta) = -\sqrt{6}\,\mathrm{tr}(\hat{\mathbf{T}}^{*\,3})/[\mathrm{tr}(\hat{\mathbf{T}}^{*\,2})]^{3/2}\,.$$

The constant φ in Eq. (3) is called critical friction angle and is related to a stationary state in triaxial compression. In contrast to the elasto-plastic concept the limit condition in hypoplasticity is embedded in the constitutive equation. Therefore, the value for $\hat{a}$ is scaled by the norm of $\hat{\mathbf{T}}^*$ and $\hat{a} = ||\hat{\mathbf{T}}^*||$ is reached for stationary stress states.
The influence of the current void ratio e and the mean pressure p on the response of the constitutive equation (1) is taken into account with the density factor f_d, i.e.

$$f_d = \left(\frac{e - e_d}{e_c - e_d}\right)^\alpha, \quad (4)$$

and the stiffness factor f_s, i.e.

$$f_s = \frac{h_s}{n}\left(\frac{e_i}{e}\right)^\beta \frac{1+e_i}{e_i}\left(\frac{3\,p}{h_s}\right)^{1-n}\left[\hat{a}_o^2 + \frac{1}{3} - \frac{\hat{a}_o}{\sqrt{3}}\left(\frac{e_{i0} - e_{d0}}{e_{c0} - e_{d0}}\right)^\alpha\right]^{-1}. \quad (5)$$

Herein α and β are dimensionless and positive constitutive constants and $\hat{a}_o = \hat{a}(\hat{\mathbf{T}}^* = \mathbf{0})$. In relations (4) and (5) the current void ratio e is related to the void ratio in the loosest state (e_i), the void ratio of maximum densification (e_d), and the void ratio in a stationary state (e_c). For granular materials the range of possible void ratios is bounded by e_i and e_d, which are pressure dependent, i.e. the limit void ratios e_i, e_d and e_c decrease with an increase of the mean pressure (Gudehus, [7], Herle and Gudehus, [8]). This is modelled by the relation

$$e_i/e_{i0} = e_d/e_{d0} = e_c/e_{c0} = \exp[\,-(3p/h_s)^n\,]\,, \quad (6)$$

wherein e_{i0} and e_{d0} and e_{c0} are the appropriate void ratios for $p = 0$, h_s has the dimension of stress, and the exponent n is a dimensionless

constant.
For state changes of the void ratio e the assumption is made that the volume change of the grains can be neglected. Then the evolution of the void ratio is directly related to the volume strain rate $\mathrm{tr}\mathbf{D}$, i.e.

$$\dot{e} = (1 + e)\,\mathrm{tr}\mathbf{D} \,. \tag{7}$$

With the density factor f_d dilatancy, strain softening and the dependence of the peak friction angle on the void ratio is included in the constitutive equation (1). In this context the assumption of strain softening as a material behaviour must not be confused with geometrical softening of the discrete mechanical structure caused by shear bands. Since the current void ratio e is related to the limit void ratios by the pressure dependent functions f_s and f_d, the constitutive constants are not restricted to a certain initial density or stress state. Thus, the mechanical behaviour of initially dense or initially loose sand can be described using one set of constitutive constants. For the numerical investigations in this paper the calibration of the constants is based on data from compression tests and triaxial tests for medium quartz sand. The following values are used: $\varphi = 30°$, $n = 0.4$, $h_s = 190$ MPa, $\alpha = 0.11$, $\beta = 1.05$, $e_{i0} = 1.2$, $e_{d0} = 0.51$, $e_{c0} = 0.82$.

2. Bifurcation analysis

Based on the general bifurcation theory (Rudnicki and Rice, [13] and Rice and Rudnicki, [12]) the possibility of a spontaneous shear band formation was analysed by Kolymbas ([9]), Chambon and Desrues ([4]), Wu and Sikora ([14]) and others using previous versions of hypoplastic models. The influence of the void ratio and the mean pressure on shear banding was first investigated without any assumption regarding the jump of the velocity gradient at the discontinuity plane by Bauer and Huang ([2]). In the following the criterion for a spontaneous shear band formation is derived and applied to the constitutive equation (1) for states which are related to plane strain compression.
The so-called discontinuity plane, i.e. the boundary of the shear band, is characterised by a different velocity gradient $\nabla\mathbf{v}$ on either side of this plane. The jump of the velocity gradient, i.e.

$$[\![\nabla\mathbf{v}]\!] = \mathbf{g} \otimes \mathbf{n} \neq \mathbf{0} \,, \tag{8}$$

can be represented by a dyadic product of the unit normal $\mathbf{n}$ of the discontinuity plane and the vector $\mathbf{g}$ defining the evolution of the plane.

Continuous equilibrium across the discontinuity requires (Rice and Rudnicki, [12])

$$[\![\dot{\mathbf{T}}]\!]\,\mathbf{n} = 0\,, \tag{9}$$

where the jump of the stress rate is related to the jump of the Jaumann stress rate according to

$$[\![\dot{\mathbf{T}}]\!] = [\![\mathring{\mathbf{T}}]\!] + [\![\mathbf{W}]\!]\mathbf{T} - \mathbf{T}[\![\mathbf{W}]\!]\,. \tag{10}$$

Together with the response of the hypoplastic model (1), i.e.

$$[\![\mathring{\mathbf{T}}]\!] = f_s\,(\mathsf{L} : [\![\mathbf{D}]\!]) + f_s\,f_d\,\mathbf{N}\,[\![\,||\mathbf{D}||\,]\!]\,, \tag{11}$$

Eq. (10) and Eq. (9) lead to the relation

$$[\,f_s\,(\mathsf{L} : [\![\mathbf{D}]\!]) + f_s\,f_d\,\mathbf{N}\,[\![\,||\mathbf{D}||\,]\!] + [\![\mathbf{W}]\!]\,\mathbf{T} - \mathbf{T}\,[\![\mathbf{W}]\!]\,]\,\mathbf{n} = \mathbf{0}\,. \tag{12}$$

Herein the jump of the stretching and the spin tensor are related to the jump of the velocity gradient (8), viz. $[\![\mathbf{D}]\!] = [\mathbf{g}\otimes\mathbf{n} + \mathbf{n}\otimes\mathbf{g}]/2$ and $[\![\mathbf{W}]\!] = [\mathbf{g}\otimes\mathbf{n} - \mathbf{n}\otimes\mathbf{g}]/2$. At the onset of shear banding the stress and the void ratio are the same on either side of the discontinuity plane so that the quantities f_s, f_d, L and $\mathbf{N}$ are also the same. Since the constitutive equation (1) is incrementally non-linear there is no need in hypoplasticity to differentiate between separate constitutive relation for loading and unloading (Kolymbas, [10]). Therefore, the possibility of different incremental stiffnesses due to a different velocity gradient on either side of the discontinuity is taken into account by the same constitutive equation. Relation (12) can be rewritten as

$$\mathbf{K}\,\mathbf{g} = \lambda\,\mathbf{r}\quad , \tag{13}$$

with $\mathbf{K} = f_s\,[\,\hat{a}^2(\mathbf{I}+\mathbf{n}\otimes\mathbf{n})/2 + \hat{\mathbf{T}}.(\mathbf{n}\otimes\mathbf{n})\,\hat{\mathbf{T}}\,] + [\,(\mathbf{n}.\mathbf{T}\mathbf{n})\,\mathbf{I} - (\mathbf{n}\otimes\mathbf{n})\mathbf{T} - -\mathbf{T} + \mathbf{T}(\mathbf{n}\otimes\mathbf{n})\,]/2$, $\mathbf{r} = -f_s\,f_d\,\mathbf{N}\,\mathbf{n}$, and $\lambda = [\![\,||\mathbf{D}||\,]\!]$.

Inserting $\mathbf{g} = \lambda\,\mathbf{K}^{-1}\,\mathbf{r}$ into the norm of $[\![\mathbf{D}]\!]$, i.e.

$$\gamma = ||[\![\mathbf{D}]\!]|| = \sqrt{[\![\mathbf{D}]\!] : [\![\mathbf{D}]\!]} = \sqrt{[\mathbf{g}\cdot\mathbf{g} + (\mathbf{g}.\mathbf{n})^2\,]/2} \tag{14}$$

leads to the bifurcation condition:

$$\sqrt{\frac{(\mathbf{K}^{-1}\mathbf{r}).(\mathbf{K}^{-1}\mathbf{r}) + (\mathbf{K}^{-1}\mathbf{r}.\mathbf{n})^2}{2}} - \frac{\gamma}{|\lambda|} = 0\,. \tag{15}$$

The inequality $|[\![\,||\mathbf{D}||\,]\!]| \le ||[\![\mathbf{D}]\!]||$, i.e. $|\lambda| \le \gamma$, is valid independent of the amount of the stretchings on either side of the discontinuity. It was found out (Bauer and Huang, [2]) that the lowest bifurcation stress ratio is obtained for $|\lambda| = \gamma$ which implies the case that the stretchings are proportional on either side of the discontinuity and the case that one stretching is zero. The latter means a rigid body motion of the material outside the shear band. For a fixed co-ordinate system the components of vector $\mathbf{n}$ can be expressed by the shear band inclination angle ϑ. Thus, relation (15) represents an equation for the unknown ϑ, whereby only real solutions to (15) indicate the possibility of a spontaneous shear band formation. Real solutions for ϑ appear in pairs which are symmetric with respect to the principal stress directions.

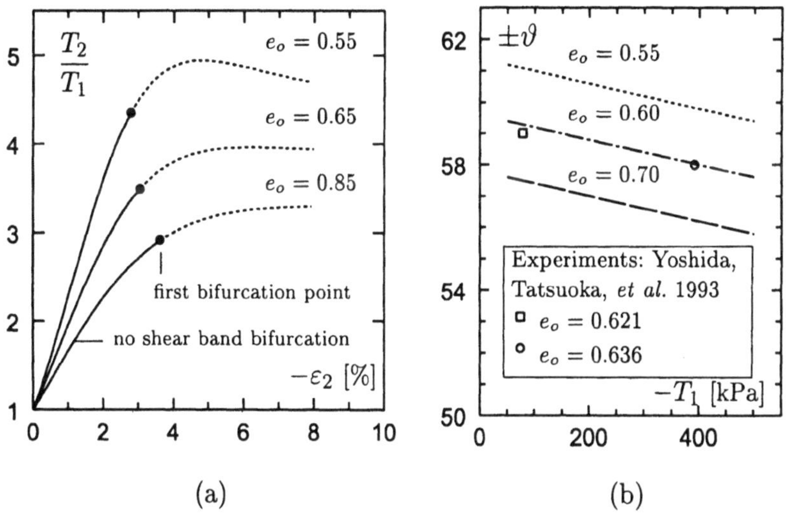

Figure 1. Onset of shear band formation in uniform biaxial compressions under constant lateral pressures T_1: (a) stress ratio T_2/T_1 versus vertical strain ε_2 for various initial void ratios and $T_1 = -100$ kPa, (b) Shear band orientation ϑ versus minimum stress T_1.

Figure 1 shows the results obtained from numerical examinations of the bifurcation condition (15) for stress paths which are related to homogeneous plane strain compressions starting from an isotropic stress state. The lowest bifurcation stress ratio detected by condition (15) is higher for an initially lower void ratio and it is reached for a smaller vertical strain than for a higher void ratio (Figure 1.a). States beyond the first bifurcation point fulfil again condition (15) in which additional real solutions are possible for $|\lambda| < \gamma$. The influence of the minimum principal stress T_1 and the void ratio on the shear band inclination ϑ is shown in Figure 1.b. For the same stress T_1 the angle ϑ increases with a decrease of the void ratio but for an increase of T_1 the shear band inclination decreases. The predictions are in accordance with the experiments performed by Yoshida *et al.* ([15]). A similar dependence of the shear band inclination on pressure and density was observed by Desrues and Hammad ([5]).

References

1. Bauer, E. Calibration of a comprehensive hypoplastic model for granular materials. *Soils and Foundations*, **36**, pp. 13–26, 1996.
2. Bauer, E. and Huang, W. The dependence of shear banding on pressure and density in hypoplasticity. *Proc. of the Int. Workshop on Localization and Bifurcation Theory for Soils and Rocks*, Gifu, Japan, eds. Adachi, Oka and Yashima, Balkema, pp. 81–90, 1997.
3. Bauer, E. Conditions for embedding Casagrande's critical states into hypoplasticity. *Mechanics of Cohesive-Frictional Materials*, **5**, pp. 125-148, 2000.
4. Chambon, R. and Desrues, J. Bifurcation par localisation et non linéarité incrémentale: un exemple heuristique d'analyse complète. *In: Plastic Instability, Presses de l'ENPC ed.*, Paris, pp. 101–113, 1985.
5. Desrues, J. and Hammad, W. Shear banding dependency on mean stress level in sand. *Proc. of the Int. Workshop on Numerical Methods for Localization and Bifurcation of Granular Bodies*, Gdansk, Poland, pp. 57–67, 1989.
6. Gudehus, G. A comprehensive constitutive equation for granular materials. *Soils and Foundations*, **36**, pp. 1–12, 1996.
7. Gudehus, G. Attractors, percolation thresholds and phase limits of granular soils. *Proc. Powder and Grains*, eds. Behringer and Jenkins, Balkema, pp. 169–183, 1997.
8. Herle, I. and Gudehus, G. Determination of parameters of a hypoplastic constitutive model from properties of grain assemblies. *Mechanics of Cohesive-Frictional Materials*, **4**, pp. 461–486, 1999.
9. Kolymbas, D. Bifurcation analysis for sand samples with non-linear constitutive equation. *Ingenieur-Archive*, **50**, pp. 131–140, 1981.
10. Kolymbas, D. (1991) An outline of hypoplasticity. *Archive of Applied Mechanics*, **61**, pp. 143–151
11. Matsuoka, H. and Nakai, T. Stress-strain relationship of soil based on the 'SMP'. *Proc. of Speciality Session 9, IX Int. Conf. Soil Mech. Found. Eng.*, Tokyo, pp. 153–162, 1977.
12. Rice, J. and Rudnicki, J. W. A note on some features on the theory of localization of deformation. *Int. J. Solids Structures*, **16**, pp. 597–605, 1980.
13. Rudnicki, J. and Rice, J. Conditions for the localization of deformation in pressure sensitive dilatant materials. *J. Mech. Phys. Solids*, **23**, pp. 371–394, 1975.
14. Wu, W. and Sikora, Z. Localized bifurcation in hypoplasticity. *Int. J. Engng. Sci.*, **29**, pp. 195–201, 1991.
15. Yoshida, T., Tatsuoka, F., Siddiquee, M. S. A., Kamegai, Y. and Park, C.S. Shear banding in sands observed in plane strain compression. *Proc. of the 3th Int. Workshop on Localisation and Bifurcation Theory for Soils and Rocks*, Grenoble, France, eds. Chambon, Desrues and Vardoulakis, Balkema, 1994, pp. 165–179, 1993.

Localization Analysis of a Saturated Elastic Plastic Porous Medium Based on Regularized Discontinuity

R. Larsson and J. Larsson
Departement of Solid Mechanics
Chalmers University of Technology
S-41296 Göteborg

Abstract. In the present paper we analyze the conditions for band-shaped localization to occur in the hydro-mechanically coupled problem. In the localization analysis, the concept of regularized discontinuity is used at the application to the conservation laws of momentum and mass. As a result, we obtain a coupled localization condition, where the situation of partly draınded conditions is discussed and compared to the extreme cases of fully drained and undrained situations.

Keywords: Porous media, plasticity, localization, discontinuity

1. Introduction

Analyses of localized failure have predominantly been restricted to onephase materials. However, when considering porous materials such as soil, the pores are iften filled with fluid whereby the behavior of the soil mass is influenced by the coupling of deformation and fluid diffusion and the drainage conditions for the characterization of a developing localization zone in the hydro-mechanically coupled problem. In the localization analysis, the concept of regularized strong discontinuity, cf. Larsson and Runesson ([1]) is extensively used at the application to the conservation laws of momentum and mass. The present work relates to Rudnicki ([2]), who presented and discussed a formulation for studying the effects of coupling between deformation and pore fluid diffusion on the development of localized deformation, aimed at fissured rock masses. In the localization analysis, he considered a band shaped zone of weakended material where the deformation is assumed to be localized.

2. Governing equations for saturated porous medium

Let us consider a fluid saturated porous solid skeleton within the small deformation theory. The constituents are denoted c^{α}, where the sup-index $\alpha = \mathrm{s}$ or $\alpha = \mathrm{f}$ denotes the solid and the fluid phases, respectively. We further assume that no voids can develop during deformation, whereby volume fractions must satisfy $\mathrm{n}_s = 1 - \mathrm{n}$ (and $\mathrm{n}_f = \mathrm{n}$ is

W. Ehlers (ed.), IUTAM Symposium on Theoretical and Numerical Methods in Continuum Mechanics of Porous Materials, 251–256.

the porosity).

Based on equilibrium consideration of the phases, we obtain the balance of momentum for the mixture per unit volume as

$$\nabla \cdot \bar{\sigma} + \hat{\rho} g = 0 \quad \text{with} \quad \bar{\sigma} = \boldsymbol{\sigma}_s + \boldsymbol{\sigma}_f \tag{1}$$

where $\bar{\sigma}$ is the total Cauchy stress, $\hat{\rho}$ is the saturated density of the soil fluid mixture, and g is the volume force per unit mass.

With the assumption of intrinsically incompressible contituents, the mass conservation for the porous medium becomes

$$\nabla \cdot \dot{\mathbf{u}} + \nabla \cdot \mathbf{v}_d = 0 \quad \text{with} \quad \mathbf{v}_d = n\mathbf{v}_r \quad \text{and} \quad \mathbf{v}_r = \mathbf{v}_f - \mathbf{v}_s \tag{2}$$

where $\mathbf{v}_d$ is the Darcian velocity which is the realive volumetric flow of fluid per unit area of the deforming soil mass, and $\dot{\mathbf{u}} \equiv \mathbf{v}_s$.

Based on thermodynamical arguments, we obtain Darcy's law for the fluidsoil interaction as

$$\mathbf{v}_d = -k\nabla p^e \tag{3}$$

where p^e is the excess pore fluid pressure and $k > 0$ is the Darcian permeability coefficient. (Henceforth, we drop the sup-index ^{e}, i.e. $p = p^e$)

Since no energy is stored or dissipated in the fluid, we retrieve the classical effective stress principle as $\boldsymbol{\sigma} = \bar{\sigma} + p1$. Considering the appropriate constitutive relation for $\boldsymbol{\sigma}$ within the plasticity framework, it may be shown in a standard fashion that the linerarized effective stress response $\dot{\boldsymbol{\sigma}} = \mathbf{E} : \dot{\boldsymbol{\varepsilon}}$ can be written in terms of the continuum tangent stiffness tensor $\mathbf{E}$ as

$$\mathbf{E} = \mathbf{E}^e \text{ if } (E) \text{ or } \mathbf{E} = \mathbf{E}^{ep} = \mathbf{E}^e - \frac{1}{h}\mathbf{E}^e : \mathbf{f}^* \otimes \mathbf{f} : \mathbf{E}^e \text{ if } (P) \tag{4}$$

where (P) and (E) denote plastic and elastic loading, respectively.

3. Localization condition

Let us investigate the condition of the existence of a regularized strong discontinutity for the hydro-mechanically coupled problem described above. To this end, we shall exploit the concept of regularized displacement discontinutiy, Larsson ([1]), within the domain $\mathbf{B}$ that is occupied by fluid saturated porous material, cf. Fig. 1. It is assumed that the discontinuities occurs across the internal surface Γ with unit normal $\mathbf{n}$.

Following the developments in e.g. Simo *et al.* ([4]), we assume, in particular, that the velocity field of the skeleton take on the structure

$$\dot{\mathbf{u}}(\mathbf{x},t) = \dot{\mathbf{u}}_c(\mathbf{x},t) + [\![\dot{\mathbf{u}}]\!](t) H_\Gamma(\mathbf{x}) \tag{5}$$

where $\mathbf{u}_c$ is the continuous portion of the solid displacement and $[\![\dot{\mathbf{u}}]\!]$ is the, spatially constant, jump of $\mathbf{u}$ across Γ, that is discontinuous with the Heaviside function H_Γ.

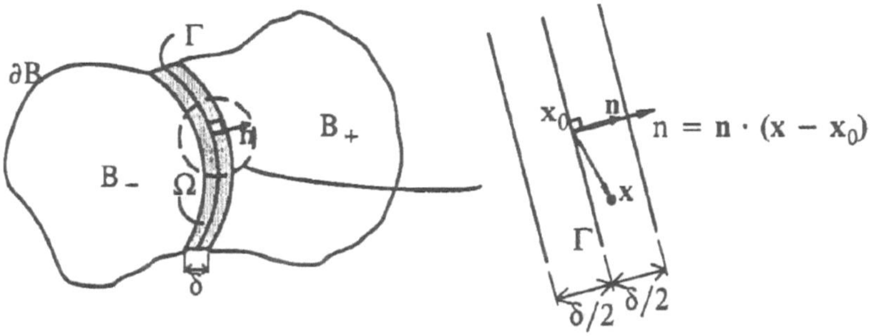

Figure 1. Body $\mathbf{B}$ with characteristic surface Γ. Regularization band of width δ, centered at Γ.

As to the gradient of H_Γ, we may formally define the Dirac delta function $\delta_\Gamma = \nabla H_\Gamma$. However, let us immendiately introduce a regularized version of δ_Γ. This is accomplished upon introducing a band shaped zone Ω along Γ with the width δ, as shown in Fig. 1. The strictly discontinuous Heaviside function is then replaced by a linear function across Ω, such that the regularization $\delta_{\nabla,r}$ becomes

$$\delta_{\Gamma,r} = \frac{f(n)}{\delta}\,\mathbf{n} \text{ with } f(n)\,\mathbf{x} = \begin{cases} 1 & \text{if } \mathbf{x} = \mathbf{x}_0 \in \Gamma \\ 0 & \text{if } \mathbf{x} = \mathbf{x}_\pm \in B\backslash\Gamma \end{cases} \tag{6}$$

where we define $\mathbf{x}_0 \in \Gamma$ and $\mathbf{x} \in \Omega$ such that $\mathbf{x} = \mathbf{x}_0 + n\mathbf{n}(\mathbf{x})$, with $-\delta/2 \le n \le \delta/2$.

We note that the regularized format $\delta_{\Gamma,r}$ is exploited at the establishment the divergence $\nabla \cdot \delta_{\Gamma,r}$ and the gradient $\nabla(\nabla \cdot \delta_{\Gamma,r})$. To this end, we consider the function $f(n)$ regularized such that

$$\nabla \cdot \delta_{\Gamma,r} := \frac{1}{\delta^2}\,g(n) \text{ with } g(n)\,\mathbf{x} = \begin{cases} 2 & \text{if } \mathbf{x} = \mathbf{x}_- \in B_-\backslash\Gamma \\ -2 & \text{if } \mathbf{x} = \mathbf{x}_+ \in B_+\backslash\Gamma \\ 0 \quad 0 & \text{otherwise} \end{cases} \tag{7}$$

Likewise, the gradient $\nabla(\nabla \cdot \delta_{\Gamma,r})$ is obtained by regularizing the function $g(n)$ such that

$$\nabla \cdot \delta_{\Gamma,r} := \frac{1}{\delta^3} h(n) \text{ with } h(n)\ \mathbf{x} = \begin{cases} -4 & \text{if } \mathbf{x} = \mathbf{x}_0 \Gamma \\ 0 & \text{if } \mathbf{x} = \mathbf{x}_\pm \in B \backslash \Gamma \end{cases} \tag{8}$$

We are now in the position to express the strain rate $\dot{\boldsymbol{\varepsilon}}$ pertinent to the proposed regularization. Hence, in view of (5) and (6) we obtain

$$\dot{\boldsymbol{\varepsilon}} = (\nabla \otimes \dot{\mathbf{u}})^{sym} = \dot{\boldsymbol{\varepsilon}}_c + \frac{f}{\delta}(\mathbf{n} \otimes [\![\dot{\mathbf{u}}]\!])^{sym} \tag{9}$$

Moreover, upon invoking (9) into the constitutive relation (4), we find that the effective stress rate field $\dot{\boldsymbol{\sigma}}$ takes the structure

$$\dot{\boldsymbol{\sigma}} = \dot{\boldsymbol{\sigma}}_c + f[\![\dot{\boldsymbol{\sigma}}]\!], \quad \dot{\boldsymbol{\sigma}}_c = \mathbf{E} : \dot{\boldsymbol{\varepsilon}}_c, \quad [\![\dot{\boldsymbol{\sigma}}]\!] = [\![\mathbf{E}]\!] : \dot{\boldsymbol{\varepsilon}}_c + \frac{1}{\delta}(\mathbf{E} \cdot \mathbf{n}) \cdot [\![\dot{\mathbf{u}}]\!] \tag{10}$$

In (10), we introduced $[\![\mathbf{E}]\!] = \mathbf{E}(\mathbf{x}_0) - \mathbf{E}(\mathbf{x}_\pm)$ as the jump in the the tangent stiffness.

As to the fluid pore pressure, we are guided by the efficive stress principle, i.e. $\bar{\boldsymbol{\sigma}} = \boldsymbol{\sigma} - p\mathbf{1}$, to employ the crucial assumption that p has the same regularity as the effective stress field. We thus assume

$$\dot{p} = \dot{p}_c + f[\![\dot{p}]\!] \quad \text{with} \quad [\![\dot{p}]\!] = \dot{p}(\mathbf{x}_0) - \dot{p}(\mathbf{x}_\pm) \tag{11}$$

where p_c is the continuous portion of the pressure field.

In order to establish the conditions that must be satisfied in order for a strong discontinuity to appear, we consider the continuity of the momentum and mass conservation relations, as expressed in eqs. (1) and (2). Assuming that the body force is continuous, it must be required due to linear momentum conservation that the *total* traction must be continuous:

$$\nabla \cdot \dot{\bar{\boldsymbol{\sigma}}} \mid_{\mathbf{x}=\mathbf{x}_\pm} - \nabla \cdot \dot{\bar{\boldsymbol{\sigma}}} \mid_{\mathbf{x}=\mathbf{x}_0} = 0 \quad \Rightarrow \quad \mathbf{n} \cdot [\![\dot{\bar{\boldsymbol{\sigma}}}]\!] = 0 \tag{12}$$

Likewise, we formulate the condition for continuity of the mass conservation across Γ in terms of eq. (2) as

$$(\nabla \cdot \dot{\mathbf{u}} + \nabla \cdot \mathbf{v}_d) \mid_{\mathbf{x}=\mathbf{x}_\pm} - (\nabla \cdot \dot{\mathbf{u}} + \nabla \cdot \mathbf{v}_d) \mid_{\mathbf{x}=\mathbf{x}_0} = 0 \tag{13}$$

which yields

$$-\frac{f}{\delta}\mathbf{n} \cdot [\![\dot{\mathbf{u}}]\!] - \frac{h}{\delta}\mathbf{n} \cdot [\![\mathbf{v}_d]\!] = 0 \tag{14}$$

Finally, upon introducing (10),(11) into (12) and (3), (10) into (14), we obtain a localization condition (=condition for the existence of regularized discontinuities) as follows:

$$\left.\begin{aligned} \tfrac{1}{\delta}\mathbf{Q}_\Gamma \cdot [\![\dot{\mathbf{u}}]\!] - [\![\dot{p}]\!]\mathbf{n} &= -\mathbf{n}\cdot[\![\mathbf{E}]\!] : \dot{\varepsilon}_c \\ \mathbf{n}\cdot[\![\dot{\mathbf{u}}]\!] + \tfrac{4}{\delta}k[\![p]\!] &= 0 \end{aligned}\right\} \text{with } \mathbf{Q}_\Gamma = \mathbf{n}\cdot\mathbf{E}(\mathbf{x}_0)\cdot\mathbf{n} \qquad (15)$$

4. Conditions for onset of localization for elasto-platicity

We consider the implications of the localization condition (15) in terms of the plasticity framework outlined in sub-section 2. To this end, it is assumed that plastic loading occurs along Γ as well as immediately outside. Only the situation at onset of localization is considered, whereby $\mathbf{E}^{ep}(\mathbf{x}_\pm) = \mathbf{E}^{ep}(\mathbf{x}_0)$. In this case we obtain

$$\begin{aligned} \tfrac{1}{\delta}\mathbf{Q}^{ep}\cdot[\![\dot{\mathbf{u}}]\!] &= 0 \\ \mathbf{n}\cdot[\![\dot{\mathbf{u}}]\!] + \tfrac{4}{\delta}k[\![p]\!] &= 0 \end{aligned} \qquad (16)$$

where $[\![\dot{\mathbf{u}}]\!]$ may be eliminated to give

$$[\![\dot{p}]\!] + c^{ep}[\![p]\!] = 0\,, \quad c^{ep} = \frac{4k}{\delta^2}\frac{1}{S^{ep}}\,, \quad S^{ep} = \mathbf{n}\cdot\mathbf{P}^{ep}\cdot\mathbf{n} \qquad (17)$$

In (17), the value of S^{ep} may be expressed more explicitly from the relation

$$S^{ep} = \frac{\mu_u}{\mu}\mathbf{n}\cdot\mathbf{P}^{ep}\cdot\mathbf{n} \qquad (18)$$

where μ and μ_u, (Runesson *et al.* [4]) represents eigenvalues of $\mathbf{Q}^{ep}$ with respect to drained and undrained conditions, respectively.

Pertinent to a ductile fracture process, as shown in Fig. 2, the value of μ will decrease from the value $\mu = 1$ (at elastic response) in a smooth fashion during plastic loading as a function of the hardening/softening parameter κ.

From (18), we note that a critical situation occurs when $\mathbf{Q}^{ep}$ becomes singular, i.e. μ traverses from positive to negative, whereby S^{ep} exhibits a singulatity. In view of eq. (17), localization is possible in the sense that $[\![\dot{p}]\!]$ may grow exponentially in time when the drained localization criterion is satisfied, i.e.when $\mathbf{Q}^{ep}$ becomes singular. We also note that if the state where the undrained localization condition is satisfied has been attained, i.e. $\mu_u = 0$, then $S^{ep} = 0$, corresponding to the situation that the discontinuity may be unbounded.

5. Concluding remarks

In the present paper we analyzed the condition for localization in the hydromechanically coupled problem. The concept of regularized discontinuity was extensively exploited, whereby a coupled localization condition was obtained. In the analytical study of the rate problem we found that if the underlying drained material signals localization, i.e. the drained localization condition is satisfied, any existing discontinuity (or perturbation thereof) may grow exponentially. On the other hand, given a homogeneous plastic state, it may be shown that the condition for localization with unloading outside the band to instanly appear, is that the undrained localization condition is satisfied.

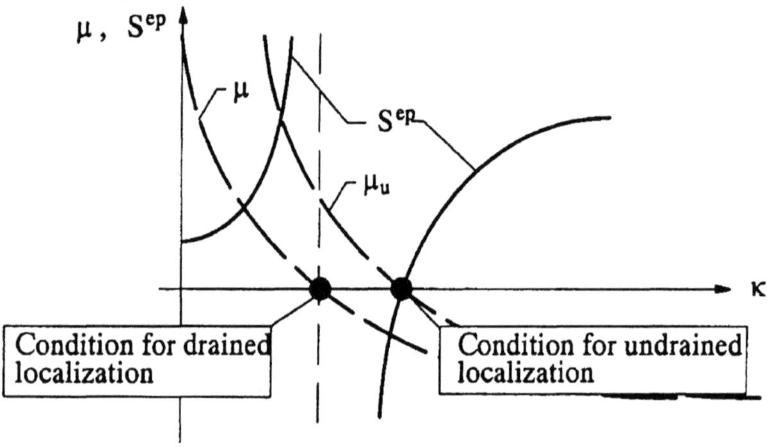

Figure 2. Schematic sketch of the development of the values of μ, μ_u and S^{ep}.

References

1. Larsson, R. and Runesson, K. Element-embedded localization band based on regularized displacement discontinuity. *J. Engrg. Mech. ASCE*, **122**, 402–411, 1996.
2. Rudnicki, J. W. A formulation for studying coupled deformation pore fluid diffusion effects on localization of deformations. *Geomechanics–AMD*, **57**, 35–44, ed. Nemat-Nasser, 1983.
3. Runesson, K., Peric, D. and Sture, S. Effect of pore fluid compressibility on localization in elastic-plastic porous solids under undrained conditions. *Int. J. Solids Structures*, **33**, 1501–1518, 1996.
4. Simo, J. C., Oliver, J. and Armero, F. An analysis of strong discontinuities induced by strain-softening in rate-independent solids, *Computational Mechanics*, **12**, 277–296, 1993.

Shear Band Localization in Frictional Geomaterials: Basic Modelling and Adaptive Computations

W. Ehlers and P. Ellsiepen
Institute of Applied Mechanics (Civil Engineering),
University of Stuttgart, D-70 550 Stuttgart, Germany
e-mail: (Ehlers, Ellsiepen)@mechbau.uni-stuttgart.de

Abstract. In saturated frictional geomaterials, shear bands occur as a result of local concentrations of plastic strains in small bands of finite width. Since in practice as well as in numerical simulations both the location of the onset and the direction of shear bands are generally unknown, time- and space-adaptive methods are an excellent tool to detect and to solve shear band problems. In the present contribution, the ill-posedness of the numerical computation of shear band phenomena is overcome by taking into account both the natural viscosity of the pore-fluid and the viscoplastic behaviour of the solid matrix.

Keywords: Frictional geomaterials, pore-fluids, localization phenomena, regularization, time and space adaptivity

1. Introduction

In the present contribution, fluid-saturated frictional geomaterials are considered within the well-founded framework of the Theory of Porous Media (TPM). Concerning a broad review of the TPM approach and the application of saturated as well as non-saturated porous media to various situations, the reader is referred to the work by Ehlers ([2]) and Diebels *et al.* ([1]). Elasto-plastic as well as elasto-viscoplastic frictional geomaterials are characterized by the fact that standard loading situations generally lead to plastic dilatation together with variations of the *Lode* angle. As a result, local concentrations of plastic strains in small bands of finite width (shear bands) occur, which have to be found and computed numerically, in the present contribution, by use of the finite element method (FEM). However, it is well known that the application of numerical methods to shear band phenomena reveals two severe problems, namely (1) the existence of a mathematically ill-posed problem and (2) the necessity to apply adaptive methods within the numerical procedures. To overcome the ill-posedness of the problem, the natural viscosity of the pore-fluid is taken into account and it is additionally assumed that the solid matrix is governed by an elasto-viscoplastic material law. Furthermore, since both the location and the development of shear bands cannot be foreseen in general, it is very convenient to proceed from time- and space-adaptive methods in order to detect the onset of the localization and to refine the FE mesh in

W. Ehlers (ed.), IUTAM Symposium on Theoretical and Numerical Methods in Continuum Mechanics of Porous Materials, 259–264.

the localization zone. In addition, using adaptive methods leads to a homogeneous distribution of the error both in the time and in the space domain ([3]). Refining the mesh also proves the regularization of the problem under study by the fact that the shear band width clearly exceeds the element size.

2. Governing Equations

By use of the Theory of Porous Media (TPM), fluid-saturated geomaterials are considered as a mixture of superimposed but immiscible constituents φ^α with particles X^α ($\alpha = S$: solid skeleton; $\alpha = F$: pore-fluid). Thus, each spatial point $\mathbf{x}$ of the current configuration at time t is simultaneously occupied by material points X^α of both constituents proceeding from different reference positions $\mathbf{X}_\alpha$ at time t_0. Furthermore, the primary kinematic variables of the binary model under study are the solid displacement vector $\mathbf{u}_S$ and the seepage velocity $\mathbf{w}_F$, viz:

$$\mathbf{u}_S = \mathbf{x} - \mathbf{X}_S\,, \qquad \mathbf{w}_F = \mathbf{v}_S - \mathbf{v}_F\,. \tag{1}$$

Therein, $\mathbf{v}_S = (\mathbf{u}_S)'_S$ and $\mathbf{v}_F$ are the macroscopic solid and pore-fluid velocities, where $(\,\cdot\,)'_\alpha$ indicates the material time derivative following the motion of the φ^α-th constituent.

Apart from the kinematical relations, the material under study is described by the balance relations of mass, momentum and moment of momentum ([2]). Associated with each φ^α is an effective (material) density $\rho^{\alpha R}$ and a partial (bulk) density ρ^α, which are related to each other by the volume fractions n^α through $\rho^\alpha = n^\alpha \rho^{\alpha R}$. Proceeding from the assumption of materially incompressible constituents ($\rho^{\alpha R} =$ const.), the partial densities can still vary through variations of the volume fractions. Following this, integration of the solid mass balance yields

$$n^S = n_0^S\,(\det \mathbf{F}_S)^{-1}\,, \tag{2}$$

where $\mathbf{F}_S = \operatorname{Grad}_S \mathbf{x}$ is the solid deformation gradient, and n_0^S is the solid volume fraction at time $t = t_0$. Excluding mass exchanges between the solid and the fluid constituents as well as inertia terms, the weak forms of the momentum and mass balance equations of the binary medium yield

$$\begin{aligned} \int_\Omega (\mathbf{T}_E^S - p\,\mathbf{I}) \cdot \operatorname{grad} \delta\mathbf{u}_S \,dv - \int_\Omega \rho\,\mathbf{b} \cdot \delta\mathbf{u}_S \,dv &= \int_{\Gamma_t} \bar{\mathbf{t}} \cdot \delta\mathbf{u}_S \,da\,, \\ \int_\Omega \operatorname{div} (\mathbf{u}_S)'_S \,\delta p \,dv - \int_\Omega n^F \mathbf{w}_F \cdot \operatorname{grad} \delta p \,dv &= -\int_{\Gamma_q} \bar{q}\,\delta p \,da\,. \end{aligned} \tag{3}$$

Therein, $\mathbf{T}_E^S = \mathbf{T}^S + n^S p\, \mathbf{I}$ is the effective part of the solid stress tensor $\mathbf{T}^S$, p is the effective fluid pressure, which, in the incompressible model under study, acts as a *Lagrange*an multiplier, whereas $\rho = n^S \rho^{SR} + n^F \rho^{FR}$ is the so-called mixture density and $\mathbf{b}$ the external volume force per unit mass (gravity). In addition, $\delta \mathbf{u}$ and δp are test functions corresponding to the respective field quantities. Furthermore, $\bar{\mathbf{t}}$ is the external load vector acting on the *Neumann* boundary Γ_t, whereas $\bar{q} = n^F \mathbf{w}_F \cdot \mathbf{n}$ is the volume flux acting and on Γ_q, wherein $n^F \mathbf{w}_F$ and $\mathbf{n}$ are the filter velocity and the outward oriented unit surface normal. Finally, grad$(\,\cdot\,)$ and div$(\,\cdot\,)$ are the gradient and divergence operators with respect to $\mathbf{x}$.

To close the problem under study, constitutive equations for the filter velocity and for the effective solid stress are needed. As was discussed, e. g., by Ehlers ([2]) and Diebels *et al.* ([1]), the filter velocity is governed by *Darcy*'s law

$$n^F \mathbf{w}_F = -\frac{K^S}{\mu^{FR}} (\operatorname{grad} p - \rho^{FR} \mathbf{b}) , \tag{4}$$

where K^S is the intrinsic permeability coefficient and μ^{FR} the effective fluid viscosity. Proceeding from an additive decomposition of the geometrically linear *Lagrange*an strain $\boldsymbol{\varepsilon}_S$ into elastic and plastic parts $\boldsymbol{\varepsilon}_{Se}$ and $\boldsymbol{\varepsilon}_{Sp}$, the elasticity law for $\mathbf{T}_E^S$ reads

$$\mathbf{T}_E^S = 2\,\mu\, \boldsymbol{\varepsilon}_{Se} + \lambda\, (\,\boldsymbol{\varepsilon}_{Se} \cdot \mathbf{I}\,)\, \mathbf{I} . \tag{5}$$

Therein, μ and λ are the *Lamé* constants of the porous skeleton material. The viscoplastic material behaviour of cohesive-frictional geomaterials is governed by the yield function

$$\begin{aligned} F &= [\,\mathrm{II}^D (1 + \gamma\vartheta)^m + \tfrac{1}{2}\alpha\, \mathrm{I}^2 + \delta^2\, \mathrm{I}^4\,]^{1/2} + \beta\, \mathrm{I} + \epsilon\, \mathrm{I}^2 - \kappa = 0 , \\ \vartheta &= \mathrm{III}^D / (\mathrm{II}^D)^{3/2} , \end{aligned} \tag{6}$$

the plastic potential

$$G = [\,\mathrm{II}^D + \tfrac{1}{2}\alpha\, \mathrm{I}^2 + \delta^2\, \mathrm{I}^4\,]^{1/2} + \beta\, \mathrm{I} + \epsilon\, \mathrm{I}^2 - g\,(\,\mathrm{I}\,) \tag{7}$$

and the flow rule

$$(\boldsymbol{\varepsilon}_{Sp})'_S = \frac{1}{\eta} \left\langle \frac{F(\mathbf{T}_E^S)}{\sigma_0} \right\rangle^r \frac{\partial G}{\partial \mathbf{T}_E^S} . \tag{8}$$

In (6)-(8), I, II^D and III^D are the principal invariants of the effective stress $\mathbf{T}_E^S$ and of the effective stress deviator $\mathbf{T}_E^{SD}$. Furthermore, the parameters α, β, γ, δ, ϵ, m, κ govern the shape of the yield function, whereas $g(\,\mathrm{I}\,)$ relates the dilatation angle to experimental data. Finally,

$\langle \cdot \rangle$ are the *Macauley* brackets, η is the relaxation time, σ_0 is the reference stress, and r is the viscoplastic exponent.

3. Time and space adaptivity

In the framework of the FEM, the spatial discretization of the field equations (3) is based on quadratic shape functions for the solid displacement $\mathbf{u}_S$ and linear shape functions for the fluid pressure p (primary variables $\boldsymbol{u}$). Furthermore, the evolution equation for the viscoplastic strain tensor (8) (internal variables $\boldsymbol{q}$) are computed at the integration points of the numerical quadrature. Following this, one obtains a semi-discrete system of first order in time that can be solved by time- and space-adaptive methods ([1], [3]).

In the time domain, one-step methods with an embedded time step control are applied, where the solution at time t_{n+1} only depends on the solution at time t_n. This choice is of essential importance with respect to space-adaptive methods, since the transfer of the numerical solution thus only includes two meshes. In particular, 2-stage singly diagonally implicit *Runge-Kutta* methods (SDIRK) are used, thus allowing for an efficient embedded time-step control through the error estimation

$$ERR \approx \|\boldsymbol{y}_{n+1} - \hat{\boldsymbol{y}}_{n+1}\|, \tag{9}$$

where $\boldsymbol{y}_{n+1}$ and $\hat{\boldsymbol{y}}_{n+1}$ are two solutions of $\boldsymbol{y} = (\boldsymbol{u}, \boldsymbol{q})^T$ at time t_{n+1} with convergence orders $p = 2$ and $\hat{p} = 1$ ([3]). Using the relative and the absolute tolerances ϵ_r and ϵ_a together with the weighted error measures

$$e_u := \left(\frac{1}{N} \sum_{k=1}^{N} \left[\frac{u_{n+1}^k - \hat{u}_{n+1}^k}{\epsilon_r \cdot |u_n^k| + \epsilon_a} \right]^2 \right)^{1/2}, \qquad e_q := \max_k \left| \frac{q_{n+1}^k - \hat{q}_{n+1}^k}{\epsilon_r \cdot |q_n^k| + \epsilon_a} \right|, \tag{10}$$

where u^k and q^k are the components of $\boldsymbol{u}$ and $\boldsymbol{q}$ and $N = \dim \boldsymbol{u}$, the time-step is accepted if $e_y := \max\{e_u, e_q\} \le 1$ and rejected otherwise. In both cases, a new step size is predicted by

$$h_{\text{new}} := h_{\text{old}} \cdot \min\left\{ f_{\text{max}}, \max\left\{ f_{\text{min}}, f_{\text{safety}} \cdot e_y^{-1/(\hat{p}+1)} \right\} \right\}. \tag{11}$$

Therein, $f_{\text{safety}} < 1$ is a safety factor, which prevents an oscillation of the time-step size, whereas $f_{\text{max}} > 1$ and $f_{\text{min}} < 1$ are used to limit the step size variation.

In the space domain, the adaptive procedure is based on an extension of the gradient-based error indicator by *Zienkiewicz* and *Zhu* ([4])

through the inclusion of the driving quantities of the fluid-saturated elasto-viscoplastic material ([3]). Proceeding from the L_2-norm $\|\cdot\|_2$ and the corresponding element-wise norm $\|\cdot\|_{2,e}$, smoothened values $(\cdot)^*$ are computed on the basis of the FE quantities $(\cdot)^h$. Following this, the error indicators

$$\eta_1^{(e)} := \|\mathbf{T}_E^{S\,*} - \mathbf{T}_E^{S\,h}\|_{2,e}\,, \quad \eta_2^{(e)} := \|\boldsymbol{\varepsilon}_{Sp}^{*} - \boldsymbol{\varepsilon}_{Sp}^{h}\|_{2,e}\,, \quad \eta_3^{(e)} := \|\mathbf{w}_F^{*} - \mathbf{w}_F^{h}\|_{2,e}\,, \tag{12}$$

are applied, where $\eta_1^{(e)}$ considers the solid elasticity through the effective stresses, $\eta_2^{(e)}$ the viscoplatic deformations through the plastic strains and $\eta_3^{(e)}$ the fluid viscosity through the seepage velocity. The domain integrals

$$W_1 := \|\mathbf{T}_E^{S\,h}\|_2, \quad W_2 := \|\boldsymbol{\varepsilon}_{Sp}^{h}\|_2, \quad W_2 := \|\mathbf{w}_F^{h}\|_2, \tag{13}$$

serve as reference quantities of the respective error indicators. Given the relative and absolute tolerances ϵ_r and ϵ_a together with the weighting factors $\alpha_i \geq 0$ to be included in the local error measures ξ_e, weighted global error measures ξ and weighted local error measures ξ_e per element can be computed on the actual mesh with N_e elements and $\sum_{i=1}^{3} \alpha_i = 1$:

$$\xi := \left(\sum_{e=1}^{N_e} \xi_e^2 \right)^{1/2}, \quad \xi_e := \left(\sum_{i=1}^{3} \alpha_i \left[\frac{\eta_i^{(e)}}{\epsilon_r \cdot W_i + \epsilon_a} \right]^2 \right)^{1/2}. \tag{14}$$

The solution on the actual mesh is accepted if $\xi \leq 1$ and not accepted else. In order to refine ($\xi > 1$) or to coarsen ($\xi \leq 1$) the mesh, a new element radius h_{new} is be computed in the frame of a hierarchical scheme via

$$h_{\text{new}} := h_{\text{old}} \cdot \xi_e^{-1/(r+1)} \cdot \left[\sum_{e=1}^{N_e} \xi_e^{2/(r+1)} \right]^{-1/(2r)}. \tag{15}$$

4. Numerical example

The numerical example exhibits the time- and space-adaptive computation of a biaxial experiment, where a side load of $t_1 = 110\,\text{kN/m}^2$ was applied to a liquid-saturated elasto-viscoplasitic specimen before a displacement driven vertical loading process took place. The specimen was fully drained on both sides and undrained on the top and at the bottom, where, to initiate a shear band, a weakened zone was assumed.

It is seen from Figure 1 that (1) a considerable mesh refinement took place in the localization zone and that (2) the included solid and fluid viscosities led to a satisfying regularization of the problem. Furthermore, as a result of the applied side load, a dilatant shear band was obtained, what is easily seen from the arrows of the seepage velocity streaming into the shear band. Finally, increasing t_1 considerably leads to a compressive shear band situation.

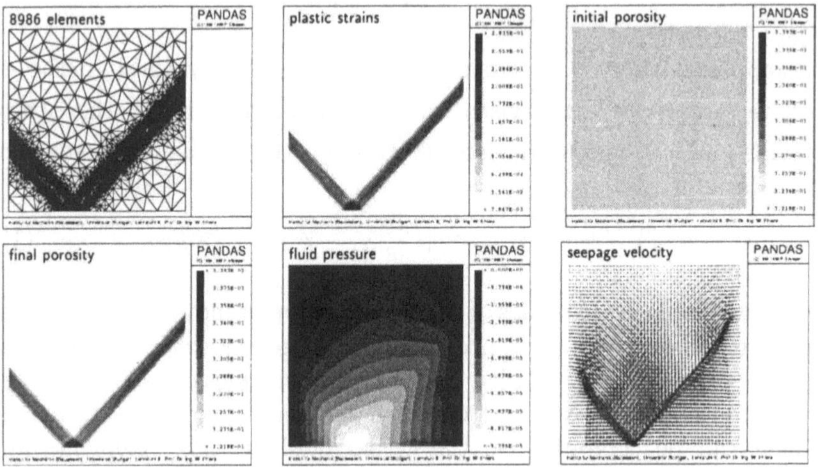

Figure 1. Dilatant shear band development ($t_1 = 110\,\mathrm{kN/m^2}$).

References

1. Diebels, S., Ellsiepen, P. and Ehlers, W. Error-controlled *Runge-Kutta* time integration of a viscoplastic hybrid two-phase model. *Technische Mechanik* **19**, 19–27, 1999.
2. Ehlers, W. Constitutive equations for granular materials in geomechanical context. In: K. Hutter (ed.): *Continuum Mechanics in Environmental Sciences and Geophysics*, CISM Courses and Lectures No. 337. Springer-Verlag, Wien, pp. 313–402, 1993.
3. Ellsiepen, P. *Zeit- und ortsadaptive Verfahren angewandt auf Mehrphasenprobleme poröser Medien.* Dissertation, Bericht Nr. II-3 aus dem Institut für Mechanik (Bauwesen), Universität Stuttgart, 1999.
4. Zienkiewicz, O. C. and Zhu, J. Z. A simple error estimator and adaptive procedure for practical engineering analysis. *Int. J. Numer. Methods Eng.* **24**, 337–357, 1987.

Analysis of Instability Conditions for Normally Consolidated Soils

R. Nova and S. Imposimato
Milan University of Technology (Politecnico) Piazza L. da Vinci, 32, 20133 Milan, Italy

Abstract. The traditional method of undrained stability analysis of geostructures is addressed first. It is shown that the reason why it is possible to predict the collapse condition by using only total stresses is linked to that this method is not a limit analysis (in the sense of plasticity theory) but an analysis of the instability condition under the constraint of no volume change. It is shown further that a normally consolidated or slightly overconsolidated soil specimen may become unstable, under more general loading , for very low values of the stress level. Under convenient loading conditions, unlimited pore water pressure generation, drained or undrained shear banding, loss of controllability of the loading programme may take place still in the hardening regime, well before the limiting stress condition is achieved.

1. Introduction

The term stability is traditionally used in Soil Mechanics with a meaning different from the same term used in Solid Mechanics and in other branches of Science. In a mathematical sense, stability is usually interpreted to mean that any possible infinitesimal disturbance will cause only infinitesimal departures from the given equilibrium configuration, while in traditional Soil Mechanics the term stability is used making reference to a limiting equilibrium state, controlled by the strength of the material. Clearly, the limit state condition is unstable in the sense that a small perturbation (a load increment for instance) causes collapse of the structure and therefore large departures from the initial equilibrium configuration. The converse is not true, however, since we may have unstable equilibrium configurations even though the limit strength of the material is nowhere reached (as in the Euler beam for instance). In the case of the soils with low permeability such as clay, the stability analysis is made in undrained conditions. This method is very effective in determining the failure conditions of a slope or the limit height of an embankment, despite the fact that the undrained strength has no actual mechanical meaning. The strength of the soil is in fact controlled by effective stresses and by the Mohr-Coulomb limit condition. The Tresca criterion in total stresses appears only as an artifice due to a misunderstanding of the role of pore water pressures in soil in the early times of the development of Soil Mechanics. The aim

W. Ehlers (ed.), IUTAM Symposium on Theoretical and Numerical Methods in Continuum Mechanics of Porous Materials, 265–272.

of this paper is to solve this apparent contradiction. It will be shown first that the traditional limit analysis in undrained conditions and total stresses coincides in fact with a proper (static) stability analysis in term of effective stresses, which is the reason why good predictions can be obtained by means of a method based on wrong premises. It will be shown in fact that the limit condition in total stresses coincides with the possibility of existence of an undrained shear band, what may happen well before the Mohr-Coulomb effective limit condition is achieved. It will be shown further that the instability condition is a function of the type of loading programme followed. Whenever an instability occurs, the second order work is zero. The locus in the stress space where the second order work becomes zero first will be determined. A locus for which unlimited pore pressure generation may occur will be also determined as well as loci for drained and undrained shear banding.

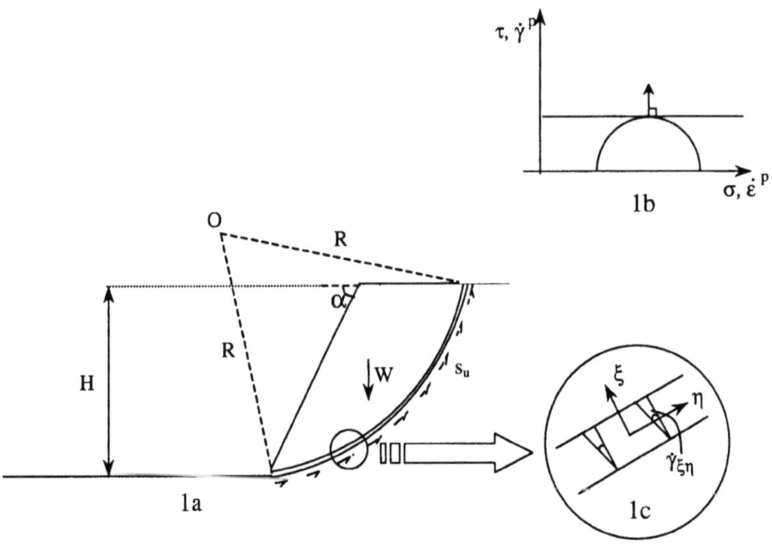

Figure 1. Slope stability analysis

2. Slope stability analysis

Consider the slope of Figure 1, which we shall assume is made of clay either normally consolidated or slightly overconsolidated. Heavily overconsolidated clays are not considered here, since their behaviour is brittle and gives rise to progressive failure. For certain combinations of the mechanical and geometrical parameters, a portion of soil can slide downhill while the rest remains in place. A thin transition zone delineates between the moving soil and the soil at rest. Compared to the size of the slope such a zone can be considered to be of infinitesimal thickness and reduces to a line (if plane strain conditions are assumed to hold true). In the traditional analysis such a line is a circular arc. The collapse condition is obtained by imposing that the overturning moment of the weight of the upper portion of soil with respect to the centre of the arc is counterbalanced by the stabilising moment due to the shear stresses along the transition line. Such stresses are assumed to be equal to the undrained strength s_u since drainage is not allowed for. The clay is therefore assumed to have a Tresca type failure condition. Such an analysis is a limit equilibrium analysis and it can be shown that it coincides with a limit analysis of rigid-plastic theory, since in this case the normality rule holds true (Fig. 1b), for the imposed undrained conditions. We can look at this problem from another standpoint, however. Consider an element volume of the transition zone and take as reference axes the axis ξ orthogonal to the line, the axis η tangent to the line, and axis χ orthogonal to the plane $\xi\eta$, Fig 1c. Such an element will be sheared by the motion of the upper part of the soil under the following constraints:

$$\dot{\varepsilon}_\chi = \dot{\gamma}_{\chi\eta} = \dot{\gamma}_{\chi\xi} = 0 \tag{1}$$

for collapse is occurring in plane strain conditions;

$$\dot{\varepsilon}_\eta = 0 \tag{2}$$

since the transition zone, infinitesimally thin in the x direction, can be assumed infinitely long in the h direction;

$$\dot{\varepsilon}_\xi = 0 \tag{3}$$

since no volume change can take place and both $\dot{\varepsilon}_\chi$ and $\dot{\varepsilon}_\eta$ are zero. For the problem at hand, therefore, the only strain rate component that can be different from zero is $\dot{\gamma}_{\xi\eta}$.Each element of the shear band of Fig. 1c undergoes therefore an undrained simple shear test. A change in $\gamma_{\xi\eta}$ will cause a change in the effective stresses according to the constitutive law

$$\{\dot{\sigma}'_\xi \;\; \dot{\sigma}'_\eta \;\; \dot{\sigma}'_\chi \;\; \dot{\tau}_{\xi\eta} \;\; \dot{\tau}_{\chi\xi} \;\; \dot{\tau}_{\chi\eta}\}^T = \{\mathrm{D}_{14} \;\; \mathrm{D}_{24} \;\; \mathrm{D}_{34} \;\; \mathrm{D}_{44} \;\; \mathrm{D}_{54} \;\; \mathrm{D}_{64}\}^T \dot{\gamma}_{\xi\eta} \tag{4}$$

where D_{ij} are elements of the stiffness matrix. If stresses and strains are assumed to be coaxial

$$\mathrm{D}_{54} = \mathrm{D}_{64} = 0 \tag{5}$$

since χ is also a principal stress axis. In order to have $\dot{\gamma}_{\xi\eta} \neq 0$, in general it is necessary to change the external loads and therefore the stresses. If, however,

$$D_{44} = 0 \tag{6}$$

a spontaneous variation of $\gamma_{\xi\eta}$ is possible without any change in external loads. Such a shear strain increment causes variation of $\dot{\sigma}'_\chi$ and $\dot{\sigma}'_\eta$, which, summed to the variation of pore pressure, u̇, generate in turn total stress variations $\dot{\sigma}_\chi$ and $\dot{\sigma}_\eta$, which are self-equilibrated. The shear components are all zero for Eqs. (5, 6). Under constant normal loads, a possible self-equilibrated solution is:

$$\dot{\sigma}'_\xi = D_{14}\dot{\gamma}_{\xi\eta} = -\dot{u} \, . \tag{7}$$

Any value of $\dot{\gamma}_{\xi\eta} \neq 0$ is therefore associated to zero load change. Equation (6) is therefore a condition for non-uniqueness of the solution, analogous to the condition we obtain in a static instability analysis of the Euler beam. Eq. (6) is an instability condition since under constant external loads unlimited shear strains and therefore displacements may occur and is expressed in term of effective stresses since the mechanical behaviour of soil and therefore the stiffness matrix depend solely on effective stresses. Condition (6) corresponds to the peak of the shear stress shear strain curve in an undrained simple shear test, as shown in Fig.2, and determines therefore the traditional undrained strength of material s_u.

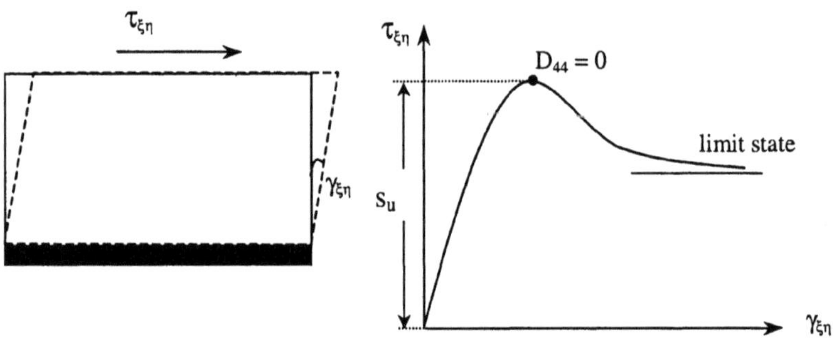

Figure 2. Undrained simple shear test

It is important to note that, as it will be shown in the following section, the stress point for which Eq. (6) is fulfilled is not at all at the limit state, which corresponds to the ordinary drained strength. If we would be able to reach the peak in undrained simple shear and then change the loading programme (for instance allowing drainage to occur) the specimen would be able to sustain a shear stress increment. Unstable states may occur, therefore, before the limit condition is reached, i.e. still in the hardening regime, and a stress condition can be unstable or not depending on the type of loading programme followed.

3. Determination of unstable states

Consider again the element of Fig.1b in the transition zone or the simple shear test, Fig.2. It is straightforward to note that when Eq. (6) is fulfilled the second order work is zero. All strain rate components other than $\dot{\gamma}_{\xi\eta}$ are zero in fact as well as $\dot{\tau}_{\xi\eta}$. The second order work is zero, however, for particular stress paths, even for stress states which are far away from any traditional failure condition. The second order work is in fact zero first when the stiffness matrix becomes positive semidefinite

$$\dot{\sigma}'_{ij}\,\dot{\varepsilon}_{ij} = \mathrm{D}_{ijhk}\,\dot{\varepsilon}_{hk}\,\dot{\varepsilon}_{ij} = 0 \tag{8}$$

what happens when the determinant of the symmetric part of the stiffness matrix is zero. Soil behaviour is often described by elasto-plastic work-hardening laws with non-associated flow rules (Lade, [3]; Nova and Wood, [4]: Pastor *et al.*, [7]), therefore the stiffness matrix is non-symmetric and by the Ostrowksy and Taussky ([6]) theorem when the determinant of the symmetric part of the stiffness matrix is zero, the determinant of the entire stiffness matrix is positive, which means that the soil is still in the hardening regime. By taking a convenient constitutive model, e.g. Nova ([5]), we can quantify when limit state and loss of positive semi-definiteness take place. Such loci are depicted in Fig.3 for loose Hostun sand. It is clear that the stress level for which the second order work can be zero first for a particular stress path is very small, even lower than the stress level characteristic of the geostatic state at rest. For clay, the deviation from normality is less marked than for sand and the locus for loss of positive definiteness approaches the limit locus, but it is still different. Only the stress points within the inner locus of Fig.3 are therefore associated to stable states. If a point belongs to the zone between the two loci of Fig.3, there is at least one special loading path for which unlimited strain response may occur (Imposimato and Nova, [1]).

Assume for instance that a minor of the stiffness matrix is nil, e.g.

$$\det \mathrm{D}_{11} = 0 \tag{9}$$

where

$$\begin{Bmatrix} \dot{\sigma}'_1 \\ \dot{\sigma}'_2 \end{Bmatrix} = \begin{bmatrix} \mathbf{D_{11}} & \mathbf{D_{12}} \\ \mathbf{D_{21}} & \mathbf{D_{22}} \end{bmatrix} \begin{Bmatrix} \dot{\varepsilon}_1 \\ \dot{\varepsilon}_2 \end{Bmatrix} \tag{10}$$

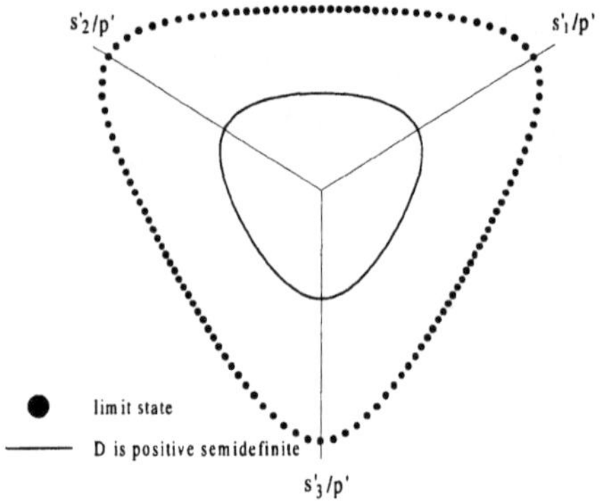

Figure 3. Loci for det $\mathbf{D} = 0$ (limit state) and det $\mathbf{D}_s = 0$ according to 'Sinfonietta classica'

is the partitioned stiffness matrix. To zero stress increment it is obviously associated a zero strain increment. However if we control partly stress ($\dot{\sigma}'_1$) and partly strain ($\dot{\varepsilon}_2$), it can be that a zero variation of the controlling quantities is associated to an infinity of responses of the type:

$$\dot{\sigma}'_2 = \beta \mathrm{D}_{21} \dot{\varepsilon}_1^* \tag{11}$$

where $\dot{\varepsilon}_1^*$ is the normalised eigenvector of matrix D_{11} and β is an arbitrary scalar. Fig.4 shows the loci for which it is fulfilled Eq. (9). In this figure it is shown that two particular stress paths (b and c) cannot be continued beyond the point for which such loci are reached, because condition (9) is fulfilled and an infinity of solutions is possible.

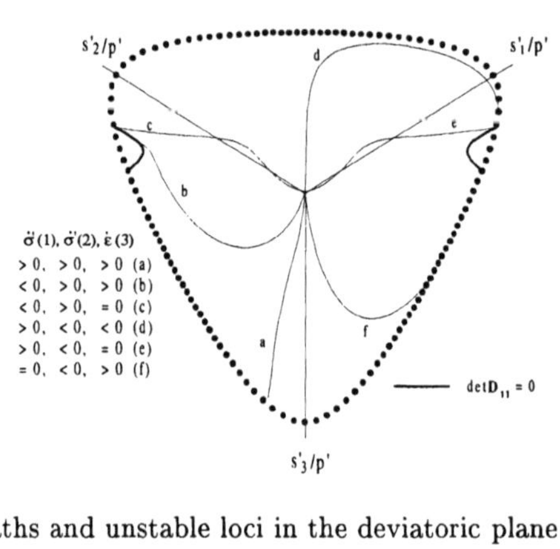

Figure 4. Stress paths and unstable loci in the deviatoric plane

Condition (6) is a special case of the more general condition (9). If we restrict our analysis to undrained conditions, for which we control the volume change and for instance two stress parameters, e.g. $\dot{\sigma}'_1 - \dot{\sigma}'_2$ and

$\dot{\sigma}'_1 - \dot{\sigma}'_3$, it can be shown (Imposimato and Nova, [2]) that when the volumetric compliance under isotropic loading is zero [$C_{pp} = 0$], unlimited spontaneous increase in pore water pressure is possible under no external load increment in undrained conditions ($\dot{\sigma}_1 = \dot{\sigma}_2 = \dot{\sigma}_3 = \dot{v} = 0$), with a consequent generation of unlimited strains. The locus for which this occurs is shown in Fig.5. If we now consider a biaxial undrained stress path, we find that the locus of the peak of the deviatoric stresses, which coincides with the locus for which an undrained shear band can be formed and therefore with condition (6) in general loading conditions, is given by a curve also shown in Fig.5. Depending on the initial state after consolidation such a peak will be reached sooner or later and therefore far or close to the limit locus. A similar analysis would allow to determine the locus for occurrence of a drained shear band, which can be seen as a special case of loss of controllability of the loading programme (Imposimato and Nova, [2]).

4. Conclusions

It is shown in this paper that the traditional method of undrained limit analysis of geostructures in terms of total stress levels gives the same results of a static stability analysis in terms of effective stresses, which is the reason why such a method, based on wrong premises, gives good predictions. Unstable states can be reached at different stress levels, but anyway in the hardening regime, depending on the loading programme followed. Unlimited pore water pressure generation, drained or undrained shear banding, loss of controllability may take place well before the limiting stress condition is achieved. This is intimately linked to the non-symmetry of the stiffness matrix, which is linked in turn to the non-associativeness of the elastic-plastic law describing the mechanical behaviour of this type of soils.

Acknowledgements

This research is performed within the framework of EU Environment Programme *Mechanisms of Catastrophic Landslides*, ENV4-CT97-0619. The financial support of CNR is also gratefully acknowledged.

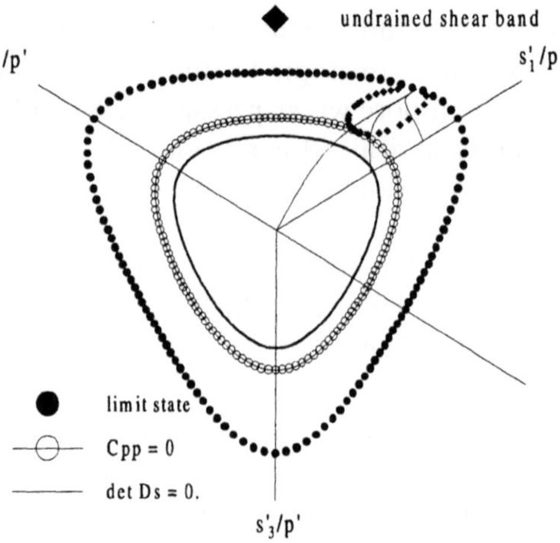

Figure 5. Locus of spontaneous pore pressure increase in undrained conditions and locus of stress points for which a locally undrained shear band can occur, according to 'Sinfonietta classica' (deviatoric plane)

References

1. Imposimato, S. and Nova, R. An investigation on existence and uniqueness of the incremental response of elastoplastic models for virgin sand. *Mechanics of Cohesive Frictional Materials*, **3**, 1–16, 1998a.
2. Imposimato, S. and Nova, R. Instability of loose sand specimens in undrained tests. Proceedings of the 4th Int. Workshop on Localization and Bifurcation Theory for Soils and Rocks, 313-322, Gifu (Japan), 1998b.
3. Lade, P. V. Elasto-plastic stress-strain theory for cohesionless soil with curved yield surfaces. *Int. J. Solids Struct.*, **13**, 1019–1035, 1977.
4. Nova, R. and Wood, D. M. A constitutive model for sand in triaxial compression. *Int. J. Num. Anal. Meth. Geomech.*, **3**, 255–278, 1979.
5. Nova, R. Sinfonietta classica: an exercise on classical soil modelling Constitutive equations for granular non-cohesive materials, Saada and Bianchini Eds., Balkema, 501–520, 1988.
6. Ostrowksy, A. and Taussky, O. On the variation of the determinant of a positive definite matrix, Nederl. Akad. Wet. Proc., (A) **54**, 333–351, 1951.
7. Pastor, M., Zienkiewicz, O. C. and Chan, A. H. C. Generalized plasticity and the modelling of soil behaviour. *Int. J. Num. Anal. Meth. Geomech.*, **14**, 150–190, 1990.

Structure and Elastic Properties of Reinforced Cellular Plastics

J. Brauns
Department of Structural Engineering, Latvia University of Agriculture, 19 Academy St., Jelgava, LV-3001, Latvia
e-mail: braun@acad.latnet.lv

Abstract. This paper presents methods on the determination of elastic properties of fibre-reinforced high-density and low-density cellular plastics in the context of the structure of composites. On the basis of effective compliance and rigidity the technical characteristics were determined. Agreement of analytical solution with experimental data can be described as reasonable showing for further investigations, particularly in the area of the interface between the fibre and matrix.

1. Introduction

The development of reinforced cellular plastics has taken place over the past 30 years with a considerable upsurge in interest in materials characterised by excellent impact properties particularly at high strain rates and under ballistic conditions. Reinforced cellular plastics are multiphase composite materials consisting of a polymeric matrix, reinforcement and a mobile, usually gaseous, phase.
The motivation to examine reinforced plastics is specifically attributable to the following reasons: 1) to reduce the high coefficient of thermal expansion, 2) to produce a flatter modulus temperature profile as well as a higher modulus, and 3) to reduce temperature creep.
The objective of this study is to present an adequate model of the reinforcement of a foamed matrix by fibres. The derivation of a model to describe the mechanical properties of a reinforced foam provides a basis for the interpretation of structure and properties relationship which in turn may be used to derive and predict design data.

2. Structure of cellular plastics

Cellular plastics have been classified in a variety of ways. The most usual are: the cellular morphology, the mechanical behaviour and the composition. Structurally cellular plastics can be described as an open or closed cell. With closed-cell materials the gas is dispersed in the form of discrete gas bubbles and a reinforced polymer matrix forms a continuous media. In open-cell foam the voids coalesce so both the

W. Ehlers (ed.), IUTAM Symposium on Theoretical and Numerical Methods in Continuum Mechanics of Porous Materials, 273–278.

solid and the gas phases are continuous. These materials are classified according to their stiffness, the two extremes being rigid and flexible. A strut-like polyhedral structure is usually a characteristic for the open-cell plastic. The base polymer is concentrated in struts and knots. There are no polymer membranes in this foam or they are thin and can be neglected. Four to six struts usually enter a knot in the plastic foam with a uniform (isotropic) structure. An additional orientation of the struts parallel to rise direction can be observed in monotropic foams [1, 6].

3. Models of reinforced cellular plastics

3.1. High-density reinforced plastics

The high-density reinforced materials can be described as tree-phase composites taking into account the relative volume fraction of each of the constituents. Assuming that pores have uniform distribution in volume the coefficient of reinforcement can be expressed in the form

$$\mu_f = \frac{A_f}{A_m + A_f + A_p}, \tag{1}$$

where A_m, A_p and A_f are the cross sections of matrix, pores and fibres of given direction, respectively. Here and bellow, the indexes m, f and p refer to the matrix, fibre and pore, respectively.
By introducing the coefficient of porosity of composite and matrix η_m the elastic modulus of porous matrix can be determined by using the formula:

$$E_{m,p} = \eta_m E_m. \tag{2}$$

Considering now the relationship for a unidirectional finite fibre-reinforced element in the direction of fibre x_1', we can determine the elastic compliance S_{1111}' by using the following expression:

$$S_{1111}' = \frac{1}{[1 + \mu_f(m_1 - 1)]E_{m,p}}, \tag{3}$$

where in the case of anisotropic fibre $m_1 = E_{f1}/E_{m,p}$.
The influence of the fibre length l_f and incomplete bonding is taken into account in finding the value of the longitudinal fibre modulus E_{f1} and Poisson's coefficients ν_{fi1} $(i = 2, 3)$. Here the reduction factors are used [3, 4]
When examining the structural element as an elastic anisotropic body, an analogous expression for transverse compliance S_{2222}' is

$$S'_{2222} = \frac{(1-\mu_f)r_2 + \mu_f^2 m_1 + (1-\mu_f)^2 m_2}{[1+\mu_f(m_1-1)]E_{f2}}, \tag{4}$$

where $r_2 = 1 + m_1 m_2 - \nu_m^2 m_1 m_2 + \nu_m m_1 + \nu_m(\nu_{f12} m_1 + \nu_{f21} m_2) - \nu_{f12}\nu_{f21}$ and $m_2 = E_{f2}/E_{m,p}$.

The elastic solution for the component S'_{1122} is

$$S'_{1122} = \frac{\nu_m(1-\mu_f) + \nu_{f12}\mu_f m_{12}}{[1+\mu_f(m_1-1)]E_{m,p}}, \tag{5}$$

where $m_{12} = E_{f1}/E_{f2}$. The remaining compliances of this group are determined in a similar way.

The detailed knowledge derived from microstructure investigation gives possibility to characterise the fabric function of material $f(\theta,\varphi)$ with spherical angles θ,φ. The lower bond of material was determined by the orientational averaging according to Reuss hypothesis [5, 8]. The components of the compliance tensor of the composite are determined using the relationship:

$$S_{\alpha\beta\gamma\delta} = \frac{1}{V*}\int_0^{\pi}\int_0^{2\pi} S'^{(n)}_{ijkl} l_{i\alpha} l_{j\beta} l_{k\gamma} l_{l\delta} f(\theta,\varphi)\sin\theta\, d\theta\, d\varphi \tag{6}$$

where $l_{i\alpha}$ are the cosines of the angles between the composite axis and the axis of the structural element of the given direction n; $i,j,k,l = 1,2,3$; $\alpha,\beta,\gamma,\delta = 1,2,3$.

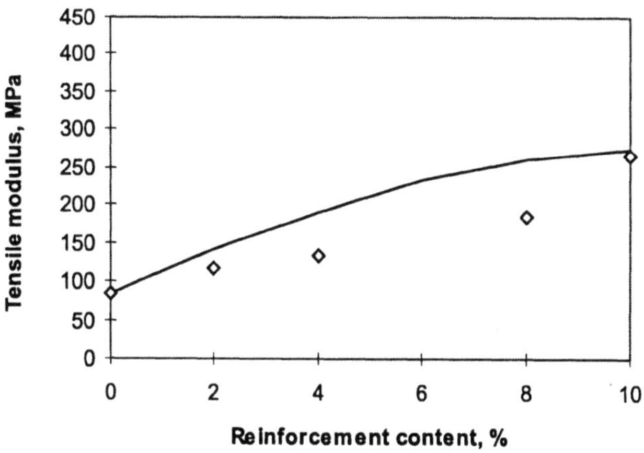

Figure 1. Tensile modulus for short-fibre reinforced high-density cellular plastic; ◇-experimental points [7].

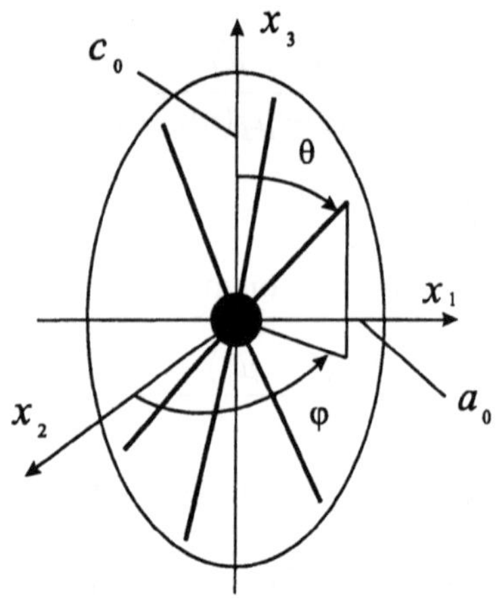

Figure 2. Ellipsoidal model cell of monotropic plastic with struts and knot; a_0, c_0 - semiaxis of model.

In a similar way, according to the assumption of uniformity of the strain field [9] the rigidity tensor is found. On the basis of compliances $S_{\alpha\beta\gamma\delta}$ the technical characteristics were determined. In Fig. 1 the variation of elastic modulus depending on reinforcement coefficient μ_f for the case of uniform fibre distribution is shown.

3.2. Low-density reinforced plastics

Fibres incorporated in the matrix are coated with resin and become in essence a composite struts. These struts bridge across several cells with the individual cell struts radiating outwards. The aim of the research is to characterise the low-density cellular plastics as integral ones with following introduction of reinforcement.

As regards deformative properties the free-rise plastic foams are monotropic materials with the isotropy plane perpendicular to the rise direction. The tensor of the effective elastic constants $A_{\alpha\beta\gamma\delta}$ of plastic characterise the deformative properties of a nonhomogeneous material. It connects stresses and strains averaged throughout the material:

$$< \sigma_{\alpha\beta} >= A_{\alpha\beta\gamma\delta} < \epsilon_{\gamma\delta} > . \tag{7}$$

The ergodic hypothesis is assumed to be valid in calculations. This permits to perform an averaging throughout a cluster of one-type situations (an ensemble). A local model cell is obtained by cutting out a rotational ellipsoid around a polymeric knot (Fig. 2).

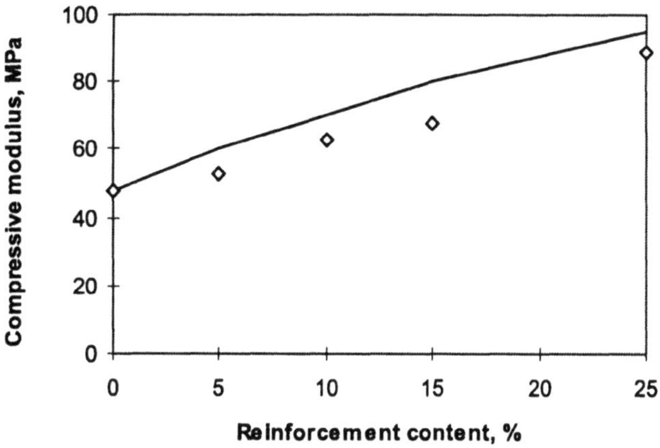

Figure 3. Effect of glass-fibre reinforcement on compressive modulus of low-density foam; ◇ - experimental points [7].

If the strain ϵ_{33} parallel to the rise direction is applied to a cell of cellular monotropic plastic:

$$\epsilon_{33} = (c - c_0)/c_0, \tag{8}$$

the effective stiffness A_{3333} can be expressed as follows:

$$A_{3333} = < \sigma_{33} > /\epsilon_{33}. \tag{9}$$

In order to calculate average stress $< \sigma_{33} >$ the stress in every microsituation of the cellular plastic structure has to be known. The post-deformation rotational ellipsoid is replaced by a circular cylinder the height and volume of which are equal to those of ellipsoid. Because the deformation of the model cell is equal to the energy W accumulated in N struts the stress σ_{33} can be calculated in every microsituation θ, φ:

$$\sigma_{33}(\theta, \varphi) = \frac{3}{2\pi(c - c_0)} \frac{\min \sum\limits_{n=1}^{N} W_n(a, \theta, \varphi)}{a^2(\theta, \varphi)}. \tag{10}$$

The calculation of average stress $< \sigma_{33} >$ can be performed in the following simplified way

$$< \sigma_{33} > = \frac{1}{4\pi} \int\limits_0^{\pi} \int\limits_0^{2\pi} \sigma_{33}(\theta, \varphi) \sin\theta \, d\theta d\varphi. \tag{11}$$

In an analogous way the elastic characteristics of cellular plastic can be found if the strain is applied perpendicular to the rise direction of

material. In Fig. 3 a compressive modulus of fibre reinforced low-density foam depending on fibre content is shown.
If the shear strain ϵ_{13} in the plane perpendicular to the plane of isotropy is applied to the local model cell the effective shear stiffness A_{1313} can be expressed as

$$A_{1313} = < \sigma_{13} > / \epsilon_{13}, \tag{12}$$

where $2\epsilon_{13} = \gamma$ is a shear angle of a model cell in the plane $x_1 x_3$.
The deformation energy depends on all the three Euler's angles and no simplification can be made in the calculation of averaged stress $< \sigma_{13} >$. A numerical value of the deformation energy in every microsituation may be determined from the local structure model taking into account the energy accumulated in a cell of hinged struts and by using a co-ordinate method [2].

4. Conclusions

Agreement of the analytical solution with the experimental data can be described as a reasonable showing for further investigations, particularly in the area of the interface between the fibre and matrix. The benefit of higher reinforcement content on modulus is negated somewhat by the problems of short-fibre reinforcement incorporation into the system.

References

1. Baxter, S. and Jones, T. T. The physical properties of foamed plastics and their dependence on structure. *Plast. and Polymer*, 69–76, 1972.
2. Beverte, I. V. and Kregers, A. F. Young's modulus of monotropic foam plastics subjected to deformation parallel to rise direction. *Mech. Compos. Mat.*, **29**, 19–26, 1993.
3. Brauns, J. A. and Rocens, K. A. Applied model of composite with incomplete bonding and finite length of reinforcing elements. *Mech. Compos. Mat.*, **27**, 567–573, 1991.
4. Brauns, J. A. and Rocens, K. A. Analysis of the effect of stress state of structural elements on the viscoelastic properties of a composite. *Mech. Compos. Mat.*, **29**, 20–26, 1993.
5. Christensen, R. M. Introduction to the Mechanics of Composites. *John Wiley & Sons, Inc.*, 1979.
6. Lederman, J. M. The prediction of the tensile properties of flexible foams. *J. Appl. Polym. Sci.*, **15**, 693–703, 1971.
7. Methven, J. M. and Dawson, J. R. Reinforced Foams. Mechanics of Cellular Plastics. Ed. N.C. Hilyard, *Macmillan Publishing Co.*, 323–358, 1982.
8. Reuss, A. Berechnung der Fliessgrenzen von Mischkristallen und Grund des Plastizittsbedingung für Eiskristalle. *ZAMM*, **9**, 49–58, 1929.
9. Voigt, W. Lehrbuch der Kristallphysik. *Berlin - Leipzig: Teubner-Verlag*, 1910.

Session D1: Micromechanics

Chairman: C. Miehe

Effective Stress and Capillary Pressure in Unsaturated Porous Media

A. K. Didwania
Department of Mechanical and Aerospace Engineering
University of California, San Diego, La Jolla, CA 92093-0411, USA

Abstract. The effective stress and capillary pressure in unsaturated porous media are analyzed from a micromechanical ansatz. The assumptions underlying the commonly used Bishop's expression for effective stress are elucidated.

1. Introduction

The concept of effective stress in a liquid saturated porous media has been the subject of several investigations since Terzaghi ([8]) proposed the principle of effective stress for porous media with rigid grains and incompressible liquid. However for unsaturated porous media, the investigations have been rather limited. For partially saturated porous media (also refered to as ternary model) Bishop suggested the following variation of Terzaghi's expression for effective stress (See Skempton [7])

$$\mathbf{T}_E^S = (p^G - \chi(p^G - p^L))\mathbf{I} + \mathbf{T} \tag{1}$$

where $\mathbf{T}_E^S$ and $\mathbf{T}$ refer to effective and total stress respectively, p^G and p^L to the pressures in gas and liquid phase and χ is a parameter related to the degree of saturation and equals unity for fully saturated soils. More recently, Bluhm and de Boer ([1]) have also obtained a similar expression for effective stress in porous media. These expressions however don't include explicit dependence of effective stress on various physical properties like surface tension, contact angle, porosity, saturation etc. even for porous materials with rigid grains and incompressible fluid constituents. In this paper, we attempt to examine the expression for effective stress for unsaturated porous media and its relation to capillary pressure from a micromechanical ansatz and elucidate the assumptions underlying Bishop's expression.

2. Idealized partially saturated porous media

We adopt a particular model of the partially saturated porous media in which the solid skeleton is assumed to consist of N identical grains of

W. Ehlers (ed.), IUTAM Symposium on Theoretical and Numerical Methods in Continuum Mechanics of Porous Materials, 281–286.

a specific configuration C^N, specified by a set of position vectors $\mathbf{r}^\alpha$ for $\alpha = 1, 2, \ldots, N$. For simplicity we restrict our discussion here to spherical grains, an incompressible wetting liquid phase and a nonwetting gas phase whose compressibility is of no significance for the problem under consideration. In general, the wetting phase is determined by the contact angle of the fluid-solid interface, for wetting contact angle θ is less than 90 whereas for nonwetting it is more than 90. Defining the capillary pressure as the discontinuity in pressure that exist across an interface between the two phases, the Laplace's equation gives

$$p^C = p^G - p^L = \gamma(\frac{1}{\rho_1} + \frac{1}{\rho_2}) \tag{2}$$

where γ is the interfacial tension and ρ_1 and ρ_2 are the principal radii of curvature of the interface. We have further assumed there is no gravitational field, no relative velocity between the phases and all phases are isothermal. The gravitational terms can be easily included in a manner outlined earlier by de Boer and Didwania ([2]).

We recall some of the definitions and manipulations commonly employed in the ensemble averaging approach detailed in (Didwania and de Boer [3], Sangani and Didwania [6]). We consider an ensemble of realizations and denote by $f(C^N)$ the probability of a specific configuration C^N. In view of the identity of the grains, the appropriate normalization is

$$\int dC^N \, f(C^N) = N! \tag{3}$$

We define an indicator function $\chi^i(\mathbf{x}; C^N)$, such that $\chi^i(\mathbf{x}; C^N) = 1$ if the point $\mathbf{x}$ is in the phase i, otherwise $\chi^i(\mathbf{x}; C^N) = 0$. Since the boundary of the phases has a measure of zero, $\chi^G + \chi^L + \chi^S = 1$. For spherical grains of radii R, the indicator function is of the form

$$\chi^S(\mathbf{x}; C^N) = \sum_{\alpha=1}^{N} H(R - |\mathbf{x} - \mathbf{r}^\alpha|), \tag{4}$$

with H the Heaviside distribution. The gradient of the indicator function is then

$$\nabla\chi^S(\mathbf{x}; C^N) = -\sum_{\alpha=1}^{N} \delta(R - |\mathbf{x} - \mathbf{r}^\alpha|) \frac{\mathbf{x} - \mathbf{r}^\alpha}{|\mathbf{x} - \mathbf{r}^\alpha|}, \tag{5}$$

where δ refers to the Dirac delta function. The phase volume fractions are then defined by

$$n^{G,L,S}(\mathbf{x}) = \frac{1}{N!} \int dC^N \, f(C^N) \, \chi^{G,L,S}(\mathbf{x}; C^N), \tag{6}$$

from which we obtain the condition, $n^G + n^L + n^S = 1$. Let $\theta^{F,S}(\mathbf{x}; C^N)$ with $F = L$ or G, be any flow quantity pertaining to the liquid or solid phase at position $\mathbf{x}$ in the presence of the configuration C^N of the grains. The phase-ensemble average of $\theta^{F,S}$ is defined by averaging over all the configurations such that the point $\mathbf{x}$ is in the appropriate phase,

$$< \theta^{F,S} > (\mathbf{x}) = \frac{1}{n^{F,S} N!} \int dC^N \, \theta^{F,S}(\mathbf{x}; C^N) \, \chi^{F,S}(\mathbf{x}; C^N) \, f(C^N) \quad (7)$$

It can be easily shown that

$$\nabla[n^F < \theta^F >] = n^F < \nabla\theta^F > + \quad (8)$$
$$+ \int dC^N \, \nabla\chi^F(\mathbf{x}; C^N) \, f(C^N)\theta^F(\mathbf{x}; C^N)$$

3. Ensemble-averaged momentum balances

Next we derive continuum momentum balance equations for each of gas and liquid phases. In the quasistatic limit with no relative velocity between the phases, as considered here, we ignore fluid viscosity and inertia contributions. The fluid pressure (for each of both phases, $F = L$ or G) at any point $\mathbf{q}$ in the fluid, is then given as:

$$\nabla p^F(\mathbf{q}; C^N) = 0 \quad (9)$$

Applying (9), we obtain

$$-\nabla[n^F < p^F >] + \int dC^N f(C^N) \nabla\chi^F \, p^F(\mathbf{x}; C^N) = 0 \quad (10)$$

Adding each of the two averaged equations for gas and liquid

$$-\nabla[n^L < p^L > + n^G < p^G >] + \quad (11)$$
$$\int dC^N f(C^N)[\nabla\chi^L p^L(\mathbf{x}; C^N) + \nabla\chi^G p^G(\mathbf{x}; C^N)] = 0$$

Noting $\nabla\chi^L + \nabla\chi^G + \nabla\chi^S = 0$ and $n^G + n^L + n^S = 1$ and assuming $p^G \approx < p^G >$ we obtain after some rearrangement

$$-\nabla[(n^L + n^G) < p^G >] = < p^G > \nabla n^S \quad (12)$$

Next we direct our attention to the solid phase. The force balance on a single grain with its center located at position $\mathbf{x}$ is

$$- \int_{|\mathbf{x}-\mathbf{q}|=R} dS \, \mathbf{n} \, p^w(\mathbf{q}; C^N) + \mathbf{F}_c(\mathbf{x}; C^N) = 0 \quad (13)$$

where dS is the differential surface element of the grain with the unit normal $\mathbf{n}$ directed out of the grain and $\mathbf{F}_c(\mathbf{x};C^N)$ is the contact force exerted by the contacting neighbors. The pressure at a point $\mathbf{q}$ on the grain surface is given by $p^w(\mathbf{q};C^N) = p^G\chi^{Gw} + p^L\chi^{Lw}$ where $\chi^{Gw} + \chi^{Lw} = 1$ with χ^{Gw} and χ^{Lw} indicating the portion of the grain surface in contact with gas or liquid respectively. Rewriting

$$-\int_{|\mathbf{x}-\mathbf{q}|=R} dS\,\mathbf{n}\,p^G(\mathbf{q};C^N) - \mathbf{F}_{cp}(\mathbf{x};C^N) + \mathbf{F}_c(\mathbf{x};C^N) = 0 \tag{14}$$

where $\mathbf{F}_{cp}(\mathbf{x};C^N) = \int_{|\mathbf{x}-\mathbf{q}|=R} dS\,\mathbf{n}\,\chi^{Lw}(p^L - p^G)(\mathbf{q};C^N)$ is the force on the grain due to surface wetting by the liquid. Ensemble averaging the force balance eq. (14)

$$\begin{aligned} &- < \int_{|\mathbf{x}-\mathbf{q}|=R} dS\,\mathbf{n}\,p^G(\mathbf{q};C^N) > \\ &- < \int_{|\mathbf{x}-\mathbf{q}|=R} dS\,\mathbf{n}\,\chi^{Lw}(p^L - p^G)(\mathbf{q};C^N) > + \nabla\cdot\mathbf{T}^S_E = 0 \end{aligned} \tag{15}$$

where $\mathbf{T}^S_E$ denotes contact stress among the grains and can be computed exactly from the knowledge of $\mathbf{F}_c$ (See Zhuang, Didwania and Goddard [9], Goddard and Didwania [4] for rigid grains) if it is assumed that $\mathbf{F}_c$ is independent of $\mathbf{F}_{cp}$ and p^G. As shown in Didwania and de Boer ([3]), for slow variation of averages on the scale of the grain, the first term on the left hand side of eq. (15) reduces to

$$\begin{aligned} < \int_{|\mathbf{x}-\mathbf{q}|=R} dS\,\mathbf{n}\,p^G(\mathbf{q};C^N) > &= n^S\nabla < p^G > \mathbf{x}) = \\ &\nabla[n^S < p^G > (\mathbf{x})] - < p^G > \nabla n^S(\mathbf{x}) \end{aligned} \tag{16}$$

Substituting the relation (16) in eq. (15)

$$\begin{aligned} &-\nabla[n^S < p^G > (\mathbf{x})] + < p^G > \nabla n^S(\mathbf{x}) \\ &- < \int_{|\mathbf{x}-\mathbf{q}|=R} dS\,\mathbf{n}\,(p^L - p^G)\chi^{Lw}(\mathbf{q};C^N) > + \nabla\cdot\mathbf{T}^S_E = 0 \end{aligned} \tag{17}$$

Thus the momentum balance eqs. (12) and (17) are

$$\nabla\cdot\mathbf{T}^F + \mathbf{I}^F = 0, \quad \nabla\cdot\mathbf{T}^{\mathbf{S}} + \mathbf{I}^{\mathbf{S}} = 0. \tag{18}$$

where

$$\begin{aligned} \mathbf{T}^F &= -[(n^L + n^G) < p^G >]\mathbf{I}, \quad \mathbf{I}^F = - < p^G > \nabla n^S \\ \mathbf{T}^S &= -n^S < p^G > \mathbf{I} - \mathbf{T}^S_{cp} + \mathbf{T}^S_E, \quad \mathbf{I}^S = -\mathbf{I}^F = < p^G > \nabla n^S \end{aligned} \tag{19}$$

assuming that $< \int_{|\mathbf{x}-\mathbf{q}|=R} dS\, \mathbf{n}\, \chi^{Lw}(p^L - p^G)(\mathbf{q}; C^N) >$ can be expressed as $\nabla \cdot \mathbf{T}^S_{cp}$. Adding the two equations in (18)

$$\nabla \cdot \mathbf{T} = \nabla \cdot \mathbf{T}^F + \nabla \cdot \mathbf{T}^S == -\nabla \cdot < p^G > \mathbf{I} - \nabla \cdot \mathbf{T}^S_{cp} + \nabla \cdot \mathbf{T}^S_E \quad (20)$$

On comparing (20) with (1), we find that for Bishop's expression to be valid χ must satisfy

$$\begin{aligned} \nabla \cdot [\chi(p^L - p^G)\mathbf{I}] &= \nabla \cdot \mathbf{T}^S_{cp} = \\ &= < \int_{|\mathbf{x}-\mathbf{q}|=R} dS\, \mathbf{n}\, \chi^{Lw}(p^L - p^G)(\mathbf{q}; C^N) > \end{aligned} \quad (21)$$

It is clear that Bishop's expression with χ as a scalar parameter can be valid only when the capillary stress $\mathbf{T}^S_{cp}$ assumes simple form.

4. Porous media at low saturation

In this section we outline an algorithm to evaluate the parameter χ for somewhat simpler case of low saturation. More detailed calculations involving other cases will be presented elsewhere. At low saturations liquid forms rings (often called pendular) around the grain contact points. For this case it is possible to compute $\mathbf{F}_{cp}$ for each particle as a sum of capillary forces arising at each contact. The capillary force at a contact between two equal size spheres α, β can be estimated as $\mathbf{F}^{\alpha\beta}_{cp} = \pi\gamma\rho_2(\frac{\rho_1+\rho_2}{\rho_1})\mathbf{n}^{\alpha\beta}$ where $\mathbf{n}^{\alpha\beta}$ is the unit contact normal (See Hotta et. al. [5]). The procedure for deducing ρ_1 and ρ_2 from geometrical considerations is detailed in Hotta et. al. ([5]). For very small moisture content the force expression reduces to $\mathbf{F}^{\alpha\beta}_{cp} = 2\pi\gamma R\cos\theta\mathbf{n}^{\alpha\beta}$. The capillary stress for porous media (of volume V) with low saturation is given by $\mathbf{T}^S_{cp} = (\frac{2R}{V})\sum_{\alpha<\beta}\sum \mathbf{F}^{\alpha\beta}_{cp}\mathbf{n}^{\alpha\beta}$ yielding an explicit expression for χ.

Acknowledgements

The present work is supported by NSF Grants INT-9605036, CTS-9510121 and NASA Grant NAG 3-1888.

References

1. Bluhm, J. and de Boer, R. Effective stresses-a clarification. *Arch. of Appl. Mech.*, **66**, 479–492, 1996.

2. de Boer, R. and Didwania, A. K. The effect of uplift in liquid-saturated porous solids - Karl Terzaghi's contributions and recent findings. *Geotechnique*, **47**, 289–298, 1997.
3. Didwania, A. K. and de Boer, R. Saturated Compressible and Incompressible Porous Solids: Macro- and Micromechanical Approaches. *Transport in Porous Media*, **34**, 101–115, 1999.
4. Goddard, J. D. and Didwania, A. K. Computations of dilatancy and yield surfaces for assemblies of rigid frictional spheres. *Q. Jl. Mech. appl. Math.*, **51**, 15–43, 1997.
5. Hotta, K., Takeda, K. and Iinoya, K. The capillary binding force of a liquid bridge. *Powd. tech.*, **10**, 231–242, 1974.
6. Sangani, A. S. and Didwania, A. K. A dispersed-phase stress tensor in flows of bubbly liquids at large Reynolds number. *J. Fluid Mech.*, **248**, 27–54, 1993.
7. Skempton, W. W. Effective stress in soils, concrete and rock. In *Symposium on pore pressure and suction in soils*, Conference organized by the British National Siciety of Foundations Engineer at ICE, March 30-31, 4–16, London, Butterworth, 1962.
8. von Terzaghi, K. The shearing resistance of saturated soils and the angle between the planes of shear. *First Int. Conf. Soil Mech.*, **1**, Harvard University, 54–56, 1936.
9. Zhuang, X., Didwania, A. K. and Goddard, J. D. Simulation of the quasi-static mechanics and scalar transport properties of granular assemblies. *J. Comp. Phys.*, **121**, 331–346, 1995.

Porous Medium Mechanics and the Skin Barrier

J. M. Huyghe
Department of Mechanical Engineering, Eindhoven University of Technology, Eindhoven, The Netherlands

P. M. van Kemenade and L. F. A. Douven
Philips Personal Care Institute, Eindhoven, the Netherlands

P. H. M. Bovendeerd
Department of Mechanical Engineering, Eindhoven University of Technology, Eindhoven, The Netherlands

Abstract. A skin-air model is developed to predict changes in transepidermal water loss following changes in ambient relative humidity and following minute damages to the skin. In vivo experiments on human subjects are used to validate the model.

1. Introduction

Our body is porous medium with a fluid volume fraction of well over 50 %, depending on age. It has a contact area with air in the order of magnitude of 5 m^2. Naturally one expects that the fluid contained within the body tends to evaporate at its surface. Transepidermal water loss is measured in the order of 5 to 20 g/m^2h at rest. This is 15 to 30 times less than the water evaporating from a pond of similar contact area with air and similar temperature. The difference is explained by the skin barrier. The fluid content of the body - and hence the skin barrier - is of vital importance for diffusion of waste materials and nutrients within the body [1]. The observation that transepidermal water loss continues unchanged after death, indicates that the phenomenon does not involve gland secretion or other complex biochemical processes and is therefore within reach of present knowledge of porous medium mechanics.

A remarkable finding that might shed light on the location of the skin barrier is the significant increase of transepidermal water loss after minute - with naked eye invisible - damage to the skin surface. Even overexposure to detergents, increases the water loss by a factor two [2]. The purpose of the present study is to develope a porous medium finite element model of the skin barrier, quantify the model parameters on the basis of our experimental data of steady state behaviour of human *in vivo* skin and verify the model using our experimental data of transient behaviour of human *in vivo* skin.

W. Ehlers (ed.), IUTAM Symposium on Theoretical and Numerical Methods in Continuum Mechanics of Porous Materials, 287–292.

2. Model

Our approach involves a skin model and an air model coupled by tyings. The tyings require the hydraulico-chemical potential of the fluid to equate the chemical potential of the vapour at the contact plane between air and skin. We deal with each model separately.

2.1. Skin

The skin is a saturated porous medium. Considering the experimental evidence that the skin barrier is localised in the outermost layer of the skin we restrict the modelling to the upper 300 μm : a fraction of the stratum corneum. In order keep the model tractable we neglect the influence of solutes in the fluid fase. Based on the ultrasound wave velocity, we infer that compressibility is not an issue in the present problem. Assuming both fluid and solid components intrinsically incompressible and excluding mass transfer between phases, the mass balance of the mixture is:

$$\nabla \cdot \boldsymbol{v}^s + \nabla \cdot \left(n^f (\boldsymbol{v}^f - \boldsymbol{v}^s) \right) = 0 \qquad (1)$$

in which n^α is the volume fraction and $\boldsymbol{v}^\alpha$ the velocity of component α. Volume change from the initial to the current state is the determinant of the deformation gradient tensor $J = det\boldsymbol{F}$. Neglecting body forces and inertia, the momentum balance takes the form:

$$\nabla \cdot \boldsymbol{\sigma}^e + \nabla p = \boldsymbol{0}, \qquad (2)$$

$\boldsymbol{\sigma}^e$ is the effective stress tensor and p is the fluid pressure, Assuming hyperelastic behaviour, the effective Cauchy stress tensor is given by

$$\boldsymbol{\sigma}^e = \frac{1}{J}\boldsymbol{F} \cdot \frac{\partial W}{\partial \boldsymbol{E}} \cdot \boldsymbol{F}^c, \qquad (3)$$

where W is the strain energy function. The hydraulico-chemical potential of the fluid (μ^f) is

$$\mu^f = \frac{\partial W}{\partial (J n^f)} + p, \qquad (4)$$

There is considerable evidence of the applicability of Darcy 's law to biological tissue [3]. Micromechanically, this fact is probably associated with the fact that the size of the pores (several nm) is larger than the size of a water molecule. Transport of fluid and ions is caused by gradients in their hydraulico-chemical potentials according to an isotropic Darcy's equation:

$$n^f \vec{v}^{fs} = K \vec{\nabla}_0 \mu^f, \qquad (5)$$

where $\vec{\nabla}_0$ is the gradient operator with respect to the initial configuration ($\vec{\nabla}_0 = \boldsymbol{F} \cdot \vec{\nabla}$), K is the permeability and $\vec{v}^{fs}$ is the relative velocity defined as $\vec{v}^{fs} = \boldsymbol{F}^{-1} \cdot (\vec{v}^f - \vec{v}^s)$. Assuming a strain energy function W of the form

$$W = W_E(\boldsymbol{E}) + W_\psi(Jn^f) \tag{6}$$

where W_E is the elastic energy function corresponding to linear isotropic elasticity and W_ψ the matric energy function. The matric potential $\psi = \partial W_\psi / \partial(Jn^f)$ appearing in eq. 4, accounts for fluid-solid interaction (capillary and adsorptive effects) . We choose on the basis of sorption curves from the literature:

$$\psi = \partial W_\psi / \partial(Jn^f) = a(e^b - e^{b/Jn^f}) \quad a = 9.84\ 10^{10}\ Pa \quad b = 9.7\ 10^{-5} \tag{7}$$

In many porous media the matric potential is fairly constant and is therefore often omitted. However, for skin in contact with air the matric potential plays an important role. Note the order of magnitude of the matric potential (1-100 MPa) compared to a typical mechanical pressure in the body, e.g. systolic blood pressure (15 kPa). The fluid volume fraction in skin has been measured as a function of depth. Its value ranges from 0.12 at the surface to 0.40 at depths of more than 15 μm. The initial fluid volume fraction in the model is chosen accordingly. From eq. 7 we can infer that the matric potential dips towards the surface to values below 50 MPa. This sharp dip can only be compensated by a sharp decrease of permeability towards the surface of the skin, in order to comply with fluid mass conservation in steady state condition. We choose a relationship permeability versus fluid volume fraction so as to comply with the constraint of water flow equal to about 10 g/m^2h :

$$K = K_0\ e^{Mn^f} \quad K_0 = 1e^{-23}\ m^4/(Ns) \quad M = 17.75 \tag{8}$$

An experimental relationship between water content and relative humidity and between relative humidity and Young's modulus - both taken from the literature - provide a clue to choose the Young's modulus as a function of depth, ranging from 1.7 10^7 Pa at a depth of 15 μm to 5 10^8 Pa at the surface. Poisson's ratio is chosen 0.15.

2.2. Air

Vapour diffusion through (non-moving) air is described by the equation:

$$\frac{1}{R^v T}\frac{\partial p^v}{\partial t} - \frac{1}{R^v T}\vec{\nabla} \cdot (D^v \vec{\nabla} p^v) = 0, \tag{9}$$

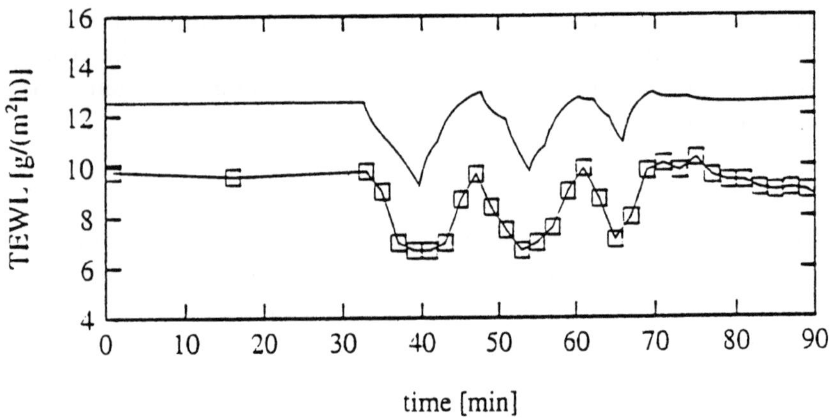

Figure 1. Simulated transepidermal water loss (–) and measured transepidermal water loss (□) .

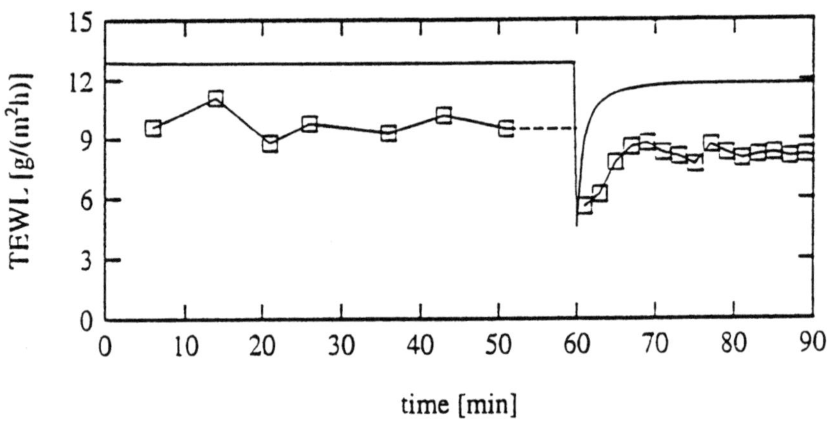

Figure 2. Simulated transepidermal water loss (–) and measured transepidermal water loss (□) .

where R^v is the specific gas constant of water vapour, T the absolute temperature, p^v the vapour pressure and D^v is the diffusion coefficient of vapour through air.

2.3. Numerical implementation

The skin model is implemented as a twodimensional geometrically and physically non-linear quadrilateral 8 node Serendipity finite element with 8 displacements and 4 hydraulico-chemical potentials as degrees of freedom. The air model is a 4-node quadrilateral element with the vapour hydraulico-chemical potential as degree of freedom. Newton

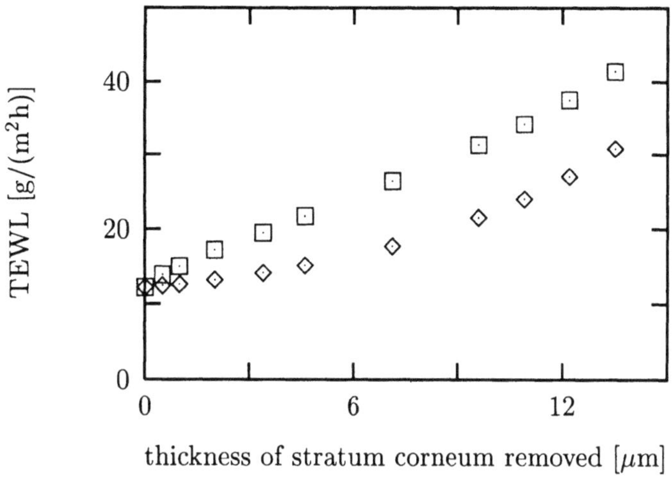

Figure 3. Simulated transepidermal water loss as a function of stratum corneum removed. □: transepidermal water loss 1 minute after damaging . ◇: transepidermal water loss 60 minutes after damaging.

Raphson iterative procedure is used within each time step. Euler backward time integration scheme is used . In this paper only 1D results are presented. The skin mesh consists of a sequence of 98 elements ranging from a thickness of 0.1 μm to 7 μm. The air mesh consists of 13 elements of 1 mm thickness each. At 13 mm distance from the skin the vapour pressure is assumed to equal ambient vapour pressure. At 300 μm depth in the skin the hydraulico-chemical potential of the water is assumed to equal the hydraulico-chemical potential of blood plasma in the capillary vessels.

3. Experiments

Transepidermal water loss (TEWL) was measured on the volar forearm of 9 healthy volunteers using TEWAmeter (Courage and Khazaka, Cologne, Germany). In Figure 1 the relative humidity was varied between 70 and 100 % in several cycles. In Figure 2 the relative humidity was changed stepwise from 53 % to 89 % after 60 minutes.

4. Results

The two experiments described above are simulated using the model. Reasonable agreement between numerical and experimental results is found (fig. 1and 2). In addition minute damages to the skin surface are simulated by removing elements from the finite element mesh. The response of the model is an instantaneous increase in water loss followed by progressive decrease to a value above the baseline (fig. 3), which is consistent with experiments reported elsewhere [4]. Indeed, the simulated increase after one minute lies between 10 and 20 g/m^2h, while the measured increase varies between 7 and 35 g/m^2h at the first measurement after damage. The simulated final increase in water loss lies between 3 and 9 g/m^2h while the measured final increase varies between -2 and 18 g/m^2h. The simulated initial and final increase are comparable to the measured values.

5. Discussion

The above example demonstrate the capabilities of a porous medium approach in simulating transepidermal water loss under different conditions. In Figure 2 the relative humidity is changed from 53 % to 89 % at t = 60 minutes. This results for both measurement and simulation in an immediate decrease in transepidermal water loss followed by a slow recovery to values below baseline. The initial increase is somewhat higher in the simulation as compared to the measurement. This might be explained by the time needed to obtain a water loss measurement (40-60 s). The present work represents to our knowledge the first *in vivo* validation of a porous medium model of a living tissue.

6. Acknowledgement

The research of Dr. J. M. Huyghe was made possible through a fellowship of the Royal Netherlands Academy of Arts and Sciences.

References

1. Guyton, A. C. and Hall, J. E. Textbook of medical physiology. 9th ed. Saunders, Philadelphia, PA, 1996.
2. Lo, L. S., Oriba, H. A., Maibach, H. I. and Bailin, P. L., Transepidermal potassium ion, chloride ion, and water flux across delipidized and cellophane tape-stripped skin. *Dermatologica* **180**, 66–68, 1990.
3. Mow, V. C., Holmes, M. H. and Lai, W. M. Fluid transport and mechanical properties of articular cartilage. *J. Biomech.* **17**, 377–394, 1984.
4. van Kemenade, P. M. Water and Ion Transport Through Intact and Damaged Skin. Ph.D.-thesis, Eindhoven University of Technology, Department of Mechanical Engineering, 1998.

Discrete Element Modelling of Compaction of Cylindrical Powder Particles

P. Redanz* and N. A. Fleck
Cambridge Centre for Micromechanics, Cambridge University Engineering Department, Trumpington Street, Cambridge CB2 1PZ, U.K.

Abstract. Cold compaction of a random array of cylindrical metal powder particles has been investigated numerically using a network model also known as the discrete element method. In the particle network, each particle center is represented by a node and each contact between neighboring particles is represented by a spring element. The random packing of the particles in the network is generated with an algorithm using the ballistic deposition method. Deformation of the particles takes place by plastic flattening of the contacts. Interparticle sliding is accounted for in the tangential force-displacement relation and in the moment-spin relation. The network model has been used to study the micromechanical behaviour of a powder aggregate undergoing compaction.

1. Introduction

In the present work, we will focus on the initial stage of the compaction process, stage I, wherein isolated contacts between particles control the macroscopic deformation. Subsequently, the contacts start to interact and towards the end of compaction porosity exists in the form of isolated voids; this is known as stage II. Ashby and co-workers (Helle *et al.*, [9] ; Artz, [3]; Artz *et al.*, [4]) introduced a macroscopic yield condition for hydrostatic stage I compaction. They assumed that the porous aggregate consists of randomly packed, rigid-perfectly plastic, equi-sized spheres joined by discrete necks. Fleck *et al.* ([8]) generalized this approach to general stress states and Fleck ([7]) included the effects of deformation induced anisotropy and inter-particle cohesion. Similar yield surfaces were found experimentally for copper powder by Akisanya *et al.* ([1]). Recently, Storåkers *et al.* ([13]) extended this model to powder composites with a power law visco-plastic material behaviour.

The material models are based on micromechanical assumptions such as the average number of contacts at each particle, the average area of each contact, and affine motion of the particles. In the present work, the validity of these assumptions through the compaction process is investigated and the yield surfaces for isostatic precompacts are

* Permanent address: Department of Solid Mechanics, Technical University of Denmark, DK-2800 Lyngby, Denmark

W. Ehlers (ed.), IUTAM Symposium on Theoretical and Numerical Methods in Continuum Mechanics of Porous Materials, 293–298.

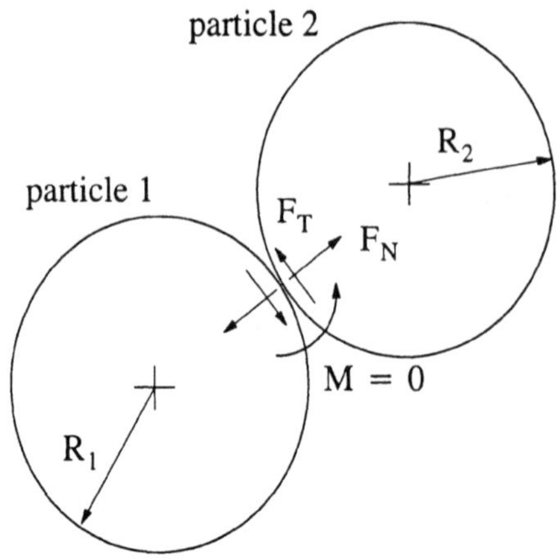

Figure 1. Two cylinders of radii R_1 and R_2 in contact. The contact forces in the normal and tangential directions are denoted F_N and F_T, respectively, and the rolling moment is denoted M.

determined and compared with the 2D version (Sridhar *et al.*, [12]) of the yield surface given by Fleck *et al.* ([8]).

2. Numerical method

The cross section of two cylinders in contact is shown in Figure 1a. Particles 1 and 2 are circular cylinders of radii R_1 and R_2, respectively. The contact force from one cylinder to the other is resolved into a normal force, F_N, acting along the normal to the plane of contact and a tangential force, F_T, due to friction in the tangential plane. The rolling moment, M, is small as long as the area of contact is small. This is the case for stage I compaction and the rolling moment is therefore set to zero. In the network each contact between two cylinders is represented by a spring element and each particle center is represented by a node.

Deformation during compaction occurs by the plastic flattening of contacts. The normal pressure at each contact in a rigid-perfectly plastic material is approximately equal to three times the yield stress of the matrix material, $\sigma = 3\sigma_y$. Hence, the normal force - normal displacement relationship has the following form

$$F_N = 3\sigma_y A_c \quad where \quad A_c = \sqrt{2(R_1 + R_2)u_N} \tag{1}$$

where u_N is the displacement of one particle with respect to the other in the normal direction and A_c is the area of the contact. Due to the non-linear structure of (1) the incremental form is needed,

$$\dot{F}_N = 3\sigma_y \sqrt{\frac{R_1 + R_2}{2u_N}} \dot{u}_N \tag{2}$$

In the tangential plane, the relative displacement is equal to the difference between the tangential displacements of the centers of particles 1 and 2 plus a contribution from rotation of the particles. As done by Parhami and McMeeking ([11]), we assume a linear relationship between the shear stress increment and the tangential displacement increment, $\dot{\tau} = \eta \dot{u}_T$, with η termed the drag coefficient. This leads to an incremental relationship between the tangential force, $\dot{F}_T$, and the tangential displacement increment, $\dot{u}_T$, in the form

$$\dot{F}_T = \eta A_c \dot{u}_T \tag{3}$$

It should be noted that (3) is a rather crude description of friction, however, as we will only study the two limits of η, zero and infinity, this formulation is sufficient.

Besides the normal and tangential forces, each node experiences a moment, $M_i = d_N F_T$, due to the tangential force, F_T, at the particle periphery. Here, d_N is the distance from the particle center to the contact. This type of network formulation has earlier been used by Newell ([10]) and Deutschmann *et al.* ([6]).

Equilibrium of forces and moments for each particle is expressed by

$$\sum_{q=1}^{N_{con,p}} f_i^{pq} = 0 \quad \forall \quad p = 1, N_{tot} \tag{4}$$

where f_i^{pq} are the generalized forces, forces and moments, transmitted between particles p and q. The number of contacts at particle p is denoted $N_{con,p}$ and the total number of contacts is N_{tot}. Body forces are neglected.

Equations (2)-(4) lead to the following set of equations

$$A_{ij} \dot{D}_i = \dot{F}_i^{applied} \tag{5}$$

where A_{ij} is the incremental global stiffness matrix, $\dot{D}_i$ are the generalized displacement increments, and $\dot{F}_i^{applied}$ are the applied generalized forces. This is solved for the unknown incremental particle displacements and rotations. After each increment, the state of the aggregate is updated and the procedure is repeated.

The macroscopic stresses are determined from

$$\Sigma_{ij} = \frac{1}{A_{box}} \sum_{p=1}^{N_{par}} F_i^{applied} x_j \tag{6}$$

as suggested by Christoffersen *et al.* ([5]) where the deformed area of the container is denoted A_{box}.

The initial packing of the particles has been generated with an algorithm using the ballistic deposition method, see for example Aparicio and Cocks ([2]).

The displacement boundary conditions are applied on a layer of particles at the exterior of the aggregate and have the following form

$$\dot{u}_1 = \dot{E}_{11}x_1 + \dot{E}_{12}x_2 \tag{7}$$
$$\dot{u}_2 = \dot{E}_{21}x_1 + \dot{E}_{22}x_2 \tag{8}$$

where $\dot{E}_{ij}$ is the incremental macroscopic strain tensor and (x_1, x_2) are the global coordinates of the centers of the particles in the boundary layer.

During the compaction process, new contacts between the particles are formed, hence, new elements are added to the network.

3. Results and discussion

A polydispersed packing with cylinders of radii $r = R[0.9, 1.1]$ is shown in Figure 2. This aggregate holds 6604 particles (nodes) with 12,999 contacts (elements) and has an initial relative density of $D_0 = 0.810$.

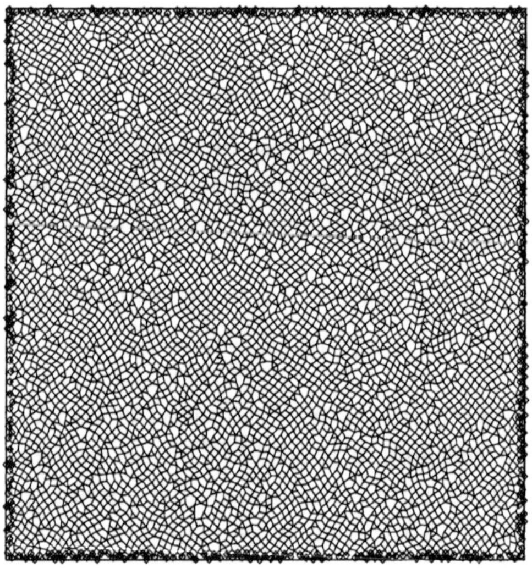

Figure 2. The initial packing of the cylinders. Each line represents a contact and each junction represents a cylinder center. The cylinders are of radii $r = R[0.9, 1.1]$, where R is $6.25 \cdot 10^{-3}$ times the initial width of the box.

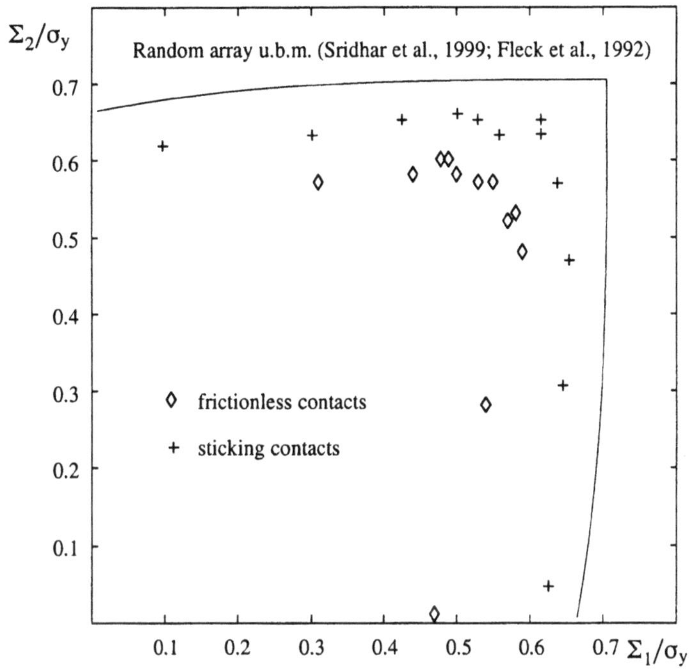

Figure 3. The 2D version (Sridhar *et al.*, [12]) of Fleck, Kuhn and McMeeking's ([8]) random array, upper bound yield surface for a hydrostatic precompact at $D = 0.85$ compared with the yield surfaces of the network shown in Figure 2. Results of both frictionless and sticking contacts are shown.

The network has been used to study the micromechanical behaviour of a particulate aggregate undergoing compaction. Yield surfaces of the network were determined as follows: The network was first compacted to a density of $D = 0.85$, then unloaded elastically and subsequently reloaded using different strain paths. During reloading, an approximately linear relationship between stress and density exists until yielding occurs. The yield stress and thus each point of the yield surface is defined as the value at which the stress-density curve displays a corner.

The yield surfaces of hydrostatic precompacts of both frictionless and sticking particles are shown in $(\Sigma_{11}\Sigma_{22})$-space in Figure 3. The particles exhibit no cohesive (tensile) inter-particle strength, thus, the lines $\Sigma_{11} = 0$ and $\Sigma_{22} = 0$ form additional boundaries for the yield surfaces. The network calculations suggest that the yield surface of an aggregate consisting of frictionless particles is approximately 15% smaller than the corresponding upper bound approximation of the yield surface (Fleck *et al.*, [8]; Sridhar *et al.*, [12]), see Figure 3. A corner exists at the random array upper bound approximation of the yield surface but this is not the case for the yield surface of the network.

For sticking inter-particle contacts the yield surface is slightly larger than in the frictionless case; this noticeable difference is larger than in the upper bound approximations in Fleck ([7]) who found that inter-particle friction plays only a minor role.

In both Storåkers *et al.* ([13]) and Fleck *et al.* ([8]) they assume a kinematic field in the particulate aggregate where the centers of the particles move compatibly with the macroscopic strain. It has been found here that the actual average relative normal displacement between two particles deviates between 6% and 24% from this assumption depending on initial particle packing and on the degree of inter-particle friction.

References

1. Akisanya, A. R., Cocks, A. C. F. and Fleck, N. A. The yield behaviour of metal powders, *International Journal of Mechanical Sciences*, **39**, 1315–1324, 1997.
2. Aparicio, N. D. and Cocks, A. C. F. On the representation of random packings of spheres for sintering simulations, *Acta metall.*, **43**, 3873–3884, 1995.
3. Arzt, E. *Acta metall.*, **30**, 1883–1890, 1982.
4. Arzt, E., Ashby, M. F. and Easterling, K. E. *Metall. Trans.*, **14A**, 211, 1983.
5. Christoffersen, J., Mehrabadi, M. M. and Nemat-Nasser, S. A micromechanical description of granular material behavior, *Journal of Applied Mechanics*, **48**, 339–344, 1981.
6. Deutschmann, G., Landis, C. M. and McMeeking, R. M. A Network Model for the Plastic Compaction of Monodispersed Spherical Powder, University of California Santa Barbara, 1998.
7. Fleck, N. A. On the cold compaction of powders, *J. Mech. Phys. Solids*, **43**, 1409-1431, 1995.
8. Fleck, N. A., Kuhn, L. T. and McMeeking, R. M. Yielding of metal powder bonded by isolated contacts, *J. Mech. Phys. Solids*, **40**, 1139–1162, 1992.
9. Helle, A. S., Easterling, K. E. and Ashby, M. F. Hot-isostatic pressing diagrams: new developments, *Acta metall.*, **33**, 2163–2174, 1985.
10. Newell, K. J. Discrete element modelling of powder consolidation and the formation of Titanium matrix composites from powder-fiber monotapes, *Ph.D. thesis*, University of California Santa Barbara, 1995.
11. Parhami, F. and McMeeking, R. M. A network model for initial stage sintering, *Mechanics of Materials*, **27**, 111–124, 1998.
12. Sridhar, I., Fleck, N. A. and Akisanya, A. R. Cold compaction of an array of cylindrical powder particles, *submitted*, 1999.
13. Storåkers, B., Fleck, N. A. and McMeeking, R. M. Viscoplastic compaction of composite powders, *J. Mech. Phys. Solids*, **47**, 785–815, 1999.

Modeling of In-Situ Solution Mining Processes

H.-B. Mühlhaus and J. Liu
CSIRO Exploration & Mining, 39 Fairway, Nedlands, WA 6009 Australia

B. E. Hobbs
CSIRO Exploration & Mining, Private Bag, PO Wembley WA 6014, Australia

Abstract. We apply a nonlinear reaction-diffusion-advection model to the simulation of a five-spot in-situ solution mining operation. The simulation is based on a finite element model of the governing equations. The porosity dependency of the hydraulic conductivity gives rise to a positive feedback, manifesting itself in porosity fingering instabilities. The deformation dependency of dispersion coefficients is also taken into consideration. Through the simulation, we demonstrate that the characteristics of mineral recovery concentration histories are affected significantly by the total injection flux or by the total ejection flux. The peak mineral recovery concentration decreases as the total injection or ejection flux increases.

Keywords: Solution Mining, permeability, porosity, transport

1. Introduction

In-situ solution mining represents a hydro-metallurgical operation usually following conventional surface or underground ore extraction. On the other hand, in-situ solution mining can also be regarded as a distinct, low environmental impact, and low cost mining method. An inexpensive reagent is injected into a mineralized rock mass and moves through the orebody. The reagent is selected in part to maximize the leaching rate of a specific mineral in a particular orebody, and ordinarily the rate constant is only one of a number of chemical factors that would be considered [5, 1, 6].

The significance of this mining technology is illustrated by problems which are becoming apparent with current practice of mineral resource engineering. Decreasing grade of near-surface deposits (i.e. within 500m of the ground surface) will in future result in mining, transport and processing of much larger volumes of rock per unit product recovered. In addition to continued exploitation of low-grade and near-surface deposits, the technology of *in-situ* leaching may be used to extract higher grade and deeper deposits. Assuming exploration technology advances sufficiently to permit economic location and delineation of deep orebodies, current practice has already confronted limitations in mining technology for large-scale and deep-level mining. Induced seismicity and rockbursts are recognized as a pervasive problem in deep-level hard-rock mining [3].

W. Ehlers (ed.), IUTAM Symposium on Theoretical and Numerical Methods in Continuum Mechanics of Porous Materials, 301–306.

Although *in situ* leaching is increasingly becoming an attractive mining method for extraction of mineral values from near-surface low-grade ore deposits or higher grade and deeper deposits, this method may not be used in mining practice until the governing sciences are fully understood. A fully coupled leach solution flow-mineral dissolution-mechanical deformation model for in-situ leaching processes has been developed in this study. The model is applied to a typical five-spot in-situ leaching operation.

2. Conceptual model

The valuable minerals may be deposited uniformly within rock matrixes (matrix-hosted) or on fracture walls (fracture-hosted). Assuming that rock matrix is functionally impermeable, and that fluid flows predominantly in fractures, the porosity of fractured pore system is equal to the effective porosity. The determination of the effective porosity for the fractured pore system is quite complex, if not impossible, because the complicated internal structure may be unknown and perhaps unknowable. A method, which links the effective porosity of a fractured rock mass to an empirical rock mass classification index, has been developed [4]. This method is based on the relation between the effective porosity and the permeability for a specific ore deposit.

Assuming the ore deposit is anisotropic with hydraulic conductivities, K_{11}, K_{22}, and K_{33}, in the x_1-, x_2-, and x_3-directions, respectively, the rock mass may be substituted by a fractured medium with three orthogonal sets of fractures. The effective porosity, ϕ_0, can be defined as

$$\phi_0 = \left[\frac{6\mu}{gs^2}(K_{22}+K_{33}-K_{11})\right]^{\frac{1}{3}} + \left[\frac{6\mu}{gs^2}(K_{11}+K_{33}-K_{22})\right]^{\frac{1}{3}} + + \left[\frac{6\mu}{gs^2}(K_{11}+K_{22}-K_{33})\right]^{\frac{1}{3}} \tag{1}$$

where s is the equivalent fracture spacing, g is gravitational acceleration, μ is kinematic viscosity. Assuming the original porous media are isotropic, substituting $K_{11} = K_{22} = K_{33} = K_0$ into equation 1 yields

$$\phi_0 = 3\left(\frac{6\mu K_0}{gs^2}\right)^{\frac{1}{3}}. \tag{2}$$

Equation 2 may be interpreted as a microscale model, assuming the material is made of cubes and s is the side length of cubes. When s is small, it represents that the medium is made of very fine grains such as soil. When s is large, it represents that the medium is made of big "grains" or rock matrixes. As shown in equation 1 or 2, the effective porosity decreases as the value of s increases.

3. Governing equations

The governing equations of the in-situ leaching processes read

$$\phi_{,t} = (\phi K(\phi) h_{,i})_{,i} \tag{3}$$

$$V_i = -K(\phi) h_{,i} \tag{4}$$

$$(\phi C_1)_{,t} = -(\phi C_1 V_i)_{,i} + (\phi D_{1ij} C_{1,j})_{,i} - \beta \frac{G \rho_{rock}}{\rho_w} \phi_{,t} \tag{5}$$

$$(\phi C_2)_{,t} = -(\phi C_2 V_i)_{,i} + (\phi D_{2ij} C_{2,j})_{,i} + \frac{G \rho_{rock}}{\rho_w} \phi_{,t} \tag{6}$$

where

$$K(\phi) = K_0 \left(\frac{\phi}{\phi_0}\right)^3 \tag{7}$$

and

$$\phi_{,t} = k \Delta\phi_{max} \left(\frac{\phi_f - \phi}{\phi_f - \phi_0}\right)^{2/3} C_1 = k(\Delta\phi_{max})^{1/3} (\phi_f - \phi)^{2/3} C_1. \tag{8}$$

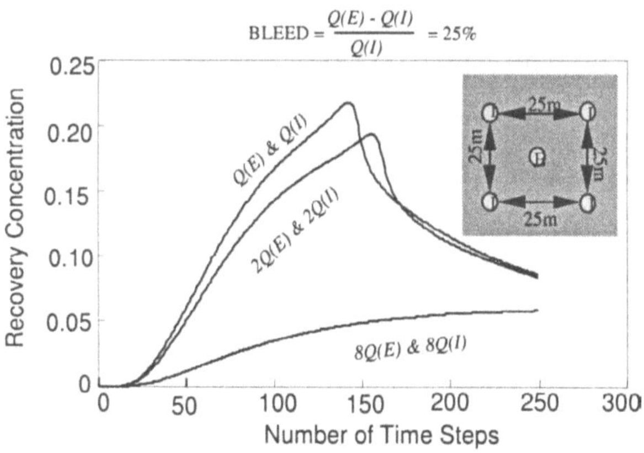

Figure 1. Operation curves for a five-spot in-situ leaching operation. E and I represent ejection well and injection well, respectively. $Q(E)$ and $Q(I)$ represent ejection flux and injection flux, respectively.

where ϕ is the effective porosity; $K(\phi)$ is the effective porosity-dependent hydraulic conductivity; h is the total hydraulic head; V_i is the ith velocity component; x_i is the coordinates; C_1 is the concentration of a

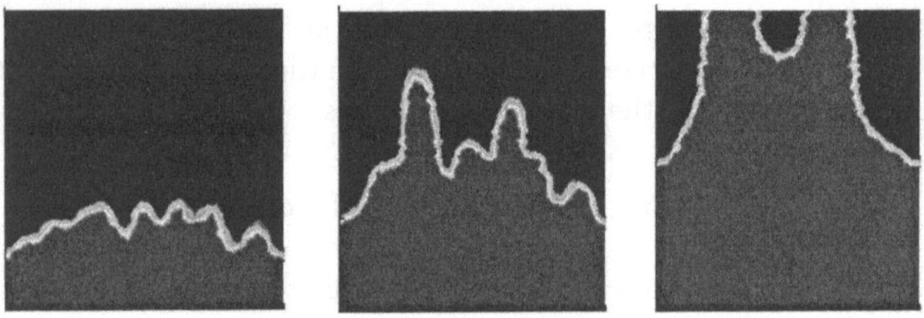

Figure 2. Evolution of porosity fingering instability; 2000 triangular elements; unstructured mesh. The initial state is shown in the left figure, the intermediate state in the central figure, and the final state in the right figure.

reagent; D_{ij} are components of the dispersion tensor; ρ_{rock} is the mass density of the orebody; ρ_w is the mass density of the solution; β is the stoichiometric coefficient; C_2 is the concentration of the dissolved mineral; k is the rate constant; G is the ore grade; K_0 is the initial hydraulic conductivity; ϕ_f is the final porosity; and ϕ_0 is the initial effective porosity. Equations 3 and 4 are fluid continuous equation and Darcy's law, respectively. Equations 5 and 6 are transport equations of the reagent and the dissolved mineral, respectively. Equation 8 describes the evolution of the effective porosity due to the dissolution/precipitation. Equation 7 describes the dependency of the hydraulic conductivity on the effective porosity. Details of equations, 5, 6 and 8, can be found in the paper [5]. These sets of coupled equations have been implemented into and solved by the, *FASTFLO* (see http://www.nag.co.uk/simulation/Fastflo/fastflo.html).

4. Numerical results

We now apply our model to the simulation of a typical five-spot in-situ solution mining, where we prescribe constant flux conditions both at the extraction well (central well) and at four injection wells (surrounding wells). The model results are shown in Figure 1. In the simulation, the so called bleed, defined as

$$bleed = \frac{Q(E) - Q(I)}{Q(I)} \tag{9}$$

is kept constant (25%). $Q(E)$ and $Q(I)$ are the total ejection flux and the total injection flux, respectively. It is apparent that the characteristics of mineral recovery concentration versus time curves are affected significantly by the total injection flux, $Q(I)$, or by the total ejection flux, $Q(E)$. The peak mineral recovery concentration decreases as the

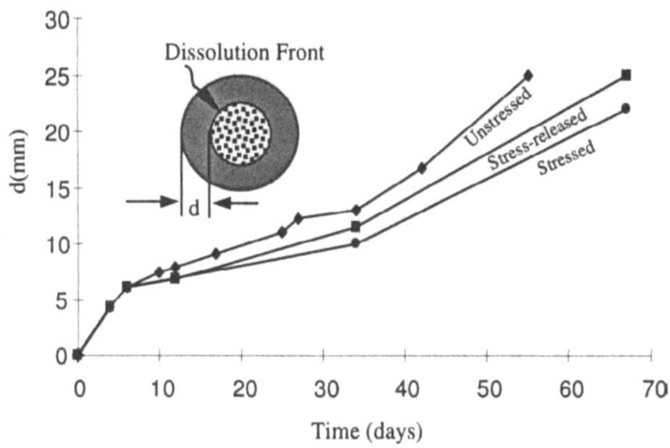

Figure 3. Migration of the dissolution front. d represents depth of the dissolution front from the sample surface (after Chigira, 1993). $pH = 3$. $\sigma = 0$ or $0.49MPa$.

total injection or ejection flux increases. This phenomenon may be explained in the following.

The pore velocity for the case with one injection well can be defined as

$$V = \frac{Q(I)}{2\pi r d\phi} \tag{10}$$

where V is the pore velocity, r is the distance away from the well center, and d is the thickness of the ore deposit. It is obvious that the pore velocity is directly proportional to the total flux. The mineral recovery concentration decreases with a higher injection flux because the higher pore velocity, resulting from the higher injection flux, reduces the contact time between leach solution and target mineral. This conclusion has been verified by the numerical results.

5. Concluding remarks

We have applied a nonlinear reaction-diffusion-advection model to the simulation of a typical unit operation of in-situ solution mining. The operation consists of a five-spot arrangement of wells, where constant flux conditions are prescribed both at the extraction well (central well) and at four injection wells (surrounding wells). It is concluded from the model results that the characteristics of mineral recovery concentration histories are affected significantly by the total injection flux or by the total extraction flux. The peak mineral recovery concentration decreases as the total injection or extraction flux increases.

We have already mentioned the positive feedback resulting from the porosity dependency of the permeability: The solution process increases the porosity and thus the permeability as well. The porosity feedback causes instability of initial straight reaction fronts. A corresponding simulation is illustrated by means of a finite element simulation (see Figure 2). The problem is defined on a rectangular domain with $h = 0$ and zero reagent flux on top and prescribed pore fluid flux and prescribed reagent concentration at the bottom. The reagent concentration is constant with small superimposed perturbation. On the lower boundary we also have assumed that $\phi = \phi_f$, where in the domain, initially we have $C_1 = 0$ and $\phi = \phi_0$. The scale of the initial concentration perturbation was reflected according to the linear instability analysis in [2]. The instability and the evaluation of the porosity "fingering" is represented in Figure 2(a-c).

Finally we would like to mention that in the present model we have neglected any effect which the chemical process might have on the mechanical properties of the material model consideration and vice-versa. In this context, an interesting experimental result is displayed in Figure 3. The dependency of the penetration, d, on the time and the state of loading reflects the dependency of permeability, conductivity and diffusivity on the state of deformation.

Acknowledgements

The work reported in this paper is supported jointly by the CSIRO Exploration & Mining and by the CSIRO Minerals. This support is gratefully acknowledged.

References

1. Brady, B. H. and Liu, J. Geomechanics and Analysis of *In-situ* Solution Mining. In: H. D. et al. (ed.): *Proceedings of the Mining Technology Conference.* Fremantle WA, Australia, 150–158, 1996.
2. Chadam, J., Hoff, D., Merino, E., Orteleva, P. and Sen, A. Reactive infiltration instabilities. *Journal of Applied Mathematics*, **36**, 207–227, 1986.
3. Gibowicz, S. J. The mechanism of seismic events induced by mining. In: *Rockbursts and Seismicity in Mines*, 3–27, 1990.
4. Liu, J., Elsworth, D. and Brady, B. H. Linking Stress-dependent Effective Porosity and Hydraulic Conductivity Field to RMR. *Int. J. Rock Mech. & Min. Sci.*, **36**, 581–596, 1999.
5. Muhlhaus, H. B., Liu, J. and Hobbs, B. A Porosity Evolution Model for In-situ Leaching Processes. In: *Proceedings of The Second Australasian Congress on Applied Mechanics.* Paper No. U-74. Canberra, Australia, 1999.
6. Wiley, K. L., Ramey, D. S. and Rex, M. J. In-situ leaching wellfield design at San Manuel. *Mining Engineering* August, 991–994, 1994.

Coupling of Damage and Fluid-Solid Interactions in Quasi-Brittle Unsaturated Porous Materials

J. Carmeliet
Laboratory of Building Physics (Civil Engineering)
University of Leuven, Belgium
e-mail: Jan.Carmeliet@bwk.kuleuven.ac.be

Abstract. In this paper constitutive equations are formulated describing the fluid-solid interactions and damage behaviour of unsaturated porous materials. The fluid-solid interactions are described by coupling coefficients dependent on the degree of saturation, strain and damage state. Simplified expressions for these dependencies are developed. The model is illustrated by analysing the swelling and compressive damage behaviour of unsaturated materials.

Keywords: Poromechanical coupling, damage, unsaturated porous media

1. Introduction

In dealing with the modelling of microscopic fluid-solid interaction forces in hydrophilic porous materials, we follow the poromechanical continuum approach associated with the pioneering work of Biot ([1]). Coussy *et al.* ([4]) extended the Biot theory to partially saturated deformable porous materials. His approach is limited to poroelastic behaviour of undamaged porous media with an incompressible matrix. In this paper, the framework of Coussy ([3]) is generalised to unsaturated damageable materials with a compressible matrix.

2. Constitutive equations

The pore space is assumed to be partially filled by an ideal mixture (mix) of water vapour (v) and dry air (a), forming together with the liquid water (l) the fluid particle. We will assume the pressure in the mixture to be constant, or, $\mathrm{d}p_{mix}$=0. In the constitutive equations the total stress σ and liquid pressures p_i are related to the unknown fields of strain ε and moisture content variation m_i. Each field can by decomposed in elastic (e) and anelastic (a) parts

$$\varepsilon = \varepsilon^e + \varepsilon^a \quad \text{and} \quad m_l = m_l^e + m_l^a \tag{1}$$

The anelastic strain ea, which remains in the material after complete unloading, results from (1) incomplete contact between crack

W. Ehlers (ed.), IUTAM Symposium on Theoretical and Numerical Methods in Continuum Mechanics of Porous Materials, 307–312.

surfaces and (2) plastic deformations of the solid matrix (figure 1b). This distinction will appear to be important since microcracking, in contradiction to plastic solid matrix deformation, leads to an increase of porosity, which highly affects the fluid-solid interaction forces. Thus, the anelastic strain increment can be written as

$$\mathrm{d}\varepsilon^a = (1-\phi)\mathrm{d}\varepsilon_s^a + \mathrm{d}\phi^a \tag{2}$$

The term in $\mathrm{d}\varepsilon_s^a$ describes the incremental plastic deformation of the solid matrix, while the term $\mathrm{d}\phi^a$ accounts for the porosity change due to incomplete closure of micocracks upon unloading.
The introduction of free energy $\Psi{=}\Psi(\varepsilon^e,\, m_i^e,\, D)$, with D denoting the internal damage variable, leads to the constitutive equations

$$\sigma = \frac{\partial\Psi}{\partial\varepsilon^e} \qquad \text{and} \qquad \frac{P_i}{\rho_l} = \frac{\partial\Psi}{\partial m_i^e} \tag{3}$$

The damage variable D can be regarded as a measure for the reduction of the material stiffness and grows from zero to one at complete loss of integrity of the porous material (see figure 1a). Following Herrmann and Kestin ([5]), the absolute value of the anelastic strain is assumed to be only a function of D

$$\mid \varepsilon^a \mid = \frac{\gamma_1 D^2}{E(1-\gamma_2 D)}1 \tag{4}$$

with E the elastic modulus, γ_1 the crack closure pressure and γ_2 a parameter ensuring finite values of the anelastic strain for the almost completely damaged material.
Following Coussy ([3]), the following incremental constitutive equations can be derived (Carmeliet, [2])

$$\mathrm{d}\sigma = (1-D)C_0 : \mathrm{d}\varepsilon^e - C_0 : \varepsilon^e \mathrm{d}D + b_l\, \mathrm{d}p_c \tag{5}$$

$$\frac{\mathrm{d}m_l}{\rho_l} = -\frac{\mathrm{d}p_c}{M} + b_l : \mathrm{d}\varepsilon^e + \mathrm{d}m_l^a \tag{6}$$

where p_c is the capillary pressure ($p_c = p_{mix}$ - p_l). In (5,6) C_0 is the drained elastic stiffness tensor, b_l and M respectively the Biot coefficient and Biot modulus.
In the view of simplifying the constitutive equations, we assume isotropic

behaviour and the drained elastic stiffness C_0 to be independent of the capillary pressures p_c, or $C_0 = C_0(\varepsilon^e)$. This assumption is also referred to as the effective stress assumption. Then, it can be shown that the coupling coefficient b_l is only a function of p_c and D. This also means that the degree of saturation S_l is independent of the volumetric strain ε^e which implies the validity of the generally postulated state equation $S_l = S_l\ (p_c)$ for constant D.
The coupling coefficients are derived from the microscopic properties of the constituents of the porous medium (Carmeliet, [2]). In this appoach, we assume that

$$\mathrm{d}\phi^a = \beta\, \mathrm{d}\varepsilon^a \tag{7}$$

holds for any anelastic evolution of the porosity. From (2) and (7), it follows

$$(1-\phi)\, \mathrm{d}\varepsilon_s^a = (1-\beta)\, \mathrm{d}\varepsilon^a \tag{8}$$

A value β=1 corresponds to a solid matrix, that undergoes no permanent deformation $\mathrm{d}\varepsilon_s^a$=0. The limiting case β=0 corresponds to a material, where all anelastic deformations are due to the anelastic behaviour of the solid matrix.

a)
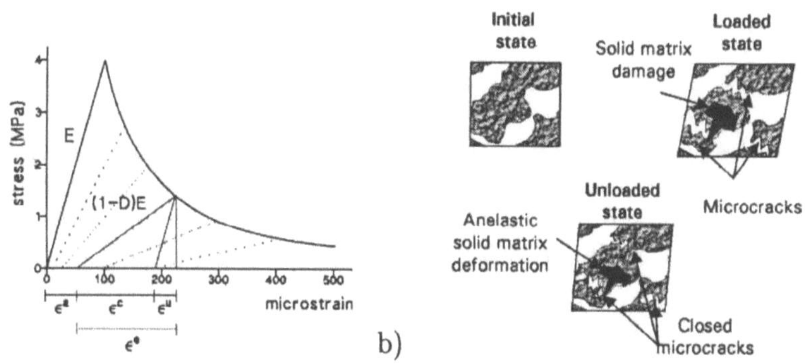

Figure 1. a) Strain softening behaviour of the porous material with permanent strain upon unloading. b) Idealised different states of the porous material.

Using the above-mentioned micro-macro approach, the poromechanical coefficient $\mathrm{b}_1 = b_l 1$ can be identified as

$$b_l = S_l\,(1 - \frac{K_0}{K_s}) + S_l D \frac{K_0}{K_s} \qquad (0 \leq S \leq 1) \tag{9}$$

with K_0 the drained bulk modulus, K_s the bulk modulus of the solid matrix. The anelastic moisture content variation dm_l^a is given by

$$\mathrm{d}m_l^a = b_l^a \mathrm{d}D \qquad (0 \leq S \leq 1)$$
$$b_l^a = S_l\,(\frac{K_0}{K_s}\,\mathrm{tr}\,\varepsilon^e + \beta\,\mathrm{tr}\gamma_D) + \frac{\partial S_l}{\partial D} \quad , \qquad \gamma_d = \frac{\partial \varepsilon^a}{\partial D} \tag{10}$$

The coupling coefficients are found to be dependent on the degree of saturation S_l, the damage state D, the crack closure pressure γ_1, the elastic opening of microcracks ε^e, the ratio K_0/K_s, the anelastic behaviour of solid martix (β).It is important to note that the coupling coefficient b_l increases with the degree of saturation as well as with the damage state of the material.
It is remarked that relation (9) reduces to the classical Biot relation $b_l = 1 - K_0/K_s$ for undamaged saturated media ($S_l = 1$ and $D = 0$). In letting $K_s \to \infty$ (incompressible matrix) relation (9) reduces to bl = Sl, which was assumed by Coussy *et al.* ([4]). The coupling coefficient b_l^a accounts for anelastic moisture content variations due to damage growth and consists of three parts: the two first terms of (10) account for the moisture content variation due to the elastic and anelastic opening of developing microcracks, while the third term describes moisture content variation due to the change of the state equation S_l (p_c, D) during damage growth.

3. Examples

In an isothermal adsorption experiment an initially dry specimen is conditioned at different capillary pressures (or relative humidity h). The specimen is not subjected to an external loading ($\mathrm{d}\sigma = 0$). We assume that no damage develops ($\mathrm{d}D = 0$). Introducing these conditions into (5) gives

$$\mathrm{d}\varepsilon = -\frac{b_l}{(1 - D_0)K_0}\,\mathrm{d}p_c \tag{11}$$

Figure 2a shows the measured uniaxial swelling strain e11 as a function of relative humidity h for a cellulose fibre cement mortar. From (11), one can notice that the Biot coefficient b_l (S_l , D_0) can be determined by measuring the volumetric strain change $\mathrm{d}\varepsilon$=3$\mathrm{d}\varepsilon_{11}$. Figure 2a gives the fitted theoretical curve using the measured saturation curve $S_l(p_c)$. The Biot coefficient is found to be equal to $b_l = 0.79$ S_l. According to (10) $K_s = 4.8$ K_0, which proves the solid matrix to compressible. It is noted that a validation of the present model requires the measurement of the bulk moduli K_s and K_0. It is shown in fig. 2b that the swelling strain significantly increases with the damage state. This is caused by two effects: a decrease of the apparent bulk modulus (1-D)K_0 (dashed line) and an increase of the coupling coefficient b_l due to damage (last term of 10).

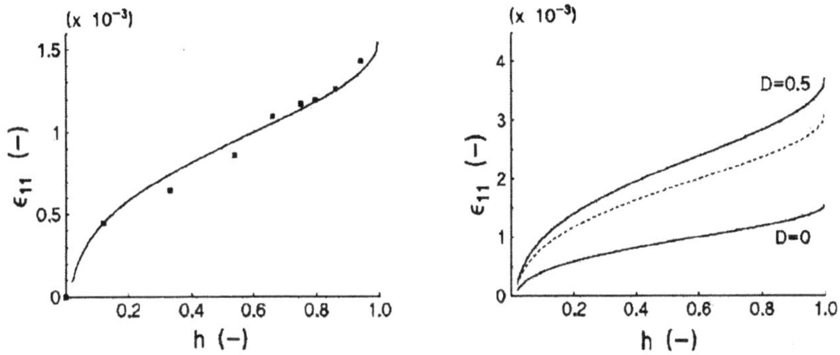

Figure 2. a) Measured (□) and simulated (-) swelling strain as a function of the relative humidity. b) Influence of damage on swelling strain.

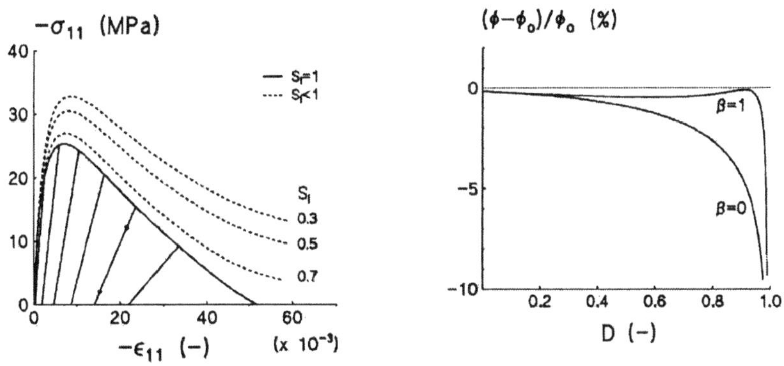

Figure 3. a) Compression test for different degrees of saturation. b) Change of porosity as a function of damage for two values of the parameter β.

In second academic example, we consider an uniaxial drained compression test (dp_c=0). The material is initially fully saturated, then dried to different degrees of saturation and loaded. The material properties are given in Carmeliet ([2]). The axial load-deformation curves for different degrees of saturation are given in figure 3a. Lower degrees of saturation indicate an increase of the compressive stresses. The strengthening effect can be explained by the fact that the radial tensile

strain ε_{22}, which governs the damage process, reduces at low degree of saturation. This reduction is due to the high compressive capillary pressure at low degree of saturation.
We now analyse the change of porosity during the damage process for two limiting values of the parameter β. For β=0, the anelastic change of porosity $d\phi^a$ is zero (see eq. 6). The porosity reduction is then only caused by the elastic squeezing of the material, which is favoured by the stiffness reduction resulting from the damage process. For β=1, the damage process not only favours the squeezing and porosity reduction, but also introduces an anelastic increase of porosity in the form of micro-cracks. As a result, the porosity reduction is found to be much lower compared to the curve for β=0.

4. Conclusions

Constitutive relations for unsaturated damageable porous materials are derived based on the thermodynamic theory of damage mechanics and mechanics of deformable porous media. Expressions for the dependence of the coupling coefficients on degree of saturation, strain and damage state, are proposed. The model is illustrated by analysing the swelling and damage behaviour of unsaturated materials. The important influence of damage on the poromechanical behaviour is clearly illustrated.

References

1. Biot M. A. General theory of three dimensional consolidation. *J. Appl. Phys.*, **12**, 155–164, 1941.
2. Carmeliet J. Poroelastic and damage coupling in unsaturated porous media, submitted to *Mechanics of Materials*, 1999.
3. Coussy, O. *Mechanics of porous continua*, Wiley and sons, Chichester, 1995.
4. Coussy, O., Eymard, R. and Lassabatere, T. Constitutive modelling of unsaturated drying deformable materials. ASCE *J. Engng. Mech.*, **124**, 658–667, 1998.
5. Hermann, G. and Kestin, J. On the thermodynamics foundations of damage theory in elastic solids. In *Cracking and Damage*, J. Mazars and Z. P. Bažant (eds.), Elsevier, 217–227, 1989.

An Influence of Initial Porosity on Damage Process in Semi-Brittle Polycrystalline Ceramics under Compression

T. Sadowski and S. Samborski
Faculty of Mechanical Engineering Technical University of Lublin
20-618 Lublin, Nadbystrzycka 36 Str., Poland
e-mail: tskms@archimedes.pol.lublin.pl

Abstract. The initial porosity inside ceramics strongly influences the material response tQ applied mechanical load. In the paper two types of porosity were considered: distributed inside grains and along grain boundaries. Both types of porosity influence elastic material properties (compliance tensor S_{ijkl}). The pores distributed along grain boundary lead to significant crack resistance reduction in that part of the material. These local weakening places create preferential conditions for microcracks growth, which are responsible for internal damage process in macro-scale. The proposed theory was illustrated by uniaxial compression loading process of polycrystalline magnesium oxide sample.

1. Motivation

The ceramic porous materials are widely used as refractory materials in steal making or cement industry as furnace lining (in the form of bricks) because of good compressive strength and thermal stress resistance. The typical material for bricks is magnesium oxide containing up to 30elliptical or spherical pores. The technology of casting process is connected with cyclic loading and reloading of furnace, what causes progressive cracking of bricks during mechanical loading. The degree of the material cracking, responsible for internal damage in macro-scale, depends on the current level of loading.

Experimental observations, e.g. [1] and [8], suggest to take into account in the polycrystalline material modelling information concerning:

• elastic and plastic properties of the material,

• internal structure of the material which can be described by: grain size distribution, pore size distribution, pore placement inside material and mode of microcracking.

The modelling of such complex material response is possible by micromechanical method.

2. Constitutive modelling of the polycrystalline porous ceramics by micromechanical approach

Let us consider a material sample subjected to uniaxial compression. Assume, that the material is porous, semi-brittle (i.e. exhibits a lim-

W. Ehlers (ed.), IUTAM Symposium on Theoretical and Numerical Methods in Continuum Mechanics of Porous Materials, 313–318.

ited plasticity) and has nucleated cracks which grow during loading process. If the defect density inside the material is dilute or at most moderate one can decompose the total macroscopic strain tensor into the following parts:

$$\varepsilon_{ij} = \varepsilon_{ij}^{e} + \varepsilon_{ij}^{po} + \varepsilon_{ij}^{pl} + \varepsilon_{ij}^{cr} , \tag{1}$$

where: ε_{ij}^{e} are purely elastic strains, ε_{ij}^{po} are pore existence-dependent strains, ε_{ij}^{pl} are plastic strains, ε_{ij}^{cr} are cracks growth-dependent strains.

In order to get particular specification of the constitutive equation (1) let us apply micromechanical approach. For two-dimensional case one can consider representative surface element (RSE) (for example [2] - [7]), which area is denoted by A. Assume that RSE contains: N - grains, N_{po}- number of pores, N_{pl}- number of grains with activated slip system ($N_{pl} \ll N$), N_{cr}- number of growing cracks. Denoting by $\varepsilon'_{ij}(\sigma_{kl})$ local microstrains created by: holes, plastic grains and cracks inside RSE, one can get averaged values of deformation tensor $\bar{\varepsilon}_{ij}(\sigma_{kl})$ describing:

- elastic deformations on the elastic or elasto-plastic part of RSE denoted by A_e

$$\bar{\varepsilon}_{ij}^{e}(\sigma_{kl}) = \frac{1}{A_e} \int\limits_{A_e} \varepsilon_{ij}^{\prime e}(\sigma_{kl})\, dA . \tag{2}$$

- additional deformations caused by pores existence in the part of RSE denoted by $A_{po} = \sum\limits_{s=1}^{N_{po}} A_{po}^{(s)}$

$$\bar{\varepsilon}_{ij}^{po}(\sigma_{kl}) = \frac{1}{A_{po}} \sum_{s=1}^{N_{po}} \int\limits_{A_{po}^{(s)}} \varepsilon_{ij}^{\prime po(s)}(\sigma_{kl})\, dA . \tag{3}$$

- additional deformations created by grains with activated slip system (in the part of RSE denoted by $A_{pl} = \sum\limits_{s=1}^{N_{pl}} A_{pl}^{(s)}$)

$$\bar{\varepsilon}_{ij}^{pl}(\sigma_{kl}) = \frac{1}{A_{pl}} \sum_{s=1}^{N_{pl}} \int\limits_{A_{pl}^{(s)}} \varepsilon_{ij}^{\prime pl(s)}(\sigma_{kl})\, dA . \tag{4}$$

- additional deformations reflecting crack growth (in the part of RSE denoted by $A_{cr} = \sum\limits_{s=1}^{N_{pl}} A_{cr}^{(s)}$)

$$\bar{\varepsilon}_{ij}^{cr}(\sigma_{kl}) = \frac{1}{A_{cr}} \sum_{s=1}^{N_{cr}} \int\limits_{A_{cr}^{(s)}} \varepsilon_{ij}^{\prime cr(s)}(\sigma_{kl})\, dA . \tag{5}$$

In the formula (3), (4), (5) symbol "(s)" is related to any inhomogenity or discontinuity "s" existing in RSE.

Introducing (2), (3), (4) and (5) to equation (1) and according to mixed rule we get particular forms of the right hand side of (1):

$$\begin{aligned} \varepsilon_{ij}(\sigma_{kl}) &= [1-\mathrm{f}_d]\varepsilon^{e}_{ij}(\sigma_{kl}) + \mathrm{f}_{po}\bar{\varepsilon}^{po}_{ij}(\sigma_{kl}) + \mathrm{f}_{pl}\bar{\varepsilon}^{pl}_{ij}(\sigma_{kl}) + \mathrm{f}_{cr}\bar{\varepsilon}^{cr}_{ij}(\sigma_{kl}) \\ &= \mathrm{S}_{ijkl}(\sigma_{mn})\sigma_{kl}\ , \end{aligned} \tag{6}$$

where f_d is the surface area density of all defects (pores and cracks): $\mathrm{f}_d = \mathrm{A}_d/\mathrm{A}$. A_d is the total area of defects in RSE. Moreover: $\mathrm{f}_{po} = \mathrm{A}_{po}/\mathrm{A}$, $\mathrm{f}_{pl} = \mathrm{A}_{pl}/\mathrm{A}$; $\mathrm{f}_{cr} = \mathrm{A}_{cr}/\mathrm{A}$ and $\mathrm{A}_e = (1\text{-}\mathrm{f}_d)\mathrm{A}$. S_{ijkl} is the actual compliance tensor characterizing the total material response.

2.1. Porosity of the material

The closed initial porosity is distributed in grains (p_g) or along grain boundaries (p_{gb}). Both types of porosity $\mathrm{p}=\mathrm{p}_g+\mathrm{p}_{gb}$ significantly influence the initial components of the compliance tensor S_{ijkl}. Assuming that the pores are spherical and initially homogeneously distributed in RSE one can introduce porosity parameter (see [2]):

$$\mathrm{p} = \sum_{s=1}^{\mathrm{N}_{po}} \left[\frac{\mathrm{r}_{(s)}}{\mathrm{R}}\right]^2 , \tag{7}$$

where r_s is the pore "s" radius and R is the radius of RSE. For the plain strain conditions under loading process we have:

$$\varepsilon^{po}_{ij} = \frac{1-(v_0)^2}{\mathrm{E}_0}[\mathrm{p}(4\sigma_{ij} - \sigma_{kk}\delta_{ij})] \tag{8}$$

for non-interacting pores or

$$\varepsilon^{po}_{ij} = \frac{1-(v_0)^2}{\mathrm{E}_0}\left[\frac{\mathrm{p}}{1-\mathrm{p}}(4\sigma_{ij} - \sigma_{kk}\delta_{ij})\right] \tag{9}$$

for interacting pores.

2.2. Plasticity of the material

A certain part of grains (N_{pl}) within RSE undergoes conjugate slip system, when the shear stress τ along the slip direction overcomes the threshold value of the shear stress τ_{so}. In order to estimate plastic deformation, the Eshelby inclusion model can by applied. Then the additional strains ε^{pl}_{ij} are equal:

$$\varepsilon^{pl}_{ij} = \frac{\mathrm{N}_{pl}}{\pi \mathrm{G}_0(1-\mathrm{b}_1)} \int_{\mathrm{D}_m}^{\mathrm{D}_M}\int_{\beta_1}^{\beta_2} \Big[\tau(\beta) - \tau_o\Big]\alpha_{ij}(\beta)\mathrm{p}_1(\beta)\mathrm{p}_2(\mathrm{D})\mathrm{d}\beta\mathrm{d}\mathrm{D}\ , \tag{10}$$

where β is the slip line inclination angle to the axis x_1, $p_1(\beta)$ is the distribution function of slip line inclination angle, D_m and D_M are the smallest and largest grain size (D) respectively, G_0 is the Kirchhoff modulus of the virgin material, b_1 is the coefficient connected with the shape of grains. $\beta_1 \leq \beta \leq \beta_2$ denotes the fan of N_{pl} grains with activated slip systems inside RSE.

2.3. Cracking of the material

In polycrystalline ceramics, cracks are created by Zener-Stroh's initiation mechanism, i.e. when the dislocations pile-up at the grain boundary. The texture of the cracking phenomenon inside the RSE of polycrystalline material is build up of the two types of defects: rectilinear cracks (number $N_{cr}^{(rec)}$, inclined to horizontal axis under angle ϕ) and wing cracks (number $N_{cr}^{(w)}$, wing is inclined in relation to rectilinear middle crack under angle θ). The initiated cracks can develop along grain facets changing their direction. This process is strongly influenced by grain boundary porosity p_{gp} . Namely, any crack (rectilinear or wing) can grow if the energy release rate G is:

$$G(\sigma_{22}^{(b)}, \phi, \theta, p_{pb}) \geq \gamma_{gb}^{cr}(p_{gb}) \ . \tag{11}$$

$\gamma_{gp}(p_{gp})$ is the critical value of the grain boundary fracture surface energy. This energy is significantly less than the grain surface energy γ_g:

$$\gamma_{gb}(p_{gb}) = (0.5 - 0.75)\gamma_g \ . \tag{12}$$

In order to describe function $\gamma_{gp}(p_{gp})$ we assume uniform distribution of holes along grain boundary $\bar{p_{gb}}$. Then the following simplest correlation can be applied

$$\gamma_{gb}(\bar{p}_{gb}) = (1 - \bar{p}_{gb})\gamma_g \ . \tag{13}$$

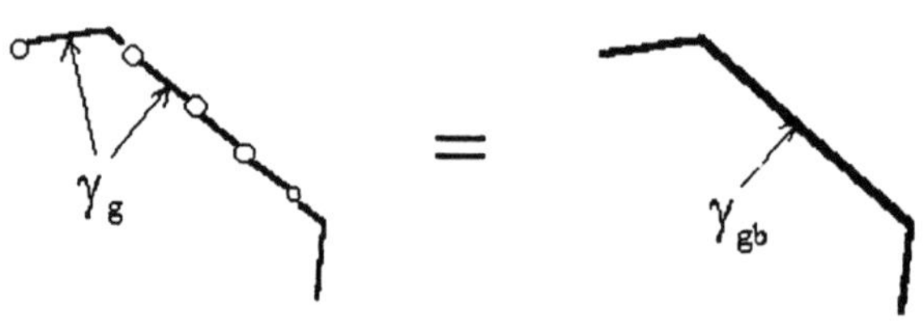

Figure 1. Homogenisation procedure of the surface energy for grain boundary

Having defined criterion of crack growth (10), the additional strains describing cracking process inside RSE consist of two parts related to rectilinear or wing defects:

$$\begin{aligned} \varepsilon_{ij}^{cr}(\sigma_{kl}) &= \varepsilon_{ij}^{cr(rec)}(\sigma_{kl}, N_{cr}^{(rec)}, \phi, \mu, D) + \\ &+ \varepsilon_{ij}^{cr(w)}(\sigma_{kl}, N_{cr}^{(w)}, \phi, \theta, \mu, D, l) \ , \end{aligned} \tag{14}$$

where l is the wing length.

2.4. Experimental estimation of the Young modulus and Poisson coefficient

Fig. 2 presents experimental results of the Young modulus(solid line) which were compared

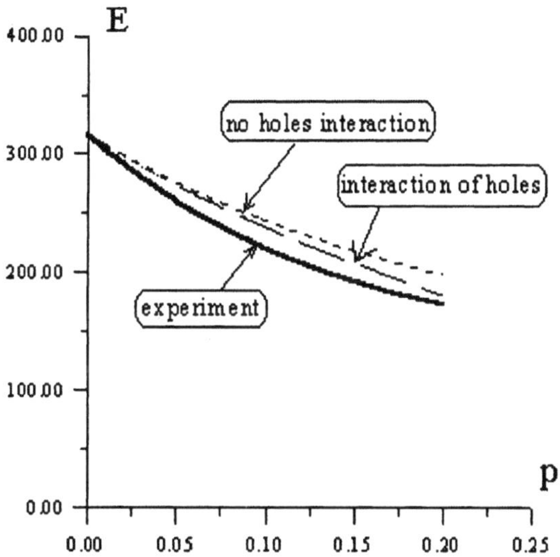

Figure 2. Experimental and theoretical estimations of the Young modulus for MgO porous ceramics.

with theoretical approach for non-interacting pores (11) and interacting ones (12). Experimental observations of the Poisson's coefficient lead to the conclusion that ν=.27 for the analysed porosity. However theoretical modelling suggests small increase of this mechanical property with the increase of porosity.

2.5. Numerical example

Numerical calculations were performed according to theoretical formulation for the following data: $E_0(p=0) = 316.4$ GPa, $G_0(p=0) = 121.4$ GPa, $\nu(p=0) = 0.27$, $\gamma_g(p=0) = 1$ J/m^2, $p \in (0;0.2)$. It was assumed that RSE contains hexagonal grains of the mean diameter $\bar{D}=45\mu$m .

The densities of pores and cracks are diluted, therefore we do not take into account interaction between them.

Obtained results were plotted in Fig. 3. One can notice the influence of the porosity on the material response. It is reflected by significant increase of the components of compliance tensor $S_{ijkl}(\sigma,p)$. The numerical calculations were performed for $\gamma_{gp}(p_{gp}) = 0.5$, assuming that pores diameter is $2\mu m$.

The presented results show capability of micromechanical approach in modelling of material with internal structure.

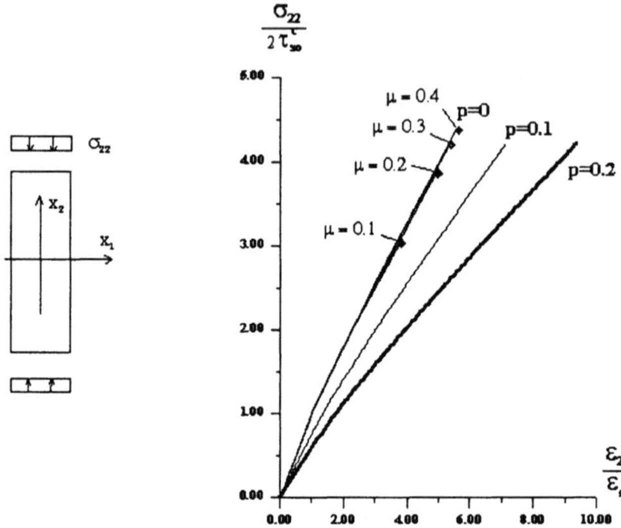

Figure 3. Stress-strain relation for different porosity level

References

1. Davidge, R. W. Mechanical behaviour of ceramics. Cambridge Univ. Press, 1979.
2. Kachanov, M. On the effective moduli of solids with cavities and cracks. *Int. J. Fracture*, **59**, R17–R21, 1993.
3. Kachanov, M. Elastic solids with many cracks and related problems. *Advances in Appl. Mech.*, **30**, 259–445, 1993.
4. Krajcinovic, D. Damage mechanics. *Mech. Materials*, **8**, 117–197, 1989.
5. Nemat-Nasser, S. and Horii, M. Micromechanics: overall properties of heterogeneous materials. *Elsevier Sci. Publ.*, 1993.
6. Nemat-Nasser, S. and Obata, M. A microcrack model of dilatancy in brittle materials. *J. Appl. Mech.*, **55**, 24–35, 1988.
7. Sadowski, T. Modelling of semi-brittle MgO ceramics behaviour under compression.*Mech. Materials*, **18**, 1–14, 1994.
8. Sadowski, T., Boniecki, M., Librant, Z. and Ruiz, C. Fracture process of monolithic polycrys-talline ceramics (Al2O3 and MgO) under quasi-static and dynamic loading. In: A.Brandt, V.C.Li, I.H.Marshall (eds.), Brittle Matrix Composites 5, BIGRAF and Woodhead Publ., Warsaw, 567–576, 1997.

Session D3: Swelling, Drying and Shrinkage

Chairman: Y. Abousleiman

The Physical Role of Crack Rate Dependence in the Long-Term Behaviour of Cementitious Materials

G. P. A. G. van Zijl and J. G. Rots
Faculty of Architecture
Delft University of Technology
P.O. Box 5043, 2600 GA Delft, The Netherlands

R. de Borst
Koiter Institute Delft / Faculty of Aerospace Engineering
Delft University of Technology
P.O. Box 5058, 2600 GB Delft, The Netherlands

1. Introduction

The behaviour of porous, cementitious materials is strongly loading rate dependent. This is not only the case for high rates, but has also been observed for low, quasi-static rates (Rüsch [6], Zhou [10], Bažant and Gettu [2]). Therefore, it is important to include rate effects for the proper modelling of fracture of cementitious materials.

Moisture migration is largely responsible for the rate/time-dependent behaviour. It is thus of paramount importance to study the moisture migration in order to understand the key phenomena originating from it, such as true material shrinkage, stress-induced shrinkage/drying creep, basic creep and the viscosity of the degradation process. A numerical model has been formulated which comprises these phenomena in a unified manner (Van Zijl [8]). It can be argued that the crack mouth opening rate is reduced by the viscosity of bond rupture, be it caused by the presence of moisture, or the viscosity of the cement paste itself. This justifies the inclusion of a cracking viscosity by Sluys ([7]) and de Borst *et al.* ([3]), who presented it as a solution to the loss of well-posedness of the governing equations in the event of softening. Alternatively, a physical rationale for the rate dependence has been sought in the activation theory (Bažant [1]). In this study a strain-softening cracking model including viscosity is presented and compared with the formulation resulting from the activation theory. Through the analyses of three-point bending creep experiments by Zhou ([10]) the physical role of the crack rate dependence is underlined.

W. Ehlers (ed.), IUTAM Symposium on Theoretical and Numerical Methods in Continuum Mechanics of Porous Materials, 321–328.

2. Crack rate dependence

To facilitate the description of the constitutive behaviour of concrete it is assumed that the strain rate can be decomposed as follows:

$$\dot{\varepsilon} = \dot{\varepsilon}_e + \dot{\varepsilon}_c + \dot{\varepsilon}_{cr} + \dot{\varepsilon}_s + \dot{\varepsilon}_T \tag{1}$$

where ε_e is the elastic strain, ε_c the creep strain, ε_{cr} represents the cracking strain, ε_s the shrinkage strain and ε_T the thermal strain. In this way the constitutive laws for the different phenomena can be elaborated independently (eg. Van Zijl [8]). For constant hygral and thermal conditions only the first three components prevail. It can be further reduced by exploiting the fading memory, visco-elastic nature of concrete. To this end, the first two components are combined

$$\dot{\varepsilon}_{ve} = \dot{\varepsilon}_e + \dot{\varepsilon}_c \tag{2}$$

and are represented by a Maxwell chain to capture the elasticity and bulk creep. However, the time-dependence of the behaviour of cementitious materials is caused not only by the bulk creep, but also by the rate dependence of the breakage of bonds in the fracture process zone. Bažant ([1]) has derived a formulation for the latter rate effect from the theory of activation energy, according to which the rupture of a bond requires that the limiting bond potential, called the activation energy, be exceeded. Departing from the Maxwell-Boltzmann distribution of the frequency of exceedence of the activation energy, the following expression for the crack opening rate $\dot{w}$ can be derived for isothermal conditions (Wu and Bažant [9]):

$$\dot{w} = \dot{w}_r \sinh\left[\frac{\sigma - \sigma_t(w)}{k_0\ [\sigma_t(w) + k_1\ f_t]}\right] \tag{3}$$

where $\dot{w}_r$ is a constant, reference crack opening velocity and $\sigma_t(w)$ describes the strength degradation with an infinitely low crack opening velocity. The material parameter k_0 is estimated to be in the range 0.01 - 0.08 from the knowledge that for a 10^4-fold increase of the loading rate a 25% increase in peak strength is found experimentally. The material parameter k_1 is an offset factor to prevent the denominator in eq. (3) from becoming zero. By approximating the crack opening displacement $w = l_b\ \kappa$, with l_b the crack band width and κ the crack strain, eq. (3) can be rewritten as

$$\sigma = \sigma_t(\kappa)\left[1 + k_0 \sinh^{-1}\left(\frac{\dot{\kappa}}{\dot{\kappa}_r}\right)\right] + k_0\ k_1\ f_t \sinh^{-1}\left(\frac{\dot{\kappa}}{\dot{\kappa}_r}\right). \tag{4}$$

A simple, alternative approach is to capture the crack rate dependence via a single variable viscosity term, following Sluys ([7]) and de Borst *et al.* ([3]). Thereby, the cracking stress is supplemented with a rate term

$$\sigma = \sigma_t(\kappa)\left[1 + \frac{m}{f_t}\,\dot{\kappa}\right], \tag{5}$$

where m is the cracking viscosity. The rate term is degraded with increasing crack width to avoid a residual strength. This formulation is of a similar form as the three-parameter model of eq. (4), except for the residual term. Both equations have been incorporated in a softening Rankine plasticity formulation to capture cracking in a phenomenological way.

3. Creep failure: a case study

To verify the model the three-point bending creep tests performed by Zhou ([10]) are analysed.

3.1. Description of the tests

The geometry of the 100 mm thick notched concrete beams which were tested by Zhou, is shown in Figure 1a. The concrete specimens were kept at hygral and thermal equilibrium, so the further complication introduced by the simultaneous drying and thermal shrinkage was avoided.

Zhou performed displacement-controlled tests to obtain the total load-deformation response. In subsequent tests he employed force control to study the time-dependent interaction of creep and cracking. In these tests the central force was increased gradually to a predefined level, which was subsequently sustained. A sustained load of larger than about 60% of the peak load eventually led to failure of the beams. Typical results of total load-deformation responses under displacement control at 5 $\mu m/s$, as well as the results of sustained load tests at 92%, 85%, 80% and 76% of the peak loads obtained in the former tests are shown in Figure 1b. The crack mode opening displacement (CMOD) at failure under each sustained load is indicated by an X.

3.2. Finite element model

The finite element mesh employed for the analyses is also shown in Figure 1a. It consists of plane-stress, four-noded quadrilateral elements. Symmetry is exploited, enabling one half of the model only to be modelled.

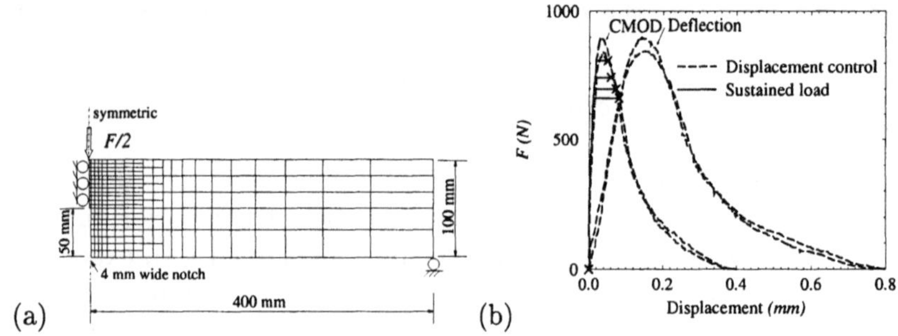

Figure 1. Three-point bending tests. (a) Specimen geometry, set-up and FE model, (b) CMOD and load-deflection measurements (Zhou [10]).

A Rankine plasticity formulation (Feenstra [4], Lourenço [5]) is employed to capture the crack initiation and propagation in a smeared fashion. A crack band width $l_m = 4\ mm$ is assumed, which is equal to the notch width. To study the cracking rate dependence, the formulation proposed by Bažant ([1]), eq. (4), as well as the simple cracking viscosity model, eq. (5), are employed. Bulk creep is considered by activating the Maxwell chain incorporated in the constitutive model. The own weight is compensated for by applying a volume load of mass density 2400 kg/m^3 in an initial step in each analysis.

Separate tests had been performed to determine the material parameters, yielding a Young's modulus 36 kN/mm^2 and the tensile strength f_t=2.8 N/mm^2 (Zhou [10]). Also, relaxation tests had been performed on cylindrical, notched tensile specimens, providing information for determining the bulk creep parameters, Figure 2. A 10-element Maxwell chain model has been fitted by a least squares method to the relaxation curve - Figure 2. Unfortunately, the relaxation was measured over short times (maximum 1 hour), calling for extrapolation. An extremely high creep coefficient $\phi = \varepsilon_c/\varepsilon_e = 5$ has been assumed after 100 days. This has been done to numerically maximise the role of the bulk creep in the creep fracture process and thereby to demonstrate that, despite such a large rate contribution by the bulk creep, the crack rate dependence must be added to simulate the measured responses.
Zhou ([10]) has also performed three-point bending tests under displacement control on smaller beams (600 mm long by 50x50 mm^2 section) to determine the fracture energy G_f. By varying the deflection rate from slow (0.05 $\mu m/s$ - peak load after about 80 minutes) to fast (50 $\mu m/s$ - peak load after about 5 s) he studied the rate influence on the fracture energy, Figure 3a, and peak strength, Figure 3b.

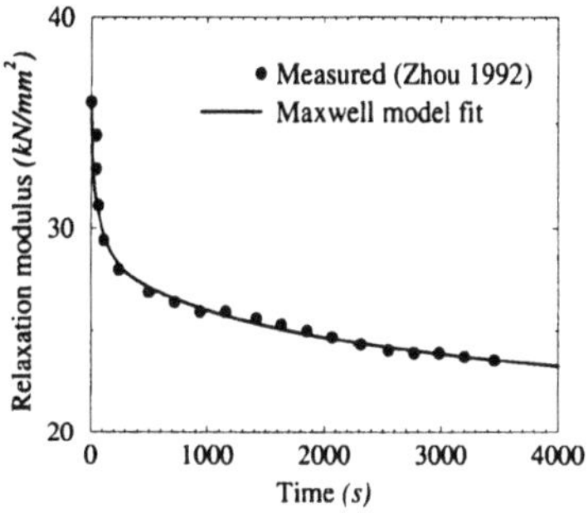

Figure 2. Maxwell chain fit to the relaxation data.

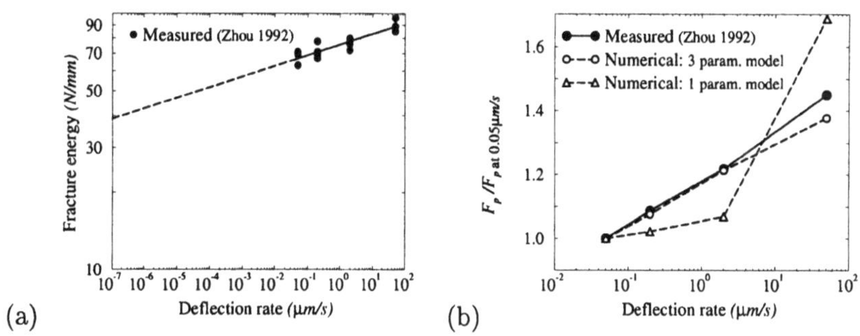

Figure 3. Rate effect on (a) fracture energy and (b) peak strength (Zhou [10]).

To derive the parameters for the crack mouth opening rate (CMOR) dependent models, eqs. (4,5), a process of inverse fitting has been followed. For these analyses the smaller beam has been modelled with the same mesh shown in Figure 1a, scaled to the small beam geometry. The parameters which have been obtained in this way are, for eq. (4): $k_0 = 0.05, k_1 = 0.1, \dot{\kappa}_r = 10^{-7}\ s^{-1}$ and for eq. (5): $m = 1500\ Ns/mm^2$. In figure 3b the normalised numerical peak strengths are compared with the measured values. Reasonable agreement is found with the three-parameter model, but with the simple one-parameter model it is impossible to fit the strength increase over the entire range of loading rates. A possible remedy is to employ a rate-dependent viscosity $m(\dot{\kappa})$. This has not been attempted. Instead, the three-parameter model has been employed for the subsequent analyses.

With regard to the apparent increase in fracture energy with loading rate, it must be noted that this follows from the numerical model although constant fracture energy is prescribed. This value can be estimated by extrapolation to the deflection rate at the reference

CMO strain rate $\dot{\kappa}_r$, Figure 3a. A constant value of G_f=0.035 N/mm has been used.

3.3. Results

The experimental results indicate that the displacement-controlled response forms an envelope for failure under a sustained load, Figure 1b. Therefore, this case is analysed first. Figure 4a compares the numerical response with the experimental responses. To obtain this agreement a 5% lower tensile strength than the reported f_t=2.8 N/mm^2 (Zhou [10]) and a Young's modulus 30 kN/mm^2 have been used. On the one hand these adjustments have been made to give reasonable agreement with the experimental responses, in order to make possible the subsequent comparison between the responses under sustained load. On the other hand the strength and stiffness are rate dependent. Thus, the reduction of the measured parameters is in line with the determination of the "rate-independent" values, as has been done in the previous section for the fracture energy.

Next, the sustained load cases are analysed, Figure 4b. As was attempted in the experiments (Zhou [10]), the initial, ascending loading rate is the same as for the displacement-controlled case. Beyond this ascending branch the load level is kept constant and the creep behaviour is analysed. During this stage the crack propagates and the deflection and CMOD increase up to a point where equilibrium can no longer be achieved for the sustained load level. Here, the load bearing capacity of the beam is exceeded. To ensure that failure under the sustained load is indeed imminent, the analyses are continued, by replacing the force control with displacement control at this point. This results in the subsequent softening responses which prove that failure would have

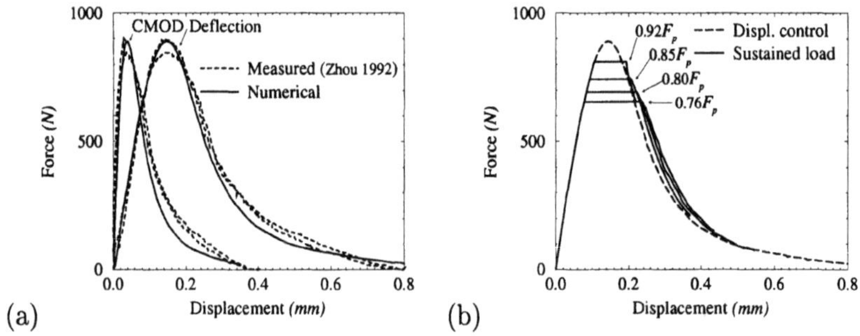

Figure 4. (a) Comparison of experimental and numerical displacement-controlled (5 $\mu m/s$) responses. (b) Numerical creep responses, with switch to displacement control at failure.

occurred under continued load control. These results confirm the experimental observation that the displacement-controlled response serves as an envelope for failure under sustained loads.

The CMODs computed numerically compare reasonably with the measured values, Figure 5a. If the crack rate dependence is ignored, this is not the case. Furthermore, apart from simulating the deformational response, it is imperative that the time scale involved is captured accurately by the model. Reasonable agreement is found with the measured times between reaching the sustained load level and failure, Figure 5b. However, if the crack rate dependence is not activated, these times are greatly overestimated. This provides strong evidence of the validity of the inclusion of the cracking rate dependence.

Table I. Three-point bending time to failure.

F/F_p	0.92	0.85	0.80	0.76
Measured CMOD (μm): (Zhou 1992)	48	57	70	77
Numerical (μm):	56	59	67	72
Numerical (μm): (Bulk creep only)	52	73	88	102

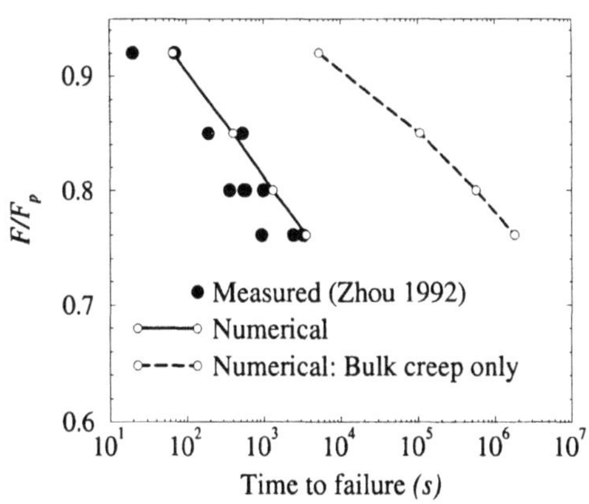

Figure 5. CMOD at the point of failure under sustained load.

4. Conclusions

Evidence has been presented for the necessity to consider the crack rate dependence as an important source of time dependence next to bulk creep in a porous, cementitious material. Only by including this contribution to the cracking resistance, the observed times to failure and the crack mouth opening displacements in three-point bending creep tests of concrete beams can be computed with reasonable accuracy.

Two formulations have been presented for the modelling of the crack rate dependence. A simple, single parameter model accounts for the viscosity of the cracking process, but cannot simulate the observed strength increase over a wide range of loading rates. A cracking strain rate-dependent viscosity may solve this problem. It has been shown that the three-parameter model proposed by Bažant ([1]), where such a dependence is incorporated, can represent the rate dependence over a wide range of loading rates.

References

1. Bažant, Z. P. Current status and advances in the theory of creep and interaction with fracture, *Creep and Shrink. of Concr.* (eds. Z. P. Bažant and I. Carol), E.&F. N. Spon, London, 291–307, 1993.
2. Bažant, Z. P. and Gettu, R. Rate effects and load relaxation in static fracture of concrete, *ACI Materials J.*, 456–68, 1992.
3. de Borst, R., Sluys, L. J., Van den Boogaard, A. H. and Van den Bogert, P. A. J. Computational issues in time-dependent deformation and fracture of concrete, *Creep and Shrinkage of Concrete* (eds. Z. P. Bažant and I. Carol), E.&F. N. Spon, London, 309–26, 1993.
4. Feenstra, P. H. *Computational aspects of biaxial stress in plain and reinforced concrete.* Dissertation, Delft Univ. of Techn., Delft, The Netherlands, 1993.
5. Lourenço, P. B. *Computational strategies for masonry structures.* Dissertation, Delft Univ. of Techn., Delft, The Netherlands, 1996.
6. Rüsch, H. Researches toward a general flexural theory for structural concrete, *ACI J.*, **57**, 1–28, 1960.
7. Sluys, L. J. *Wave propagation, localisation and dispersion in softening solids.* Dissertation, Delft Univ. of Techn., Delft, The Netherlands, 1992.
8. Van Zijl, G. P. A. G. *Computational Modelling of Masonry Creep and Shrinkage.* Dissertation (to be published), Delft Univ. of Techn., Delft, The Netherlands, 1999.
9. Wu, Z. S. and Bažant, Z. P. Finite element modelling of rate effect in concrete fracture with influence of creep, *Creep and Shrink. of Concr.* (eds. Z. P. Bažant and I. Carol), E.&F. N. Spon, London, 427-32, 1993.
10. Zhou, F. P. *Time-dependent Crack Growth and Fracture in Concrete.* Dissertation, Lund Univ., Lund, Sweden, 1992.

Macroscopic Swelling of Clays Derived from Homogenization

C. Moyne and M. Murad
LNCC/MCT and IPRJ/UERJ
Rua Getulio Vargas 333
25651-070 Petrópolis, RJ, Brazil

1. Introduction

Swelling clay consists of large flat sheets of deformable silicate structures separated from one-another by an aqueous layer (adsorbed water) [1]. Each clay mineral consists of a 2:1 layer consisting of an octahedral aluminia sheet between two silica tetrahedral sheets. Crystal imperfections and isomorphous substitutions in the smectitic minerals produce negative surface charge density which is neutralized by exchangeable cations to form a diffuse positively ionic atmosphere around the clay mineral surfaces. Swelling phenomena result from change in the crystal dimension when water is incorporated into the lattice structure. When a phyllislicate crystal is put in contact with water, the water penetrates between the superimposed layers and intermolecular forces (hydration and electrical repulsive forces) operate to disjoin the stacked silicate layers. During water uptake, the volume of montmorillonite increases by absorbing water. Water will flow osmotically into regions of higher ionic concentration and particles will separate causing swelling. For low moisture content range (interstices smaller than $50\AA$) swelling is dominated by hydration forces, which consist of bonding forces between the mineral surfaces and the water which arise from the hydrophilic structure of the platelets resulting from the ordering of polar water molecules near the clay minerals (Israelachvili [2]). In contrast, for long-range interactions, swelling is dominated by the electrostatic effect, which is classically governed by the conventional Gouy-Chapman theory of diffuse double layer wherein the equilibrium charge distribution and the electrical field are governed by a Poisson-Boltzman equation [3].

To the authors knowledge a comprehensive microscopic theory capable of generalizing Gouy-Chapman theory to incorporate non-equilibrium effects (such as ion transport and fluid movement) and their correlation with the overall macroscopic response of the clay clusters has not been developed yet. The paper aims to fill this gap. Here we

W. Ehlers (ed.), IUTAM Symposium on Theoretical and Numerical Methods in Continuum Mechanics of Porous Materials, 329–334.

adopt the homogenization procedure to upscale the non-equilibrium version of the Gouy-Chapman theory whose governing equations in the fluid domain incorporate fluid motion and ion transport. This system is coupled with the elasticity problem governing the deformation of the clay platelets in the solid phase. Homogenized equations are derived by a rigorous upscaling of the microstructure. Among other effects, the homogenized results include additional physico-chemical terms, such as disjoining pressure. Unlike experimental results, which have pursued a direct macroscopic view for the constitutive behavior of these quantities (see e.g. Low [4]), the upscaling provides a micromechanical representation for them.

2. Microscopic description of the problem

In what follows we consider swelling clays with a monomodal distribution, in which all the water is essentially adsorbed to the clay surface (e.g. highly compacted clay). Consider at the microscopic scale the clay material composed of a solid phase "s" and an adsorbed liquid one "f". For simplicity the adsorbed liquid phase is considered a mixture of water and ions of an entirely dissociated monovalent salt: anions "$-$" and cations "$+$" with volumetric concentrations n^- and n^+. If e designates the charge of a positron, ϵ_0, ϵ the permittivity of free space and the relative one of the mixture, the electric potential ϕ is given by the Poisson equation (1) (with $\eta = 1$). The movement of the fluid phase (assumed newtonian of dynamic viscosity μ_f) is given by the Stokes equations (2, 3) supplemented by an electric force term represented in terms of the divergence of the Maxwell tensor. Here $\mathbf{v}$ is the velocity of the fluid, p the pressure and $\mathbf{E} = -\nabla\phi$ denotes the electric field. The movement of the ions is due to a convective movement and a diffusion one. The diffusion term is the sum of a fickian term ($\mathcal{D}_{w\pm}$ are the binary water–ions diffusion coefficients, k the Boltzman constant and T the temperature) and a forced diffusion one accounting for the effect of the electric forces acting on the ions (4, 5). The solid phase is supposed to be isotropic elastic with Lamé parameters λ_s and μ_s and its deformation is denoted by $\mathcal{E}$. The equation for the displacement $\mathbf{u}$ is given by (6). Boundary conditions are prescribed on the solid–fluid interfaces (A_{fs}). Since the solid surface is charged with a surface charge density σ, the electric field at the interface is given by (7) where $\mathbf{n}_{fs}$ is the normal to the interface oriented from the fluid to the solid phase. The usual non-slip boundary condition (8) is for the Stokes problem. For the transport of the ions, the solid is assumed impervious (9). Finally, the stress continuity is given by (10).

3. Order of magnitude analysis

In order to derive the equations governing the phenomena at the macroscopic scale, the homogenization method is used [5]. Two length scales are introduced, a microscopic one ℓ of the order of the pore size and a macroscopic one L corresponding to the macroscopic scale of variation of the parameters such as ions concentrations, fluid velocity and solid displacement. Their ratio $\eta \equiv \ell/L$ is a small parameter. The range of the pore scale ℓ is usually lower than 0.1 microns (dozens of namometers for practical purposes). This corresponds to the range where intermolecular forces begin to operate in a colloidal system. In addition, since we have assumed that no bulk phase water is present in the model, here we shall refer to as L the homogenized scale of bentonites, i.e. highly compacted clays where the hydrodynamics of macro-voids is often neglected and their structure is primarily dominated by the narrow spaces occupied by the electrolyte solution. Typically, L is the macroscopic scale where Darcy's law holds when the hydrodynamics of the very dense smectite clay is mainly dictated by the adsorbed water flow. The derivation of models which incorporate both adsorbed and bulk water movements in narrow and macro voids respectively can be obtained within the framework of three-scale models for loose clays [1]. In this framework L would correspond to the intermediate (mesoscale) of the adsorbed water flow and the new macroscale would be of order of centimeters, where Darcy's law dictates the bulk water flow in the macro-void system.

In what follows the microscopic governing equations are written in dimensionless form thoroughly using references values in the spirit of [6]. The order of magnitude of the electric field is given by the imposed value for the normal component on the solid-fluid interface, $E_{ref} = \sigma/(\epsilon\,\epsilon_0)$. Therefore, $\phi_{ref} = \ell\, E_{ref}$. The electric charge are supposed to counterbalance the surfaces charge and $n_{ref} = \sigma/(e\,\ell)$. The classical theory based on Darcy's law allows to assume $v_{ref} = \ell^2\, p_{ref}/(\mu_f\, L)$. Three dimensionless quantities appear: the Péclet number $Pe = v_{ref}\,\ell/\mathcal{D}$; $N = e\,\sigma\,\ell/(\epsilon\,\epsilon_0\,k\,T)$ which is the ratio of electric energy to the thermal energy of an ion; $M = (\epsilon\,\epsilon_0/2)\, E^2_{ref}\, L/u_{ref}\,(3\,\lambda_s + +2\,\mu_s)$ which is a measure of the importance of the electric effects on the stress acting on the solid phase. When swelling effects are important, it seems reasonable to consider that $M = \mathcal{O}(1)$ and $N = \mathcal{O}(1)$. The Péclet number is assumed $\mathcal{O}(\eta^m)$ with $m \geq 1$. Using the macroscopic scale L to normalize the operator $\boldsymbol{\nabla}$ and the time scale $t_{ref} = L^2/\mathcal{D}$ for the time derivatives, the following equations are listed below in dimensional form with a formal η^n factor to indicate the order of magnitude of each term.

– Poisson equation

$$\eta^2\,\nabla^2\phi = -\frac{(n^+ - n^-)\,e}{\epsilon\,\epsilon_0} \tag{1}$$

– Stokes equation

$$0 = \nabla\,.\,\mathbf{v} \tag{2}$$

$$0 = \nabla\,.\left[-p\,\mathbf{I} + \frac{\epsilon\,\epsilon_0}{2}\left(2\,\mathbf{E}\,\mathbf{E} - E^2\,\mathbf{I}\right)\right] + \eta^2\,\mu_f\nabla^2\mathbf{v} \tag{3}$$

– Transport equation for cations

$$\frac{\partial n^+}{\partial t} + \eta^{m-1}\,\nabla\,.\,\left(n^+\,\mathbf{v}\right) = \nabla\,.\left[\mathcal{D}_{w+}\left(\nabla n^+ + \frac{e\,n^+}{k\,T}\nabla\phi\right)\right] \tag{4}$$

– Transport equation for anions

$$\frac{\partial n^-}{\partial t} + \eta^{m-1}\,\nabla\,.\,\left(n^-\,\mathbf{v}\right) = \nabla\,.\left[\mathcal{D}_{w-}\left(\nabla n^- - \frac{e\,n^-}{k\,T}\nabla\phi\right)\right] \tag{5}$$

– Elasticity equation for the solid phase

$$\mu_s\nabla^2\mathbf{u} + (\lambda_s + \mu_s)\,\nabla\,(\nabla\,.\,\mathbf{u}) = 0 \tag{6}$$

The boundary conditions on the solid–fluid interface (A_{fs}) are:

$$\eta\,\nabla\phi\,.\,\mathbf{n}_{fs} = \frac{\sigma}{\epsilon\,\epsilon_0} \tag{7}$$

$$\mathbf{v} - \frac{\partial\mathbf{u}}{\partial t} = 0 \tag{8}$$

$$-\mathcal{D}_{w\pm}\left(\nabla n^\pm \pm \frac{e\,n^\pm}{k\,T}\nabla\phi\right)\,.\,\mathbf{n}_{fs} = 0 \tag{9}$$

$$\left[-p\,\mathbf{I} + \frac{\epsilon\,\epsilon_0}{2}\left(2\,\mathbf{E}\,\mathbf{E} - E^2\,\mathbf{I}\right) + \eta^2\left(\nabla\mathbf{v} + \nabla\mathbf{v}^T\right)\right]\,.\,\mathbf{n}_{fs} = \boldsymbol{\sigma}_s\,.\,\mathbf{n}_{fs} \tag{10}$$

where $\boldsymbol{\sigma}_s$ is the stress tensor in the solid phase. Two independent variables are introduced $\mathbf{x}$ for the macroscopic scale and $\mathbf{y}$ for the microscopic scale, the medium being spatially periodic in $\mathbf{y}$. With the choice of the macroscopic length as reference value, $\nabla = 1/\eta\,\nabla_y + \nabla_x$. For the time derivatives, the phenomena at the macroscopic scale are described by two time scales: a diffusion time t_d and a convection time t_c respectively with time scales $(L^2/\mathcal{D})$ and (L/v_{ref}). Therefore $\partial/\partial t = \partial/\partial t_d + Pe/\eta\;\partial/\partial t_c$. The homogenization method consists in searching a development in series of the small parameter η for each variable $\psi = \sum \eta^k\,\psi^k$ of the problem.

4. Homogenization

4.1. Transport of ions

We shall work with the equivalent "bulk" concentration (i.e. the equilibrium value in the absence of electric field) $n_b^+ = n^+ \exp(e\,\phi/k\,T)$ (and the corresponding Boltzman distribution for the anions) depend only of the macroscopic variable $\mathbf{x}$ and t. In the case $m = 2$, we have

$$\frac{\partial}{\partial t_d}\left[\left\langle \exp\left(-\frac{e\,\phi^0}{k\,T}\right)\right\rangle^f n_b^{+\,0}\right] =$$

$$= \boldsymbol{\nabla}_x \,.\, \left[\mathcal{D}_{w+}\left\langle \exp\left(-\frac{e\,\phi^0}{k\,T}\right)(\mathbf{I} + \boldsymbol{\nabla}_y\,\mathbf{f}^+)\right\rangle^f .\, \boldsymbol{\nabla}_x\, n_b^{+\,0}\right] \qquad (11)$$

where $\langle\ \rangle$ designates the volume average over the unit cell and $\mathbf{f}^+$ is solution of the closure problem (similar result holds for the anions)

$$\boldsymbol{\nabla}_y \,.\, \left[\mathcal{D}_{w+}\exp\left(-\frac{e\,\phi^0}{k\,T}\right)(\boldsymbol{\nabla}_y\,\mathbf{f}^+ + \mathbf{I})\right] = 0 \qquad (12)$$

$$-\mathcal{D}_{w+}\exp\left(-\frac{e\,\phi^0}{k\,T}\right)\mathbf{n}_{fs} \,.\, (\boldsymbol{\nabla}_y\,\mathbf{f}^+ + \mathbf{I}) = 0 \quad \text{on} \quad \partial Y_{fs} \qquad (13)$$

4.2. Poisson equation

The Poisson equation (1) at order $\mathcal{O}(1)$ leads to a local problem in Y_f

$$\nabla_y^2\,\phi^0 = -\left(n^{+\,0} - n^{-\,0}\right)\,e/\epsilon\,\epsilon_0 \qquad (14)$$

$$\boldsymbol{\nabla}_y\,\phi^0 \,.\, \mathbf{n}_{fs} = \sigma/\epsilon\,\epsilon_0 \quad \text{on} \quad \partial Y_{fs}. \qquad (15)$$

4.3. Fluid movement

Introducing $P^0 = p^0 + \int_0^{\phi^0} e\left(n^{+\,0} - n^{-\,0}\right)d\phi$, the Stokes equations show that P^0 is only a function of $\mathbf{x}$ and t. Following standard arguments used in the homogenization of the Stokes problem (with $m = 2$) leads to the following modified form of Darcy's law

$$\langle\mathbf{v}^0\rangle - n_f\frac{\partial\mathbf{u}^0}{\partial t} = -\frac{\mathbf{K}}{\mu_f}\,.\,\boldsymbol{\nabla}_x\,P^0 - \frac{k\,T}{\mu_f}\left(\langle\boldsymbol{\kappa}^+\rangle .\boldsymbol{\nabla}_x\,n_b^{+\,0} + \langle\boldsymbol{\kappa}^-\rangle \,.\, \boldsymbol{\nabla}_x\,n_b^{-\,0}\right) \qquad (16)$$

where $\mathbf{K}$ is the standard permeability tensor, n_f the volume fraction of the fluid phase and $\boldsymbol{\kappa}^+$ is the solution of the following closure problem

$$0 = \boldsymbol{\nabla}_y \,.\, \boldsymbol{\kappa}^+ \qquad (17)$$

$$0 = -\boldsymbol{\nabla}_y\,\delta^+ + \nabla_y^2\,\boldsymbol{\kappa}^+ - k\,T\left[\exp\left(-\frac{e\,\phi^0}{k\,T}\right) - 1\right](\mathbf{I} + \boldsymbol{\nabla}_y\,\mathbf{f}^+) \qquad (18)$$

with $\boldsymbol{\kappa}^+ = 0$ on ∂Y_{fs}, $\boldsymbol{\delta}^+$ a Lagrange multiplier and similar results for $\boldsymbol{\kappa}^-$. The latter two terms in Darcy's law show that, in addition to a pressure gradient, flow is also driven by electro-osmotic effects.

4.4. Total stress tensor

The total averaged stress tensor ($\boldsymbol{\sigma}_t = \boldsymbol{\sigma}_f$ in Y_f and $\boldsymbol{\sigma}_s$ in Y_s) can be written

$$\langle \boldsymbol{\sigma}_t^0 \rangle = P^0(-n_f\mathbf{I} + \langle \mathbf{C} \,.\, \boldsymbol{\mathcal{E}}_y(\boldsymbol{\zeta})\rangle) + \langle \mathbf{C} \,.\, (\mathbf{I} + \boldsymbol{\mathcal{E}}_y(\boldsymbol{\xi}))\rangle \,.\, \boldsymbol{\mathcal{E}}_x(\mathbf{u}^0) - \langle \boldsymbol{\Pi} \rangle_S \quad (19)$$

$$\text{with} \quad \langle \boldsymbol{\Pi} \rangle_S = -\left\langle \mathbf{C} \,.\, \boldsymbol{\mathcal{E}}_y(\mathbf{u}^1_{part})\right\rangle + \langle \boldsymbol{\Pi} \rangle \quad (20)$$

where the disjoining "pressure" $\boldsymbol{\Pi}$ (in fact a tensor) is the sum of the osmotic pressure and of the electric forces one

$$\boldsymbol{\Pi}(\mathbf{x}, \mathbf{y}, t) = -\int_0^{\phi^0} e\left(n_0^+ - n_0^-\right) d\phi\, \mathbf{I} - \frac{\epsilon\,\epsilon_0}{2}\left(2\mathbf{E}\mathbf{E} - E^2\,\mathbf{I}\right) \quad (21)$$

The problems for $\boldsymbol{\zeta}$ and $\boldsymbol{\xi}$ are classical [7]. $\mathbf{u}^1_{part}$ satisfies

$$\mu_s \nabla_y^2 \mathbf{u}^1_{part} + (\lambda_s + \mu_s)\boldsymbol{\nabla}_y(\boldsymbol{\nabla}_y \,.\, \mathbf{u}^1_{part}) = 0 \quad (22)$$

$$\text{and on} \partial Y_{fs} : -\boldsymbol{\Pi} \,.\, \mathbf{n}_{fs} = \left[\lambda_s \boldsymbol{\nabla}_y \,.\, \mathbf{u}^1_{part}\,\mathbf{I} + 2\,\mu_s\,\boldsymbol{\mathcal{E}}_y(\mathbf{u}^1_{part})\right] \,.\, \mathbf{n}_{fs} \quad (23)$$

The relation (19) is a generalization of the Terzaghi's effective stress principle adding a swelling "pressure" term $\langle \boldsymbol{\Pi} \rangle_S$ defined by (20).

References

1. Murad, M. A. A Thermomechanical Model of Hydration Swelling in Smectite Clays: I. Two-Scale Mixture-Theory Approach; II. Inter-Phase Mass Transfer: Homogenization and Computational Validation. *International J. for Numer. and Analytical Methods in Geomechanics*, **23**, 673–719, 1999.
2. Israelachvili, J. Intermolecular and Surfaces Forces. *Academic Press*, New York, 1991.
3. Van Olphen, H. An Introduction to Clay Colloid Chemistry: for Clay Technologists, Geologists, and Soil Scientists. *Wiley*, New York, 1977.
4. Low, P. F. Structural Component of the Swelling Pressure of Clays. *Langmuir*, **3**, 18–25, 1987.
5. Sanchez–Palencia, E. Non-Homogeneous Media and Vibration Theory. Lecture Notes in Physics, *Springer - Verlag*, 1980.
6. Auriault, J. L. Is an Equivalent Homogeneous Description Always Possible? *Int. Journal of Engineering*, **29**, 785–795, 1991.
7. Auriault, J. L. and Sanchez-Palencia, E. Étude du comportement macroscopique d'un milieu poreux saturé déformable. *Journal de Mécanique*, **16**, 575–603, 1977.

Extending Griffiths Theory to Cohesive Types of Dried Materials

S. J. Kowalski and K. Rajewska
Poznań University of Technology, Pl. Marii Sklodowskiej-Curie 2, 60-695 Poznań, Poland

1. Introduction

The paper is concerned with condensed dispergate systems, as for example, solid suspensions, ceramic pastes used for production of electronic elements, clay for brick production, etc. These systems, after drying and calcination become porous with relatively high mechanical strength. The characteristic feature of the dispergate systems is their great amount of developed interfacial surfaces. It means that they contain a great amount of the surface energy. As a consequence of this, the particles (crystalline grains) tend to join each other and create agglomerates. By growth of the agglomerates the surface energy decreases but the particle bonds become thermodynamically stable.

Two items are discussed in this paper: first, the maximal value of stress that can be carried by the body in a given stage of drying; and second, the criterium of fracture. Some knowledge concerning fracture of ceramics, like materials, is presented in monographs by Cottrell ([1]), Pampuch ([8]). General knowledge concerning drying problems is given, for example, in Kowalski ([5]), Kowalski and Strumillo ([6]).

2. Cohesion forces in drying processes

There exist two kinds of interactions between particles: electrostatic forces (repulsive), resulting from existence of monomial charges on the particle surfaces, and London's dispersive forces (attractive), caused by a instantaneous asymmetry of the charge distribution as a result of electron fluctuation. The dispersive forces are of long range. They are responsible for the attraction of the particles and creation of the agglomerates.

The potential energy of atom interactions U results from the attraction and repulsion effects between atoms

$$U = -\frac{A}{L^n} + \frac{B}{L^m} \tag{1}$$

where A and B are the proportionality constants referring to the attraction and repulsion, respectively, m and n - the exponents of the

W. Ehlers (ed.), IUTAM Symposium on Theoretical and Numerical Methods in Continuum Mechanics of Porous Materials, 335–340.

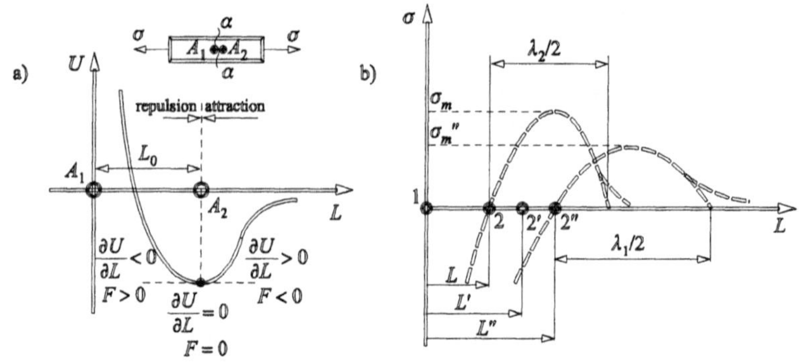

Figure 1. Condon-Morse curve: a) variation of potential energy in dependence on the interatomic distance, b) stress σ resulting from tension of the sample

current interatomic distance referring to the attraction and repulsion, respectively.

The variation of potential energy for interaction of two atoms is illustrated in the Condon-Morse curve presented in figure 1a. Minimum of the potential energy corresponds to the interatomic distance L_0, at which the interaction force is equal to zero (equilibrium state). If the interatomic distance is changed ($L \neq L_0$), a force F arises, and tends to bring the system to equilibrium. It will be a repulsive force if $L < L_0$, or attractive one if $L > L_0$. This force is equal to alteration of the potential energy with respect to the interatomic distance

$$F = -\frac{\partial U}{\partial L} = n\frac{A}{L^{n+1}} - m\frac{B}{L^{m+1}} \tag{2}$$

The minus sign results from the fact that the potential energy increases and the force decreases with the increase of the interatomic distance. The approximate values of the cohesion forces for the particles of oxides dispergate in water are in the range of $2.5 \cdot 10^{-15}$ to $6 \cdot 10^{-8}$N.

The negative capillary pressure in liquid bridges between adjoining particles makes additional forces creating the aggregates. The value of this pressure is determined by the Laplace's formula (see e.g. Kirkham and Powers, 1972)

$$p = \gamma \left(\frac{1}{R_1} - \frac{1}{R_2} \right) \tag{3}$$

where γ denotes the liquid surface tension, and R_1 and R_2 are the main radii of curvature of the bridge concave surface.

The necessary condition for the negative pressure is that $R_2 < R_1$. For the water surface tension $\gamma = 7.2 \cdot 10^{-2}$N/m and the particle dimension ranges from 10^{-6} to 10^{-8}m, the magnitude of capillary forces F_k is in the range from $2 \cdot 10^{-7}$ to 10^{-9}N. These values are comparable with those for the dispersive forces at the atomic contacts. So, the capillary forces may bring the particles to achieve contacts of atomic distance at the final stage of drying process.

Let us imagine two adjoining atoms placed on the opposite sides of a mental cross-section $\alpha - \alpha$ of a tensed bar (Fig.1a). If the force interacting between two atoms is F, then the stress σ is approximately F/L_0^2, and the strain is equal to $\varepsilon = (L - L_0)/L_0$. For stretching of the two adjoining atoms by dL, the increase of stress is

$$d\sigma = \frac{1}{L_0^2}\left(\frac{\partial F}{\partial L}\right)_{L=L_0} dL = \frac{1}{L_0}\left(\frac{\partial F}{\partial L}\right)_{L=L_0} d\varepsilon = E d\varepsilon \tag{4}$$

where $E = (\partial F/\partial L)_{L=L_0}/L_0$ denotes the modulus of elasticity.

Figure 1b illustrates the displacement of two particles of a dried material. In a saturated state the distance between particles is denoted by L_1''. Due to the shrinkage, the particles displace nearer to each other, for example, from the position 2" to 2 for the free shrinkage, or from position 2" to 2' for the constrained (not free) shrinkage. The constrained shrinkage takes place in the presence of the stresses that counteract the shrinkage phenomenon. Let us denote X as the current moisture content (m.c.) in a dried material, that is the ratio of the liquid mass to the mass of the bone-dry material, $L(X)$ – the distance between two atoms at the m.c. X and the stress free state, and $L'(X)$ - the distance between these two atoms at the same m.c. X but additionally at some state of stress. Mechanical strain caused by this stress reads

$$\varepsilon^M = \frac{L'(X) - L(X)}{L(X)} = \frac{1}{E(X)}\sigma \tag{5}$$

where Young's modulus $E(X)$ is a function of the current moisture content.

Let us assume the strain-stress relation to be still a linear one, but different for various moisture contents, (see Fig. 2a). In such a case the compliance (inverse of Young's modulus) is a linear function of moisture content, and the Young's modulus depends on the moisture content in the following way

$$E(X) = \frac{E_0}{1 + aE_0(X - X_0)} \tag{6}$$

where E_0 denotes the Young's modulus of a dry material at the final equilibrium humidity X_0, and a is a coefficient of influence of the moisture content on the compliance. Figure 2b illustrates this relation for various values of coefficient a.

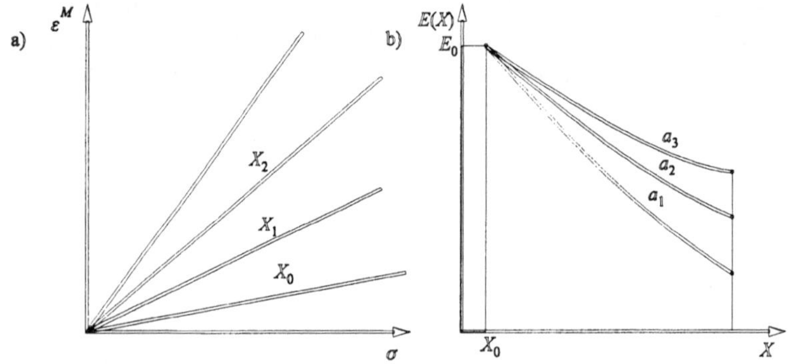

Figure 2. Influence of humidity on: a) relation between strain and stress, b) Young's modulus

When a wet material dries, the dried surface attempts to shrink but is restrained by the wet core. The surface is stressed in tension and the core in compression. The tension stresses can cause a damage of the surface structure. The tension curves for dried material of different humidity are shown in figure 1b. The curve has a shape that makes it possible to approximate with the sine function (see e.g. Cottrell, [1] or Pampuch, [8])

$$\sigma = \sigma_m sin\frac{2\pi u}{\lambda} \tag{7}$$

where σ_m denotes the theoretical (maximal) strength of material, $u = L' - L$ - displacement equivalent to the applied stress, and λ - the period of the sine function being dependent on the moisture content.

The magnitude of σ_m may be determined by the amount of work necessary to apply to a material to stretch atoms at a distance $\lambda/2$, at which their separation takes place. This work is expressed by the following integral

$$U = \int_0^{\lambda/2} \sigma du = \int_0^{\lambda/2} \sigma_m sin\frac{2\pi u}{\lambda} du = \frac{\lambda\sigma_m}{\pi} = \frac{2L(X)}{E(X)}[\sigma_m(X)]^2 \tag{8}$$

where λ in the above formula was replaced by that derived from the limit conditions $\frac{d\sigma}{du}\mid_{u=0} = \sigma_m\frac{2\pi}{\lambda} = \frac{E}{L}$.

If we assume the energy lost for plastic deformations and reconstruction of crystal lattice during the fracture to be negligibly small, then the energy of the two newly created surfaces 2γ is, according to Griffith's theory, equal to the work U. This statement allows us to write the theoretical strength of dried material as a function of moisture content

$$\sigma_m(X) = \sqrt{\frac{\gamma E(X)}{L(X)}} \tag{9}$$

where $L(X) = L_0[1 + \alpha_X(X - X_0)$ is the interatomic distance in a wet material of m.c. X, and α_X is the coefficient of linear shrinkage.

The increase of the theoretical strength during drying process results from two reasons: an increasing Young's modulus, and the diminishing of distances between atoms.

3. Fracture

Generally, cracks appear at the tips of flaws or pores, where the stress concentration takes place. The stress that causes the fracture is not a macroscopic stress σ that acts on the network, but it is the stress σ_c concentrated at the flaw tip of length c, having radius ρ at the tip, (see e.g. Siedov, [9], or Pampuch, [8]): $\sigma_c = \sigma(1 + 2\sqrt{c/\rho})$.

One assumes that a fracture occurs when the stress σ_c is greater than the theoretical strength σ_m. The fracture criterium formulated at the level of macroscopic stresses reads

$$\sigma > \sigma_{cr} \text{ where } \sigma_{cr} = \sqrt{\frac{\gamma E(X)}{L(X)}}/\sigma\left(1 + 2\sqrt{c/\rho}\right) \approx \sqrt{\frac{\gamma E(X)\rho}{2cL(X)}} \tag{10}$$

The above presented fracture criterium is suitable for drying processes, where the surface of the dried body is in tensile stress state and the fracture takes place on the boundary surface.

Let us consider a sample in the form of thin plate having a flaw in the middle. Let us imagine first a chosen fragment of a similar sample without flaw, and assume that it has some amount of an accumulated energy U_0 =const. The same fragment with flaw in the middle has the total energy U equal to

$$U = U_0 - \frac{\pi\sigma^2 c^2}{E} + 2\gamma(2c) \tag{11}$$

The second term expresses the negative energy of the flaw, and the third one the surface energy of the flaw. The slit will then increase its size spontaneously, but only when a decrease of energy during this process takes place, that is $\frac{\partial U}{\partial c} \leq 0$.

Differentiating (11) one can find the critical length of the slit c_{cr} for a given stress

$$c_{cr} = \frac{2\gamma E}{\pi\sigma_c^2} \tag{12}$$

The fracture process will proceed spontaneously at a given stress σ_c for all slits with $c > c_{cr}$.

4. Final remarks

Every suitable theory of cracking during drying should account for the common observations that cracking is more likely if the body is thick or the drying rate is high. This is because such conditions involve great inhomogeneities in moisture distribution and these induce great stresses. The cohesion force, on the other hand, depends evidently on the moisture content, which causes the porous body to swell.

Using the interaction model of atoms, expressed by the Condone-Morse curves, a theoretical strength of drying materials was determined. If the stress concentrated at the flaw tip is greater than the theoretical one, the fracture of the body can take place. Thus, the fracture criterion is derived from the comparison of the concentrated stress and the theoretically admissible stress. Besides, based on the Griffith's concept, a critical length of the flaw is determined, which grows spontaneously at a given stress.

Acknowledgements

This work was carried out as a part of research project No 3 T09C 0015 12 sponsored by the Polish State Committee for Scientific Research in the year 1999.

References

1. Cottrell, A. H. The mechanical properties of matter. *PWN, Warszawa*, (in Polish), 1970.
2. Griffith, A. A. The phenomena of rupture and flow in solids. *Engng. Royal Aircraft Establishment*, 163–198, 1920.
3. Kirkham, D. and Powers, W. L. Advanced soil physics. *John Wiley &Sons*, New York, 1972.
4. Kneule, F. Drying. *ARKADY*, Warszawa, (in Polish), 1970.
5. Kowalski, S. J. Toward a thermodynamics and mechanics of drying processes. *Chemical Engng. Science*, (in print), 1999.
6. Kowalski, S. J. and Strumillo, C. Moisture transport, thermodynamics and boundary conditions in porous materials in presence of mechanical stresses. *Chemical Engng. Science*, **52**, 1141–1150, 1997.
7. Lykov, A. W. Theory of drying. *ENERGIA*, Moskow, (in Russian), 1968.
8. Pampuch, R. Ceramics materials. *PWN*, Warszawa (in Polish), 1988.
9. Siedov, A. I. Mechanics of continua. *Nauka*, Moscow, (in Russian), 1984.

Session D4: Waves in Porous Media II

Chairman: S. Kowalski

Reflection and Transmission of Waves at a Fluid/Porous-Medium Boundary

A. I. M. Denneman, G. G. Drijkoningen, D. M. J. Smeulders and C. P. A. Wapenaar
Centre for Technical Geoscience
Delft University of Technology
P.O. Box 5028, 2600 GA Delft
The Netherlands

Abstract. Relatively simple closed-form expressions are derived for the reflection and transmission coefficients belonging to a fluid/porous-medium interface with open-pore boundary conditions. The wave propagation in the fluid-saturated porous-medium is described using Biot's theory and it is assumed that both the porous skeleton and the pore fluid are much more compressible than the skeletal solid grains themselves. The obtained results find their application in forward and inverse surface wave analysis.

1. Introduction

To calculate the reflection and transmission coefficients belonging to a fluid/porous-medium interface, one normally uses the boundary conditions of Deresiewicz and Skalak [1]. In a number of papers [2, 3, 4, 5, 6, 7] it has been shown that these boundary conditions lead to a set of four linear equations with the reflection and transmission coefficients as the four unknowns.

It is rather easy to solve this set of equations numerically, while closed-form expressions for the reflection and transmission coefficients can be obtained by applying the well-known Cramer's rule (each coefficient is then equal to the ratio of two determinants of two different 4×4 matrices). Simply applying Cramer's rule results in closed-form expressions for the reflection and transmission coefficients that are very complicated. Due to this complexity it is rather difficult to acquire a good physical insight in the dependencies of these coefficients on the many measurable quantities defining the fluid/porous-medium interface.

We will show in this paper that it is possible to derive expressions for the reflection and transmission coefficients that are much simpler than the ones already known. We note, however, that we have assumed in this derivation that both the porous skeleton and the pore fluid are much more compressible than the skeletal solid grains themselves. The availability of simpler expressions for the reflection and transmission

W. Ehlers (ed.), IUTAM Symposium on Theoretical and Numerical Methods in Continuum Mechanics of Porous Materials, 343–350.

coefficients contributes to a better physical understanding of the reflection and transmission properties of waves at a fluid/porous-medium boundary.

We finally note that we restrict ourselves to a fluid/porous-medium interface with open-pore boundary conditions, while results associated with sealed-pore boundary conditions can be found in a different paper [8].

2. Waves at a fluid/porous-medium boundary

According to Biot's theory [9] three different types of waves may propagate through a porous material: a fast P-wave, a slow P-wave, and a S-wave. Consequently, at the fluid/porous-medium boundary an incident P-wave in the fluid is converted simultaneously into (i) a reflected P-wave, (ii) a transmitted fast P-wave, (iii) a transmitted slow P-wave, and (iv) a transmitted S-wave (see Fig. 1).

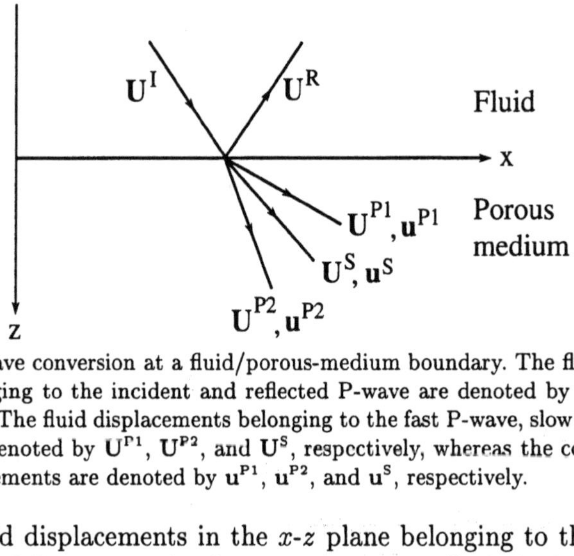

Figure 1. Wave conversion at a fluid/porous-medium boundary. The fluid displacements belonging to the incident and reflected P-wave are denoted by $\mathbf{U}^{\mathrm{I}}$ and $\mathbf{U}^{\mathrm{R}}$, respectively. The fluid displacements belonging to the fast P-wave, slow P-wave, and S-wave are denoted by $\mathbf{U}^{\mathrm{P1}}$, $\mathbf{U}^{\mathrm{P2}}$, and $\mathbf{U}^{\mathrm{S}}$, respectively, whereas the corresponding solid displacements are denoted by $\mathbf{u}^{\mathrm{P1}}$, $\mathbf{u}^{\mathrm{P2}}$, and $\mathbf{u}^{\mathrm{S}}$, respectively.

The fluid displacements in the x-z plane belonging to the incident and reflected P-wave are in the space-frequency domain given by

$$\mathbf{U}^{\mathrm{I}} = \begin{pmatrix} U_{\mathrm{x}}^{\mathrm{I}}(x,z,\omega) \\ U_{\mathrm{z}}^{\mathrm{I}}(x,z,\omega) \end{pmatrix} = \begin{pmatrix} p \\ q \end{pmatrix} A^{\mathrm{I}} \exp\left[-j\omega(px+qz)\right], \tag{1}$$

$$\mathbf{U}^{\mathrm{R}} = \begin{pmatrix} U_{\mathrm{x}}^{\mathrm{R}}(x,z,\omega) \\ U_{\mathrm{z}}^{\mathrm{R}}(x,z,\omega) \end{pmatrix} = \begin{pmatrix} p \\ -q \end{pmatrix} A^{\mathrm{R}} \exp\left[-j\omega(px-qz)\right], \tag{2}$$

where ω is the angular frequency, p the horizontal slowness, q the vertical slowness with a positive real part and a negative imaginary part, and A^{I} and A^{R} are wave-amplitudes. The slownesses p and q are related to the propagation velocity c as

$$p^2 + q^2 = \frac{1}{c^2} = \frac{\rho}{K}, \tag{3}$$

where ρ is the fluid density and K the fluid bulk modulus. Furthermore, by combining the deformation equation $P = -K\boldsymbol{\nabla}\cdot(\mathbf{U}^{\mathrm{I}} + \mathbf{U}^{\mathrm{R}})$ with Eqs. (1)–(3) one finds that the fluid pressure P is given by

$$P(x,z,\omega) = j\omega\rho\left[A^{\mathrm{I}}\exp\left[-j\omega(px+qz)\right] + A^{\mathrm{R}}\exp\left[-j\omega(px-qz)\right]\right]. \tag{4}$$

The displacements of the solid skeleton in the x-z plane belonging to the fast P-wave, slow P-wave, and S-wave are given by

$$\mathbf{u}^{\mathrm{P1}} = \begin{pmatrix} u_{\mathrm{x}}^{\mathrm{P1}}(x,z,\omega) \\ u_{\mathrm{z}}^{\mathrm{P1}}(x,z,\omega) \end{pmatrix} = \begin{pmatrix} p \\ q_{\mathrm{P1}} \end{pmatrix} A^{\mathrm{P1}}\exp\left[-j\omega(px+q_{\mathrm{P1}}z)\right], \tag{5}$$

$$\mathbf{u}^{\mathrm{P2}} = \begin{pmatrix} u_{\mathrm{x}}^{\mathrm{P2}}(x,z,\omega) \\ u_{\mathrm{z}}^{\mathrm{P2}}(x,z,\omega) \end{pmatrix} = \begin{pmatrix} p \\ q_{\mathrm{P2}} \end{pmatrix} A^{\mathrm{P2}}\exp\left[-j\omega(px+q_{\mathrm{P2}}z)\right], \tag{6}$$

$$\mathbf{u}^{\mathrm{S}} = \begin{pmatrix} u_{\mathrm{x}}^{\mathrm{S}}(x,z,\omega) \\ u_{\mathrm{z}}^{\mathrm{S}}(x,z,\omega) \end{pmatrix} = \begin{pmatrix} -q_{\mathrm{S}} \\ p \end{pmatrix} A^{\mathrm{S}}\exp\left[-j\omega(px+q_{\mathrm{S}}z)\right], \tag{7}$$

where A^{P1}, A^{P2}, and A^{S} are wave-amplitudes. The vertical slownesses q_{P1}, q_{P2}, and q_{S} (all with a positive real part and a negative imaginary part) are related to the horizontal slowness p and the propagation velocities c_{P1}, c_{P2}, and c_{S} as

$$p^2 + q_{\mathrm{P1}}^2 = \frac{1}{c_{\mathrm{P1}}^2}, \qquad p^2 + q_{\mathrm{P2}}^2 = \frac{1}{c_{\mathrm{P2}}^2}, \qquad p^2 + q_{\mathrm{S}}^2 = \frac{1}{c_{\mathrm{S}}^2}. \tag{8}$$

According to Biot's theory [3, 9] the propagation velocities c_{P1}, c_{P2}, and c_{S} are given by

$$c_{\mathrm{P1}}^2 = \frac{c_1 + \sqrt{c_1^2 - 4c_0c_2}}{2c_0}, \qquad c_{\mathrm{P2}}^2 = \frac{c_1 - \sqrt{c_1^2 - 4c_0c_2}}{2c_0}, \qquad c_{\mathrm{S}}^2 = \frac{G\rho_{22}}{c_0} \tag{9}$$

with

$$c_0 = \rho_{11}\rho_{22} - \rho_{12}^2, \qquad c_2 = R(A+2G) - Q^2, \tag{10}$$

$$c_1 = R\rho_{11} + (A+2G)\rho_{22} - 2Q\rho_{12}, \tag{11}$$

in which G is the shear modulus of the porous material. The generalized elastic coefficients A, Q, and R are related to measurable quantities by the following expressions [3, 10]

$$A = \frac{(1-\phi)^2K_{\mathrm{s}}K_{\mathrm{f}} - (1-\phi)K_{\mathrm{b}}K_{\mathrm{f}} + \phi K_{\mathrm{s}}K_{\mathrm{b}}}{K_{\mathrm{f}}(1-\phi-K_{\mathrm{b}}/K_{\mathrm{s}}) + \phi K_{\mathrm{s}}} - \frac{2}{3}G, \tag{12}$$

$$Q = \frac{\phi K_{\mathrm{f}}(K_{\mathrm{s}}(1-\phi) - K_{\mathrm{b}})}{K_{\mathrm{f}}(1-\phi-K_{\mathrm{b}}/K_{\mathrm{s}}) + \phi K_{\mathrm{s}}}, \tag{13}$$

$$R = \frac{\phi^2 K_{\mathrm{f}}K_{\mathrm{s}}}{K_{\mathrm{f}}(1-\phi-K_{\mathrm{b}}/K_{\mathrm{s}}) + \phi K_{\mathrm{s}}}, \tag{14}$$

where K_s is the skeletal grain bulk modulus, K_f the pore fluid bulk modulus, K_b the "jacketed" bulk modulus of the porous material, and ϕ the porosity (pore fluid volume divided by bulk volume).

The density terms ρ_{11}, ρ_{22}, and ρ_{12} in Eqs. (9)–(11) are defined as

$$\rho_{11} = (1-\phi)\rho_s - \rho_{12}, \qquad \rho_{22} = \phi\rho_f - \rho_{12}, \qquad \rho_{12} = -(\alpha-1)\phi\rho_f, \quad (15)$$

where ρ_s and ρ_f are the densities of the solid skeleton and the pore fluid, respectively. According to Johnson *et al.* [11] the drag coefficient α belonging to a fluid-saturated porous material can be defined as

$$\alpha = \alpha_\infty \left(1 - j\frac{\omega_c}{\omega}\sqrt{1 + j\frac{M}{2}\frac{\omega}{\omega_c}}\right) \quad \text{with} \quad \omega_c = \frac{\eta\phi}{k_0\rho_f\alpha_\infty}, \quad (16)$$

where $M \approx 1$ is the so-called similarity parameter, α_∞ the inertial drag at infinite frequency, η the pore fluid viscosity, k_0 the permeability of the porous material, and the critical frequency ω_c is the frequency at which the inertial and viscous drag are of comparable magnitude.

According to Biot's theory the pore fluid displacements $\mathbf{U}^{P1}$, $\mathbf{U}^{P2}$, and $\mathbf{U}^{S}$ are related to the solid skeleton displacements $\mathbf{u}^{P1}$, $\mathbf{u}^{P2}$, and $\mathbf{u}^{S}$ as

$$\mathbf{U}^{P1} = G_{P1}\mathbf{u}^{P1} \text{ with } G_{P1} = \frac{Q - c_{P1}^2\rho_{12}}{c_{P1}^2\rho_{22} - R} = \frac{A + 2G - c_{P1}^2\rho_{11}}{c_{P1}^2\rho_{12} - Q}, \quad (17)$$

$$\mathbf{U}^{P2} = G_{P2}\mathbf{u}^{P2} \text{ with } G_{P2} = \frac{Q - c_{P2}^2\rho_{12}}{c_{P2}^2\rho_{22} - R} = \frac{A + 2G - c_{P2}^2\rho_{11}}{c_{P2}^2\rho_{12} - Q}, \quad (18)$$

$$\mathbf{U}^{S} = G_S\mathbf{u}^{S} \text{ with } G_S = \frac{-\rho_{12}}{\rho_{22}} = \frac{\alpha - 1}{\alpha}. \quad (19)$$

The pore fluid stress $\boldsymbol{\tau}_f$ and solid skeleton stress $\boldsymbol{\tau}_s$ are defined as

$$\boldsymbol{\tau}_f = -\phi P_f\boldsymbol{\delta} = Q(\boldsymbol{\nabla}\cdot\mathbf{u})\boldsymbol{\delta} + R(\boldsymbol{\nabla}\cdot\mathbf{U})\boldsymbol{\delta}, \quad (20)$$

$$\boldsymbol{\tau}_s = -\boldsymbol{\sigma} - (1-\phi)P_f\boldsymbol{\delta} = G\left[\boldsymbol{\nabla}\mathbf{u} + (\boldsymbol{\nabla}\mathbf{u})^T\right] + A(\boldsymbol{\nabla}\cdot\mathbf{u})\boldsymbol{\delta} + Q(\boldsymbol{\nabla}\cdot\mathbf{U})\boldsymbol{\delta} \quad (21)$$

with $\mathbf{u} = \mathbf{u}^{P1} + \mathbf{u}^{P2} + \mathbf{u}^{S}$ and $\mathbf{U} = \mathbf{U}^{P1} + \mathbf{U}^{P2} + \mathbf{U}^{S}$, and where P_f is the pore fluid pressure, $\boldsymbol{\delta}$ a unit tensor, and $\boldsymbol{\sigma}$ the intergranular stress tensor.

3. Reflection and transmission coefficients

The reflection and transmission coefficients R^F, T^{P1}, T^{P2}, and T^S are related to the wave-amplitudes A^I, A^R, A^{P1}, A^{P2}, and A^S as

$$A^R = R^F A^I, \qquad A^{P1} = T^{P1}A^I, \qquad A^{P2} = T^{P2}A^I, \qquad A^S = T^S A^I. \quad (22)$$

To solve R^{F}, T^{P1}, T^{P2}, and T^{S} we use boundary conditions that are valid for a fluid/porous-medium boundary with open pores [1, 3, 12]. Hence, at the boundary $z=0$:

$$\phi(U_z^{\mathrm{P1}} + U_z^{\mathrm{P2}} + U_z^{\mathrm{S}}) + (1-\phi)(u_z^{\mathrm{P1}} + u_z^{\mathrm{P2}} + u_z^{\mathrm{S}}) = U_z^{\mathrm{I}} + U_z^{\mathrm{R}}, \tag{23}$$

$$P_{\mathrm{f}} = P, \qquad \sigma_{\mathrm{xz}} = 0, \qquad -\sigma_{\mathrm{zz}} - P_{\mathrm{f}} = -P. \tag{24}$$

By combining these boundary conditions with Eqs. (1), (2), (4)–(8), and (17)–(21) in an appropriate way one obtains the following set of equations

$$\begin{bmatrix} q & q_{\mathrm{P1}}(1-\phi+\phi G_{\mathrm{P1}}) & q_{\mathrm{P2}}(1-\phi+\phi G_{\mathrm{P2}}) & p(1-\phi+\phi G_{\mathrm{S}}) \\ -\rho\phi & \dfrac{Q+RG_{\mathrm{P1}}}{c_{\mathrm{P1}}^2} & \dfrac{Q+RG_{\mathrm{P2}}}{c_{\mathrm{P2}}^2} & 0 \\ 0 & pq_{\mathrm{P1}} & pq_{\mathrm{P2}} & \left(p^2 - \dfrac{1}{2c_{\mathrm{S}}^2}\right) \\ 0 & \left(p^2 - \dfrac{N_1}{2Gc_{\mathrm{P1}}^2}\right) & \left(p^2 - \dfrac{N_2}{2Gc_{\mathrm{P2}}^2}\right) & -pq_{\mathrm{S}} \end{bmatrix} \begin{bmatrix} R^{\mathrm{F}} \\ T^{\mathrm{P1}} \\ T^{\mathrm{P2}} \\ T^{\mathrm{S}} \end{bmatrix} = \begin{bmatrix} q \\ \rho\phi \\ 0 \\ 0 \end{bmatrix} \tag{25}$$

with

$$N_1 = A + 2G - \frac{1-\phi}{\phi}Q + G_{\mathrm{P1}}\left(Q - \frac{1-\phi}{\phi}R\right), \tag{26}$$

$$N_2 = A + 2G - \frac{1-\phi}{\phi}Q + G_{\mathrm{P2}}\left(Q - \frac{1-\phi}{\phi}R\right). \tag{27}$$

To obtain relatively simple closed-form expressions for R^{F}, T^{P1}, T^{P2}, and T^{S} we assume that both the porous skeleton and the pore fluid are much more compressible than the skeletal solid grains themselves. Consequently, the substitution of $K_{\mathrm{s}} \gg K_{\mathrm{b}}$ and $K_{\mathrm{s}} \gg K_{\mathrm{f}}$ in Eqs. (12)–(14) leads to

$$A = \frac{(1-\phi)^2}{\phi}K_{\mathrm{f}} + K_{\mathrm{b}} - \frac{2}{3}G, \qquad Q = (1-\phi)K_{\mathrm{f}}, \qquad R = \phi K_{\mathrm{f}}. \tag{28}$$

By combining Eq. (28) with Eqs. (9), (10), (15), (17)–(19), (26), and (27) one obtains

$$1 - \phi + \phi G_{\mathrm{S}} = \gamma_0 = 1 - \frac{\phi}{\alpha}, \tag{29}$$

$$1 - \phi + \phi G_{\mathrm{P1}} = \gamma_0(1+\gamma_1) \quad \text{with} \quad \gamma_1 = \frac{K_{\mathrm{f}}}{\rho_{\mathrm{f}}\alpha c_{\mathrm{P1}}^2 - K_{\mathrm{f}}}, \tag{30}$$

$$1 - \phi + \phi G_{\mathrm{P2}} = \gamma_0(1+\gamma_2) \quad \text{with} \quad \gamma_2 = \frac{K_{\mathrm{f}}}{\rho_{\mathrm{f}}\alpha c_{\mathrm{P2}}^2 - K_{\mathrm{f}}}, \tag{31}$$

$$c_{\mathrm{P1}}^2 c_{\mathrm{P2}}^2 = \frac{K_{\mathrm{p}}K_{\mathrm{f}}}{G\alpha\rho_{\mathrm{f}}}c_{\mathrm{S}}^2 \quad \text{with} \quad K_{\mathrm{p}} = N_1 = N_2 = K_{\mathrm{b}} + \frac{4}{3}G, \tag{32}$$

where K_p is the constrained modulus. The substitution of Eqs. (28)–(32) in the four boundary conditions represented by Eq. (25) leads to

$$\begin{bmatrix} q & q_{\mathrm{P1}}\gamma_0(1+\gamma_1) & q_{\mathrm{P2}}\gamma_0(1+\gamma_2) & p\gamma_0 \\ -\rho\phi & \rho_\mathrm{f}(\alpha-\phi)\gamma_1 & \rho_\mathrm{f}(\alpha-\phi)\gamma_2 & 0 \\ 0 & pq_{\mathrm{P1}} & pq_{\mathrm{P2}} & \left(p^2-\frac{1}{2c_\mathrm{S}^2}\right) \\ 0 & \left(p^2-\frac{K_\mathrm{p}}{2Gc_{\mathrm{P1}}^2}\right) & \left(p^2-\frac{K_\mathrm{p}}{2Gc_{\mathrm{P2}}^2}\right) & -pq_\mathrm{S} \end{bmatrix} \begin{bmatrix} R^\mathrm{F} \\ T^{\mathrm{P1}} \\ T^{\mathrm{P2}} \\ T^\mathrm{S} \end{bmatrix} = \begin{bmatrix} q \\ \rho\phi \\ 0 \\ 0 \end{bmatrix}. \tag{33}$$

By solving this set of linear equations one finds that the closed-form expressions for R^F, T^{P1}, T^{P2}, and T^S are given by

$$R^\mathrm{F} = \frac{R_1-R_2}{R_1+R_2}, \qquad T^{\mathrm{P1}} = \frac{\Delta_{\mathrm{R2}}^{\mathrm{S,P2}}}{R_1+R_2}, \qquad T^{\mathrm{P2}} = \frac{-\Delta_{\mathrm{R1}}^{\mathrm{S,P1}}}{R_1+R_2}, \tag{34}$$

$$T^\mathrm{S} = \frac{pq_{\mathrm{P2}}\Delta_{\mathrm{R1}}^{\mathrm{S,P1}} - pq_{\mathrm{P1}}\Delta_{\mathrm{R2}}^{\mathrm{S,P2}}}{R_1+R_2}\left(p^2-\frac{1}{2c_\mathrm{S}^2}\right)^{-1}, \tag{35}$$

where R_1 and R_2 are defined as

$$\begin{aligned} R_1 &= \frac{\rho_\mathrm{f}}{2\rho}\left(1-\frac{\alpha}{\phi}\right)\left(\gamma_2\Delta_{\mathrm{R1}}-\gamma_1\Delta_{\mathrm{R2}}\right), \\ &= \frac{\rho_\mathrm{f}}{2\rho}\left(1-\frac{\alpha}{\phi}\right)\left(\gamma_2\Delta_{\mathrm{R1}}^{\mathrm{S,P1}}-\gamma_1\Delta_{\mathrm{R2}}^{\mathrm{S,P2}}\right), \end{aligned} \tag{36}$$

$$R_2 = \frac{1}{2q}\left(\frac{\phi}{\alpha}-1\right)\left(q_{\mathrm{P2}}\gamma_2\Delta_{\mathrm{R1}}^{\mathrm{P1,P1}} - q_{\mathrm{P1}}\gamma_1\Delta_{\mathrm{R2}}^{\mathrm{P2,P2}}\right), \tag{37}$$

while Δ_{R1}, $\Delta_{\mathrm{R1}}^{\mathrm{S,P1}}$, $\Delta_{\mathrm{R1}}^{\mathrm{P1,P1}}$, Δ_{R2}, $\Delta_{\mathrm{R2}}^{\mathrm{S,P2}}$, and $\Delta_{\mathrm{R2}}^{\mathrm{P2,P2}}$ are given by

$$\Delta_{\mathrm{R1}} = p^2q_\mathrm{S}q_{\mathrm{P1}} + \left(p^2-\frac{1}{2c_\mathrm{S}^2}\right)^2, \tag{38}$$

$$\Delta_{\mathrm{R1}}^{\mathrm{S,P1}} = p^2q_\mathrm{S}q_{\mathrm{P1}} + \left(p^2-\frac{1}{2c_\mathrm{S}^2}\right)\left(p^2-\frac{K_\mathrm{p}}{2Gc_{\mathrm{P1}}^2}\right), \tag{39}$$

$$\Delta_{\mathrm{R1}}^{\mathrm{P1,P1}} = p^2q_\mathrm{S}q_{\mathrm{P1}} + \left(p^2-\frac{K_\mathrm{p}}{2Gc_{\mathrm{P1}}^2}\right)^2, \tag{40}$$

$$\Delta_{\mathrm{R2}} = p^2q_\mathrm{S}q_{\mathrm{P2}} + \left(p^2-\frac{1}{2c_\mathrm{S}^2}\right)^2, \tag{41}$$

$$\Delta_{\mathrm{R2}}^{\mathrm{S,P2}} = p^2q_\mathrm{S}q_{\mathrm{P2}} + \left(p^2-\frac{1}{2c_\mathrm{S}^2}\right)\left(p^2-\frac{K_\mathrm{p}}{2Gc_{\mathrm{P2}}^2}\right), \tag{42}$$

$$\Delta_{\mathrm{R2}}^{\mathrm{P2,P2}} = p^2q_\mathrm{S}q_{\mathrm{P2}} + \left(p^2-\frac{K_\mathrm{p}}{2Gc_{\mathrm{P2}}^2}\right)^2. \tag{43}$$

Note that Δ_{R1} (multiplied by 4) is the Rayleigh function $\mathsf{R}(p)$, which is associated with surface waves traveling along a solid/vacuum boundary [13].

In Fig. 2 results are shown belonging to an interface between water and a water-saturated sand-layer. These results can be easily transformed into figures showing R^F, T^{P1}, T^{P2}, and T^S as a function of the incident angle θ by using the relation $\theta = \arcsin(pc)$ for the region $|pc| \leq 1$. In the region $|pc| > 1$ the P-waves in the fluid are evanescent; in this region one observes that the reflection and transmission coefficients are very large for $pc \approx 1.44$. The reciprocal of this p-value is equal to the propagation velocity of the surface wave traveling along the interface between water and a water-saturated sand layer, i.e., it is 0.7 times the P-wave velocity in water.

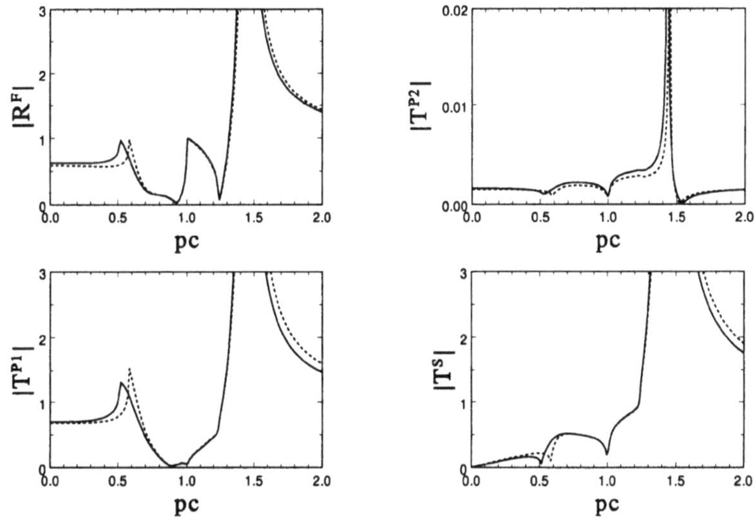

Figure 2. The coefficients R^F, T^{P1}, T^{P2}, and T^S as a function of the horizontal slowness p times the propagation velocity c of the P-wave in the fluid. The solid lines correspond to the case $K_s/K_b = \infty$: the coefficients R^F, T^{P1}, T^{P2}, and T^S are calculated by using Eqs. (34)–(43). The dashed lines correspond to the case $K_s/K_b = 5$: the coefficients R^F, T^{P1}, T^{P2}, and T^S are obtained by solving numerically the set of equations given by Eq. (25) and by using the expressions for A, Q, and R given by Eqs. (12)–(14) instead of the ones given in Eq. (28). The parameters used to obtain these results are: $\omega = 10000\,\mathrm{rad\,s^{-1}}$, $\phi = 0.24$, $\eta = 0.001\,\mathrm{Pa\,s}$, $k_0 = 0.3910^{-12}\,\mathrm{m^2}$, $\rho = \rho_f = 1000\ \mathrm{kg\,m^{-3}}$, $\rho_s = 2760\ \mathrm{kg\,m^{-3}}$, $K = K_f = 2.22\ \mathrm{G\,Pa}$, $K_b = 5.8\ \mathrm{G\,Pa}$, $G = 3.4\ \mathrm{G\,Pa}$, $\alpha_\infty = 2.3$, and $M = 1$.

4. Concluding remarks

In general, the surface wave velocity can be obtained as follows: find the horizontal slowness p for which the denominator $R_1 + R_2$ of the reflection and transmission coefficients given by Eqs. (34) and (35) is zero. Note, however, that the obtained $p = p_0$ for which $R_1 + R_2 = 0$ is complex-valued. Actually, $\mathrm{Re}(p_0)^{-1}$ is the propagation velocity of the surface wave traveling along the fluid/porous-medium interface, whereas $\mathrm{Im}(-\omega p_0)$ is its attenuation in the propagation direction.

It is clear that the availability of closed-form expressions for R_1 and R_2, as given by Eqs. (36) and (37), will facilitate our research in solving the inverse problem: "given a measured surface wave, find the physical parameters describing the porous medium".

References

1. Deresiewicz, H. and Skalak, R. On Uniqueness in Dynamic Poroelasticity. *Bull. Seism. Soc. Am.*, **53**, 783–788, 1963.
2. Rosenbaum, J. H. Synthetic Microseismograms: Logging in Porous Formations. *Geophysics*, **39**, 14–32, 1974.
3. Feng, S. and Johnson, D. L. High-Frequency Acoustic Properties of a Fluid/Porous Solid Interface. I. New Surface Mode and II. The 2D Reflection Green's Function. *J. Acoust. Soc. Am.*, **74**, 906–924, 1983.
4. Wu, K., Xue, Q. and Adler, L. Reflection and Transmission of Elastic Waves from a Fluid-Saturated Porous Solid Boundary. *J. Acoust. Soc. Am.*, **87**, 2349–2358, 1990.
5. Santos, J., Corbero, J., Ravazzoli, C. and Hensley, J. Reflection and Transmission Coefficients in Fluid-Saturated Porous Media. *J. Acoust. Soc. Am.*, **91**, 1911–1923, 1992.
6. Albert, D. G. A Comparison between Wave Propagation in Water-Saturated and Air-Saturated Porous Materials. *J. Appl. Phys.*, **73**, 28–36, 1993.
7. Cieszko, M. and Kubik, J. Interaction of Elastic Waves with a Fluid-Saturated Porous Solid Boundary. *J. Theor. Appl. Mech.*, **36**, 561–580, 1998.
8. Denneman, A. I. M., Drijkoningen, G. G., Smeulders, D. M. J. and Wapenaar, C. P. A. Reflection and Transmission of Waves at an Impermeable Interface between a Fluid and a Porous Medium. In: *Proceedings of Fifth International Conference on Mathematical and Numerical Aspects of Wave Propagation (July 10–14, Santiago de Compostela, Spain)*. Dordrecht, The Netherlands: Kluwer Academic, 2000.
9. Biot, M. A. Theory of Propagation of Elastic Waves in a Fluid-Saturated Porous Solid. I. Low-Frequency Range and II. Higher Frequency Range. *J. Acoust. Soc. Am.*, **28**, 168–191, 1956.
10. Biot, M. A. and Willis, D. G. The Elastic Coefficients of the Theory of Consolidation. *J. Appl. Mech.*, **24**, 594–601, 1957.
11. Johnson, D. L., Koplik, J. and Dashen, R. Theory of Dynamic Permeability and Tortuosity in Fluid-Saturated Porous Media. *J. Fluid Mech.*, **176**, 379–402, 1987.
12. Gurevich, B. and Schoenberg, M. Interface Conditions for Biot's Equations of Poroelasticity. *J. Acoust. Soc. Am.*, **105**, 2585–2589, 1999.
13. Aki, K. and RichardsP. G. *Quantitative Seismology: Theory and Methods*, Vol. 1. San Fransisco, USA: W. H. Freeman and Company, 1980.

Poroelasto-Electric Longitudinal Waves in Porous Wet Long Bones - a Transmission Line Model

R. Uklejewski
Department of Environmental Mechanics Pedagogical University of Bydgoszcz 85-064 Bydgoszcz, Chodkiewicza 30 and Paediatric Endocrinology Clinic, Medical University of Poznań, Poland
e-mail: mkk@rose.man.poznan.pl

Abstract. Porous long bone under longitudinal harmonic load is shown - on the basis of the earlier author's papers (Uklejewski [5], [6], [7]) - to can be modelled as a poroelasto-electric transmission line. The mechano-electric parameters per unit length of long bone transmission line are introduced and the fundamental set of equations is presented.

Keywords: Cortical bone tissue, extracellular and intracellular strain generated potentials (SGPs), poroelasticity, electrokinetics, piezoelectricity

1. Introduction

The extracellular strain-generated electric potentials (SGPs) of porous cortical bone filled with a physiological fluid can be used not only to the experimental study of the local bone pore fluid flow, but - together with the intracellular stain-generated potentials (ic-SGPs, Zhang *et al.* [2]) - they will contribute to the bone cells physiology and bone tissue bioengineering. Bone cells (osteoblasts, osteoclasts) possess the hormonal receptors (for PTH, Vit.D, calcitonin and growth factors), the voltage-activated ionic channels (Ferrier *et al.* [3]; Duncan, Misler [1]) and probably also the fluid shear-stress-activated ionic channels (Duncan, Misler [1]). Salzstein, Pollack and coworkers ([4]) were the first who applied the combination of electrokinetics together with a poroelasticity formalism derived for non-compressible solid and fluid phases from the Bowen's mixture theory to explain the SGPs phenomenon in wet cortical bone. In (Uklejewski [5], [6]) the author has presented an explaining hypothesis on the physiological and bioelectromechanical role of the bone matrix piezoelectricity (osteonic lamella acts as an mechanoelectret detector of cortical bone tissue growth) and has gived an explanation *why* the electrokinetics (without piezoelectricity) together with the poroelasticity give an effective theoretical tool in cortical bone tissue bioengineering.

W. Ehlers (ed.), IUTAM Symposium on Theoretical and Numerical Methods in Continuum Mechanics of Porous Materials, 351–356.

2. Long bone shaft as a transmission line

A long bone shaft has the shape of an irregular hollow cylinder and the interesting structural properties resulting from the existence of the remodelling (strain-history adaptive) processes in living bone. In (Lazenby [10]) it was shown that the cross-section geometry of a long bone is strictly connected with the distribution of bone porosity in the cross-section region, and both these parameters are remodelled under the influence of bone bending loads. We can assume that the value of the product: $S_x(1-f_x)$ (where S_x - the long bone shaft cross section area at the distance x; $0 \leq x \leq L$, L - length of the shaft, f_x - the mean bone porosity in element $S_x \Delta x$) is the same along the bone shaft length

$$\forall x_1, x_2 \in \langle 0, L \rangle \qquad S_{x_1}(1-f_{x_1}) = S_{x_2}(1-f_{x_2}). \tag{1}$$

If, moreover, for a given long bone the cross-section area of its shaft varies very small along the shaft length, as it occurs usually (Gies, Carter [9]; Piekarski [11]), then also the following approximate equality holds

$$\forall x_1, x_2 \in \langle 0, L \rangle \qquad S_{x_1} f_{x_1} \approx S_{x_2} f_{x_2}, \tag{2}$$

i.e. the mean bone porosity f_x in the bone shaft element $S_x \Delta x$ is approximately the same along the length of long bone shaft. The properties (1), (2) allow us to introduce in a natural way the mechanical and the electrical parameters per unit length of the long bone shaft, and to treate the long bone shaft as a mechano-electric transmission line with macroparameters distributed practically uniformly along the line length.

3. Equations of the Biot theory for a long bone shaft transmission line under a longitudinal harmonic load

We consider a wet porous long bone shaft and assume that at $x = 0$ on the cross-section of the shaft acts a longitudinal harmonic load $P(t)$ with the pulsation ω, $P(t) = P_o \sin \omega t = Im\left(P_o e^{j\omega t}\right)$. The cortical bone of long bone shaft wall can be treated in these conditions as an *isotropic* material (Huiskes [12]). We assume that the deformation of the porous cortical bone of a long bone shaft wall is described by the one-dimensional dynamical state equations of the Biot theory ([8]), cf. (Williams [13]). Because our system is linear and stationary (i.e. its parameters values do not change in time), thus the state variables: $\sigma_{11}(x,t)$ - the normal component of the stress tensor of the solid frame

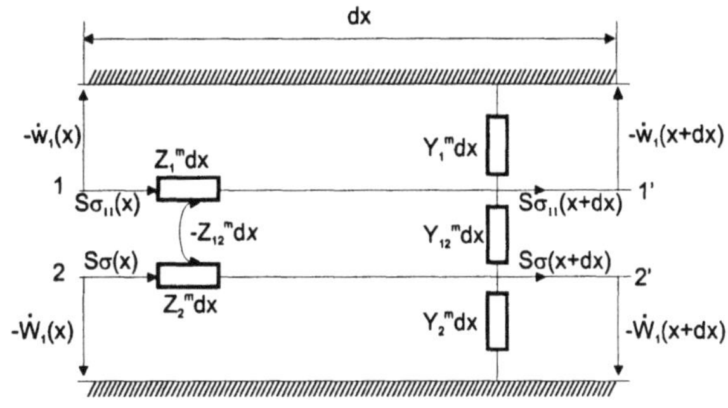

Fig.1

Figure 1. The formal analog scheme of an element dx of a shaft of a wet long bone during propagation of longitudinal harmonic elastic waves, on the basis of the Biot theory equations and the electro-mechanical analogy (Uklejewski [7]). Z_1^m , Z_2^m , Z_{12}^m - the mechanical impedances (Y_1^m , Y_2^m , Y_{12}^m - the admitances) per unit length of a bone shaft of cross-section area S; $\dot{w}_1$, $\dot{W}_1$ - velocities of displacements of the solid and the fluid phase, σ_{11}, σ - stresses in the solid and the fluid phase of wet porous cortical bone of the shaft.

in the x direction, $\sigma(x,t)$ - the fluid stress, $\dot{w}_1(x,t)$ - the velocity of the solid frame displacement in the x direction, $\dot{W}_1(x,t)$ - the velocity of the fluid displacement in the x direction vary also sinusoidally. We will use in the following the complex amplitudes: $\sigma_{11}(x)$, $\sigma(x)$, $\dot{w}_1(x)$, $\dot{W}_1(x)$ of the above mentioned state variables.

In Fig.1 we present schematically - on the basis of the electromechanical analogy (Uklejewski [7]) - an element dx of a bone transmission line, i.e. of a wet porous long bone shaft during propagation of longitudinal harmonic poroelastic waves. The transmission line 1 corresponds to the solid frame of wet long bone shaft, whereas the transmission line 2 corresponds to the fluid phase of wet long bone shaft.

The costitutive relations in complex form **for a wet long bone shaft** under longitudinal harmonic load one can obtain on the basis of the scheme given in Fig.1 applying formally the second Kirchhoff's law to the elementary circuits formed by the element 1 - 1' of the line 1 and "the earth", and by the element 2 - 2' of the line 2 and "the earth". We have

$$\begin{aligned} \frac{d\dot{w}_1(x)}{dx} &= Z_1^m S\sigma_{11}(x) - Z_{12}^m\sigma(x) \\ \frac{d\dot{W}_1(x)}{dx} &= -Z_{12}^m S\sigma_{11}(x) - Z_2^m\sigma(x)\,. \end{aligned} \tag{3}$$

The Eqs. (3) are isomorphic with the constitutive equations of the Biot theory ([8]). The mechanical impedances Z_1^m , Z_2^m , Z_{12}^m per unit length should be determined experimentally, or they may be calculated if the Biot-Willis poroelastic coefficients for cortical bone and the geometry of long bone shaft are known.

The equations of poroelastodynamics in complex form **for a wet long bone shaft** under longitudinal harmonic load can be obtained by applying formally the first Kirchhoff's law to the nodal points of the circuit element given in Fig.1; we have

$$\begin{aligned} \frac{d}{dx}[S\sigma_{11}(x)] &= Y_1^m \dot{w}_1(x) + Y_{12}^m[\dot{w}_1(x) - \dot{W}_1(x)] \\ \frac{d}{dx}[S\sigma(x)] &= Y_2^m \dot{W}_1(x) + Y_{12}^m[\dot{W}_1(x) - \dot{w}_1(x)] \,. \end{aligned} \tag{4}$$

The Eqs.(4) are isomorphic with the equations of motion of the Biot theory. The mechanical admitances Y_1^m , Y_2^m , Y_{12}^m per unit length of wet long bone shaft play similar role as the inertial coefficients ρ_1, ρ_2, ρ_a . The admitances should be determined experimentally, or they may be calculated if the material coefficients of cortical bone and the geometry of long bone shaft are known. We can say that the Eqs. (3) and (4) constitute *the matrix state equation* for a wet long bone shaft transmission line under longitudinal harmonic load.

4. Electric potentials associated with propagation of harmonic longitudinal poroelastic waves in wet long bone shafts

During the propagation of harmonic longitudinal poroelastic waves in a wet long bone shaft the oscillatory flow of ionic physiological fluid in cortical bone pores occur. This flow - described by the fluid relative velocity $\dot{W}_1(x) - \dot{w}_1(x)$ obtained from the state equations (3) and (4) - produces the electrokinetic streaming currents and streaming potentials. In Fig.2 the electric scheme of an element dx of wet long bone shaft under harmonic longitudinal mechanical load, placed in the air over a conducting layer, is presented.

In Fig. 2 $E_{str}^0(x)$ represents the electromotive force of voltage sources of mechanical (strain) origin, distributed along the wet long bone shaft. We define $E_{str}^0(x)$ as

$$E_{str}^0 = Z_e S J_{str}, \tag{5}$$

where J_{str} - the streaming current density produced by the strain generated pore fluid flow in long bone shaft wall, S - the cross-section area of long bone shaft, and Z_e ($= R + j\omega L$; R - resistance, L - inductance)

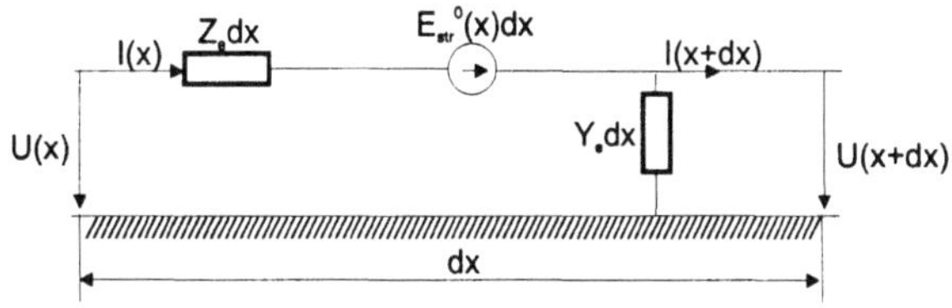

Fig.2

Figure 2. The electric scheme of an element dx of wet long bone shaft under harmonic longitudinal mechanical load, $U(x)$ - electric voltage associated with propagation of longitudinal elastic waves in wet long bone shaft, E^0_{str} - electromotive force of voltage sources of mechanical origin, per unit length; Z_e $(= R + j\omega L)$ - electric impedance of bone transmission line per unit length, Y_e $(= G + j\omega C)$ - electric admitance of bone transmission line per unit length.

is the electric impedance per unit length of wet long bone shaft.
The streaming current density J_{str} can be determined from the linear equations of classical electrokinetics, which are here presented in the vector form

$$\dot{\mathbf{W}} - \dot{\mathbf{w}} = -A_{11}\nabla p - A_{12}\nabla V \ , \quad \mathbf{J} = -A_{21}\nabla p - A_{22}\nabla V \ , \tag{6}$$

where $\dot{\mathbf{W}} - \dot{\mathbf{w}}$ is the vector of relative velocity of ionic fluid in bone pores, p - the fluid pressure, V - the electric potential, $\mathbf{J}$ - the vector of the electric current density, A_{11} - the hydraulic permeability of porous bone, A_{22} - the electric conductivity, A_{12} - the electrokinetic coefficient of electroosmosis, A_{21} - the electrokinetic coefficient of streaming current. Substituting in Eqs. (6) $\nabla V = 0$ (streaming current phenomenon) we obtain the following formula for the streaming current density vector $\mathbf{J}_{str}$

$$\mathbf{J}_{str} = -A_{21}\nabla p = A_{21}A_{11}^{-1}(\dot{\mathbf{W}} - \dot{\mathbf{w}}). \tag{7}$$

The relative velocity $\dot{\mathbf{W}} - \dot{\mathbf{w}}$ of ionic fluid in bone pores for the considered 1D problem of wet long bone shaft under harmonic longitudinal load can be obtained by solving the state equations (3) and (4). On the basis of the scheme given in Fig.2 and the Kirchhoff's laws we can write the following non-homogeneous electric state equation

$$\frac{d}{dx}\begin{bmatrix} U(x) \\ I(x) \end{bmatrix} = \begin{bmatrix} 0 & -Z_e \\ -Y_e & 0 \end{bmatrix} \begin{bmatrix} U(x) \\ I(x) \end{bmatrix} + \begin{bmatrix} E^o_{str}(x) \\ 0 \end{bmatrix}, \tag{8}$$

where: $I(x)$ is the complex amplitude function of the electric current in wet long bone shaft under harmonic longitudinal load ($I = JS$, where J - the current density, S - the cross-section area) and $U(x)$ is the complex amplitude function of the electric voltage associated with the propagation of harmonic longitudinal waves along a wet long bone shaft. The state equations (3), (4) and (8) together with Eqs. (5) and (7) constitute *the complete set of equations* for poroelasto-electric longitudinal harmonic waves in porous long bones filled with physiological fluid.

References

1. Duncan, R. and Misler, S. Voltage-activated and Stretch-activated Conducting Channels in an Osteoblast-like Cell Line (UMR 106). *FEBS Letters*, **251**, 17–21, 1989.
2. Zhang, D., Weinbaum, S. and Cowin, S. Electrical Signal Transmission in a Bone Cell Network: the Influence od a Discrete Gap Junction. *Annals of Biomedical Engineering*, **26**, 644–659, 1998.
3. Ferrier, J., Ross, S. M., Kanehisa, J. and Aubin, J. E Osteoblasts and Osteoclasts Migrate in Opposite Directions in Response to a Constant Electrical Field. *J. Cell. Physiol.*, **129**, 283–288, 1986.
4. Salzstein, R. A., Pollack, S. R., Mak, A. F. T. and Petrov, N. Electromechanical Potentials in Cortical Bone. I - A Continuum Approach. *J. Biomech.*, **20**, 261–270, 1987.
5. Uklejewski, R. Electromechanical Potentials in a Fluid-Filled Cortical Bone: Initial Stress State in Osteonic Lamellae, Piezoelectricity and Streaming Potential Roles - A Theory. *Biocybern. Biomed. Engineering*, **13**, 97–112, 1993.
6. Uklejewski, R. Initial Pizoelectric Polarization of Cortical Bone Matrix as a Determinant of the Electrokinetic Potential Zeta of that Bone. Osteonic Lamella as a Mechanoelectret. *J. Biomech.*, **27**, 991–993, 1994.
7. Uklejewski, R. *On the Electromechanical Properties of Porous Cortical Bone Filled with Physiological Fluid and on the Acousto-Electrical Effects in Wet Long Bone Shafts.* Habiltastion thesis, Inst.Biocybern.Biomed.Engng. Polish Acd. Sci., Warszawa, 1994. Biomedical Engineering Habil.
8. Biot, M. A. Theory of Propagation of Elastic Waves in a Fluid-Saturated Porous Solid. I. Low Frequency Range. *J. Acoust. Soc. Am.*, **28**, 168–178, 1956.
9. Gies, A. A. and Carter, D. R. Experimental Determination of Whole Long Bone Sectional Properties. *J. Biomech.*, **15**, 297–303, 1982.
10. Lazenby, R. Porosity-Geometry Interaction in the Conservation of Bone Strength. *J. Biomech.*, **19**, 257–259, 1986.
11. Piekarski, K. K. Biomechanics of Bone. In A. Morecki and K. Fidelus and K. Kedzior and A. Wit, editors, *Biomechanics VII A, International Series on Biomechanics*, Univ.Park Press - Polish Sci.Publ., Baltimore-Warszawa, 1981.
12. Huiskes, R. On the Modelling of Long Bones in Structural Analysis. *J. Biomech.*, **15**, 55–69, 1982.
13. Williams, J. L. Ultrasonic Wave Propagation in Cancellous and Cortical Bone: Prediction of Some Experimental Results by Biot's Theory. *J. Acoust. Soc. Am.*, **91**, 1106–1112, 1992.

Scattering of SH-Waves by a Porous Slab - Approximate Solution

Y. C. Angel* and A. R. Aguiar
Department of Mechanical Engineering and Materials Science - MS 321
Rice University, Houston, Texas 77251-1892, USA
e-mail: (angel, aguiar)@rice.edu
**Now at: Laboratoire MECAL, Université Claude Bernard Lyon 1 - Bât 721*
43 Boulevard du 11 Novembre 1918, 69622 Villeurbanne Cedex, France
e-mail: angel@mecamat.univ-lyon1.fr; aguiar@aem.umn.edu

Abstract. This work is concerned with the propagation of time-harmonic waves in an unbounded elastic solid containing empty cylindrical cavities that are randomly distributed in a slab region. It is known that, for an incident antiplane (SH) wave, the average wave motion in the solid is governed by an integro-differential equation. An approximate solution to this equation is obtained by assuming that the average wave motion inside the slab is the sum of a forward attenuated wave and a backward attenuated wave. It is shown that this approximation yields accurate results for the attenuation, velocity, reflection, and transmission in the solid.

Keywords: Seismic, ultrasonic, multiple scattering, reflection, transmission

1. Introduction

The seismic or ultrasonic evaluation of materials containing multiple scatterers, such as porous or fractured solids, or composites containing fibers, is a difficult problem. To predict the scattering of waves, several methods have been proposed. These methods, however, are valid in general in a specific frequency range and do not yield the reflection and transmission from a bounded region containing a distribution of scatterers. In this paper, we consider the interaction between antiplane waves and a large number of cylindrical cavities contained in a slab, and our objective is to derive simple formulae for the reflection and transmission on either side of the slab. We first review recent results of Aguiar and Angel [1, 2].

2. Multiple scattering by a random distribution of cavities

Consider an unbounded, homogeneous, and isotropic linear elastic solid of density ρ and shear modulus μ that contains a random distribution of cylindrical cavities of radius a, as shown in Figure 1. The cavities are centered in the slab region $\mathcal{S} = \{\mathbf{x} \in \mathbb{R}^3 : |y_1| < h, |y_2| < \infty, |y_3| < \infty\}$. The number n of cavities per unit area is constant on average in $\mathcal{S}$.

W. Ehlers (ed.), IUTAM Symposium on Theoretical and Numerical Methods in Continuum Mechanics of Porous Materials, 357–362.

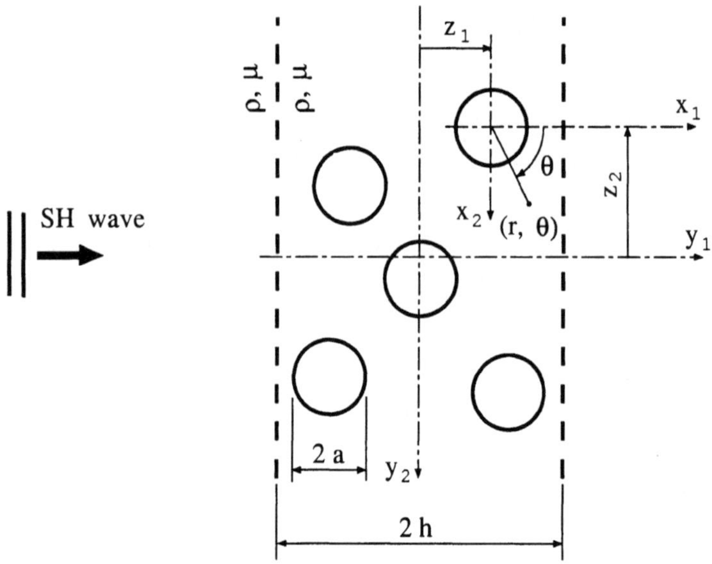

Figure 1. Incidence of an antiplane wave on cavities of diameter $2a$.

The SH wave of Figure 1 has a displacement $u^I(y_1) = u_0 \exp(i\,k\,y_1)$ in the y_3-direction, where the time factor $\exp(-i\,\omega\,t)$ has been omitted, u_0 and ω are the amplitude and the frequency, $k = \omega/c_T$ is the wavenumber, and c_T is the transverse sound speed. We average the displacement and stress fields over all possible configurations of cavities in S. In the geometry of Figure 1, the average total displacement $< u^T >_\infty$ is a function of y_1 only. Next, we define $y = y_1/a$, $\tilde{h} = h/a$, $\tilde{k} = k\,a$, $\varepsilon = n\,a^2$, $u(y) =< u^T >_\infty (a\,y)/u_0$, where ε is the cavity density or porosity. When the average exciting motion near a fixed cavity is equal to the average total motion [4], Aguiar and Angel [1] show that $u(y)$, for y in the interval $(-\infty, \infty)$, is given by the integro-differential equation

$$u(y) = \exp(i\,\tilde{k}\,y) - \frac{2\,\varepsilon}{\pi\,\tilde{k}^2} \int_{-\tilde{h}}^{\tilde{h}} s(z, y - z)\, \exp(i\,\tilde{k}\,|y - z|)\, d\,z\,, \quad (1)$$

$$s(z, y) = \int_0^\pi u'(\eta)\Big|_{\eta = z + \cos\zeta} \cos\zeta\, [\,S_0(\zeta) - (\operatorname{sgn} y)\, S_1(\zeta)\,]\, d\,\zeta\,, \quad (2)$$

$$S_j(\zeta) = i^j \sum_{m=0}^{\infty} (-1)^m \frac{\varepsilon_{2m+j}}{h_{2m+j}} \cos(\,(2\,m + j)\,\zeta\,)\,, \quad j = 0, 1\,. \quad (3)$$

In (3), $h_m = (H_m^{(1)})'(\tilde{k})$, where $H_m^{(1)}$ is the Hankel function of the first kind and order m and either $\varepsilon_m = 1$ for $m = 0$ or $\varepsilon_m = 2$ for $m > 0$.

Outside the slab, in the range $|y| > \tilde{h}$, equation (1) has the form

$$\begin{aligned} u(y) &= T \exp(i \tilde{k} y), & y > \tilde{h}, \\ u(y) &= R \exp(-i \tilde{k} y) + \exp(i \tilde{k} y), & y < -\tilde{h}, \end{aligned} \tag{4}$$

where T and R are, respectively, transmission and reflection coefficients.

Equation (1) is solved by using a Fourier series with complex coefficients, as discussed in Aguiar and Angel [1, 2]. This method gives exact numerical results for the coefficients R and T and for the ratio $u''(y)/u(y)$ inside the slab region. For a fixed value of the porosity ε, this ratio has been found numerically to be nearly constant in the interval $|y| < \tilde{h} - 2$, and independent of $\tilde{h}$ in that interval. Further, the ratio lies in the lower complex half-plane. Thus, we select $y = 0$ and define a complex number K such that

$$\frac{u''(0)}{u(0)} = -K^2, \qquad K = \frac{\tilde{k}}{c} + i\alpha, \qquad c > 0, \ \alpha > 0. \tag{5}$$

3. Approximate analysis by a semi-inverse method

We substitute for u in (1) the approximation $\hat{u}$ such that

$$\hat{u}(y) = C^+ \exp(i \hat{K} y) + C^- \exp(-i \hat{K} y), \qquad |y| < \tilde{h} + 1, \tag{6}$$

where $C^\pm$ and $\hat{K}$ are complex constants. Equation (6) corresponds to the sum of a forward attenuated wave and a backward attenuated wave, and is only an approximation to the solution $u(y)$ of (1), because it does not agree well with $u(y)$ for $\tilde{h} - 2 < |y| < \tilde{h} + 1$ [2].

Direct substitution of (6) into (1) yields a linear combination of the exponential functions $\exp(\pm i \hat{K} y)$ and $\exp(\pm i \tilde{k} y)$, from which we infer that

$$1 = \frac{2 i \varepsilon \hat{K}}{(\hat{K}^2 - \tilde{k}^2)\, \tilde{k}^2} \left[(\hat{K} + \tilde{k})\, E^+ - (\hat{K} - \tilde{k})\, E^- \right], \tag{7}$$

$$C^\pm = \frac{(\hat{K}^2 - \tilde{k}^2)\, \tilde{k}^2}{2 i \varepsilon \hat{K}} \, \frac{\Delta^\pm}{\Delta \exp(i \tilde{k} \tilde{h})}, \tag{8}$$

where $E^{\pm}$, $\Delta^{\pm}$, and Δ are functions of $\hat{K}$ and are given by

$$E^{\pm} = \sum_{m=0}^{\infty} (\pm 1)^m \frac{\varepsilon_m}{h_m} J'_m(\hat{K}), \qquad \Delta = \left[\Delta^+\right]^2 - \left[\Delta^-\right]^2, \tag{9}$$

$$\Delta^{\pm} = (\hat{K} \pm \tilde{k})\, E^{\pm} \exp(\mp i \hat{K} \tilde{h}). \tag{10}$$

In (9), J_m is the Bessel function of the first kind and order m, and the prime superscript denotes the derivative. We have found that, given a set of parameter values, the transcendental equation (7) has a unique simple complex root for the wavenumber $\hat{K}$, and the root lies in the first quadrant. Thus, we write

$$\hat{K} = \frac{\tilde{k}}{\hat{c}} + i\,\hat{\alpha}, \qquad \hat{c} > 0, \ \hat{\alpha} > 0. \tag{11}$$

Approximate transmission and reflection coefficients $\hat{T}$ and $\hat{R}$ are obtained by using (1), (4), and (6) with $y > \tilde{h}$ and $y < -\tilde{h}$, respectively. Using $\hat{K}$ and $C^{\pm}$ of (7) and (8), we find that

$$\begin{aligned} \hat{T} &= \frac{\left[(\hat{K} + \tilde{k})\, E^+\right]^2 - \left[(\hat{K} - \tilde{k})\, E^-\right]^2}{\Delta \exp(2 i \tilde{k} \tilde{h})}, \\ \hat{R} &= \frac{(\hat{K}^2 - \tilde{k}^2)\, E^- E^+}{\Delta \exp(2 i \tilde{k} \tilde{h})} \left[\exp(2 i \hat{K} \tilde{h}) - \exp(-2 i \hat{K} \tilde{h})\right]. \end{aligned} \tag{12}$$

4. Numerical results

We present numerical results in Figures 2-3 which are obtained from both the exact solution presented in Aguiar and Angel [1, 2] and the approximation (6). The exact results are represented by solid lines and the approximate results by dashed lines .

In Figure 2a, we show the attenuations α and $\hat{\alpha}$ versus the wavenumber $\tilde{k} = \omega a / c_T$ for porosities $\varepsilon = 0.01,\ 0.03$, and 0.05. The attenuation vanishes at zero frequency and increases with frequency. Observe that the exact and approximate results coincide with each other .

In Figure 2b, we show the velocities c and $\hat{c}$ versus $\tilde{k}$ for porosities $\varepsilon = 0.01,\ 0.03$, and 0.05, where c and $\hat{c}$ are the ratios between the corresponding wave velocities inside the slab and c_T. The velocity first decreases and then increases with $\tilde{k}$ toward the asymptotic value 1.

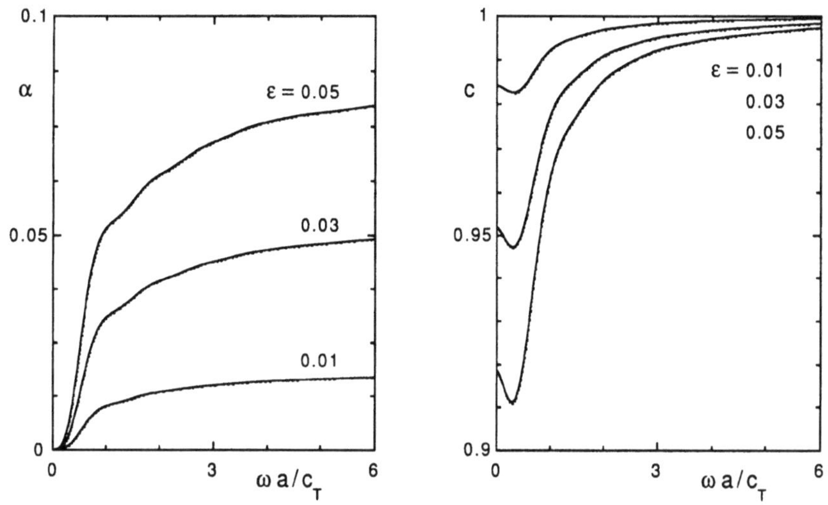

Figure 2. The attenuation α and velocity c versus the wavenumber $\tilde{k}$.

In Figures 3a, b we show, respectively, the moduli of the reflection and transmission coefficients versus $\tilde{k}$ for porosities $\varepsilon = 0.01,\ 0.03,\ 0.05$ and thickness $\tilde{h} = 3$. In both figures, approximate and exact results are close to each other. We have also evaluated the reflection and transmission for $\tilde{h} = 5,\ 30$, and the agreement between the approximate and exact results is as good as that in Figures 2-3.

The curves in Figure 3a have cyclic variations. The first local minimum occurs approximately at $\tilde{k} = \pi/(2\tilde{h})$. The following minima occur at integer multiples of $\pi/(2\tilde{h})$. The reflection generated by the cavities of this work is about twice as large as that generated by the flat cracks of Angel and Koba [3]. In Figure 3b, the transmission is less than 1 for all values of the frequency, and has an asymptote as the frequency becomes large. The asymptotic limit depends on the porosity and on the slab thickness.

5. Conclusion

We have evaluated the reflection and transmission on either side of a slab region containing cylindrical cavities by using a simple approximation for the mean displacement inside the slab. The resulting formulae (12) are easy to implement numerically. Further, they are valid for all frequency and yield numerical results that in very good agreement with exact results obtained previously.

These formulae are of interest for applications in seismic or ultrasonic evaluation, where it is important to develop reliable approxima-

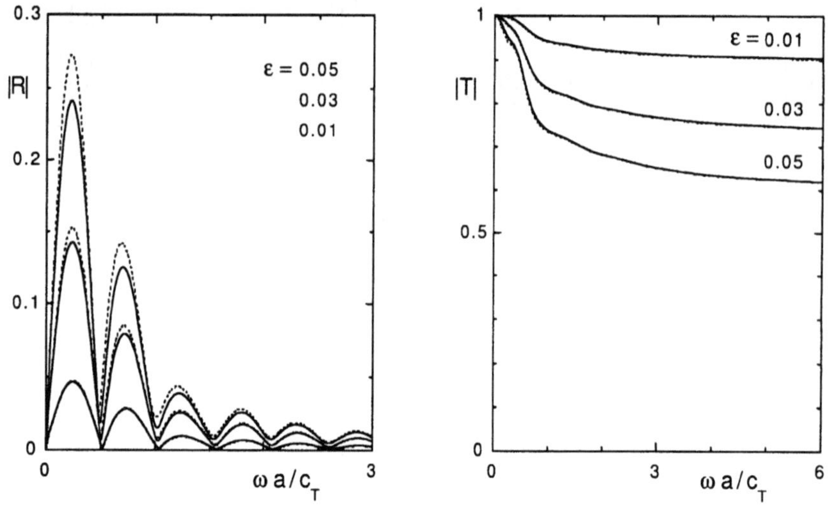

Figure 3. Moduli of R and T versus the wavenumber $\tilde{k}$ for $\tilde{h} = 3$.

tions that can be run at a low cost and used as a tool in the design of real-time control systems. The analytical predictions presented in this paper are in need of a detailed experimental comparison, but the method opens new possibilities in the seismic or ultrasonic evaluation of materials containing multiple scatterers.

Acknowledgements

This material is based in part upon work supported by the Texas Advanced Technology Program under Grant No. 003604-004.

References

1. Aguiar, A. R. and Angel, Y. C. Antiplane coherent scattering from a slab containing a random distribution of cavities. *Proceedings of the Royal Society of London, series A*. Accepted for publication, 1999a.
2. Aguiar, A. R. and Angel, Y. C. Singular solution of an integro-differential equation governing the multiple scattering of antiplane waves. *SIAM Journal on Applied Mathematics*. Submitted for publication, 1999b.
3. Angel, Y. C. and Koba, Y. K. Complex-valued wavenumber, reflection and transmission in an elastic solid containing a cracked slab region. *International Journal of Solids and Structures*, **35**, 573-592, 1998.
4. Waterman, P. C. and Truell, R. Multiple scattering of waves. *Journal of Mathematical Physics*, **2**, 512-537, 1961.

Session D5: Applications

Chairman: Y. C. Angel

A Three-Dimensional Nonlinear Model for Soil Consolidation

R. Lancellotta and G. Musso
Dipartimento di Ingegneria Strutturale e Geotecnica
Politecnico di Torino
Corso Duca degli Abruzzi 24, Torino, 10129, Italy

L. Preziosi
Dipartimento di Matematica
Politecnico di Torino
Corso Duca degli Abruzzi 24, Torino, 10129, Italy

Keywords: Consolidation, deformable porous media, drains

1. Introduction

This paper is aimed at presenting a three-dimensional mathematical model for soil consolidation. The starting point is the theory of mixtures (see Bowen ([6])), which models the macroscopic behavior of composite systems. A departure from this theory is represented in this paper by a unified Lagrangian description for both the soil skeleton and the fluid phase. In addition, it is also shown how the three-dimensional model suggested by Biot ([3]) and the axisymmetric model introduced by Barron ([1]) can be both derived from this general formulation. Finally, the assumptions under which the coupled flow-deformation problem can be simplified into an uncoupled one and the practical relevance of soil nonlinearity for the axisymmetric problem are discussed.

2. The Eulerian formulation of a deformable porous media model

In a previous paper (see Lancellotta and Preziosi ([10])) a non-linear model for soil consolidation has been deduced. Under the assumptions of soil saturation, isothermal process, negligible inertia, and validity of Darcy's law, the model was based on:

- The solid mass balance equation

$$\frac{\partial \phi_s}{\partial t} + \nabla \cdot (\phi_s \mathbf{v}_s) = 0 \, , \tag{1}$$

W. Ehlers (ed.), IUTAM Symposium on Theoretical and Numerical Methods in Continuum Mechanics of Porous Materials, 365–378.

where ϕ_s is the solid volume fraction, i.e. the volume occupied by the soil grains over the total volume, and $\mathbf{v}_s$ is the velocity of the soil grains. The saturation assumption implies that liquid and solid volume fraction sum up to 1, i.e. $\phi_\ell = 1 - \phi_s$. The void ratio e is then defined as $e = \phi_\ell / \phi_s$;

- The linear combination of the solid and liquid mass balance,

$$\nabla \cdot \mathbf{v}_c = 0 \,, \tag{2}$$

which identifies a divergence-free velocity field in terms of composite velocity

$$\mathbf{v}_c = \phi_s \mathbf{v}_s + (1 - \phi_s)\mathbf{v}_\ell \,,$$

where $\mathbf{v}_\ell$ is the liquid velocity;

- The Darcy's law

$$\mathbf{v}_\ell - \mathbf{v}_s = -\frac{\mathrm{K}(\mathrm{F}_s)}{\mu(1 - \phi_s)}(\nabla P_\ell - \rho_\ell \mathbf{g}) \,, \tag{3}$$

where P_ℓ is the pore pressure, μ is the liquid viscosity, $\mathbf{g}$ is the gravitational acceleration, and ρ_ℓ is the "true" density of water. The symmetric tensor K is the "effective permeability" tensor, and depends on the deformation state of the soil skeleton, which can be identified by the related deformation gradient F_s

$$F_{iL} = \frac{\partial x_i}{\partial X_L} \,;$$

- The momentum balance equation for the saturated soil

$$\nabla P_\ell + \nabla \cdot \mathrm{T}' = \rho_m \mathbf{g} \,, \tag{4}$$

where both pore pressure P_ℓ and effective stress T' are assumed positive in compression. Moreover, $\rho_m = \rho_s \phi_s + \rho_\ell (1 - \phi_s)$ is the density of the wet soil as a whole.

By expressing $\mathbf{v}_\ell$ from (3) in terms of other state variables

$$\begin{aligned} \mathbf{v}_\ell &= \mathbf{v}_s - \frac{\mathrm{K}(\mathrm{F}_s)}{\mu(1 - \phi_s)}(\nabla P_\ell - \rho_\ell \mathbf{g}) \\ &= \mathbf{v}_s - \frac{\mathrm{K}(\mathrm{F}_s)}{\mu(1 - \phi_s)}[-\nabla \cdot \mathrm{T}' + (\rho_s - \rho_\ell)\phi_s \mathbf{g}] \,, \end{aligned} \tag{5}$$

and by substitution of (5) in the definition of $\mathbf{v}_c$, it follows that

$$\nabla \cdot \left[\mathbf{v}_s - \frac{\mathrm{K}(\mathrm{F}_s)}{\mu} (\nabla P_\ell - \rho_\ell \mathbf{g}) \right] = 0 \ , \tag{6}$$

or

$$\nabla \cdot \left\{ \mathbf{v}_s - \frac{\mathrm{K}(\mathrm{F}_s)}{\mu} [-\nabla \cdot \mathrm{T}' + (\rho_s - \rho_\ell)\phi_s \mathbf{g}] \right\} = 0 \ . \tag{7}$$

The Eulerian formulation of the model can then be reduced to Eqs.(1), (4), and (6), or (7), which give the evolution of the state variables ϕ_s, $\mathbf{v}_s$, and P_ℓ, provided that the constitutive law linking the Cauchy stress tensor T' to the deformation gradient is specified.

At this stage it can be noticed how Biot's model (see Biot ([3])) can be obtained in the framework of small deformations and assuming that soil behavior is linear.

Being

$$\nabla \cdot \mathbf{v}_s = \frac{d_s \epsilon_v}{dt} \ , \tag{8}$$

where ϵ_v is the volumetric strain of the soil skeleton and

$$\frac{d_s f}{dt} = \frac{\partial f}{\partial t} + \mathbf{v}_s \cdot \nabla f \ , \tag{9}$$

is the total derivative following the soil particles, by inserting (8) into (6) we obtain

$$\nabla \cdot \left[\frac{K}{\mu} (\nabla P_\ell - \rho_\ell \mathbf{g}) \right] = \frac{d_s \epsilon_v}{dt} \ , \tag{10}$$

where the permeability tensor is assumed isotropic.
If P_ℓ and T' are the pore pressure and the effective stress tensor in excess to the initial stationary values, i.e. in excess to the values in equilibrium with the self-weigth and stationary flow, the equilibrium equation reduces to

$$\nabla P_\ell + \nabla \cdot \mathrm{T}' = \mathbf{0} \ . \tag{11}$$

The linear constitutive law is expressed in the form

$$\mathrm{T}' = -G[\nabla \mathbf{s} + \nabla \mathbf{s}^T] - \lambda(\nabla \cdot \mathbf{s})\mathrm{I} \ , \tag{12}$$

being $\mathbf{s}$ the displacement vector of the soil skeleton and G and λ are Lamè's constants.

Substitution of (12) into (11) gives the equilibrium equation

$$G\nabla^2\mathbf{s} + (\lambda + G)\nabla\epsilon_v - \nabla P_\ell = \mathbf{0} \,, \tag{13}$$

which combined with Eq. (10), rewritten in the following form according to the introduced hypothesis,

$$\frac{K}{\mu}\nabla^2 P_\ell = \frac{d_s\epsilon_v}{dt} \,, \tag{14}$$

represent Biot's formulation of the three-dimensional linear consolidation model.

3. The Lagrangian formulation

In this section we will formulate the model using the Lagrangian coordinate system related to the solid constituent.

The continuity equation of the solid constituent (1) in Lagrangian form writes

$$\frac{d}{dt}(\phi_s \det \mathrm{F}_s) = 0 \,, \tag{15}$$

being

$$J_s \overset{\text{def}}{\equiv} \det \mathrm{F}_s = \frac{1+e}{1+e_0} \,, \tag{16}$$

where e_0 is the void ratio in the reference configuration.

On the other hand, the balance of mass for the liquid constituent, taking as control volume a material volume $\mathcal{V}$ fixed on the solid constituent writes

$$\frac{d}{dt}\int_{\mathcal{V}} \rho_\ell \phi_\ell \, dV = -\int_{\partial\mathcal{V}} \rho_\ell\phi_\ell(\mathbf{v}_\ell - \mathbf{v}_s)\cdot \mathbf{n}\, d\Sigma \,, \tag{17}$$

where, as already stated, $\phi_\ell = 1 - \phi_s$.

Transforming the flux term in (17) first in an integral over the surface of the reference material volume $\mathcal{V}_0$, and then in a volume integral gives

$$\begin{aligned}\int_{\partial\mathcal{V}} \rho_\ell\phi_\ell(\mathbf{v}_\ell - \mathbf{v}_s)\cdot \mathbf{n}\, d\Sigma &= \int_{\partial\mathcal{V}_0} \rho_\ell\phi_\ell(\mathbf{v}_\ell - \mathbf{v}_s)\cdot\left[J_s\left(\mathrm{F}_s^{-1}\right)^T\mathbf{N}\right] d\Sigma_0 \\ &= \int_{\partial\mathcal{V}_0} \rho_\ell\phi_\ell J_s\left[\mathrm{F}_s^{-1}(\mathbf{v}_\ell - \mathbf{v}_s)\right]\cdot\mathbf{N}\, d\Sigma_0 \qquad (18) \\ &= \int_{\mathcal{V}_0} \rho_\ell \mathrm{Div}\left[\phi_\ell J_s \mathrm{F}_s^{-1}(\mathbf{v}_\ell - \mathbf{v}_s)\right] dV_0\end{aligned}$$

where Div(·) is the divergence operator in Lagrangian coordinates. After transforming the left hand side to an integral over the reference volume, one can write

$$\int_{\mathcal{V}_0} \rho_\ell \left\{ \frac{d}{dt}(\phi_\ell J_s) + \text{Div}\ \left[\phi_\ell J_s \mathrm{F}_s^{-1}(\mathbf{v}_\ell - \mathbf{v}_s)\right] \right\} dV_0 = 0\ , \tag{19}$$

which under suitable smoothness assumptions gives the balance of mass for the liquid constituent in a Lagrangian framework

$$\frac{d}{dt}(\phi_\ell J_s) + \text{Div}\ \left[\phi_\ell J_s \mathrm{F}_s^{-1}(\mathbf{v}_\ell - \mathbf{v}_s)\right] = 0\ . \tag{20}$$

In addition, Darcy's law can be written in a Lagrangian formulation as

$$\phi_\ell(\mathbf{v}_\ell - \mathbf{v}_s) = -\frac{\mathrm{K}(\mathrm{F}_s)}{\mu}\left[(\mathrm{F}_s^{-1})^T \text{Grad}\, P_\ell - \rho_\ell \mathbf{g}\right]\ , \tag{21}$$

where Grad(·) is the gradient operator in Lagrangian coordinates, which can then be used to simplify (20) to

$$\frac{\mu}{1+e_0}\frac{de}{dt} = \text{Div}\ \left\{J_s \mathrm{F}_s^{-1}\mathrm{K}(\mathrm{F}_s)\left[(\mathrm{F}_s^{-1})^T \text{Grad}\, P_\ell - \rho_\ell \mathbf{g}\right]\right\}\ . \tag{22}$$

In order to obtain the Lagrangian formulation of the momentum equation for the mixture, it is convenient to introduce (see Biot([5]) and Coussy ([7])) the change m of fluid mass referred to the initial volume, so that the quantity $m dV_0$ indicates the difference of fluid mass passing from the initial to the current configuration

$$m = J_s \rho_\ell \phi_\ell - \rho_{\ell 0}\phi_{\ell 0}\ .$$

With such an assumption and by introducing the second Piola-Kirkhoff stress tensor, the momentum balance equation writes

$$\text{Div}\,(\mathrm{F}_s \mathrm{T}_*) = (\rho_{m0} + m)\mathbf{g}\ , \tag{23}$$

where

$$\mathrm{T}_* = J_s \mathrm{F}_s^{-1}\mathrm{T}(\mathrm{F}_s^{-1})^T \tag{24}$$

4. Coupled and uncoupled soil models

In the following we will try to simplify the model. This is done assuming that shear deformations are negligible compared to the diagonal components, which in particular implies that horizontal planes remain essentially horizontal.

We then split F_s as

$$\mathrm{F}_s = \mathrm{D} + \hat{\mathrm{F}}_s \,, \tag{25}$$

where $\mathrm{D} = \mathrm{Diag}(D_x, D_y, D_z)$ is a diagonal matrix which is not uniquely identified and at the moment is left unspecified. It may be the diagonal part od F_s, i.e.

$$D_x = \frac{\partial x}{\partial X} \,, \qquad D_y = \frac{\partial y}{\partial Y} \,, \qquad D_z = \frac{\partial z}{\partial Z} \,, \tag{26}$$

(for which however the determinant is usually different from J_s) or the horizontal components may be modified to force $\det \mathrm{D} = J_s$, e.g.

$$D_x = D_y = \sqrt{\frac{J_s}{D_z}} \,, \qquad D_z = \frac{\partial z}{\partial Z} \,. \tag{27}$$

It need be mentioned, however, that in our assumption these two choices differ for quantities of the order of the off-diagonal part, which is assumed small.

Expanding (22) to first order one can then write

$$\frac{\mu}{1+e_0}\frac{de}{dt} = \mathrm{Div}\left[J_s \mathrm{D}^{-1}\mathrm{K}(\mathrm{D})\mathrm{D}^{-1}\mathrm{Grad}\, P_\ell - \rho_\ell J_s \mathrm{D}^{-1}\mathrm{K}(\mathrm{D})\mathbf{g}\right] \,, \tag{28}$$

Assuming that the medium is initially transversely isotropic with respect to the vertical direction, and observing that under the above specified assumptions this property is preserved during the evolution we can explicit (28) as

$$\begin{aligned} \frac{\mu}{1+e_0}\frac{de}{dt} = {} & \frac{\partial}{\partial X}\left(J_s \frac{K_\perp}{D_x^2}\frac{\partial P_\ell}{\partial X}\right) + \frac{\partial}{\partial Y}\left(J_s \frac{K_\perp}{D_y^2}\frac{\partial P_\ell}{\partial Y}\right) + \\ & + \frac{\partial}{\partial Z}\left[J_s \frac{K_z}{D_z^2}\left(\frac{\partial P_\ell}{\partial Z} + D_z \rho_\ell g\right)\right] \,, \end{aligned} \tag{29}$$

where $K_\perp$ is the permeability in the horizontal plane and the z-axis is directed upward.

Assuming (27), Eq. (29) rewrites

$$\frac{\mu}{1+e_0}\frac{de}{dt} = \frac{\partial}{\partial X}\left(\frac{K_\perp}{D_z}\frac{\partial P_\ell}{\partial X}\right) + \frac{\partial}{\partial Y}\left(\frac{K_\perp}{D_z}\frac{\partial P_\ell}{\partial Y}\right) + \\ + \frac{\partial}{\partial Z}\left[\frac{1+e}{1+e_0}\frac{K_z}{D_z^2}\left(\frac{\partial P_\ell}{\partial Z} + D_z\rho_\ell g\right)\right], \qquad (30)$$

The equation above needs to be joined to the stress equation

$$J_s \mathrm{D}^{-1}\mathrm{Grad}\, P_\ell + \mathrm{Div}\,(J_s \mathrm{T}'\mathrm{D}^{-1}) = (\rho_{m0} + m)\mathbf{g}\,, \qquad (31)$$

which under our hypothesis can be approximated as

$$\begin{cases} \frac{J_s}{D_x}\frac{\partial P_\ell}{\partial X} + \frac{\partial}{\partial X}\left(\frac{J_s}{D_x}T'_{xx}\right) = 0\,, \\ \frac{J_s}{D_y}\frac{\partial P_\ell}{\partial Y} + \frac{\partial}{\partial Y}\left(\frac{J_s}{D_y}T'_{yy}\right) = 0\,, \\ \frac{J_s}{D_z}\frac{\partial P_\ell}{\partial Z} + \frac{\partial}{\partial Z}\left(\frac{J_s}{D_z}T'_{zz}\right) + (\rho_{m0} + m)g = 0\,, \end{cases} \qquad (32)$$

as in the expansion introduced above shear stresses can be neglected if compared to normal stresses.

According to the assumption that displacements occur only in the vertical direction one obtains:

$$\frac{\delta\varepsilon}{1+e_0} = -\frac{T'_{zz}}{\lambda + 2G} = -m_v T'_{zz}\,, \qquad (33)$$

being m_v the soil compressibility in one-dimensional conditions (no lateral strains).

By combining Eq. (33) with Eqs.(29) and (32), an uncoupled model can be derived, which allows to compute the evolution of the system first in term of pore pressure and to determine then the distribution δ_ε throught Eq. (4.9) (see Verruijt ([13]) and Bear and Verruijt ([2])).

Also note that in the one-dimensional case, combination of (30) with the third equation in (32) gives the nonlinear model

$$\frac{\mu}{1+e_0}\frac{de}{dt} = \frac{\partial}{\partial Z}\left[K_z(e)\left(\frac{1+e_0}{1+e}\frac{\partial T'_{zz}}{\partial Z} + \frac{\rho_s - \rho_\ell}{1+e}g\right)\right], \qquad (34)$$

dealt with in Lancellotta and Preziosi (cit.) and therefore all the linearizations discussed therein.

5. The axisymmetric problem

In geotechnical engineering, it is a rather diffuse practice to accelerate the consolidation of fine-grained soils by using vertical drain wells (see Lancellotta ([8]) and Lee *et al.*([9])). The length of vertical drains is typically up to 20–30 meters and the spacing among them is of the order of 1.5 to 3.5 meters.

On these drains and on the draining part of the boundary one can then impose a Dirichlet boundary condition on the pressure. For instance, if the initial depth of the layer is H and the bottom layer $z = 0$ is assumed draining and fixed, one can set on it

$$P_\ell(x, y, z = 0) = \rho_\ell g H \ .$$

and on the vertical wall relative to the drains

$$P_\ell(\mathbf{x}) = \rho_\ell g H \left(1 - \frac{z}{h(t)}\right) \ , \qquad \mathbf{x} \in \partial\Omega_d \ ,$$

where $h(t)$ is the depth of the layer near the drain.

In addition, in general, there is a portion $\partial\Omega_{\rm imp}$ of the boundary which behaves like an impervious surface, e.g. a vertical surface sufficiently far away, or a symmetry surface. On this boundary the no-flux condition need be imposed, corresponding to a Neumann boundary condition, which by (21) writes

$$\mathbf{n} \cdot \mathrm{K}(\mathrm{F}_s^{-1} \mathrm{Grad} P_\ell - \rho_\ell \mathbf{g}) = \mathbf{0} \ ,$$

where $\mathbf{n}$ is the outward normal to $\partial\Omega_{\rm imp}$.

To be more specific we here consider a consolidation process driven by a draining rigid load of infinite extent placed on the top surface $z = h(t)$. This means that

$$P_\ell(x, y, z = h(t)) = 0 \ .$$

In this case the motion of the top surface can be determined using a body force diagram on the foundation which yields

$$M\ddot{h} = -Mg + \int_A T'_{zz}(x, y, z = h(t))\, dx\, dy \ , \qquad (35)$$

where, of course, it is possible to substitute the gravitational force Mg of the load with any vertical force F applied to the solid foundation.

Note that, in addition to (35), considering tilting of the foundation would require an analysis of the resulting moments acting on the foundation and an angular momentum balance. However if the applied load and the soil beneath satisfy a symmetry condition the top surface stays horizontal and Eq. (35) is sufficient to describe its motion.

The inertial term in (35) is usually so small that it can be reasonably assumed that the soil near the top surface readily compresses to balance the applied

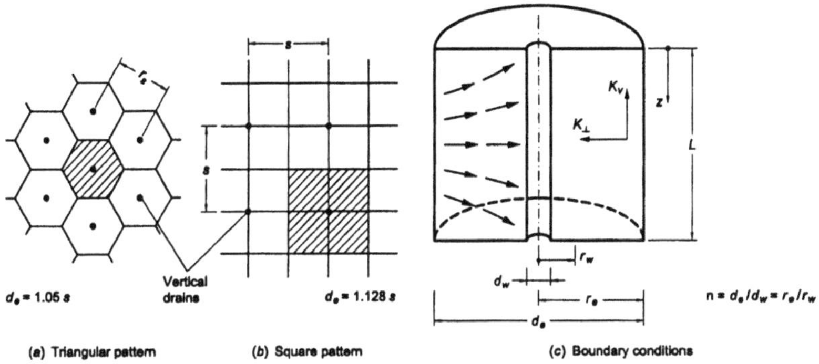

Figure 1. Drain and typical drain dipositions

The most economical pattern to study the consolidation with drains is the one shown in Fig. 1 (see Barron ([1])), where it is assumed that the zone of influence of each drain is a circle of radius R_e, and that consolidation and flow are axisymmetric.

In this geometrical configuration, assuming that displacements occur only in the vertical direction,

$$D_z = \frac{1+e}{1+e_0},$$

and from Eq. (10) one can write

$$\begin{aligned}\frac{\mu}{1+e_0}\frac{de}{dt} = & \frac{1}{R}\frac{\partial}{\partial R}\left(RK_\perp(e)\frac{\partial P_\ell}{\partial R}\right) + \\ & +\frac{\partial}{\partial Z}\left[\frac{1+e}{1+e_0}\frac{K_z(e)}{D_z^2}\left(\frac{\partial P_\ell}{\partial Z} + \rho_\ell g D_z\right)\right].\end{aligned} \tag{36}$$

which still need to be coupled with

$$\begin{cases}\frac{1}{R}\frac{\partial}{\partial R}\left[R(T'_{RR}+P_\ell)\right] - \frac{1}{R}(T'_{\theta\theta}+P_\ell) = 0\,, \\ \frac{\partial}{\partial Z}\left[\frac{1+e}{1+e_0}\frac{1}{D_z}(T'_{zz}+P_\ell)\right] + (\rho_{m0}+m)g = 0\,,\end{cases} \tag{37}$$

and

$$\frac{\partial z}{\partial Z} = D_z\,. \tag{38}$$

As already stated the cylindrical outer surface $R = R_e$ of this zone is taken to be so that no flow occurs across it, i.e. it behaves as an impervious boundary. As already mentioned, the pore pressure at the drain surface R_d is assumed to be hydrostatic (here the atmospheric pressure is considered as the reference value).

In these circumstances, the boundary conditions are

$$\begin{cases} P_\ell = \rho_\ell g H \left(1 - \frac{z}{h(t)}\right) & \text{on } R = R_d; \\ P_\ell = 0 & \text{on } Z = H; \\ P_\ell = \rho_\ell g H & \text{on } Z = 0; \\ \dfrac{\partial P_\ell}{\partial R} = 0 & \text{on } R = R_e; \end{cases} \tag{39}$$

where H is the initial depth of the layer and the settlement $z(Z)$ is determined integrating (35).

Under the additional assumption that hydraulic gradients in the z-direction can be considered negligible, or that K_z is negligible compared to $K_\perp$, the two-dimensional equation can be reduced to the one-dimensional one

$$\frac{\mu}{1+e_0}\frac{de}{dt} = \frac{1}{R}\frac{\partial}{\partial R}\left[R\left(K_\perp(e)\frac{\partial P_\ell}{\partial R}\right)\right] . \tag{40}$$

If now the soil compressibility is introduced, which allows to write

$$\frac{\delta e}{1+e_0} = m_v \delta P_\ell , \tag{41}$$

one has

$$m_v \mu \frac{\partial P_\ell}{\partial t} = \frac{1}{R}\frac{\partial}{\partial R}\left[R K_\perp \frac{\partial P_\ell}{\partial r}\right] . \tag{42}$$

Remark 5.1. Equation (42) is the nonlinear equation of the consolidation due to radial flow toward a vertical well in terms of pore pressure. Both soil compressibility m_v and hydraulic conductivity K are functions of the radial distance, due to their dependence on the void ratio and therefore on the pore pressure.

Remark 5.2. The equation obtained by Barron ([1]) can be derived from (42) under the assumptions of small strains and that both m_v and

K are constant during the consolidation process. In this circumstance one gets

$$\frac{\partial P_\ell}{\partial t} = \frac{K_\perp}{\mu m_v} \frac{1}{R} \frac{\partial}{\partial R} \left[R \frac{\partial P_\ell}{\partial R} \right] . \tag{43}$$

An analysis of the differences deriving from the use of (43) instead of (42) are still open to investigation and is presented in the following.

In order to compare the results of nonlinear analysis with the conventional ones (linear model), let us define the consolidation coefficient for radial flow as

$$c_r = \frac{K_\perp}{\mu m_v} , \tag{44}$$

so that Eq. (43) writes

$$\frac{\partial P_\ell}{\partial t} = c_r \frac{1}{R} \frac{\partial}{\partial R} \left[r \frac{\partial P_\ell}{\partial R} \right] . \tag{45}$$

The coefficient of consolidation can be regarded as defining a characteristic time

$$t = \frac{d_e^2}{c_r} , \tag{46}$$

provided that a convenient length scale d_e (i.e. the diameter of the draining cylinder in Figure 1) is identified.

This characteristic time can be regarded as a measure of the time can be regarded as a measure of the time required in order to appreciate significative changes of the process.

According to this position, a dimensionless time factor is introduced in the presentation of the results

$$T_r = \frac{c_r t}{d_e^2} . \tag{47}$$

In nonlinear analysis both the soil conductivity K and the compressibility m_v change during the process, and their dependence with stress level or void ratio can be expressed according to the following empirical rules

$$K = K_0 \exp \left(\frac{e_0 - e}{C_k} \right) , \tag{48}$$

$$m_v = \frac{0.435 C_c}{(1+e) T'_{zz}} , \tag{49}$$

where K_0 and e_0 are the initial values of conductivity and void ratio

$$C_c = \frac{\Delta e}{\Delta \log T'_{zz}}, \tag{50}$$

is the change in void ratio for a tenfold change of vertical stress and

$$C_k = \frac{\Delta e}{\Delta \log K}, \tag{51}$$

is the change in void ratio for any log cycle of conductivity.

According to these well established empirical rules, the resulting coefficient of consolidation C_r can increase or decrease during the process, depending on the ratio C_k/C_c, as shown in Figure 2.

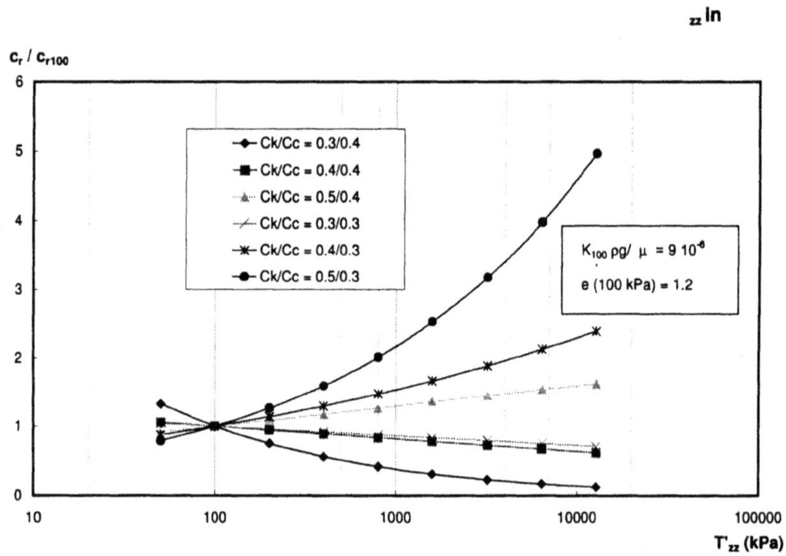

Figure 2. Dependence of consolidation coefficient on vertical stress T'_{zz} in nonlinear analysis

As a consequence, the following two integral measures of the evolution of the process,

- the degree of consolidation settlement

$$U_s = \frac{\text{settlement at time } t}{\text{final settlement}}; \tag{52}$$

- the degree of pore pressure dissipation

$$U_p = \frac{\int_{r_w}^{r_e} [P_\ell(t=0) - P_\ell(t)]\, dr}{\int_{r_w}^{r_e} P_\ell(t=0)\, dr}, \tag{53}$$

which are coincident in linear analysis, can have significantly different values in the nonlinear case.

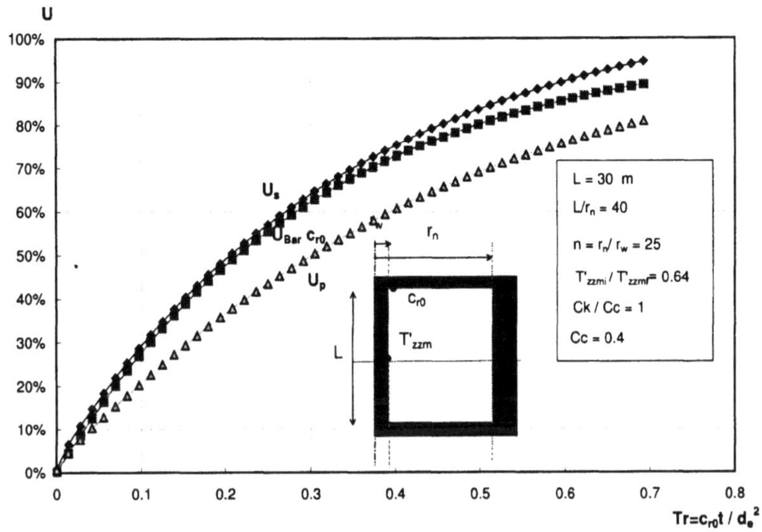

Figure 3. Comparison between Barron solution and nonlinear analysis

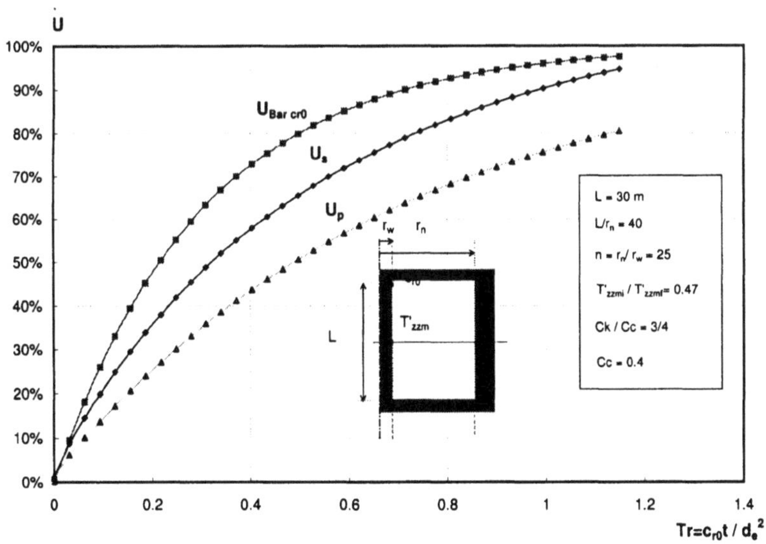

Figure 4. Comparison between Barron solution and nonlinear analysis

A practical implication is that the predictions based on Barron model (shown as Ubar in Figures 3 and 4) can be unconservatives, i.e. they tend to overpredict the evolution rate, depending on the ratio C_k/C_c.

Acknowledgements

The Authors are grateful to the Italian National Research Council (C.N.R.) Contract n° 98.01024.ct01, and to the Italian Ministry for the University and Scientific Research (M.U.R.S.T.) for funding the present research.

References

1. Barron, R. A. Consolidation of fine-grained soils by drain wells, *Trans. ASCE*, 713–754, 1948.
2. Bear, J. and Verruijt, A. Modeling Groundwater Flow and Pollution, D. Reidel Publ. Co. (Kluwer group), 414, 1987.
3. Biot, M. A. General theory of three-dimensional consolidation, *J. Appl. Phys.*, **12**, 155–164, 1941.
4. Biot, M. A. A theory of finite deformations of porous solids, *Indiana Univ. Math. J.*, **21**, 597, 1972.
5. Biot, M. A. Variational Lagrangian thermodynamics of nonisothermal finite strain mechanics of porous solids and thermomolecular diffusion, *Int. J. Solids Structures*, **13**, 579–597, 1977.
6. Bowen, R. M. Theory of Mixtures, in Continuum Physics, A. C. Eringen Ed., **3**, Academic Press, 1976.
7. Coussy, O. Mechanics of Porous Continua, Wiley, 1995.
8. Lancellotta, R. Geotechnical Engineering, Balkema, 1995.
9. Lee, I. K., White, W., and Ingles, O. G. Geotechnical Engineering, Pitman, 1983.
10. Lancellotta, R., and Preziosi, L. A general nonlinear mathematical model for soil consolidation problems, *Int. J. Engng. Sciences*, **35**, 1045–1063, 1997.
11. Munaf, D., Wineman, A. S., Rajagopal, K. R. and Lee, D. W. A boundary value problem in groundwater motion analysis — Comparison of predictions based on Darcy's law and the continuum theory of mixtures, *Math. Models Methods Appl. Sci.*, **3**, 231–248, 1993.
12. Rajagopal, K. R., and Tao, L. Mechanics of Mixtures, World Scientific, 1995.
13. Verruijt, A. Elastic storage of aquifiers, in Flow through Porous Media, R. J. M. de Wiest Ed., Academic Press, 331–376, 1969.

Drying Induced Stresses in Viscoelastic Sphere

J. Banaszak[1] and S. J. Kowalski
Poznań University of Technology
Institute of Technology and Chemical Engineering
Pl. M. Sklodowskiej-Curie 2, 60-965 Poznań, Poland

[1]*Polish Academy of Sciences*
Institute of Fundamental Technological Research

1. Introduction

The main subject of this paper is to study how the constitutive model of saturated porous body influences the numerically estimated drying induced stresses and deformations. Mechanical properties of porous materials are changing during drying. At the beginning they are viscous and in the course of moisture removing become viscoelastic, elastic or even brittle.To authenticate the calculations, the material of the sphere is asumed to be both elastic and viscoelastic. One states a significant difference between the results obtained for the elastic sphere and viscoelastic one, particularly for stresses.

The thermodynamical background of the model used in this paper is presented in Kowalski ([4]) and Kowalski and Strumillo ([3]). The solution of the problem was obtained making use of both the Laplace transformations and the finite difference method. The considerations are confined to the constant drying rate period in which the temperature of the saturated body is constant in the whole cross-section and equal to the wet-bulb temperature. Phase transitions inside the dried material are ignored and the whole evaporation of the moisture is assumed to proceed on the boundary of the drying material.

2. Model presentation

To solve the problem of convective drying of the sphere the following assumptions have to be satisfied:

1) The sphere consists of a porous elastic or viscoelastic material (Maxwellian model), whose pores are filled with water.

2) The drying of the sphere proceeds symmetrically with respect to the middle point, so that only the displacement in radial direction is different from zero, that is $u_r(r,t) \neq 0, u_\varphi = 0$, $u_\theta = 0$.

3) The external surface $r = R$ is free of external loading.

W. Ehlers (ed.), IUTAM Symposium on Theoretical and Numerical Methods in Continuum Mechanics of Porous Materials, 381–386.

4) The analysis is confined to the constant drying rate period, which is characterised by the uniform temperature in the whole body, equal to the wet-bulb temperature.

5) The sphere is assumed to be isotropic and continuos.

The equation of equilibrium for the moist porous sphere is:

$$\frac{\partial \sigma_{rr}}{\partial r} + 2\frac{\sigma_{rr} - \sigma_{\theta\theta}}{r} = 0. \tag{1}$$

The strains in spherical coordinates are:

$$\varepsilon_{rr} = \frac{\partial u_r}{\partial r}, \quad \varepsilon_{\theta\theta} = \varepsilon_{\varphi\varphi} = \frac{u_r}{r}. \tag{2}$$

The viscoelastic properties of the material are simulated by Maxwell's model:

$$\dot{\sigma}_{ij} + \frac{M}{\eta}\sigma_{ij} = 2M\,\dot{\varepsilon}_{ij} + \left(\frac{A}{K}\,\dot{\sigma} + \frac{\xi}{\chi}\frac{M}{\eta}\sigma\right)\delta_{ij} - 2M\,\dot{\phi}\,\delta_{ij} \tag{3}$$

where $\phi = \alpha_\vartheta\vartheta + \alpha_\Theta\Theta$, expresses the volumetric deformation caused by the temperature and the moisture content, with $\vartheta = T - T_r$ and $\Theta = X - X_r$, being the relative temperature and the relative moisture content, and α_ϑ and α_Θ are coefficients of thermal expansion swelling respectively. Relation (3) is simplified in further considerations through an assumption that the ratio of volumetric modules for elastic and viscoelastic materials is equal to the ratio of shear modules of these materials, that is $\frac{K}{\chi} = \frac{M}{\eta}$. This simplification still preserves the volumetric changes of the drying body.The physical relation (3) is reduced to

$$\dot{\sigma}_{ij} + \frac{M}{\eta}\sigma_{ij} = 2M\,\dot{\varepsilon}_{ij} + 3A\,\dot{\varepsilon}\,\delta_{ij} - 3K\,\dot{\phi}\,\delta_{ij} \tag{4}$$

The alternation of the moisture content in the dried body is described by the mass balance equation :

$$\rho_0\,\dot{\Theta} = -\eta_{k,k} \tag{5}$$

The moisture mass transport equation, which relates the moisture flux η_k with the gradient of moisture potential:

$$\eta_k = -\Lambda_m\mu_{,k} \tag{6}$$

where $\Lambda_m \geq 0$ is the moisture transport coefficient. The moisture potential μ is a function of the temperature ϑ, the body volume deformation ε, and the moisture content Θ (Kowalski [4]).

As the temperature ϑ of the dried body does not alter during the constant drying rate period, the gradient of moisture potential is equal to:

$$\rho_0\mu_{,k} = -\gamma_\Theta\varepsilon_{,k} + c_\Theta\Theta_{,k} \tag{7}$$

where $\gamma_\Theta = -\rho_0 (\partial\mu/\partial\varepsilon)_{\vartheta,\Theta} = \alpha_\Theta(2M+3A)$ is the volumetric stiffness, and $c_\Theta = c_\vartheta (\partial\mu/\partial\varepsilon)_{\vartheta,\varepsilon}$- an averaged "Leverett function" connected with capillary rise of wetting fluid in porous medium (see Scheidegger [5]).

The mass transport equation takes the form:

$$\dot{\Theta}= \frac{\Lambda_m}{\rho_0^2}\nabla^2\left[-\gamma_\Theta\varepsilon + c_\Theta(\Theta-\Theta_r)\right], \tag{8}$$

The boundary conditions for the moisture mass transfer stipulate the symmetry of the moisture potential at the center point of the sphere, and convective exchange of moisture on the external surface of the moist body:

$$-\Lambda_m \frac{\partial\mu}{\partial r}\,|_{r=0}= 0 \quad \text{and} \quad -\Lambda_m \frac{\partial\mu}{\partial r}\,|_{r=R}= \alpha_m\left(\mu\,|_{r=R} -\mu_a\right) \tag{9}$$

where α_m is the convective mass transfer coefficient and μ_a - the potential of the vapour in the air (drying medium).

3. Displacements and stresses in viskoelastic sphere

The solutions for dispacements and stresses for a full elastic sphere dried convectively are:

$$u_r = \frac{4Mr}{3KR^3}\int_0^R \phi r^2 dr + \frac{1}{r^2}\int_0^r \phi r^2 dr, \tag{10}$$

$$\sigma_{rr} = 4M\left(\frac{1}{R^3}\int_0^R \phi r^2 dr + \frac{1}{r^3}\int_0^r \phi r^2 dr\right), \tag{11}$$

$$\sigma_{\varphi\varphi} = \sigma_{\theta\theta} = 2M\left(\frac{1}{r^3}\int_0^r \phi r^2 dr + \frac{2}{R^3}\int_0^R \phi r^2 dr - \phi\right). \tag{12}$$

Maxwell's physical relation (4) in Laplace transforms takes the form:

$$\overline{\sigma}_{ij}^V = \frac{s}{s+\frac{M}{\eta}}\left[2M\overline{\varepsilon}_{ij} + 3(A\overline{\varepsilon} - K\overline{\phi})\delta_{ij}\right]. \tag{13}$$

Note that if the initial values for the stresses and strains are assumed to be zero,then the physical relation for viscoelastic body expressed in Laplace transforms is proportional to that of elastic one, that is:

$$\overline{\sigma}_{ij}^V = \frac{s}{s+\frac{M}{\eta}}\overline{\sigma}_{ij}^E = 2M^V\overline{\varepsilon}_{ij} + 3(A^V\overline{\varepsilon} - K^V\overline{\phi})\delta_{ij} \tag{14}$$

where $\overline{\sigma}_{ij}^{E} = 2M\overline{\varepsilon}_{ij} + 3(A\overline{\varepsilon} - K\overline{\phi})\delta_{ij}$ and $M^V = M\frac{s}{s+M/\eta}$, $A^V = A\frac{s}{s+M/\eta}$, $K^V = K\frac{s}{s+M/\eta}$ are the constants of viscoelastic material expressed in Laplace transforms (see Alfrey ([1])).

The above statements justify construction of the solution for the viscoelastic sphere with using the solution for the elastic one.The solutions are similar to that given by (10), (11) and (12):

$$\overline{u}_r^V = \frac{4M^V r}{3K^V R^3}\int_0^R \overline{\phi} r^2 dr + \frac{1}{r^2}\int_0^r \overline{\phi} r^2 dr, \tag{15}$$

$$\overline{\sigma}_{rr}^V = 4M^V\left(\frac{1}{R^3}\int_0^R \overline{\phi} r^2 dr + \frac{1}{r^3}\int_0^r \overline{\phi} r^2 dr\right), \tag{16}$$

$$\overline{\sigma}_{\varphi\varphi}^V = \overline{\sigma}_{\theta\theta}^V = 2M^V\left(\frac{1}{r^3}\int_0^r \overline{\phi} r^2 dr + \frac{2}{R^3}\int_0^R \overline{\phi} r^2 dr - \overline{\phi}\right), \tag{17}$$

It is easy to notice that radial displacement of viscoelastic model is the same as that of elastic one, that is

$$\overline{u}_r^V = \overline{u}_r^E \qquad \text{and} \qquad u_r^V = u_r^E \tag{18}$$

The inverse transformation for stresses gives

$$\sigma_{rr}^V(r,t) = \sigma_{rr}^E(r,t) - a\int_0^t e^{-a(t-\tau)}\sigma_{rr}^E(r,\tau)d\tau, \tag{19}$$

$$\sigma_{\varphi\varphi}^V(r,t) = \sigma_{\varphi\varphi}^E(r,t) - a\int_0^t e^{-a(t-\tau)}\sigma_{\varphi\varphi}^E(r,\tau)d\tau, \tag{20}$$

$$\sigma_{\varphi\varphi}^V(r,t) = \sigma_{\theta\theta}^V(r,t) \tag{21}$$

where $a = M/\eta$ is termed as the viscosity coefficient.

4. Calculation of moisture distribution

One can obtain the moisture distribution across the sphere using the mass balance equation (5) and the moisture mass transport equation (6) with help of physical relation. Finally, the equation of diffusion type for the moisture content distribution is derived

$$\dot{\Theta} = L\nabla^2\Theta \tag{22}$$

with the coefficient $L = \frac{\Lambda_m}{\rho_0^2}(c_\Theta - \gamma_\Theta\alpha_\Theta)$,containing material constants. The drying process begins with constant moisture distribution in whole sphere $\Theta(r, 0) = \Theta_0$. Using this initial condition, the physical relationsand boundary conditions (9), one derives the moisture distribution. The Crank-Nicolson method was applied for numerical solution of the problem.

5. Results

Figure 1 presents the distribution of the moisture content versus the sphere radius in several instants of drying time. It is visible that region close to the surface (r=0,03 m) is dried relative quickly but the center of the sphere reaches the reference moisture content after 2,5 hours drying time.

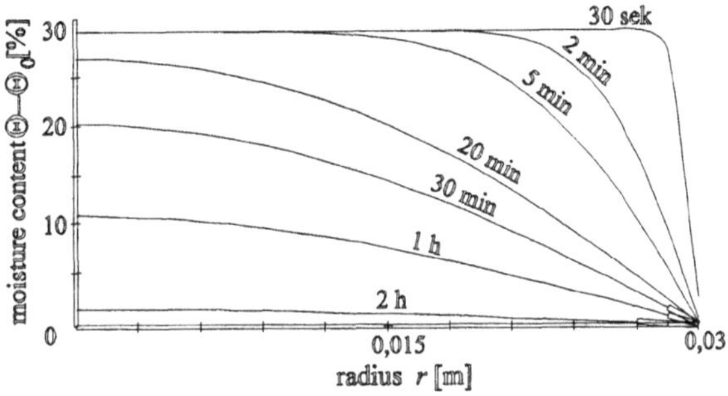

Figure 1. Moisture content distribution in the sphere.

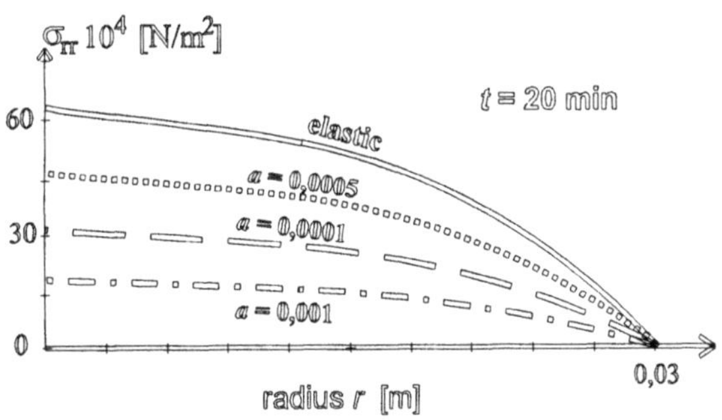

Figure 2. Comparison of stress distributions in the elastic and viscoelastic sphere after 20 minutes of drying process.

Viscosity influences significantly the magnitude of stress in dried materials. For great viscosity, represented by parameter a, the stress distribution is more uniform and reaches smaller values(Fig. 2).

Figure 3 presents the radial stresses in the drying sphere as a function of time for various radii. The stresses are the highest at the center and zero on the surface. The shape of the curve is typical for Maxwell's model. The stresses reach their maxima after three minutes of heating

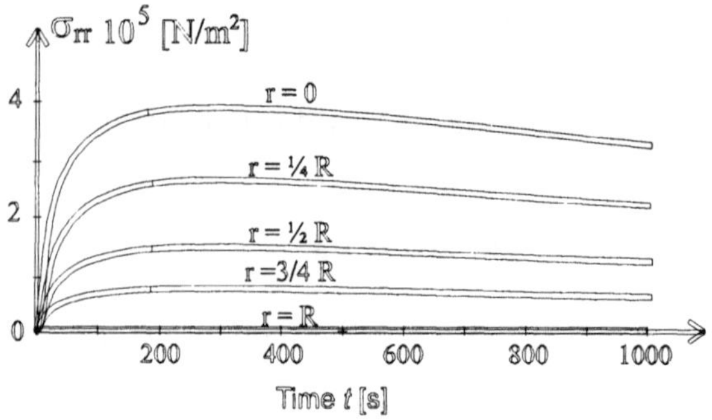

Figure 3. Radial stresses as a function of time for different radii.

and then relax slowly, approaching stress free conditions after a long period of time.

6. Final remarks

1. The rheological properties of dried material influence considerably the drying induced stresses and insignificantly the moisture distribution.

2. The stresses reach smaller magnitude (depended on viscosity) in viscoelastic material than in elastic one.

3. The stresses reach theirs maxima in the first period of drying and then relax slowly approaching stress free state after a long period of time.

4. The solution of the viscoelastic problem agrees well with the elastic one if the dependence on time is revoked.

Acknowledgements

This work was carried out as part of the research project No. 3 T09C 0015 12 sponsored by the State Committee for Scientific Research.

References

1. Alfrey T. Mechanical Behaviour of High Polymers., N. Y.- London, Interscience, 1948.
2. Kowalski, S. J. Thermomechanics of Constant Drying Rate Period, *Arch. Mech.*, **39**, 157–176, 1987.
3. Kowalski S. J. , Strumillo Cz. Moisture Transport in Dried Materials. Boundary Conditions, *Chem. Engng. Sci.*, **52**(7), 1141-1150, 1997.
4. Kowalski S. J., Musielak G. and Rybicki A. Drying Processes - Thermomechanical Approach, IFTR - PSP, P-ń - W-wa, (in Polish), 1996.
5. Scheidegger A. E. The Physics of Flow Through Porous Media, University of Toronto Press, 1957.

Settlements of Sand due to Cyclic Twisting of a Tube

R. Cudmani and G. Gudehus
Institute of Soil and Rock Mechanics, University of Karlsruhe
Postfach 6980, D-76128 Karlsruhe, Fax +49 721/696096
e-mail: (cudmani, gudehus)@ibf-tiger.bau-verm.uni-karlsruhe.de

1. Introduction

Bored piles are constructed by filling concrete into an excavated borehole with the purpose of transferring load to bearing strata below the building. Large diameter bored piles are used with increasing frequency and confidence in urban areas where the reduction of noise and vibrations during installation plays a primary role. The construction of bored piles in cohesionless soils usually requires a casing to support the borehole during its excavation. The casing consisting of a steel pipe is usually driven into the soil by a simultaneous axialy push and cyclic torsion with constant amplitude and low frequency. As it is empirically well known, cyclic torsion eases penetration considerably, but it has the negative side effect of generating settlements at the surface around the pile.
In this paper an experimental and numerical investigation of the penetration of a tube in a sand with the aid of cyclic twisting is presented.

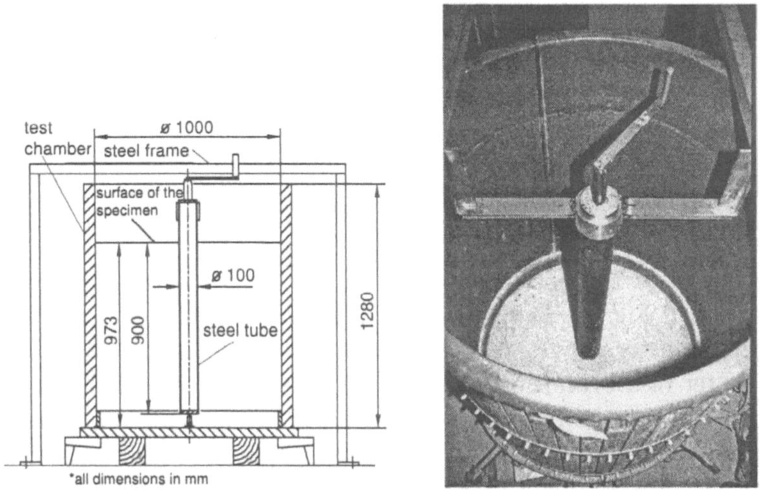

Figure 1. Testing device

W. Ehlers (ed.), IUTAM Symposium on Theoretical and Numerical Methods in Continuum Mechanics of Porous Materials, 387–396.

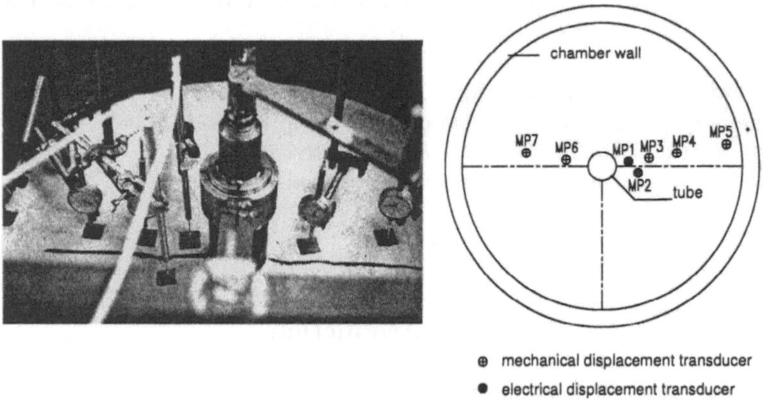

Figure 2. Transducers used to monitor the surface settlements

In the experimental part a steel pipe was installed in a cylindrical chamber, and loose sand was deposited into the chamber under controlled conditions. Afterwards, cyclic rotation with constant amplitude was applied to the tube while monitoring settlements at the surface. In the numerical part of the investigation a 3D-FE model was used to simulate the experiment and to compare the penetration resistance with and without cyclic twisting. The behaviour of the soil was modelled by a hypoplastic constitutive law, and interface friction was assumed to follow the Mohr-Coulomb friction law. The material constants were determined by means of conventional laboratory tests.

2. Experimental investigations

2.1. Testing device

The design of the equipment used in the experimental investigation is shown in Fig. 1. The kernel of the device consists of a steel tube (outer diameter D=100 mm, length L=1120 mm), which is placed in the center of a cylindrical chamber (D=1000 mm, H=1280 mm) and guided both at the top and at the bottom to eliminate all degrees of freedom except for the rotational one around the vertical axis. The upper guide is attached to a steel frame, while the lower one is fixed to the bottom of the chamber. Twist is applied manually with a crank (lever arm L=260 mm) welded onto the top of the tube.

2.2. PREPARATION OF THE SPECIMEN, INSTRUMENTATION AND TESTING

After installing the tube in the chamber the sand deposition was carried out following a procedure similar to the method proposed by DIN 18126 to generate a high void ratio. A filling funnel was attached to a container containing the granular material. Then the hopper was lifted upwards very slowly causing the sand to pour through the funnel into the chamber. Thus, a loose (relative density I_D=0.223) and relatively homogeneous sand body was attained, as cone penetration tests showed. To perform the test the tube was twisted cyclically with an angular amplitude of $\alpha = 0.20$ rad.
The evolution of settlements as a function of the number of rotation cycles was measured with five mechanical and two electrical displacement transducers, which were positioned at different distances from the tube wall (s. Fig. 2). Additionally, the electrical transducers allowed the continuous recording of the vertical displacements.

2.3. GRANULAR MATERIAL AND CALIBRATION OF THE HYPOPLASTIC PARAMETERS

A fine uniform quartz sand has been used in the experiment (uniformity coefficient: 2.35, mean grain size D_{50}: 0.25 mm, density of the grains: 2.65 g/cm^3, max. void ratio e_{max}: 0.975, min. void ratio e_{min}: 0.595). Following Herle and Gudehus [3] the material parameters for the sand were determined by performing two triaxial compression and two isotropic compression tests, the first on a dense and the second on a loose specimen (for the evaluation of φ_c, α, h_s, n und β) as well as index tests (for the evaluation of e_{c0}=e_{max}, e_{d0}=e_{min}, e_{i0}= 1,15 e_{max}). The additional five parameters of the extended hypoplastic model with intergranular strain have been estimated by using the diagrams given by Herle and Niemunis [4]. The parameters used in the numerical simulations are listed below.

φ_c [°]	e_{d0}	e_{c0}	e_{i0}	h_s [MPa]	n	α	β
33.0,0	0.60	0.98	1.15	2600	0.3	0.10	1.00

m_2	m_s	R_{max}	β	χ
2.0	5.0	$1.0 \cdot 10^{-4}$	0.5	1.0

2.4. Experimental results

The evolution of vertical displacements during the first twist cycle as well as during the first 16 cycles at positions MP1 and MP2 are shown in Fig. 3 (rotation cycles are characterized through a cycle beginning, a rotation reversal and a cycle end). During cycles the free surface only settles, indicating that the granular material behaves contractant. During subsequent cycles settlements are preceded by a small heaving of the free surface as a result of the dilatant - contractant behaviour of the sand. The additional settlement per cycle diminishes with the number of cycles. The rotation angle during which settlements takes place is a small fraction of the rotation amplitude. Since settlements decrease with increasing distances from the tube a settlement funnel develops, whose radius grows with the number of rotation cycles. Localization of deformation at the interface was not observed in the experiment. An influence of the grain size is expected to play a role when the wall roughness is at least of the same order of magnitude as the mean grain diameter, specially for angular grain shapes at dense states (Tejchman [5], Wehr [6]). Since localization did not take place the experiment the results are not expected to depend on the grain size/tube diameter ratio if the diameter of the tube is large than about 50 grain diameters, which is the thereshold at which the continuum theory is applicable.

3. Numerical simulation

3.1. Numerical model

In order to simulate numerically the experiment presented in the last section a 3D FE-model was used. The calculations were performed using the FE-code ABAQUS together with a subroutine for our constitutive law. The FE-calculation was intended to be a "Class A" prediction, i.e., the hypoplastic parameters and the parameters defining the friction law on the interface were evaluated without considering the experimental results. The FE-mesh is sketched in Fig. 5.
The hypoplastic constitutive law after von Wolffersdorf [7] with the extension proposed by Herle and Niemunis [4] to account for the so-called intergranular strain was used to model the mechanical behaviour

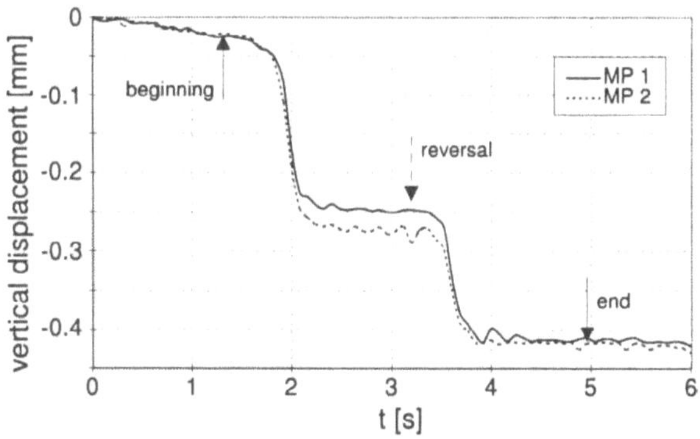

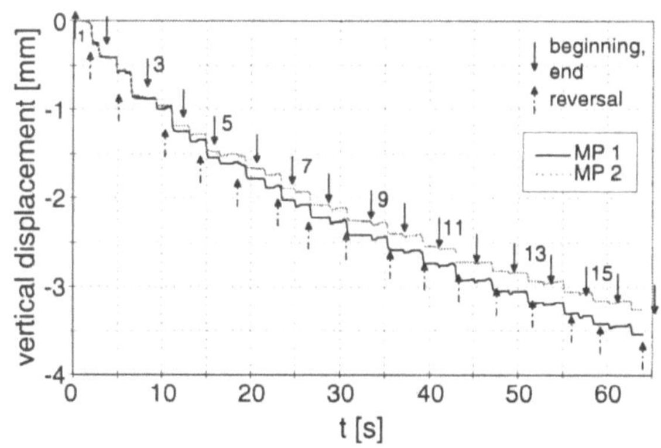

Figure 3. Evolution of settlements during the first cycle (top) and during the first 16 cycles (bottom)

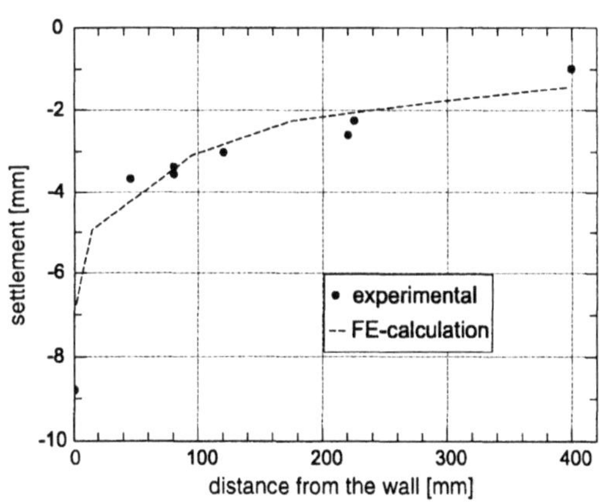

Figure 4. Comparison of measured and calculated settlements at different distances from the tube wall after 16 cycles

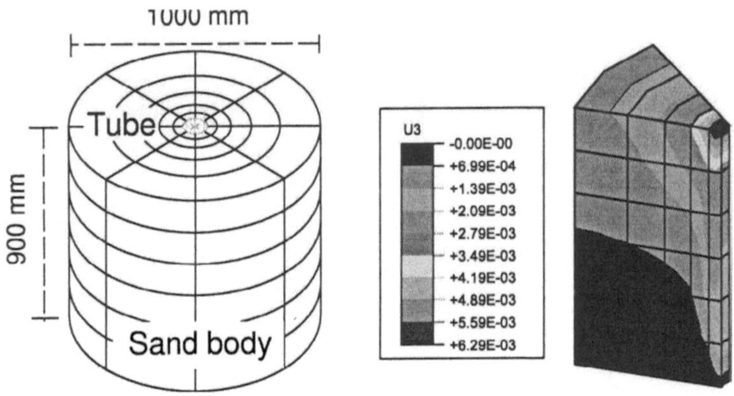

Figure 5. Finite Element mesh used in the numerical analysis (left), Distribution of vertical displacement along the specimen after 16 cycles (right

of the soil. The steel tube was regarded as rigid body. The boundary conditions were (with 1, 2, 3 for radial, tangential and vertical directions respectively):

- outer boundary of the sand body: radial displacement u_1=0

- base of the sand body: all displacement components $u_1=u_2=u_3=0$

- at the surface of the sand body a small ficticious vertical stress was applied to avoid the appearance of tension

- all degrees of freedom of the tube were fixed except for the rotation around the vertical axis. For this a sinusoidal rotation with amplitude $\varphi_0 = 0.2$ rad. was prescribed .

The initial state of the material was defined by the void ratio $e_0 = 0.89$, which was constant within the specimen, and the initial stresses $\sigma_3 = \gamma h$, $\sigma_1 = \sigma_2 = K_0 \sigma_3$. The value of K_0 was estimated using Jaky's empirical relationship $K_0 = 1 - \sin\varphi$, which is valid for normal consolidated sedimentary soils. Since the sand was in a loose state it was assumed that $\varphi=\varphi_c=33^o$, i.e, $K_0 \approx 0.5$. All shear components of the initial stress tensor were set to zero.
In order to model the friction at the contact interface between soil and tube, no sticking after a twist reversal was assumed, i.e., both surfaces were free to slip during the rotation. Vertical and tangential shear stresses at the interface were modelled by the friction law proposed by Dierssen [1] to describe the shaft friction developed during vibro-driving of piles:

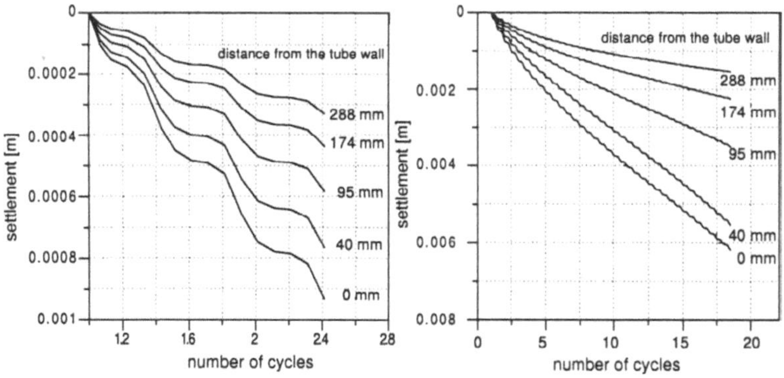

Figure 6. Calculated evolution of vertical displacement during the first 2 cycles (left) and during the first 16 cycles (right) at different distances from the tube wall.

$$\tau(t) = \exp\left[-C_1/C_2(z - z_0)\text{sgn}(\dot{z})\right](\tau_0 - \tau_{max}(t)) + \tau_{max}(t) \quad (1)$$

Herein, τ_0 is the shear stress and z_0 the shear reached at the last reversal of the motion, respectively. τ_{max}(t)=$p_r(t)$ tan δ is the Mohr-Coulomb shear resistance. p_r(t) is the pressure normal to the wall, which takes the value $p_r(0)= K_0\,\gamma\,z$ for t=0. tan δ is the friction angle of wall surface $\delta = \tan(2/3\varphi_c)$, z the displacement and $\dot{z}$ the velocity. The constants C_1 and C_2 depend on $\tau_{max}(0)$ and on the displacement needed to mobilize 95 % of $\tau_{max}(0)$, respectively.

3.2. Results of the FE-simulation

Fig. 5 illustrates the distribution of vertical displacements in the sand body obtained after 16 rotation cycles. Shear cycles induced by the cyclic rotation of the tube lead to densification of the soil, which decreases with depth and distance from the tube. Thus, a cycle-wise increasing settlement funnel at the surface develops, just as observed in the experiment (s. Fig. 4). The evolution of vertical displacements with cycles is also satisfactorily predicted by the numerical model as shown in the Fig. 6. That in the experiment the decay of settlement with cycles was faster than that in the numerical modelling indicates that the maximal friction stress in the first case was already achieved after a smaller rotation than assumed for the evaluation of the constant C_1 in eq. (1). A faster decay could be obtained by adopting a greater value of C_1.

The evolution of normal stresses and shear stresses in a soil element near the wall in the first 16 cycles is presented in Fig. 7. The mean value of

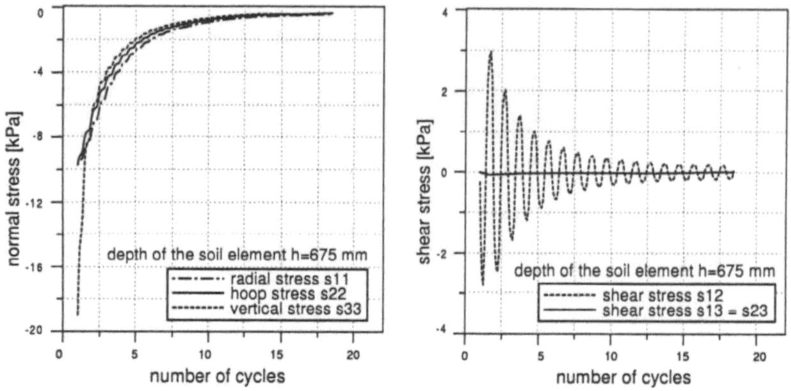

Figure 7. Evolution of normal and shear stress components in a soil element near the tube wall

the normal stresses as well as the amplitude of shear stress components in the surrounding soil decrease. The reduction of the radial stresses with the number of cycles causes a decay of the pressure normal to the wall and thus, a decay of the maximal shear stress $\tau_{max}(t)$ until a nearly cyclic response is reached (s. Fig. 8)

Finally, Fig. 9 presents a comparison of the evolution of the shear force developed on the shaft as a function of the vertical displacements when pushing the tube with and without applying cyclic twisting. Without cyclic torsion the axial force increases until the static penetration resistance is reached. On the other hand, if the tube is rotated cyclically the axial force reaches a maximal value after few cycles , which is lower than the static penetration resistance, and decays afterwards progressively with further cycles. Quantitative experimental evidence of the decay of the axial force with cycles could no be found in the literature. However, that cyclic torsion is widely used in the geotechnical practice to ease penetration indicates that in fact a decay of the axial force takes place. Measuring the evolution of torsion with cycles in future experiments will permit us to calculate the friction force acting on the interface as a function of time.

The results allow a better understanding of the mechanism leading to the development of settlements as well as to the reduction of soil resistance when pushing a casing into the soil with the aid of cyclic rotation.

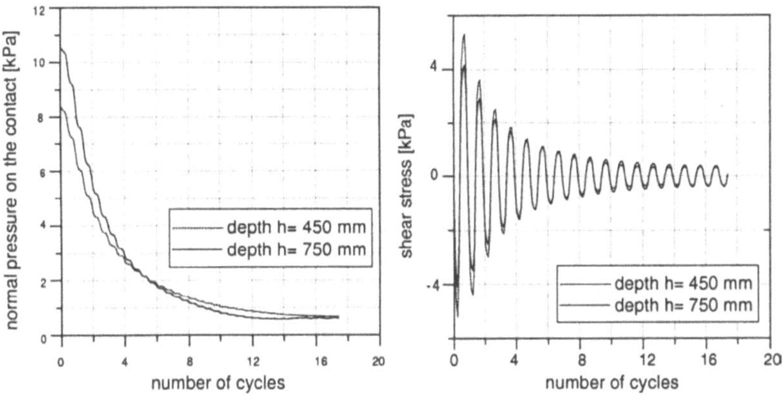

Figure 8. Evolution of normal pressure and shear stresses on the contact surface in the first 16 cycles

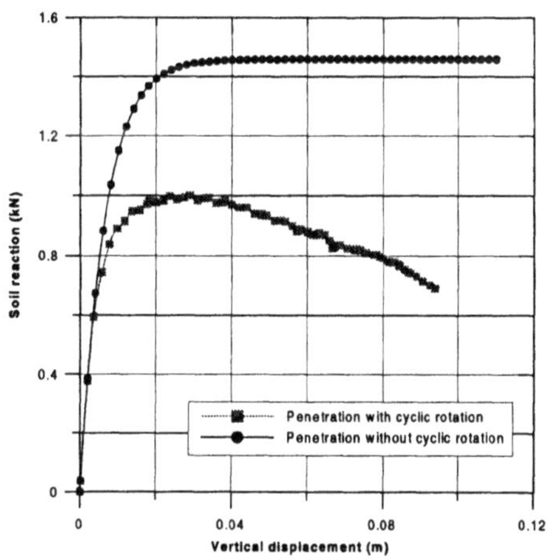

Figure 9. Comparison of the friction forces developed with and without cyclic rotation

Acknowledgements

The Autors wish to thank Dipl.-Ing. Ralf Engel for the help by performing the experiment and the numerical simulations.

References

1. Dierssen, G. Ein bodenmechanisches Modell zur Beschreibung des Vibrationsrammens in körnigen Böden. *Publication of the Institute of Soil and Rock Mechanics, Univ. Karlsruhe*, Nr. 133, 1994.

2. Gudehus, G. A comprehensive hypoplastic model for granular materials. *Soils and Foundations*, **36**, 1–12, 1996.
3. Herle, I. and Gudehus, G. Determination of parameters of a hypoplastic constitutive model from grain properties. *Mech. Cohes.-Frict. Mater.*, 1999.
4. Niemunis, A. and Herle, I. Hypoplastic model for cohesionless soils with elastic strain range. *Mech. Cohes.-Frict. Mater.*, **4**, 1997.
5. Tejchman, J. Modelling of shear localisation and autogeneous dynamic effects in granular bodies. Veröffentlichung des Institutes für Bodenmechanik und Felsmechanik der Universität Fridericiana in Karlsruhe, Heft 140, 1997.
6. Wehr, W. C. S. Granulatumhüllte Anker und Nägel -Sandanker-. Veröffentlichung des Institutes für Bodenmechanik und Felsmechanik der Universität Fridericiana in Karlsruhe, Heft 146, 1999.
7. Wolffersdorff, P. A hypoplastic relation for granular materials with a predefined limit state surface. *Mech. Cohes.-Frict. Mater.*, **1**, 1996.

Studies of Rayleigh Scattering of Longitudinal Waves in Saturated Porous Materials *

J. Kubik, M. Kaczmarek and J. Kochanski
Department of Environmental Mechanics, Pedagogical University in Bydgoszcz, 85-064 Bydgoszcz, Chodkiewicza 30, Poland
e-mail: kubik@rose.man.poznan.pl

Abstract. The paper presents experimental results and a model for propagation of ultrasonic waves in fluid saturated porous material. The model aimes at development of a two-phase description of the media assuming the presence of pore and solid scattering represented by complex and frequency dependent elastic moduli. A simple procedure allowing for determination of scattering parameters is proposed and comparison of model prediction and experimenatl data for attenuation of longitudinal waves is performed.

Keywords: Scattering, saturated porous materials, ultrasonic studies, modeling

1. Introduction

Theory of wave propagation in saturated porous materials developed by Biot [2] predicts attenuation and dispersion of waves primarily associated with the macroscopic relative motion of fluid and solid and deformation of phases. It is well known however that in order to explain some experimental data, especially for wave propagation in soil and rock like materials, number of other mechanisms must be included into two phase model of the media and particular mechanisms considered are: intergranular friction [13], local intercrack squirt flow, intracrack flow [8], [9], micro- scattering [14] and macroscopic inhomogeneity of materials [4]. Although the existing experimental results for ultrasonic frequencies show characteristic for Rayleigh scattering negative dispersion of phase velocity and frequency dependence of attenuation coefficient [11], [6] both experimental and theoretical studies of the role of scattering have numerous limitations. Experimental data concern relatively narrow frequency ranges and the scattering of slow wave is usually disregarded. Theoretical studies are concentrated on determination of frequency independent effective material parameters of porous materials [1], and explanation of the dependence of propagation parameters of shear or fast longitudinal waves on frequency [14]. The theoretical analysis usually assumes single or multiple scattering model of solid which contains gas or liquid inclusions assuming that

* The work was supported by the State Committee for Scientific Research under grant No. 7T 07A 05115.

W. Ehlers (ed.), IUTAM Symposium on Theoretical and Numerical Methods in Continuum Mechanics of Porous Materials, 397–402.

the wave carried in solid is scattered by isolated pores filled with gas or pore liquid (pore scattering). As a result the scattering of wave carried in fluid by solid skeleton is not taken into consideration and the model cannot predict scattering of waves in fluid saturating rigid porous solid or in fluid with solid inclusions. The aim of this paper is to present experimental data and develope a two-phase model for waves propagation in saturated porous media in Rayleigh scattering frequency range. The experimental results for longitudinal waves are obtained through broadband ultrasonic spectroscopy. The theoretical considerations refer to pore and solid scattering.

2. Experimental results

The studies are performed for porous materials made of sintered glass beads with average grain diameters 275 μm and vacuum saturated with water.The pulse transmission method combined with the immersion technique, similar to that used by Plona [10], is applied. The wave transducers are wide band with center frequency of 1 and 2.25 MHz. The transmitting transducer is excited through pulser by electric pulse of 350 V with a duration of 4 μs. The receiving transducer is connected to the digital storage oscilloscope which is triggered to sample the data by a synchronization signal generated by the pulser. The time records received by oscilloscope, sampled with frequency 100 MHz, and digitized are sent to FFT. In order to determine wave parameters the tests for samples of the same material but different thickness are made assuming that the energies of reflected waves at the boundaries of both samples are the same. Given the amplitude, real and imaginary components of Fourier transform of the measured signals passing through thinner and thicker samples the attenuation coefficient, α, and phase velocity, v, are determined, [7]. In order to check the accuracy of the experimental assembly the attenuation and phase velocity were determined for water. The maximum deviations of attenuation coefficient and phase velocity from the exact values equal to 0.001 Neper/mm for attenuation, and 1 % for phase velocity. In Figures 1 and 2 the coefficients of attenuation and phase velocity as functions of frequency for slow and fast longitudinal waves are given. The log-log plot of attenuation shows characteristic for Rayleigh scattering dependence on frequency to the power close to 4. Due to the shorter wavelength the scattering of slow wave is stronger and precedes in frequency domain the scattering of fast wave. The results for phase velocities indicate that the fast wave is about three times faster than the slow wave and both waves exhibit characteristic for scattered waves negative dispersion.

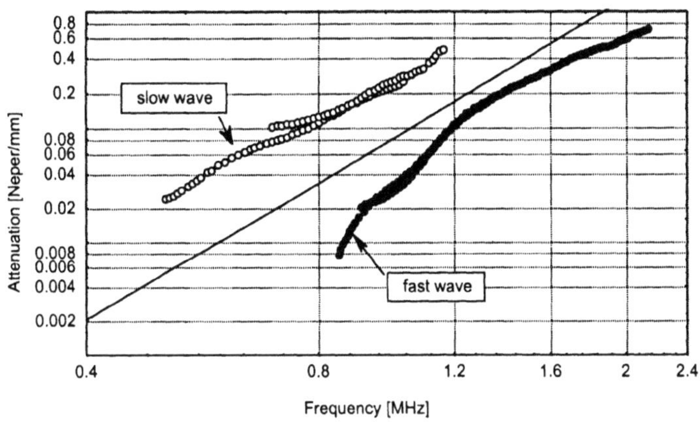

Figure 1. Attenuation of slow and fast longitudinal wave in water saturated porous glass for transducers 1 and 2.25 MHz. The solid line indicates slope of 4.

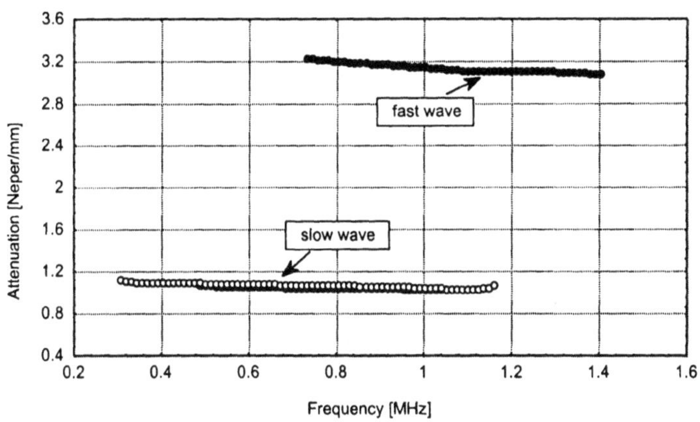

Figure 2. Phase velocity of slow and fast longitudinal wave measured in water saturated porous glass using transducers 1 and 2.25 MHz.

3. Modeling and discussion

The linear model describing propagation of two longitudinal waves in saturated porous material proposed by Biot [2] is

$$\varrho^s(1-f_v)\frac{\partial^2 e}{\partial t^2} = P\nabla^2 e + Q\nabla^2\varepsilon - b\left(\frac{\partial e}{\partial t} - \frac{\partial \varepsilon}{\partial t}\right) + \varrho_{12}\left(\frac{\partial^2 e}{\partial t^2} - \frac{\partial^2 \varepsilon}{\partial t^2}\right) \quad (1)$$

$$\varrho^f f_v \frac{\partial^2 \varepsilon}{\partial t^2} = R\nabla^2\varepsilon + Q\nabla^2 e + b\left(\frac{\partial e}{\partial t} - \frac{\partial \varepsilon}{\partial t}\right) - \varrho_{12}\left(\frac{\partial^2 e}{\partial t^2} - \frac{\partial^2 \varepsilon}{\partial t^2}\right) \quad (2)$$

where e and ε are dilatations of solid skeleton, and fluid, P, R, and Q refer to elastic or viscoelastic properties of the system, f_v is the porosity, ϱ^f and ϱ^s are mass densities of fluid, and solid material, ϱ_{12} denotes the coefficient which expresses inertial coupling between fluid and skeleton, and the coefficient b determines viscous interaction between phases (viscous drag). In general, both coefficients that define interaction between phases: ϱ_{12} and b are functions of frequency which describe a history dependence of interaction between phases. When one neglects the history dependence of the interaction force the parameters ϱ_{12} and b can be approximated by values that correspond to the flow of ideal fluid (ideal fluid approximation) and the flow with constant velocity (quasi-static approximation), i. e. $\varrho_{12} = f_v\varrho^f(1 - \alpha_T)$ and $b = \frac{f_v^2\mu}{k}$, where μ denotes the viscosity of fluid and tortuosity α_T and permeability k are parameters describing the structure of porous material.

In order to incorporate into two phase model of porous material the effect of scattering it is assumed now that the wave carried in skeleton is scattered by pore fluid (pore scattering) and the wave carried in fluid is scattered by solid (solid scattering). Moreover the concept elaborated for single phase model of inhomogeneous materials with complex elastic moduli [12] is used to include the effect of scattering. Thus, it is assumed that the elastic moduli of the Biot's model that relate stress in solid skeleton to dilatation of skeleton, P, and the stress in fluid to dilatation of fluid, R, are complex functions of frequency. The starting point of the proposed scheme are the relations obtained for inhomogeneous materials [12] which relate the real and imaginary parts of the complex elastic modulus, M, and wave parameters:

$$Re\{M\} = \frac{(\frac{1}{v^2} - \frac{\alpha^2}{\omega^2})}{v\omega(\frac{1}{v^2} + \frac{\alpha^2}{\omega^2})} \qquad Im\{M\} = \frac{2\varrho\alpha}{v\omega(\frac{1}{v^2} + \frac{\alpha^2}{\omega^2})} \tag{3}$$

where α and v denote coefficient of attenuation and phase velocity, ϱ is mass density and ω stands for angular frequency. Assuming that the coefficient of attenuation responsible for scattering is given as $\alpha = A\omega^4$ (the model assumes Rayleigh scattering) and that $\frac{\alpha^2}{\omega^2} \ll \frac{1}{v^2}$ one obtaines an approximate form for the real and imaginary parts of the elastic modulus: $Re\{M\} = \varrho v^2$ and $Im\{M\} = 2\varrho A\omega^3 v^3$. Taking into account the above expressions the real components of P and R are assumed to be constant and the imaginary parts of P and R are defined as following

$$Im\{P\} = 2\varrho^s(1 - f_v)A_1\omega^3 {v_d}^3 \qquad Im\{R\} = 2\varrho^f f_v A_2\omega^3 {v_f}^3 \tag{4}$$

where v_d and v_f are velocities of longitudinal waves in dry skeleton and fluid saturating rigid solid skeleton, and A_1 and A_2 are constants

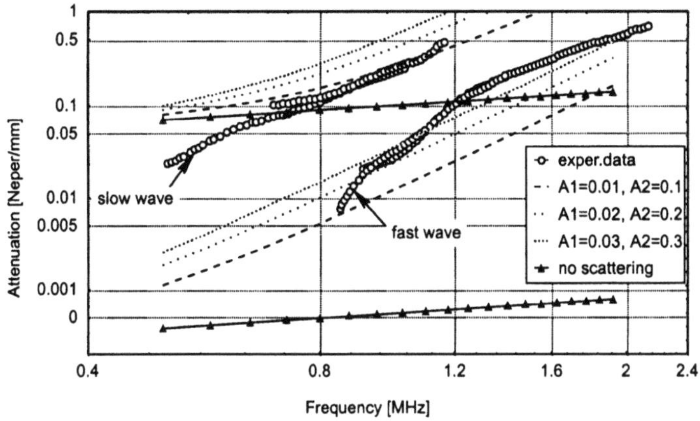

Figure 3. Attenuation of slow and fast waves predicted by models and obtained from experiments. The values of parameters A_i are given in $N(Ms)^4/m$.

which determine attenuation of fast and slow waves due to scattering. The appropriate values of the two latter parameters can be found from experimentally determined coefficients of attenuation of fast and slow wave assuming that $\alpha_i = A_i\omega^4$.

To analyze the predictions of the above proposed description the material parameters of Biot's model for water saturated porous glass were taken from the paper by Johnson *et al.* [5], and values of coefficients A_1 and A_2 are found from experimental data shown in Figure 1. The model predictions along with experimental data are gathered in Figure 3. Additionally in Figure 3 the attenuation given by the Biot's model (no scattering) are plotted. The results show good qualitative agreement of the proposed model and of experimental data, and illustrate the essential sensitivity of the model to changes of parameters which describe scattering. The sensitivity analysis proved that the changes of parameters A_1 and A_2 influence almost exclusively the attenuation of fast and slow wave, respectively. Finally, it is worth noticing that the above discussion does not refer to the dispersion of waves due to scattering. The development of a model to include the influence of scattering on dispersion will require further analysis and should take into account Kramer- Kronig relations.

4. Conclusions

The proposed two-phase model of wave propagation in saturated porous materials extends the predictive capability of the Biot's model allowing for a description of effects of scattering on attenuation of longitudinal waves. The simple procedure useful for evaluation of the model parameters responsible for scattering is developed. Numerical data prove that there is a good qualitative agreement of the model predictions and experimental results. The performed sensitivity analysis indicates that the proper choice of the parameters is essential for accurate modeling of the effects of scattering.

References

1. Berryman, J. G. Confirmation of Biot's theory. *Appl. Phys. Lett.*, **37**, 898–900, 1980.
2. Biot, M. A. Theory of elastic waves in a fluid saturated porous solid. *J. Acoust. Soc. Am.*, **28**, 168–191, 1956.
3. Bourbie, T. O. and Zinszner, B. Acoustics of porous media. Gulf Publ. Comp.Houston, 1987.
4. Chaban, A. Sound attenuation in sediments and rock. *Acoustical Physics*, **39**, 190–193, 1993.
5. Johnson, D. L., Plona, T. and Kojima, H. Probing porous media with first and second sound sound. II. Acoustic properties of water-saturated porous media. *J. Appl. Phys.*, **76**, 115–125, 1994.
6. Jungman, G., Quentin, A., Adler, L. and Xue, Q. Elastic property measurements in fluid-filled porous materials. *J. Applied Phys.*, **66**, 5179–5184, 1989.
7. Kaczmarek, M. and Kubik, J. Ultrasonic waves in saturated porous materials. Discussion of modeling and experimental results. *J. Theor. Appl. Mech.*, **36**, 597–618, 1998.
8. Mavko, F. M. and Nur, A. Wave attenuation in partially saturated rocks. *Geophysics* , **44**, 161–178, 1979.
9. O'Connell, R. J. and Budiansky, B. Viscoelastic properties of fluid-saturated cracked solids. *J. Geoph. Res.*, **82**, 5719–5736, 1977.
10. Plona, T. J. Observation of a second compressional wave in a porous medium at ultrasonic frequencies. *Appl. Phys. Lett.*, **36**, 259–261, 1980.
11. Sayers, C. M. Ultrasonic velocity dispersion in porous materials. *J. Appl. Phys.*, **14**, 413–420, 1981.
12. Schmidt, E. J. Wideband acoustic responce of fluid -saturated porous rocks: Theory and preliminary results using waveguided samples. *J. Acoustic. Soc. Am.*, **83**, 2027–2042, 1988.
13. Stoll, R. D. Sediment acoustics. Springer-Verlag, 1990.
14. Winkler, K. W. Frequency dependent ultrasonic properties of high-porosity sandstones. *J. Geoph. Res.*, **88**, 9493–9499, 1983.

A Thermomechanical Model for Partially Saturated Expansive Clay

A. J. Lempinen
Research Scientist
Helsinki University of Technology, P.O.Box 1100, FIN-02015 HUT
e-mail: antti.lempinen@hut.fi

Abstract. A thermodynamically consistent model for elastic behaviour of swelling clay is presented here. It is based on general Clausius-Duhem inequality for under-saturated porous media. In the state equations, decomposition of strain – as in moisture swelling method – will lead to inconsistencies with observed behaviour. Therefore, decomposition of stress into deformation stress, hydraulic stress and swelling stress is used. Also, the conduction equations for heat flow and moisture flow are presented. They are based on theoretical conduction equations for under-saturated porous media, taking swelling of material into account.

Keywords: Thermomechanics, porous media, swelling, bentonite

1. Introduction

Compressed bentonite, which is planned to be used in the engineered clay buffer system of high level nuclear waste depositories, has great swelling capacity: the average pressure stress in the buffer is about 10 MPa. Therefore, the possibility of mechanical instability of the buffer-canister system arises. The most important aim is that the canister should remain inside the bentonite buffer. For the performance assessment calculations of the buffer system, a thermomechanical model for expansive clay within small deformation framework is required. This model has to retain thermodynamic consistency [4].

2. Definitions

2.1. Constituents

Expansive clays are assumed to consist of four distinct phases: solid matrix (subscript s), which mostly consists of montmorillonite mineral platelets; free liquid (l) in the intergranular porous space; trapped liquid (t), which saturates the interlamellar pores in the granules; and gas, which together with the free liquid saturates the intergranular pores. The gas is moist air, i.e. a mixture of dry air (d) and water vapour (v), and the liquid is water or brine.

W. Ehlers (ed.), IUTAM Symposium on Theoretical and Numerical Methods in Continuum Mechanics of Porous Materials, 403–408.

2.2. State variables

State variables are physical variables, which describe the equilibrium energy state of elementary systems. One of these is the absolute temperature T. Because deformation of an elementary particle requires energy, the strain tensor $\boldsymbol{\epsilon}$ is postulated to be a state variable.

Work is required to change the fluid contents of an elementary particle. Therefore, it is postulated that the fluid mass content of fluid $K \in \{\mathrm{l,t,v,d}\}$ per initial volume m_K is a state variable.

Evolutions are assumed to be elastic, so no internal variables are used.

3. Balance equations

When it is assumed that there are phase changes between trapped water, free water and water vapour, the overall mass balance leads to local mass balance equations

$$\frac{dm_\mathrm{l}}{dt} + \operatorname{div}\mathbf{w}_\mathrm{l} = -\dot{m}_{\mathrm{l}\to\mathrm{t}} - \dot{m}_{\mathrm{l}\to\mathrm{v}} \qquad \frac{dm_\mathrm{t}}{dt} = \dot{m}_{\mathrm{l}\to\mathrm{t}} \tag{1}$$

$$\frac{dm_\mathrm{v}}{dt} + \operatorname{div}\mathbf{w}_\mathrm{v} = \dot{m}_{\mathrm{l}\to\mathrm{v}} \qquad \frac{dm_\mathrm{d}}{dt} + \operatorname{div}\mathbf{w}_\mathrm{d} = 0 \tag{2}$$

where $\dot{m}_{K\to L}$is the mass transform rate from fluid phase L to fluid phase K and $\mathbf{w}_K$ is the mass flow of fluid K [4] .

When all acceleration terms are neglected, the momentum balance is given by the usual equilibrium equation

$$\operatorname{div}\boldsymbol{\sigma} + r\mathbf{g} = 0 \tag{3}$$

where $\boldsymbol{\sigma}$ is the stress tensor and $r\mathbf{g}$ is weight of unit volume of the whole material.

4. State equations

Local equilibrium implies that all the water phases have equal free specific enthalpies. Therefore, they are indistinguishable in terms of free energy, and total water content $m_\mathrm{w} = m_\mathrm{l} + m_\mathrm{t} + m_\mathrm{v}$ can be used as a state variable.

When the material is isotropic, the incremental drained state equations are [3, 4]

$$d\boldsymbol{\sigma} \;=\; (K^\mathrm{o} - 2\mu/3)d\varepsilon\mathbf{1} + 2\mu\, d\boldsymbol{\epsilon} - b_\mathrm{w}\, dp_\mathrm{l} - b_\mathrm{d}\, dp_\mathrm{d} - \alpha^\mathrm{o} K^\mathrm{o}\, dT\mathbf{1} \tag{4}$$

$$dp_K = \sum_{L\in\{\mathrm{w,d}\}} \mathsf{M}_{KL}\left(-b_L\, d\varepsilon + \frac{dm_L}{\rho_L} + 3\alpha_L^{\mathrm{m}}\, dT\right) \quad K\in\{\mathrm{w,d}\} \tag{5}$$

$$dS = \alpha K\, d\varepsilon - \sum_{L\in\{\mathrm{w,d}\}}\left[3(\alpha^{\mathrm{m}}\mathsf{M})_L\frac{dm_L}{\rho_L} - S^{\mathrm{m}}\, dm_L\right] + \frac{C_m}{T}\, dT \tag{6}$$

where $p_{\mathrm{w}} \equiv p_{\mathrm{l}}$. The drained coefficients above are

K^{o}	isothermal bulk modulus	μ	shear modulus
M_{KL}	Biot moduli	b_K	Biot coefficients
α^{o}	drained thermal dilation coefficient	C_p	heat capacity
α_K^{m}	thermal expansion coefficients of fluids		

and ρ_K is density of fluid K and $\varepsilon = \mathrm{tr}\,\boldsymbol{\epsilon}$ is volume dilation of material.

5. Conduction equations

Assuming that conduction of heat and conduction of fluid mass are independent of other dissipative mechanisms, the simplest forms for fluid mass flow and heat flow $\mathbf{q}$ that satisfy the fundamental inequality are

$$\frac{1}{\rho_K}\mathrm{grad}\, p_K - \mathbf{g} = -k_K \mathbf{w}_K \quad K\in\{\mathrm{l,d,v}\} \tag{7}$$

$$\mathbf{q} = -\lambda \mathrm{grad}\, T \tag{8}$$

where k_{KL} and λ are conduction coefficients. Substitution of state equation (5) into (7) with definitions of moisture diffusion coefficients

$$\mathcal{D}_m = \sum_{K\in\{\mathrm{l,d,v}\}} \frac{k_l^{-1}}{\rho_K\rho_{\mathrm{l}}}\mathsf{M}_{K\mathrm{w}} \tag{9}$$

$$\mathcal{D}_\varepsilon = \sum_{K\in\{\mathrm{l,d,v}\}} \frac{k_l^{-1}}{\rho_K}\mathsf{M}_{K\mathrm{w}}b_{\mathrm{w}} \quad \mathcal{D}_T = \frac{k_{\mathrm{v}}^{-1}}{\rho_{\mathrm{v}}}\frac{dp_{\mathrm{v}}}{dT} \tag{10}$$

gives moisture diffusion equation

$$\mathrm{div}\left[\mathcal{D}_m\, \mathrm{grad}\, m_{\mathrm{w}} + \mathcal{D}_\varepsilon\, \mathrm{grad}\,\varepsilon + \mathcal{D}_T\, \mathrm{grad}\, T - \left(k_{\mathrm{l}}^{-1} + k_{\mathrm{v}}^{-1}\right)\mathbf{g}\right] = \frac{dm_{\mathrm{w}}}{dt} \tag{11}$$

where the effects of dry air and fluid expansion are neglected and the vapour pressure is assumed to be a function of temperature only.

When material is non-swelling, the effect of dilation gradient can be neglected in (11) when deformations are small since $b_{\mathrm{w}} \leq 1$, but with swelling material this is not neccessarily the case.

6. Material parameters

6.1. Degree of saturation

The wetting liquid content of porous material is often described using degree of saturation. In this context the degree of saturation is defined as the fraction of the porous space occupied by water

$$S_r = \frac{\phi_{\mathrm{l}} + \phi_{\mathrm{t}}}{\phi} \tag{12}$$

There are experimental problems when considering the degree of saturation: since the density of the adsorbed water is not accurately known in partially saturated state, the volume fraction of water cannot be deduced from water mass content. Therefore, degree of saturation of swelling porous medium is not a very useful quantity. Furthermore, because of equation (5), degree of saturation cannot be used as a potential for fluid conduction equations in the case of expansive media.

6.2. Effective stress

If the Biot coefficients are functions of fluid pressures only,

$$b_K = b_K(p_{\mathrm{l}}, p_{\mathrm{g}}) \quad K \in \{\mathrm{w,d}\} \tag{13}$$

the effective stress $\boldsymbol{\sigma}^*$ can be defined as

$$\boldsymbol{\sigma}^* = \boldsymbol{\sigma} + \Pi \mathbf{1} \tag{14}$$

where $\Pi = \Pi(p_{\mathrm{l}}, p_{\mathrm{d}})$ is the swelling pressure, and the Biot coefficients are

$$b_K = \frac{\partial \Pi}{\partial p_K} \quad K \in \{\mathrm{w,d}\} \tag{15}$$

and state equation (4) becomes

$$d\boldsymbol{\sigma}^* = (K^{\mathrm{o}} - 2\mu/3) d\varepsilon \mathbf{1} + 2\mu\, d\boldsymbol{\epsilon} - \alpha^{\mathrm{o}} K^{\mathrm{o}}\, dT \mathbf{1} \tag{16}$$

From symmetry relations [4] it is also concluded that

$$K^{\mathrm{o}} = K^{\mathrm{o}}(\boldsymbol{\epsilon}, T) \quad \mu = \mu(\boldsymbol{\epsilon}, T) \tag{17}$$

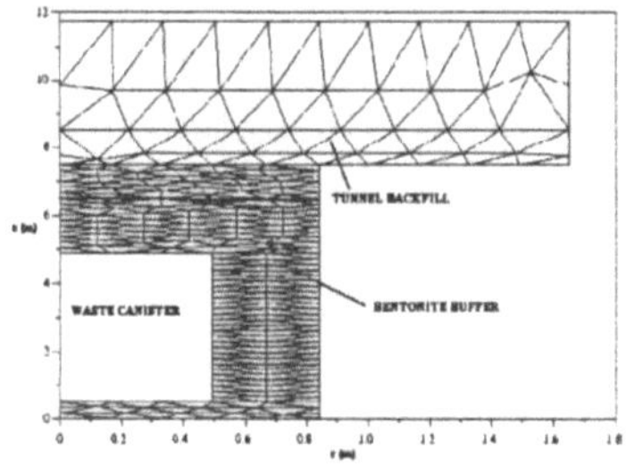

Figure 1. Finite element mesh of bentonite buffer system [6]- 386 elements

This approach including swelling in the constitutive equations is preferred to the "moisture swelling" -method [1], because swelling dilation should be independent of deformation, which would lead to too high pressure in constant volume evolutions when the swelling strain is measured by stress-free experiments.

6.3. Constant volume experiments

The coefficients related to swelling can be measured by simple constant volume experiments. In these experiments the pressure caused by swelling is measured when the sample is held in constant volume and water content of the sample is increased stepwise until saturation. When the measurement is repeated with different initial loading, coefficients b_K and $(\mathsf{M}b)_K = \sum_L \mathsf{M}_{LK} b_K$ can be calculated.

7. Numerical examples

To show the influence of the way the buffer is saturated with water, two finite element analyses of elastic deformation of the bentonite buffer system were calculated. The mesh is shown in Figure 7. The geometry in the problems is axially symmetric. Since there is little knowledge of plastic properties of the backfill material in the tunnel, it is assumed to be like the buffer material, but with no swelling pressure. This simplification can be assumed to be a stabilizing factor, because it reduces the differences in material parameters. The numerical method is explained more closely in [5]. The material parameters are calculated from data in [2].

One simulation was run with uniform saturation and the other with a linear swelling front moving upwards. In both cases, the total swelling pressure of bentonite was 14 MPa. As a result, the canister was elevated

8.9 mm with uniform swelling and 8.3 mm with a swelling front. The difference is not very great, but cannot be neglected when the long-time behaviour of the system is analysed. However, this result suggests that feedback from deformation to water conduction does not have significant influence on calculations of deformation and stress fields.

8. Conclusions

The model that is described here can be used to simulate the elastic behaviour of a bentonite buffer system of high level nuclear waste. The required parameters can be calculated from simple constant volume experiments, with no need for measuring pore water pressure. The numerical simulations show that the effect of non-linearity in material parameters is not great, but it may be significant in long-time behaviour. The connected model can be simplified by neglecting the influence of deformation in moisture diffusion.

References

1. Börgesson, L. and Hernelind, J. DECOVALEX - Test Case 3: Calculation of the Big Ben Experiment - Coupled Modelling of thermal, mechanical and hydraulic behaviour of water-unsaturated buffer material in a simulated deposition hole. Technical Report 95–29, Swedish Nuclear Fuel and Waste Management Co, Stockholm, 1995.
2. Börgesson, L., Johannesson, L.-E., Sandén, T. and Hernelind, J. Modelling of the physical behaviour of water saturated clay barriers. Technical Report 95–20, Swedish Nuclear Fuel and Waste Management Co, Stockholm, 1995.
3. Coussy, O. *Mechanics of Porous Continua.* Chichester: John Wiley & Sons Ltd. Translated from the French, 1995.
4. Lempinen, A. Thermomechanical model for expansive clays: I. Reversible processes. Technical Report 51, Helsinki University of Technology Department of Engineering Physics and Mathematics Laboratory of Theoretical and Applied Mechanics, Espoo, 1998.
5. Lempinen, A. Numerical Model for Water Unsaturated Expansive Clay. In: *Proceedings of European Conference on Computational Mechanics.* To be published, 1999.
6. Oy, P. Final disposal of spent nuclear fuel in the Finnish bedrock, technical research and developement in the period 1993–1996. Technical Report POSIVA-96-14, Posiva Oy, Helsinki. In Finnish, 1996.

Constitutive Modeling of Charged Porous Media

M. M. Molenaar*, J. M. Huyghe and F. P. T. Baaijens
Eindhoven University of Technology, Department of Materials Technology, P.O. Box 513, 5600 MB Eindhoven, The Netherlands

Abstract. As osmotic, chemical and hydrational swelling in shales are essential mechanisms in many borehole stability related problems, an electro-chemo-mechanical formulation of quasi-static finite deformation of swelling of charged porous media has been derived from the theory of mixtures. Four phases following different kinematic paths are defined: solid, fluid, anions and cations. The model includes compressibility of both the solid and the fluid component. In the limiting case without ionic effects, the resulting model from the theory of mixtures is consistent with Biot's theory.

Keywords: Swelling, hydration, osmosis, compressibility, shale

1. Introduction

The main operational parameters controlling borehole stability in drilling shales with environmentally acceptable water-based muds (WBM's) are the density and the chemical composition of the drilling fluid.

When using WBM's, the density of the fluid provides control of the support and stability of the borehole, while it also acts as a driving force for the invasion of the drilling fluid into the formation, which can lead to instabilities. By adopting the chemical composition of the drilling fluid, the infiltration rate can be reduced and a prolonged exposure time of the drilling fluid to the shale is achieved. Inevitably, the invasion of fluid and chemicals into the shale, induced by the pressure overbalance of the fluid, changes in the course of time the effective stress state around the borehole. These stress changes may exceed the elastic limits and borehole instability can occur.

To describe the changes in the 'effective' stress-strain behaviour due to the shale-drilling fluid interaction, an electro-chemo-mechanical model is formulated, using the theory of mixtures. The model presented in this paper, which is applicable to biological tissues and synthetic hydrogels also, consists of an electrically charged porous solid matrix of silt and clay minerals, saturated with an ionic solution. For the ionic solution, three types of streaming components are identified, the fluid, the cations, and the anions.

* currently holds a position as parttime contractor for Shell Int. E & P, Rijswijk.

W. Ehlers (ed.), IUTAM Symposium on Theoretical and Numerical Methods in Continuum Mechanics of Porous Materials, 409–414.

2. Kinematics of porous media

In mixture theory formulations of porous materials, it is customary to assign each constituent its own continuum:

$$\mathbf{x} = \chi_\alpha (\mathbf{X}_\alpha, t) \qquad \alpha = \mathrm{s}, \mathrm{f}, +, - \tag{1}$$

The index α in this paper denotes: the solid, the fluid, the cation and the anion constituents. It is assumed that at any time t, particles of these four constituents occupy the same representative elementary 'bulk' volume (REV) of the mixture V_b, with mass M, which is defined as the sum of all the partial masses M_α of the constituents:

$$M = \sum_\alpha M_\alpha \tag{2}$$

The fluid and ions are single species constituents, their molar volume is defined as:

$$\bar{V}_\beta = \frac{\partial V_\beta}{\partial N_\beta} \qquad \beta = \mathrm{f}, +, - \tag{3}$$

herein, N_β denotes the number of mol of constituent β and index β all the components other then the solid. Assuming, the ionic constituents incompressible (index $i = +, -$), equation (3) can be written as:

$$V_i = N_i \bar{V}_i \qquad i = +, - \tag{4}$$

where, V_i represents the volume added onto V_b, by increasing χ_i from zero to its present value. The volume fractions of the ionic constituents are defined as:

$$n_i = \frac{V_i}{V_\mathrm{b}} \tag{5}$$

Assuming the solid immiscible with the saturating ionic solution, the part V_s of the elementary bulk volume V_b occupied by the solid, is physically distinguishable from the solution $(V_\mathrm{s} + V_+ + V_-)$. The solid volume fraction, n_s, thus reads:

$$n_\mathrm{s} = \frac{V_\mathrm{s}}{V_\mathrm{b}} \tag{6}$$

Saturation of the porous medium requires:

$$\sum_\mathrm{c} n_\mathrm{c} = 1 \qquad \mathrm{c} = \mathrm{s}, \mathrm{f} \tag{7}$$

herein, the volume of the ionic species is neglected. In porous media applications it is customary to formulate the velocity of the β component relative to the solid component :

$$\mathbf{v}_{\beta\mathrm{s}} = \mathbf{v}_\beta - \dot{\mathbf{x}} \tag{8}$$

Table I. Conservation equations for the constituents.

$\dot{m}_\alpha + J\nabla \cdot (n_\alpha \rho_\alpha \mathbf{v}_{\alpha s}) = 0$	J Jacobian ρ_α intrinsic density
$\nabla \cdot (\sigma_\alpha)^{\mathrm{T}} + \hat{\pi}_\alpha = \mathbf{0}$ and $\sigma_\alpha = \sigma_\alpha^{\mathrm{T}}$	σ_α partial stress e_α internal energy
$n_\alpha \rho_\alpha (\dot{e}_\alpha + \mathbf{v}_{\alpha s} \cdot \nabla e_\alpha - h_\alpha) - \sigma_\alpha : \nabla \mathbf{v}_\alpha + \nabla \cdot \mathbf{q}_\alpha = \hat{e}_\alpha$	h_α partial heat $\mathbf{q}_\alpha$ heat flux

herein, the dot accent indicates the material time derivative 'following' the motion of the solid.

3. Conservation and constraint equations

The conservations equations for each individual constituent are listed in Table I, herein $\hat{\pi}_\alpha$ and $\hat{e}_\alpha$ denote the interaction of, momentum and energy exchange between the different constituents, respectively. It is assumed that no mass transfer and no moment of momentum interaction occurs between the constituents, and inertial terms and body forces are neglected.

In Table I the Lagrange mass m_β of a streaming constituent is introduced, which was first defined by Biot ([1]). It represents the mass variation experienced by a elementary reference volume at $t = 0$:

$$m_\beta = J n_\beta \rho_\beta - (n_\beta \rho_\beta)^0 \quad (\; > 0 \text{ if added}, \quad < 0 \text{ if extracted} \;) \qquad (9)$$

herein, the superscript 0 refers to initial values. The equations are supplemented with the second law of thermodynamics:

$$n_\alpha \rho_\alpha (\dot{\eta}_\alpha + \mathbf{v}_{\alpha s} \cdot \nabla \eta_\alpha - \frac{h_\alpha}{\theta_\alpha}) + \nabla \cdot \frac{\mathbf{q}_\alpha}{\theta_\alpha} = \xi_\alpha \qquad (10)$$

in which, η_α is the entropy density, ξ_α is the entropy production and θ_α the partial temperature of the αth constituent. The total entropy production of the mixture as a whole must be positive or zero:

$$\sum_\alpha \xi_\alpha \geq 0 \qquad (11)$$

Next, the Helmholtz free energy, Ψ_α and the strain energy function W are defined:

$$\Psi_\alpha = (e_\alpha - \theta_\alpha \eta_\alpha) \;\text{ and }\; W = J \sum_\alpha n_\alpha \rho_\alpha \Psi_\alpha \qquad (12)$$

Table II. Conservation equations for the total mixture.

$\frac{n_s}{\rho_s}\dot{\rho}_s + \nabla\cdot\mathbf{v}_s + \frac{n_f}{\rho_f}\dot{\rho}_f + \sum_\beta \frac{1}{\rho_\beta}\nabla\cdot(n_\beta\rho_\beta\mathbf{v}_{\beta s}) = 0$	C^{fc} fixed charges
$C^{fc} + \sum_{i=+,-} \frac{z_i}{V_i} J n_i = 0$	z_i valences

The mixture as a whole has to obey two additional conditions: the saturation and the electro-neutrality constraint, shown in Table II. Rewriting the entropy inequality (11), using the strain energy function (12) and substitution of the volume fraction restriction and the electro-neutrality condition by means of the Lagrange multipliers p and λ, respectively, yields the extended form of the entropy inequality:

$$\begin{aligned}
-\dot{W} + J\sum_{c}\left[p\tfrac{n_c}{\rho_c}\dot{\rho}_c + (\boldsymbol{\sigma}_c + n_c(p - \rho_c\Psi_c)\mathbf{I}) : \nabla\mathbf{v}_{cs}\right] \\
+J\sum_{c}\left[\mathbf{v}_{cs}\cdot(\nabla\cdot\boldsymbol{\sigma}_c + \tfrac{p}{\rho_c}\nabla(n_c\rho_c) - \nabla(n_c\rho_c\Psi_c)\right] \\
+J[\boldsymbol{\sigma} + p\mathbf{I}] : \nabla\mathbf{v}_s + J\sum_{i}\left[(\boldsymbol{\sigma}_i + n_i(p + \tfrac{z_i}{V_i}\lambda - \rho_i\Psi_i)\mathbf{I}) : \nabla\mathbf{v}_{is}\right] \\
+J\sum_{i}\left[\mathbf{v}_{is}\cdot(\nabla\cdot\boldsymbol{\sigma}_i + (p + \tfrac{z_i}{V_i}\lambda)\nabla n_i - \nabla(n_i\rho_i\Psi_i)\right] \geq 0
\end{aligned} \tag{13}$$

herein, isothermal condition of the mixture is assumed.

Using this Lagrangian form of the entropy inequality (13) leads to equations consistent with Biot's porous media theory in a more straightforward way than the Eulerian approach of e.g. Bowen ([2]).

4. The constitutive laws

The theory of constitution is dictated by the choice of independent variables. It is assumed that the behaviour of the mixture is dictated by the following list of independent variables [1, 2]:

$$\{\mathbf{E}_s,\ \mathbf{u}_{fs},\ \mathbf{u}_{+s},\ \mathbf{u}_{-s},\ \phi_s,\ \phi_f,\ \zeta_f,\ \zeta_+,\ \zeta_-\} \tag{14}$$

For the satisfaction of the principle of objectivity, the strain tensor $\mathbf{E}_s$, instead of the displacement gradient tensor $\mathbf{F}_s$, and the Lagrangian forms of the relative velocities $\mathbf{u}_{\beta s}$ of the constituents β are used [4]. The Lagrange volume fraction ϕ_c for the compressible constituents c is defined as:

$$\phi_c = J n_c \tag{15}$$

Table III. Constitutive laws of the mixture.

$\sigma + p\mathbf{I} = \frac{1}{J}\mathbf{F}_s \cdot \frac{\partial W}{\partial \mathbf{E}_s} \cdot \mathbf{F}_s^T$	$t_c = -\frac{\partial W}{\partial \phi_c} = p$	$\mu_\beta = \frac{\partial W}{\partial \zeta_\beta}$	$\tilde{\mu}_i = \mu_i + p + \frac{z_i}{V_i}\lambda$

The Biot terminology and notation for the 'change in constituent content', ζ_β, is adopted here and used in the list of independent variables:

$$\zeta_\beta = \frac{m_\beta}{{\rho_\beta}^0} \tag{16}$$

After the principle of equipresence is applied, the chain rule for time differentiation of strain energy function yields:

$$\dot{W} = \frac{\partial W}{\partial \mathbf{E}_s} : \dot{\mathbf{E}}_s + \sum_c \frac{\partial W}{\partial \phi_c} \dot{\phi}_c + \sum_\beta \left[\frac{\partial W}{\partial \mathbf{u}_{\beta s}} \cdot \dot{\mathbf{u}}_{\beta s} + \frac{\partial W}{\partial \zeta_\beta} \dot{\zeta}_\beta \right] \tag{17}$$

This result for $\dot{W}$ is substituted into the extended entropy inequality (13). Skipping lengthy notation, the final form of the entropy inequality for the model, the total dissipation, Φ is the defined:

$$\Phi \geq 0 \tag{18}$$

According to inequality (18), the rate of entropy production must be positive. Next the total dissipation is divided into two groups, a group Φ_R associated with the reversible processes and a group Φ_I associated with the irreversible processes:

$$\Phi = \Phi_R + \Phi_I \ \geq 0 \ \text{ with } \Phi_R = 0 \text{ and } \Phi_I \geq 0 \tag{19}$$

There is a flood of information in the equation (19) and the constitutive laws that result from the investigation of the reversible part are briefly presented in Table III. The results of the investigation of irreversible part are discussed in Molenaar *et al.* ([5]).

5. Discussion of the constitutive model

In the previous section it is demonstrated that the reversible constitutive laws are entirely described by the strain energy function W. Considering infinitesimal transformations, which allows replacement of the Green-Lagrange strain tensor $\mathbf{E}_s$ by the linearized porous matrix strain tensor ε_s, the constitutive laws from Table III recasted in their differentiated form, assuming smoothness of the W, read:

$$\begin{bmatrix} (\dot{\sigma + p\mathbf{I}}) \\ -\dot{\mathrm{t}}_{\mathrm{c}} \\ \dot{\mu}_\beta \end{bmatrix} = \mathbf{C} \begin{bmatrix} \dot{\varepsilon}_{\mathrm{s}} \\ \dot{\phi}_{\mathrm{c}} \\ \dot{\zeta}_\beta \end{bmatrix} \tag{20}$$

herein, $\mathbf{C}$ is a symmetric matrix containing the second order derivatives of the strain energy function W.

Shown in equation (20) (first row), the use of the principle of equipresence results in an 'effective' stress-strain relationship, which depends not only on the compressibility of the solid and the fluid through ϕ_{s}, ϕ_{f} and ζ_{f} but also on the ion concentration in the medium via ζ_i.

The quadriphasic model presented here is consistent with the physico-chemical model for porous solids developed by Biot ([1]). In the quadriphasic model, the fluid pressure is replaced by a chemical potential, to account for the physico-chemical effects (capillary and hydration swelling). In addition Donnan swelling is incorporated via the presence of fixed charges and a chemical swelling term is introduced, which depends on the local ionic concentration.

Acknowledgements

This research is supported by the STW and the research of J.M. Huyghe was made possible through a fellowship of the KNAW. The first author wishes to thank the members of the Shell global hole stability team for stimulating discussions, support and encouragement.

References

1. Biot, M. A. Theory of Finite Deformations of Porous Solids. *Indiana Univ. Math. J.*, **21**, 597–735, 1972.
2. Bowen, R. M. *Compressible Porous Media Models by Use of the Theory of Mixtures.* In *Int. J. Engng. Sci.*, **20**, 697–735, 1982.
3. Huyghe, J. M. and Janssen, R. R. Quadriphasic Mechanics of Swelling Incompressible Porous Media. *Int. J. Engng. Sci.*, **35**, 793–802, 1997.
4. Molenaar, M. M., Huyghe, J. M. and Baaijens, F. P. T. A Constitutive Model for Swelling of Compressible Porous Media. In J.-F. Thimus *et al.* (eds.),*Poromechanics – A Tribute to Maurice A. Biot*, Louvain-la-Neuve, Sept. 14-16, 105–110, 1998.
5. Molenaar, M. M., Huyghe, J. M. and Baaijens, F. P. T. Constitutive Modeling of Coupled Flow Phenomena in Shales. *Submitted to the Int. Symposium on Coupled Phenomena in Civil, Mining and Petroleum Engineering*, China, Nov. 1-3, 1999.

Damage of Porous Materials During Drying

G. Musielak
Poznań University of Technology
Institute of Technology and Chemical Engineering
pl. Marii Sklodowskiej-Curie 2, 60-965 Poznań, Poland
e-mail: Grzegorz.Musielak@put.poznan.pl

1. Introduction

Drying is understood here as a thermal process of moisture removal from porous material due to evaporation. The moisture evaporates from the boundary, where it is supplied from the interior of the body due to capillary forces. The heating and diffusion processes make the reasons for the non-uniform temperature and moisture content fields inside of the dried material. These non-uniformities involve thermal and shrinkage stresses, which can cause damage of dried materials.

The convective drying of plate made of clay is considered in the paper. The model of drying of capillary-porous material with material coefficients dependent on the moisture content and the temperature is presented. The material is assumed to be brittle-elastic (non-linear elasticity). The strength of the material, estimated by bending tests, is moisture dependend.

Through the comparison of the stresses arising during drying with the strength of the material the possibility of dried body damage is shown.

2. Model presentation

The dried material is assumed to be a non-linear elastic capillary-porous body saturated with liquid (water). The deformation of the dried body is caused by the alteration of the moisture content X, temperature ϑ ($\vartheta = T - T_R$ is the relative temperature and T – absolute temperature) and the dried induced stresses σ_{ij}. The one-dimensional boundary-initial problem of symmetrically dried plate (see fig.1) is considered in the paper. Therefore moisture content, temperature and stresses are functions of time t and z co-ordinate.

The moisture content and the temperature are determined by the equations of moisture transfer

$$\varrho_0 \dot{X} = \frac{\partial}{\partial z}\left[\Lambda_X\left(\frac{c_X}{\varrho_0}\frac{\partial X}{\partial z} + \frac{c_\vartheta}{\varrho_0}\frac{\partial \vartheta}{\partial z}\right)\right] \tag{1}$$

W. Ehlers (ed.), IUTAM Symposium on Theoretical and Numerical Methods in Continuum Mechanics of Porous Materials, 415–420.

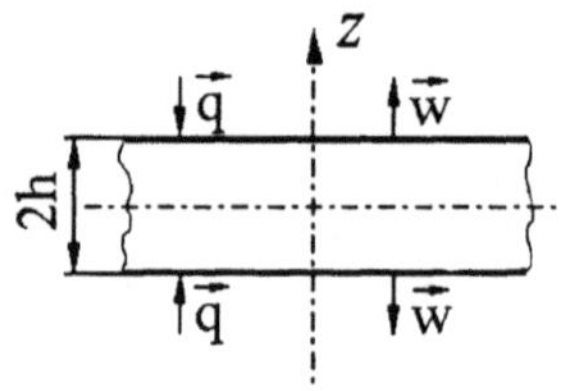

Figure 1. Dried plate; $\vec{q}$ – heat flux, $\vec{w}$ – moisture flux

and heat transfer

$$c_v \varrho_0 \dot{\vartheta} = \frac{\partial}{\partial z}\left[\Lambda_\vartheta \frac{\partial \vartheta}{\partial z}\right] \tag{2}$$

together with the boundary conditions for the moisture transfer

$$\left.\frac{\partial X(z,t)}{\partial z}\right|_{z=0} = 0,$$
$$\Lambda_X \left.\frac{\partial X(z,t)}{\partial z}\right|_{z=h} =$$
$$= \alpha_m \left[B\left(T_a - T(z,t)|_{z=h}\right) + R\left(T(z,t)|_{z=h} \ln\frac{p_s^V}{p} - T_a \ln \varphi_a \frac{p_{as}^V}{p}\right)\right] \tag{3}$$

and the heat transfer

$$\left.\frac{\partial \vartheta(z,t)}{\partial z}\right|_{z=0} = 0,$$
$$\Lambda_\vartheta \left.\frac{\partial \vartheta(z,h)}{\partial z}\right|_{z=h} = \alpha_t\left(T_a - T(z,t)|_{z=h}\right) + Q\Lambda_X \left.\frac{\partial X(z,t)}{\partial z}\right|_{z=h}, \tag{4}$$

respectively (see e.g. Kowalski, [1], [2], Musielak, [4]), where ϱ_0 is the mass density of dry body, c_X and c_ϑ are the moisture coefficient and temperature coefficient of the moisture potential, c_v is the specific heat, Λ_X and Λ_ϑ are the coefficient of the moisture flow and the thermal conductivity, R is the gas constant, B and Q are the specific entropy and the latent heat of vaporisation (both temperature dependent), α_m and α_t are the coefficients of the convective mass and heat transfer, T_a and φ_a are the temperature and the relative humidity of the drying medium, p is the total pressure and p_s^V and p_{as}^V are the partial pressures of the saturated vapour close to the boundary and in the atmosphere, respectively.

These equations are non-linear (see Musielak, [4]). The non-linear coefficient of the moisture flow Λ_X is equal

$$\Lambda_X = \frac{K(\varrho_L^R)^2}{\mu} \tag{5}$$

where K is the permeability, ϱ_L^R is the real density of liquid (water) and μ is the dynamic viscosity, temperature dependent.

The non-linear thermal conductivity Λ_ϑ is assumed to be consisted of thermal conductivity for the skeleton $\Lambda_{\vartheta S}$ (constant) and thermal conductivity for the liquid (water) $\Lambda_{\vartheta L}$ (temperature dependent) multiplied by porosity f and liquid saturation, defined as the ratio of actual to initial moisture content:

$$\Lambda_\vartheta = \Lambda_{\vartheta S}(1-f) + \Lambda_{\vartheta L} f \frac{X}{X_0}. \tag{6}$$

The dried plate is assumed to be free of external forces. In such a case the stresses tensor components are: $\sigma_{xx} = \sigma_{yy} = \sigma$ and $\sigma_{zz} = \sigma_{xy} = \sigma_{xz} = \sigma_{yz} = 0$ (see Kowalski *et al.*, [3]). These stresses are equal

$$\sigma = \frac{2+\nu^2}{1+\nu^3} E(C-\eta) \tag{7}$$

where η is a function of temperature ϑ and moisture content X

$$\eta = \alpha_\vartheta \vartheta + \alpha_X (X - X_r) H(X - X_r) \tag{8}$$

and C equal to

$$C = \frac{\displaystyle\int_0^h E\eta \, dz}{\displaystyle\int_0^h \eta \, dz}. \tag{9}$$

In equations (7–9) E is the moisture dependent Young modulus, ν is the Poisson ratio (constant), α_X, α_ϑ are the moisture and temperature expansion coefficients, X_r is the moisture content at which the shrinkage ends, H is the Heaviside's function.

3. Strength of the material

In the considered one-dimensional problem it is enough to know strength criterion for one-dimensional tension.

The three-point bending tests are performed. The bended samples are the beams (with sides of square cross-section equal to 1.5 cm and 15 cm long, initially 18–20% of moisture content) made of Poznań clay. They were dried slowly to various final moisture content and then broken in the bending test. Finally they were dried in temperature 150°C and relative humidity of the dried medium equal to 4.7%. There were carried out 12 series of the test each of 15 samples. The results are performed on the Fig.2.

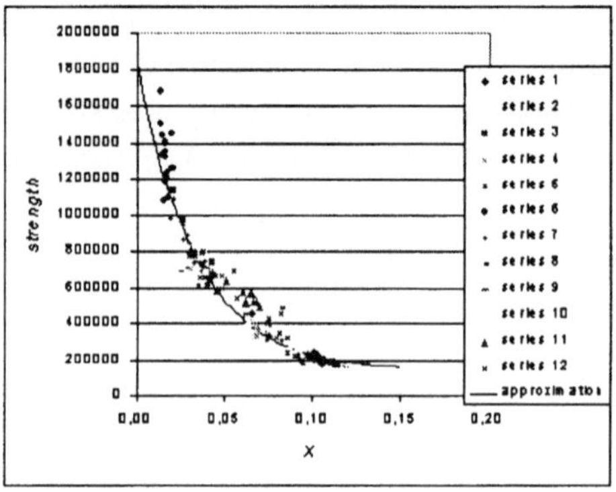

Figure 2. Strength σ_{per} of the Poznań clay versus moisture content

The permissible stresses σ_{per} appointed in the experiments are approximated by the curve

$$\sigma_{per} = A_\sigma + B_\sigma \exp(-C_\sigma X). \tag{10}$$

The coefficients $A_\sigma = 142858$ [N/m^2], $B_\sigma = 1688320$ [N/m^2] and $C_\sigma = 29.6534$ are calculated with the help of minimalisation of square deviation.

4. Numerical simulation

The convectional drying of a plate (see fig.1) was solved as the numerical example. All temperature dependent parameters were taken from Ražnjevič, [5]. Constant coefficients were assumed to be the same as in Musielak, [4]. Young modulus E was assumed to be an expotential function of moisture content:

$$E = A_E + B_E \exp(-C_E X) \tag{11}$$

similar to the strength (permissible stresses) σ_{per} (10). The coefficients $A_E = 1.0 * 10^8$ [N/m^2], $B_E = 20.0 * 10^9$ [N/m^2] and $C_\sigma = 29.6534$ were applied.

The governing set of equations (1–4) were solved with the help of explicit finite difference method. Then constant C (9) was calculated with the trapesium integration method. Resulting stresses σ (7) were compared with the strength of the material (permissible stress) σ_{per}. Distributions of stresses σ and strength σ_{per} are presented in fig.3. The results are plotted for two different drying conditions: fig.3a) – relative humidity of drying medium $\varphi_a = 0.8$ and temperature $\vartheta_a = 60°$C; fig.3b) – $\varphi_a = 0.82$, $\vartheta_a = 90°$C. The resulting stresses are very

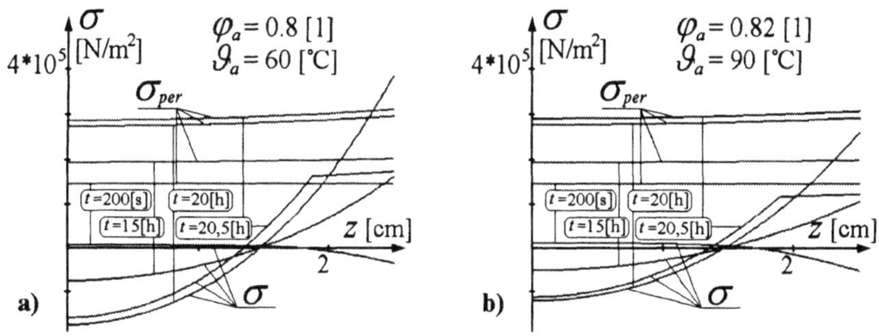

Figure 3. Distributions of stresses σ and strength σ_{per} of the dried plate at different time of drying t

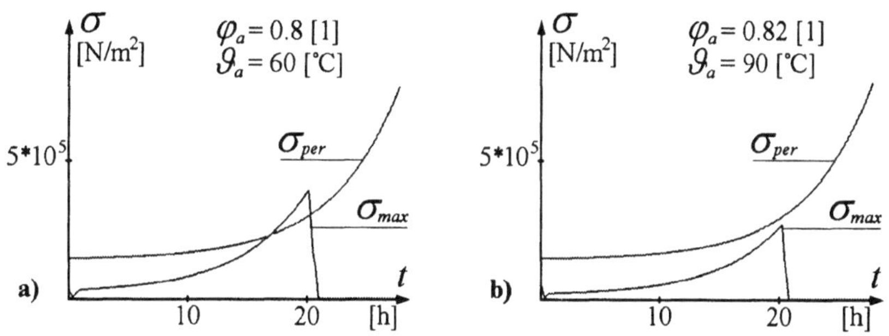

Figure 4. Evolution of maximal stresses σ_{max} and strength σ_{per} of the dried plate

similar in both cases. Thermal stresses arise mainly at the beginning of drying, compressional near the boundary and tensional inside the plate (see time $t = 200$s). However, the temperature becomes very quickly uniform and equal to the wet bulb temperature within the plate. Then the shrinkage stresses arise, tensional near the boundary and compressional inside the plate (see time $t = 15$h, and $t = 20$h). They increase with time at the beginning due to increase of moisture content gradient and next due to increase of Young modulus. When the moisture content inside of the plate becomes lower than X_r (the moisture content at which the shrinkage ends, see (8)), the stresses decreases very quickly (see also fig.4). The strength of the material σ_{per} increases all the time together with the decreasing of the moisture content X (see figs.3,4).

In the fig.4 the evolution of maximal stresses σ_{max} and strength σ_{per} during the drying are shown. The maximal stresses are possibly lower than the permissible stresses all the time (see fig.4b). But it can also happen that the stresses exceed the strength of the material (see fig.4a).

In such a case the damage of the material (fracturing) is expected. The fracture is expected in the layers near the surface of the material. One can see (fig.3a) that the stresses exceed the strength of the material only in these layers.

5. Conclusions

The shrinkage stresses during drying can exceed the strength of the material. In such a case the fracturing of the layers near the surface of the material is expected. The fracture does not happen in the middle of the plate because of the compressional stresses in this region.

The stresses caused by two various drying processes are shown. The stresses arising during the first proces exceed the strength of the material, but those arising during the second one do not. This is the consequence of increase of moisture flow coefficient Λ_X together with the temperature (see (5). The greater is this coefficient the greater rate of moisture transfer inside of material takes place and consequently the smaller moisture gradient inside of material occurs. Smaller gradient involves smaller stresses.

The shrinkage of the clay ends at some critical value of moisure content and at this moment the second drying period for this material begins. When the moisture content inside of the plate becomes smaller than X_r (the moisture content at which the shrinkage ends, see (8)), the stresses decreases very quickly. So the damage of the material is not expected during this period of drying.

Acknowledgements

This work was carried out as part of the research project No. 3 T09C 0015 12 sponsored by the State Committee for Scientific Research.

References

1. Kowalski, S. J. Thermomechanics of Constant Drying Rate Period, *Arch. Mech.*, **39**, 157–176, 1987.
2. Kowalski, S. J. Thermomechanics of Dried Materials, *Arch. Mech.*, **42**, 123–149, 1990.
3. Kowalski, S. J., Musielak, G. and Rybicki, A. Shrinkage Stresses in Dried Materials, *Engng. Trans.*, **40**, 115–131, 1992.
4. Musielak, G. Influence of the Drying Medium Parameters on Drying Induced Stresses, *Drying Technology*, (sent to the journal), 1999.
5. Ražnjevič, K. Heat Tables with Diagrams, *WNT Warszawa*, (in Polish), 1966.

Author Index

Mechanics

SOLID MECHANICS AND ITS APPLICATIONS

Series Editor: G.M.L. Gladwell

49. J.R. Willis (ed.): *IUTAM Symposium on Nonlinear Analysis of Fracture*. Proceedings of the IUTAM Symposium held in Cambridge, U.K. 1997 ISBN 0-7923-4378-6
50. A. Preumont: *Vibration Control of Active Structures*. An Introduction. 1997 ISBN 0-7923-4392-1
51. G.P. Cherepanov: *Methods of Fracture Mechanics: Solid Matter Physics*. 1997 ISBN 0-7923-4408-1
52. D.H. van Campen (ed.): *IUTAM Symposium on Interaction between Dynamics and Control in Advanced Mechanical Systems*. Proceedings of the IUTAM Symposium held in Eindhoven, The Netherlands. 1997 ISBN 0-7923-4429-4
53. N.A. Fleck and A.C.F. Cocks (eds.): *IUTAM Symposium on Mechanics of Granular and Porous Materials*. Proceedings of the IUTAM Symposium held in Cambridge, U.K. 1997 ISBN 0-7923-4553-3
54. J. Roorda and N.K. Srivastava (eds.): *Trends in Structural Mechanics*. Theory, Practice, Education. 1997 ISBN 0-7923-4603-3
55. Yu.A. Mitropolskii and N. Van Dao: *Applied Asymptotic Methods in Nonlinear Oscillations*. 1997 ISBN 0-7923-4605-X
56. C. Guedes Soares (ed.): *Probabilistic Methods for Structural Design*. 1997 ISBN 0-7923-4670-X
57. D. François, A. Pineau and A. Zaoui: *Mechanical Behaviour of Materials*. Volume I: Elasticity and Plasticity. 1998 ISBN 0-7923-4894-X
58. D. François, A. Pineau and A. Zaoui: *Mechanical Behaviour of Materials*. Volume II: Viscoplasticity, Damage, Fracture and Contact Mechanics. 1998 ISBN 0-7923-4895-8
59. L.T. Tenek and J. Argyris: *Finite Element Analysis for Composite Structures*. 1998 ISBN 0-7923-4899-0
60. Y.A. Bahei-El-Din and G.J. Dvorak (eds.): *IUTAM Symposium on Transformation Problems in Composite and Active Materials*. Proceedings of the IUTAM Symposium held in Cairo, Egypt. 1998 ISBN 0-7923-5122-3
61. I.G. Goryacheva: *Contact Mechanics in Tribology*. 1998 ISBN 0-7923-5257-2
62. O.T. Bruhns and E. Stein (eds.): *IUTAM Symposium on Micro- and Macrostructural Aspects of Thermoplasticity*. Proceedings of the IUTAM Symposium held in Bochum, Germany. 1999 ISBN 0-7923-5265-3
63. F.C. Moon: *IUTAM Symposium on New Applications of Nonlinear and Chaotic Dynamics in Mechanics*. Proceedings of the IUTAM Symposium held in Ithaca, NY, USA. 1998 ISBN 0-7923-5276-9
64. R. Wang: *IUTAM Symposium on Rheology of Bodies with Defects*. Proceedings of the IUTAM Symposium held in Beijing, China. 1999 ISBN 0-7923-5297-1
65. Yu.I. Dimitrienko: *Thermomechanics of Composites under High Temperatures*. 1999 ISBN 0-7923-4899-0
66. P. Argoul, M. Frémond and Q.S. Nguyen (eds.): *IUTAM Symposium on Variations of Domains and Free-Boundary Problems in Solid Mechanics*. Proceedings of the IUTAM Symposium held in Paris, France. 1999 ISBN 0-7923-5450-8
67. F.J. Fahy and W.G. Price (eds.): *IUTAM Symposium on Statistical Energy Analysis*. Proceedings of the IUTAM Symposium held in Southampton, U.K. 1999 ISBN 0-7923-5457-5
68. H.A. Mang and F.G. Rammerstorfer (eds.): *IUTAM Symposium on Discretization Methods in Structural Mechanics*. Proceedings of the IUTAM Symposium held in Vienna, Austria. 1999 ISBN 0-7923-5591-1

Mechanics

SOLID MECHANICS AND ITS APPLICATIONS

Series Editor: G.M.L. Gladwell

69. P. Pedersen and M.P. Bendsøe (eds.): *IUTAM Symposium on Synthesis in Bio Solid Mechanics.* Proceedings of the IUTAM Symposium held in Copenhagen, Denmark. 1999 ISBN 0-7923-5615-2
70. S.K. Agrawal and B.C. Fabien: *Optimization of Dynamic Systems.* 1999 ISBN 0-7923-5681-0
71. A. Carpinteri: *Nonlinear Crack Models for Nonmetallic Materials.* 1999 ISBN 0-7923-5750-7
72. F. Pfeifer (ed.): *IUTAM Symposium on Unilateral Multibody Contacts.* Proceedings of the IUTAM Symposium held in Munich, Germany. 1999 ISBN 0-7923-6030-3
73. E. Lavendelis and M. Zakrzhevsky (eds.): *IUTAM/IFToMM Symposium on Synthesis of Nonlinear Dynamical Systems.* Proceedings of the IUTAM/IFToMM Symposium held in Riga, Latvia. 2000 ISBN 0-7923-6106-7
74. J.-P. Merlet: *Parallel Robots.* 2000 ISBN 0-7923-6308-6
75. J.T. Pindera: *Techniques of Tomographic Isodyne Stress Analysis.* 2000 ISBN 0-7923-6388-4
76. G.A. Maugin, R. Drouot and F. Sidoroff (eds.): *Continuum Thermomechanics.* The Art and Science of Modelling Material Behaviour. 2000 ISBN 0-7923-6407-4
77. N. Van Dao and E.J. Kreuzer (eds.): *IUTAM Symposium on Recent Developments in Non-linear Oscillations of Mechanical Systems.* 2000 ISBN 0-7923-6470-8
78. S.D. Akbarov and A.N. Guz: *Mechanics of Curved Composites.* 2000 ISBN 0-7923-6477-5
79. M.B. Rubin: *Cosserat Theories: Shells, Rods and Points.* 2000 ISBN 0-7923-6489-9
80. S. Pellegrino and S.D. Guest (eds.): *IUTAM-IASS Symposium on Deployable Structures: Theory and Applications.* Proceedings of the IUTAM-IASS Symposium held in Cambridge, U.K., 6–9 September 1998. 2000 ISBN 0-7923-6516-X
81. A.D. Rosato and D.L. Blackmore (eds.): *IUTAM Symposium on Segregation in Granular Flows.* Proceedings of the IUTAM Symposium held in Cape May, NJ, U.S.A., June 5–10, 1999. 2000 ISBN 0-7923-6547-X
82. A. Lagarde (ed.): *IUTAM Symposium on Advanced Optical Methods and Applications in Solid Mechanics.* Proceedings of the IUTAM Symposium held in Futuroscope, Poitiers, France, August 31–September 4, 1998. 2000 ISBN 0-7923-6604-2
83. D. Weichert and G. Maier (eds.): *Inelastic Analysis of Structures under Variable Loads.* Theory and Engineering Applications. 2000 ISBN 0-7923-6645-X
84. T.-J. Chuang and J.W. Rudnicki (eds.): *Multiscale Deformation and Fracture in Materials and Structures.* The James R. Rice 60th Anniversary Volume. 2001 ISBN 0-7923-6718-9
85. S. Narayanan and R.N. Iyengar (eds.): *IUTAM Symposium on Nonlinearity and Stochastic Structural Dynamics.* Proceedings of the IUTAM Symposium held in Madras, Chennai, India, 4–8 January 1999 ISBN 0-7923-6733-2
86. S. Murakami and N. Ohno (eds.): *IUTAM Symposium on Creep in Structures.* Proceedings of the IUTAM Symposium held in Nagoya, Japan, 3-7 April 2000. 2001 ISBN 0-7923-6737-5
87. W. Ehlers (ed.): *IUTAM Symposium on Theoretical and Numerical Methods in Continuum Mechanics of Porous Materials.* Proceedings of the IUTAM Symposium held at the University of Stuttgart, Germany, September 5-10, 1999. 2001 ISBN 0-7923-6766-9

Kluwer Academic Publishers – Dordrecht / Boston / London